Ecology and Human Well-Being

Ecology and Human Well-Being

Edited by

Pushpam Kumar
B. Sudhakara Reddy

SAGE Publications
Los Angeles ▪ London ▪ New Delhi ▪ Singapore

First published in 2007 by

Sage Publications India Pvt Ltd
B1/I1, Mohan Cooperative Industrial Area
Mathura Road
New Delhi 110 044
www.sagepub.in

Sage Publications Inc
2455 Teller Road
Thousand Oaks, California 91320

Sage Publications Ltd
1 Oliver's Yard, 55 City Road
London EC1Y 1SP

Sage Publications Asia-Pacific Pte Ltd
33 Pekin Street
#02-01 Far East Square
Singapore 048763

Published by Vivek Mehra for Sage Publications India Pvt Ltd, typeset in 10/12 pt Book Antiqua by Star Compugraphics Private Limited, Delhi and printed at Chaman Enterprises, New Delhi.

Library of Congress Cataloging-in-Publication Data

Ecology and human well-being / edited By Pushpam Kumar, B. Sudhakara Reddy.
p. cm.
Includes bibliographical references and index.
1. Human ecology—India. 2. Human ecology—Economic aspects—India. 3. Social ecology—Economic aspects—India. 4. Well-being—India. 5. Quality of life—India. 6. India—Social conditions. 7. India—Economic conditions. I. Kumar, Pushpam. II. Sudhakara Reddy, B.

GF661.E25 304.20954—dc22 2007 2007020323

ISBN: 978-0-7619-3553-7 (HB) 978-81-7829-712-5 (India-HB)

The Sage Team: Sugata Ghosh and Janaki Srinivasan

Contents

List of Tables

List of Figures

List of Abbreviations

BCR	Benefit Cost Ratio
BCR	Benefit Cost Ratio
BIS	Bureau of Indian Standards
BWDB	Bangladesh Water Development Board
CAI	Current Annual Increment
CDM	Clean Development Mechanism
CETP	Common Effluent Treatment Plants
CFC	Chloro Fluro Carbon
CFE	Consent for Establishment
CFO	Consent for Operation
CPCB	Central Pollution Control Board
CPLR	Common Property Land Resources
CPR	Common Property Resource
CSE	Centre for Science and Environment
CSIR	Council for Scientific and Industrial Research
CVM	Contingent Valuation Method
DoIPH	Department of Irrigation and Public Health
DPCC	District Pollution Control Committees
DRDA	District Rural Development Agency
DSS	Decision Support System
EAM	Ecological Aquaculture Model
EAW	Economic Aspects of Welfare
EDP	Eco-Development Projects
EIA	Environmental Impact Assessment
EPA	Entry Point Activities
EPA	Environmental Protection Agency
ETP	Effluent Treatment Plants
FAO	Food and Agriculture Organisation
FDI	Foreign Direct Investment
FFS	Farmers' Field Schools
GDI	Gender-related Development Index
GDP	Gross Domestic Product
GEF	Global Environment Facility
GEM	Gender Empowerment Measure
GHI	Gross Happiness Index

GM	Genetically Modified
GMOs	Genetically Modified Organisms
GNP	Gross National Product
GoG	Government of Gujarat
GoHP	Government of Himachal Pradesh
GoI	Government of India
GPI	Geniune Program Indicator
HDI	Human Development Index
HLE	Happy Life Expectancy
HPI	Human Poverty Index
ICAR	Indian Council of Agriculture Research
IIFM	Indian Institute of Forest Management
IMGLP	Interactive Multiple Goal Linear Programming
INSEE	Indian Society for Ecological Economics
IPM	Integrated Pest Management
ISEW	Index of Sustainable Economic Welfare
ISP	Index of Social Progress
IUCN	International Union for Conservation of Nature
IWDP	Integrated Wasteland Development Programme
LPG	Liquefied Petroleum Gas
MAI	Mean Annual Increment
MC	Marginal Cost
MEA	Multilateral Environment Agreement
MEW	Measure of Economic Welfare
MGLP	Multiple Goal Linear Programming
MINAS	Minimum National Standards
MoEF	Ministry of Environment and Forests
MR	Marginal Revenue
MU	Marginal Utility
NEERI	National Environmental Engineering Research Institute
NGO	Non-Governmental Organisation
NIOH	National Institute of Occupational Health
NPV	Net Present Value
NRAS	Natural Resource Accounting System
NTFP	Non Timber Forest Products
NTFP	Non Timber Forest Products
OECD	Organisation of Economic Cooperation and Development
PA	Protected Area
PCA	Principal Component Analysis
PCB	Pollution Control Boards
PDS	Public Distribution System
PSU	Public Sector Unit
RPF	Resource Poor Farmers
SHG	Self Help Group

SNA	System of National Accounts
SPCB	State Pollution Control Board
SSI	Small-Scale Industries
TDS	Total Dissolved Solids
TRIPS	Trade Related Intellectual Property Rights
UNDP	United Nations Development Programme
UNEP	United Nations Environment Programme
UNFCCC	United Nations Framework Convention on Climate Change
UPDASP	Uttar Pradesh Diversified Agricultural Project
VFC	Village Forest Committees
WCED	World Conference on Environment and Development
WHO	World Health Organisation
WISP	Weighted Index of Social Progress
WRO	Water Resource Organisations
WTO	World Trade Organisation
WTP	Willingness to Pay

Preface

Through its activities like Biennial Conference, workshop, policy round table and seminar, the Indian Society for Ecological Economics (INSEE) has been making concerted effort towards furthering the understanding of ecological economics. Issues in measurement of income and wealth including green accounting, man-made capital for natural capital, valuation of ecological functions, efficiency, scale (physical size not the economies of scale), thresholds and uncertainty and the need to bridge the gap of knowledge and epistemology are some of the issues which have commanded a central place on INSEE's priority list. The theme of the Fourth Biennial Conference was 'Ecology and Human Well-Being'. Ecology and ecosystem provide an array of goods and services to the humans but their contribution remains blurred in terms of accounting and valuation, although the impact of those contributions happens to be profound. This on the one hand shows the lacuna of measurement tools, on the other hand it conveys the great need for a new and improved framework that existing 'mainstream' economics has been unable to provide. In this endeavour, one should not be oblivious to the dynamic nature of ecosystems where the interactions among four basic system functions: exploitation, conservation, release and reorganisation are critical and are always in a changing mode. Resilience or ability of the ecosystem to come back to its original position after disturbances is something one should always take into account while designing the interventions and responses to the problems of ecology and environment. Social and ecological resilience could work in different directions with varying degrees in terms of impact. Exploring and analyzing the thresholds and irreversibility of these systems could bring unprecedented excitement and be of much relevance for researchers.

Valuation of incremental change in ecosystem services is of special relevance for the decision makers who invariably face various types of trade-offs in day-to-day life. The debate on valuation is still very alive but the issue now is centered on 'how to do it' instead of 'whether we should do it'. The consensus seems to be emerging that valuation instead of being based on static and myopic preference of the individual, should be done on long-term sustainable criteria and indicators through a participative process. Valuation enables the decision makers to have better and more informed choices in the face of problems of natural resource management spreading out/spilling over into different areas especially in a country like India where forestry, common grazing land, water bodies (wetlands), coastal and marine resources, etc., need intervention at community levels through collective action at different levels of the decision making body.

The Indian economy at present is going through a process of much transformation. The process of reforms in domestic and external sectors is continuing but its impact on various aspects of ecosystem and well-being needs extra attention from academic fraternity as well as development practitioners, as the discipline of economics has never been monolithic or is the impact of changing economic policies on natural resources, their management and the overall impact on different constituents of human well-being. The issue of tiger conservation and rights of tribal people is debated and discussed all around and in this context, it is of extremely contemporary relevance that INSEE had chosen this theme of 'Ecology and Human Well-Being' for its Fourth Biennial Conference held on June 3–4 June, 2005 at the Indira Gandhi Institute of Development Research, Mumbai. The several sub themes, which had added focus during the Conference include:

- Ecological and Social Resilience
- Ecosystem Services and Quality of Life
- Policy Reform for Sustainable Development: Governance and Institutions
- Valuation for Ecosystem Changes
- Communities and Collective Action

Out of a large number of chapters presented during the conference, twenty one chapters have been carefully selected on the basis of their quality of content, relevance to the theme, methodological innovation and policy relevance. Researchers, teachers and development practitioners, in this broad area of Environment and Development, are the target group for the book.

Pushpam Kumar
B. Sudhakara Reddy

Introduction

Pushpam Kumar and *B. Sudhakara Reddy*

Ecology provides an insight into the functioning and conditions of ecosystems. A fine/fully functioning ecosystem in less perturbed conditions can yield numerous services so critical for mankind as these services sustain and enrich the human for their well-being. Human well-being comprises much more than the usual basket of material goods (income). Physical security, social harmony, health and freedom to make choices and take action in a participative and democratic way, are integral parts of human well-being (MA 2003). Obviously, well-being is different from well off. Multiple services derived from ecosystems like freshwater, forest, mountain, cultivated land, etc., contribute quite significantly towards human well-being. But their contribution remains mired in accounting and valuation, although the impact of those contributions happens to be quite profound. This on the one hand, shows the lacuna of measurement tools, on the other hand it conveys the great need for a new and improved framework that existing 'mainstream' economics has been unable to provide. In this endeavour, one should not be oblivious to the dynamic nature of ecosystems where interactions among four basic system functions: exploitation, conservation, release and reorganisation are critical and are always changing (Holling 1987). Resilience or ability of the ecosystem to come back to its original position after perturbations is something one should always take into account while designing the interventions and responses to the problems of ecology and environment. Social and ecological resilience could work in different directions with varying degrees of intensity in their impact. Exploring and analysing the thresholds and irreversibility of these systems could be of unprecedented excitement and relevance for researchers. Here, economists can learn meaningful insights about ecological production function from ecologists the way they learn from engineers about the flow of input and output in manufacturing and industries. Trans-disciplinarity and methodological pluralism are a must to understand the ecosystem services and their valuation yielding different incremental benefits to the society.

Valuation of incremental change in ecosystem services is of special relevance for the decision makers who invariably face various types of trade-offs in day-to-day life. There are numerous examples of valuation of ecosystem services, which have enabled efficient decision making. The instance of Catskill in New York is often cited where preservation of Catskill watershed yields far more benefits in terms of water supply to the city than any other treatment method to procure the same quantity and quality of water. There are other examples of ecosystem valuation and transaction for different ecosystems. A recent review has been done by Kumar (2005). In valuation, sometimes a single function valuation may be inadequate and only a multifunction valuation in an integrated ecology economy framework would provide a holistic view (Heal and Barbier 2006). The debate on valuation is still very alive but now the issue is centered on 'how to do it' instead of 'whether we should do it'. The consensus seems to be emerging that valuation instead of being based on static and myopic preference of the individual, should be done on long-term sustainable criteria and indicators through a participative process. Valuation enables the decision makers for better informed choices; problems of natural resource management transcends into a different arena especially in a country like India where forestry, common grazing land, water bodies (wetland), coastal and marine resources, etc., need intervention at community levels through collective action at different levels of decision making.

The Indian economy at present is passing through a process of transformation. The process of reforms in the domestic and external sector continues but their impact on various aspects of ecosystem and well-being needs special attention from academic fraternity as well as development practitioners. As the discipline of economics has never been monolithic, so is the impact of changing economic policies on natural resources, their management and the overall impact on different constituents of human well-being (MA 2003, 2005). The issue of tiger conservation and rights of tribal people is debated and discussed all around and in this context, it was of extreme contemporary relevance that the Indian Society for Ecological Economics (INSEE) chose this theme of 'Ecology and Human Well-Being' for its Fourth Biennial Conference. The objective is to explore the much-neglected ecology and to provide a framework for understanding its relevance to human well-being. It is aimed at providing the goals and strategies and policies that can help to provide the framework for ecological sustainability. It also brings to the participants the most up-to-date information and an appraisal of worldwide ecological issues. The conference presented an array of articles ranging from ecological and social resilience to sustainable land use management. These well-researched studies suggest that we must plan for a major restructuring of the criteria used by various agencies and others in the selection and design of development processes if we are to avoid committing future ecological blunders. The studies indicated the need for careful ecological research at various levels of the development process. With the help of a large number of technical chapters selected by the expert committee, INSEE deliberated on this theme on 3–4 June 2005 at IGIDR, Mumbai. Several sub themes got added focus during the Conference. They include ecological and social resilience, ecosystem services and quality of life, policy reform for sustainable development, governance and institutions, valuation for ecosystem changes, and communities and collective action. The majority of these chapters around the theme represent some of the latest developments in multidisciplinary/integrated research in ecological economics in India. Some of these chapters develop new tools while others demonstrate the application of existing tools for empirical observation. Invariably they also suggest how to evaluate the effects and effectiveness of different policy options on conservation of ecosystems and its impact on human well-being. Several chapters employ case studies to demonstrate the use of various models, concepts, and methodologies. As is often the case, the objective in many of them is to maintain, or at times improve, the integrity of the ecosystem under scarce resource conditions and multiple demands on them. Different drivers of change, working on ecosystems and the necessary response policy especially in the Indian context, also remain the focus of some of the chapters.

The selected chapters have been divided into six categories. The first section looks at ecological and social resilience, including riverbank erosion, vulnerability of dry ecosystems, valuation of ecosystem changes due to salinity and ground water depletion and reconciliation of weak and strong sustainability. The concept of resilience in ecological systems was introduced by G. S. (Buzz) Holling (1973), who published a classic chapter in the *Annual Review of Ecology and Systematics* on the relationship between resilience and stability. The section on resilience in social-ecological systems analyses the application of theories and ideas about resilience on interlinked systems of people and ecosystems with expansions into some areas of social science. Section four presents some of the insights gained from social-ecological systems that have been the subjects of research by the authors. The authors examined the feedbacks within and between the social and ecosystem domains. The broader issues of human well-being and sustainability with attempts to focus on operation criterion for sustainable development have also been examined. The first chapter in this section (Zulfiquar Ali Islam) expatiates the social resilience of the riverbank erosion displacees in Bangladesh and shows the sheer corrective type of measures they formulate and undertake for adaptation to unsafe riparian habitat before, during, and after their displacement from their original homestead plots. The next (R. Nallathiga) takes a different perspective on degradation of environmental resources that is conditioned by natural factors and exacerbated by anthropogenic factors. The author analyses the patterns and trends in the degradation of resource components—land and water—in the light of socio-economic pressures. The third chapter (J. Sathyapalan and S. Iyengar) provides an overview of changes in the ecosystem due to human action, induced salinity and groundwater depletion on the lives of people living in the coastal areas of the state of Gujarat. It emphasises that over-exploitation of groundwater by private farmers has imposed an environmental externality on the society. The last of the main chapters (B.S. Sri and M.S.V. Prasad) offers a new insight into the conceptual and analytical approach to reconcile weak and strong sustainability. It involves a reconsideration of the conception of total capital from an ecological-economic system perspective. The chapter stresses the need for improved institutional frameworks with support from governing bodies, local institutions and through policy reforms.

The second theme presented here is the issue of ecosystem services and quality of life. In this section, the authors explore attempts to define different 'indicators' and criteria of human well-being. The use of ecosystem related indicators is to measure the vulnerability of systems, the quality of human or ecosystem health or the level of development. In particular, this is the result of growing interest in human development. It also results from a growing understanding of the limitations of the traditional measures, such as gross national product used during the twentieth century to track quality of life. It sought to broaden public understanding about the natural ecosystems. The section on ecosystem services and quality of life (the benefits people obtain from ecosystems), focuses on how changes in these services have affected human well-being, and response options that might be adopted at the local, national or global level to improve ecosystem management and thereby contributing to human well-being. The chapter demonstrates how, starting from a condition of diversity of ideas and interests among the stakeholders, systematic dialogue and mutual learning could be generated, leading to identification of options for more sustainable resource management practices. The impact on fish harvested in Digha fishery has been addressed by modelling an aggregated Gordon-Schaefer while integrating economic biodiversity index and an environmental quality variable under different biodiversity scenarios by Anita Chattopadhyay Gupta and Rabindra N. Bhattacharya. The results show that there exists a trade-off between economic biodiversity conservation and profit maximisation and there is a need for policy measures to minimise the level of conflict between them. The research findings from a bio-economic model (K. Gupta et al.) concluded that technological change leads to an expansion of aquaculture industry and contraction of the wild fishery. This result is important from the point of view of policy makers since it emphasises the need for defining more socially and ecologically responsible aquaculture industries that enhance traditional fishery and reduce user conflicts that are currently in existence. A. Shah presents a discourse on management of protected areas from purely conservationist strategies to participatory approaches with a wide range of options that combine different elements of resource sharing, market regulation and privatisation. This made way to an analysis that the cost of biodiversity loss and the development of appropriate institutions and incentives should primarily be a local exercise. Another chapter by A. Singh et al. holds an investigation exploring the relationship of current land use and crop productivity with external factors like climate, fertiliser use and soil quality. The indices have been developed for land degradation through ranking method, index method and Principal Component Analysis (PCA). It is concluded that there is a need for the adoption of integrated pest management practices for sustainability and cost effectiveness that help in making positive environmental impact.

Various methods have been developed in the past that allow users to express the value of goods and services in quantitative and monetary units. However, these methods are often complicated and demanding in terms of the time, expertise and data required. The section on valuation of ecosystems takes a new approach, exploring how to improve the connection between the well-documented analytical efforts to place a value on forests wetlands and mangroves. The chapters presented here provide the concepts on resource management and valuation with practical experiences from case studies in which a stakeholder-oriented approach has been used as a source of empirical data. The discussion of human dimensions helps understand how to select measurable objectives that allow for the appropriate assessment of the benefits of ecosystem services to human communities and economies. D. Mondal, S. Singh and J.V. Dhameliya provide a study detailing the economic valuation of some selected wetlands in the Burdwan district of West Bengal. It presents the estimated indirect user values of wetland resources in terms of the environmental and ecological services they provide to support current production and consumption of fisheries. The detailed use of various valuation techniques for ecosystem services rendered by different ecosystems ranging from mangroves to wetlands has also been presented here. It also reflects the need for carefully using the valuation techniques to prevent bias and narrowing down the wide variability of results in the valuation findings. The chapter by S. Bandyopadhyay, K. Narayanan and A. Ramanathan focuses on urban wetlands and the need for prioritising the process of urbanisation which generates the greatest volumes of wastes and pollutants as also the large scale conversion for land use. The study attempts to explore people's perceptions and preferences regarding the wetlands of Kolkata. The chapter by Saudamini Das focuses on the question of justifying the benefits that would accrue from the initiatives taken to generate degraded lands. It also speaks about looking for mechanisms to value the forests in entirety. The study by B. Bleys advocates the use of new approaches for measurement of welfare, discussing at a great length the Index of Sustainable Economic Welfare (ISEW). The last chapter suggests some practical tips for overcoming barriers to limits of valuation of ecosystems in developing countries in particular, besides some recommendations for the same.

There are some pertinent questions about collective action and natural resource management. Why do private landowners and public land users initiate collective action for Natural Resource Management (NRM)? Whether, and how, collective action and NRM planning changed resource decision-making and management? To get the answers, one has to understand the context in which private landowners and public land users make the decisions, the factors that affect their decision-making and the reasons for initiating a collective action effort. It is also important to analyse how the regional and national political economy affects individual landowners' and public land users' decision-making. The section on natural resource management offers a general discussion of the diverse and dynamic social values that people place on natural resources and the role these values play in natural resource policy and management. The first chapter in this section by P. Sudha et al. focuses on the framework for prioritising ecological issues over forest management through people's participation in conserving the ecology both at the micro as well as at macro levels. The next chapter by H.N. Chanakya et al. takes up a broader view and raises a pertinent question -- whether our policies provide enough incentives to promote community participation? It tries to study the status of food security and vulnerability among RPF members of Self–Help Groups (SHG) who have adapted and used appropriate technology and eco-friendly inputs in agriculture. Z. Husain discusses the collective participation for conservation and environment protection with examples from cooperative fisheries in Kolkata, forest reserve in Karnataka to Sariska tiger reserve in Rajasthan. Various issues related to human-forest interaction, dependence and management are detained to bring out clarity on the underlined issues. Concerns are raised even with the careful use of radical choice model in congruence with field observation for minimising the error from the research findings. The last chapter in this section by S.C. Srivastava finds that over the years, cropping pattern under shifting cultivation has undergone significant changes mostly in favour of market economy. It provides information on the communities' preference of programmes and policies for sustainable development including planning for land use and reforestation.

Despite unprecedented levels of concern regarding ecological problems, most thinking about solutions still fail to come to terms with the 'limits to growth' argument and therefore produce strategies that have no possibility of achieving sustainability. This means that a sustainable society cannot have affluent living standards, extensive industrialisation or a growth economy. There is no possibility of all the world's people rising to the levels of production and consumption characteristic of the rich countries today. We must eventually move to ways of providing a high quality of life from a small fraction of present levels of production and consumption, within a zero growth economy. Although in literature this argument has been in place for more than three decades, when it comes to official policy discussion, the planners completely ignore this model of Ecologically Sustainable Development. The chapters in this section insist that such a debate is of fundamental importance for people concerned about resource use. The section on policy reforms and sustainable development sets the stage for what follows and concludes with several pieces of advice. The first chapter by P. Panth discusses various facets of unsustainabiltity of economic globalisation, development of pollution heavens and irony of Kuznet's curve for the development case in countries such as India, marching ahead on the path of development. Issues of equivalence of economic globalisation and sustainable development are raised with a need for more empirical evidence for holding market forces solely responsible for environmental degradation. The next chapter by P.M. Prasad stresses the need for a comprehensive management strategy of protected areas, which include purely conservationist strategies as well as participatory approaches with a wide range of options that combine different elements of resource sharing, market regulation and privatisation. In her chapter, Lekha Mukhopadhyay examines the potential of successful voluntary participation in effort regulation programme to check over utilisation and thus degradation of Common Property Resources (CPR) in a society where CPR is nested under private property regime with inequality. With the help of a simple model of agro-pastoral village society producing milk with cattle (private property, which is unequally distributed) and fodder (which is collected from forest, a common property), this chapter examines the potential of successful voluntary participation in effort regulation programme. The chapter by Saravanan shows that the management of water resources is influenced by diverse forces, but the institutional options available are divergent and do not match with the ground realities. It calls for strengthening the distributive governance of existing sectoral departments. The last chapter in the volume is by Vasisht in which a methodology for exploratory land use analysis and planning has been developed using Interactive Multiple Goal Linear Programming (IMGLP) approach. The model assumes that all water and capital within a land unit can be shared which implies that groundwater resources available within a farm can be transported to other farms without cost, irrespective of the distance involved. This would require that the small farmers who cannot use alternative, efficient and capital intensive technologies should not cultivate their land and that their water and other resources be made available to other farmers who presumably could use these more efficiently.

We have tried to present the chapters in a comprehensible manner so as to be intelligible to the technical as well as non-technical readers. We sincerely hope that this effort will be of interest to development practitioners and of particular usefulness to the students of ecological economics, policy makers, conservationists and other professionals concerned with conservation and development.

REFERENCES

Heal, Geoffrey M. and E.B. Barbier. 2006. 'Valuation of Ecosystem Services', *Economist's Voice,* January. (www.bepress.com/ev)

Holling, C.S. 1973. 'Resilience and Stability of Ecological Systems', *Annual Review of Ecology and Systematics,* 4: 1–23.

———. 1987. 'Simplifying the Complex: The Paradigms of Ecological Function and Structure', *European Journal of Operational Research*, 30: 139–46.

Kumar, Pushpam. 2005. *Market for Ecosystem Services*. Winnipeg, Canada: IISD.

Millennium Ecosystem Assessment (MA). 2003. *Ecosystem and Human Well Being: A Framework for Assessment*. Washington DC: Island Press.

———. 2005. Ecosystem and Human Well Being: Findings of the Conditions and Trends Working Group. Washington DC: Island Press.

Section 1

Ecological and Social Resilience

1

Social Resilience of the Riverbank Erosion Displacees in Bangladesh

M. Zulfiquar Ali Islam

Abstract: This chapter expatiates the social resilience of the riverbank erosion displacees in Bangladesh. It is intended to explore the prodigious needs of this riparian community induced by the catastrophic ferocity of riverbank erosion displacement. Concurrently, the chapter tries to mirror the pattern of responses that are received from different sources to the pattern of their needs. The present ecological chapter is designed to spotlight the unflinching courage and resilience the displacees show in confronting the critical and uncertain situations faced by them before, during, and after their displacement from their original homestead plots. In addition to this catalogue of findings, the chapter distils through indigenous mechanisms the displacees of a Bangladesh village—mechanisms designed and undertaken for coping with their socio-economic losses and the consequential immense sufferings in the absence of organisational support and responses. Finally, the chapter frames some recommendations, which the policy planners and development organisations may consider in their future programme content while planning for the development of the riverbank erosion displacees.

INTRODUCTION

Social resilience is interrelated with ecological resilience, as the members of human society have to be dependent on the ecological resources to meet their enormous needs induced by different environmental disasters. The riparian ecosystem of Bangladesh has to cope with many environmental disasters, such as floods and riverbank erosion, with shifting into a qualitatively different state. In this ecosystem, the process of rebuilding, after floods and erosion, promotes renewal and innovation. It may be noted that in extreme cases, the ecosystem becomes vulnerable to the effects of flood and erosion attacks that previously could be absorbed. The desolate state of the riparian ecosystem of Bangladesh renders it not only biologically and economically impoverished, but also irremediable.

The riverbank erosion displacees in Bangladesh show enormous social resilience and unflinching courage in order to withstand and recover from the environmental change and/or social, economic or political upheavals caused by such geomorphological phenomenon. They have to be socially resilient as they are the users of the riparian tract in their everyday life. The riparian ecosystem and its inhabitants are interacting with and interdependent on each other and their reciprocity is dynamic. This chapter is concerned with the social resilience of the riverbank erosion displacees and intends to find out the amount of shock they can absorb and still remain in a normal state, the degree to which they are capable of self-organisation in the sheer absence of government and non-government based organisational support and the degree to which they can build capacity for learning and adapting to the wobbly and critical riparian environment.

CONCEPTUAL FRAMEWORK

The riverbank erosion displacees are those who are displaced from their riparian homestead plots due to riverbank erosion attacks at least once in their lifetime. The shifting of major rivers of Bangladesh and their unstable character causes this environmental disaster. Consequently, thousands of people are compelled to leave the erosion-threatened areas every year. The geomorphological phenomenon of riverbank erosion displaces the riverine people from their original homestead plots and therefore their livelihood is devastated. The displacees are categorised as 'displaced once', 'displaced twice', 'displaced thrice' and 'displaced more than thrice' in the present study.

The present study uses the concept of 'social resilience' in order to explore the ability and adaptability of the displacees in the pre-displacement period, during the onslaught of riverbank erosion, as well as the post-displacement period. The resilient strategies at different levels of their adaptation to the precarious riparian habitat are focused upon in this chapter. Also, it deals with their resilience in meeting their phenomenal needs induced by the catastrophic attack of riverbank erosion and the resultant displacement. Their social resilience also encompasses their endeavours and unflinching courage in formulating and undertaking the corrective strategies for bridging the gap between their enormous needs and immense sufferings and the scarce resources and indigenous technologies at their disposal.

STUDY LOCALE AND DATA SOURCES

Sehala—a medium-sized village of Nawabganj District in the northwestern region of Bangladesh—is selected as the locus of study. It is located in Nawabganj Sadar Upazila of the district. The geographical features indicate that more or less half of the upazila (sub-district) area amounting to 91,039 acres is *char* land.[1] The principal rationale for selecting Sehala as the locus of study is that a sizeable number of displacees from different erosion-affected areas of Nawabganj district have settled in this peri-urban area in search for food, shelter and employment. It adjoins Nawabganj Town and the Barind tract[2] as well. The locale provides the displacees with access to the labour market of Nawabganj town and also to the avenues for agricultural employment in the Barind tract. The catastrophic effects of riverbank erosion in Nawabganj and the lack of social studies on this problem in the Ganges-Mahananda floodplain establish another important reason for selecting it as locus of study.

A household level survey was conducted to explore the displacee households settled in the locale. And it was followed by a sample survey for investigating the resilience of the displacees in the face of catastrophic situations they experience in the riparian tract. The sample size was 140 displacee households (100 per cent of the total) with its displacee population being 766 (100 per cent of the total). All the displacee households are considered as the appropriate primary sampling units here. The displacee household heads were directly interviewed and in this way the respective household head represents each sampling unit. In addition to the two-tier survey, the local government officials, local elites and/or public representatives and the non-displacees were interviewed. The research also collected data by conducting seventeen Focus Group Discussions (FGDs) with the displacees and their community.

The principal tools for collecting the primary data were questionnaire and interviews—two techniques of survey method. The major sources of primary data, in addition, include observation, informal interviews, case histories and case studies of selected persons and notable issues. Additional sources of data used in this study are local level official reports, evaluation of governmental and semi-governmental projects, programmes and census reports, published reports and articles, etc. The nature of this study is based on an extensive fieldwork conducted during the period June to December 2004.

RESILIENT STRATEGIES FOR LOSS REDUCTION

Riverbank erosion is a recurrent environmental disaster in Bangladesh. It contributes directly to the process of rapid pauperisation of the riparian people. It displaces millions of people from their riparian tracts every year (Elahi and Rogge 1990) and claims many lives and properties as well. The disaster often dislocates cultivable land—the principal but scarce resource to the riparian people—and human settlements, and also destroys standing crops, roads and communication systems. The displacees generally formulated and undertook corrective rather than preventive strategies in reducing their socio-economic losses as they were threatened by the riverbank erosion attacks. As a matter of fact, the displacees' position in the social hierarchy and their low-level technological know-how force them to do what is corrective in nature for minimising their losses and the consequent immense sufferings. In spite of such adversity, they were resilient and importunate in formulating and undertaking multiple remedial loss reduction strategies.

Use of Movable Housing Materials

The displacees tenaciously use movable housing materials and it is a widely practiced precautionary measure for minimising their economic losses. More than one-quarter of the displacee households of the study area reported that they had economic capability for building concrete houses on their original homestead plot but could not do so because of riverbank erosion attack. According to them, '*nodikaataar bhoye paakaa ghar korini, kakhan sarbonaash hoy ke jane*' (the threat of riverbank erosion prevented them from building concrete houses). They were compelled by the threat of riverbank erosion to use movable housing materials. They were resilient enough to adapt to such vulnerable habitat. Their strategy of using movable housing materials is not incidental but rather one of the purposive adaptation strategies. However, it does not support the findings of some studies on riparian people (Haque 1991).

The original housing structure of displacee households, prior to displacement, was usually traditional. This housing structure was constructed of materials like mud-dough, thatch, bamboo, *tarja* (fence made of bamboo), wooden plank, burnt tile, corrugated iron sheet, etc. These materials are easily movable and less susceptible to the damage caused by riverbank erosion. Not only that, the displacees of Sehala procured the resalable value of these housing materials immediately after displacement. They used this money in rebuilding their hut on the Bangladesh Water Development Board (BWDB) embankment, on khasland,[3] on any other land owned by anybody or even beside the roads.

More than half of the displacees (50.72 per cent; n=71 of 140) used corrugated iron sheets as roof material in their pre-displacement period (Table 1.1). The use of this roof material was followed by thatch (25.71 per cent; n=36 of 140) and burnt tile (17.86 per cent; n=25 of 140). These roof materials have salvageable, resalable, and reusable values in the aftermath of a riverbank erosion attack. The wall materials used by the displacees of Sehala village prior to their displacement include bamboo and/or thatch (40.71 per cent; n=57 of 140), mud-dough (35.00 per cent; n=49 of 140) and brick (24.29 per cent; n=34 of 140). Except for mud-dough, all the wall materials have salvageable and reusable values. Moreover, only the wall material of brick has salvageable, reusable and resalable values.

Investment Pattern

The displacees during the pre-displacement period invested their capital in purchasing land in Sehala for resettlement (72.86 per cent; n=102 of 140), movable assets (52.14 per cent; n=73 of 140) and livestock (30.71 per cent; n=43 of 140) (Table 1.1). They purchased these assets purposefully because they were supported by these assets in the event of desolation caused by riverbank erosion. They purchased land in Sehala for their resettlement after displacement. This type of investment was nothing but their precautionary measure to adapt to the uncertain and unsafe riparian environment.

Erosion Preventing Technology

The displacees of Sehala found their indigenous technology of piling sandbag (53.57 per cent; n=75 of 140) and building bamboo crates (52.14 per cent; n=73 of 140) partially effective in protecting their land (Table 1.1). The strategy of land protection minimises the displacees' losses but this indigenous technology may be partially effective in one season and may get eventually subjected to erosion in the next season.

Loss Acceptance

The displacees were forced to accept their losses due to riverbank erosion displacement. They did not have any alternative choice other than that of loss acceptance as they failed in protecting their cultivable land, homestead plots and other valuable properties from the cataclysmic riverbank erosion.

Table 1.1 Social Resilience of the Displacees for Minimising their Socio-economic Losses

		Sehala	
		Households	*N=140*
	Resilient Strategies for Loss Reduction	*n*	*%*
Use of Movable Housing Materials			
Roof Materials	Corrugated Iron Sheet	71	50.72
	Thatch	36	25.71
	Burnt Tile	25	17.86
	Rod Cement Concrete	7	0.05
Wall Materials	Bamboo/Thatch	57	40.71
	Mud Dough	49	35.00
	Brick	34	24.29
Investment Pattern			
	Land Purchase for Resettlement	102	72.86
	Moveable Assets	73	52.14
	Livestock	43	30.71
Erosion Preventing Technology			
	Using Sand Bag	75	53.57
	Building Bamboo Crates	73	52.14
Loss Acceptance			
	Homestead Plot Desertion	140	100.00
	Pray to Allah	50	35.71
Reducing Economic Loss			
	Salvaging Housing Structure	135	96.43
	Moving Properties	95	67.86
	Sale of the Title of Eroded Land	40	28.57
	Sale of Livestock	32	22.86
	Cutting Standing Crops	13	9.29
	Cutting and Selling Trees	13	9.29
Shift of Lives and Properties			
	Family	140	100.00
	Assets	88	62.86
	Livestock	26	18.57
Means of Transportation			
	Bullock Cart	40	28.57
	Bicycle	20	14.29
	Country Boat	18	12.86
	No Transport	63	45.00

Note: Multiple responses considered.

Their local initiative and indigenous technology of sandbag piling and building bamboo crates for preventing erosion-attack had ended in a failure. Their preventive strategies were found to be useless and the erosion-attacks had gone out of their control. Eventually, though the displacees accepted their losses due to riverbank erosion, they set about formulating some corrective measures. All the displacees (100 per cent; n=140 of 140) deserted their original homestead plots which is one of their indigenous strategies for accepting the loss (Table 1.1). Also a considerable proportion of them (35.71 per cent; n=50 of 140) prayed to Allah for preventing the erosion which is nothing but a negative acceptance of their losses.

Reducing Economic Loss

The displacees were found to be resilient in formulating and undertaking a number of mechanisms to reduce their economic losses. They salvaged their housing structures (96.43 per cent; n=135 of 140) (Haque 1991; Rogge 1991), moved properties from erosion-threatened areas to safer places (67.86 per cent; n=95 of 140), sold the titles of their eroded land (28.57 per cent; n=40 of 140) to the wealthy landowners who could afford to wait for the reemergence of their dislocated land (Rogge 1991; Zaman 1987), and also sold their livestock (22.86 per cent; n=32 of 140) which is empirically supported by Haque (1991) and Rogge (1991) (Table 1.1). In rural Bangladesh, it is easier to sell livestock than other assets in order to have some cash in any event of immediate requirement. In addition to these, they cut their standing crops (9.29 per cent; n=13 of 140) and even cut down trees and sold them (9.29 per cent; n=13 of 140) in some cases.

Shifting of Lives and Properties

The shifting of lives and properties from erosion-threatened homesteads to a safer place is another corrective strategy undertaken by the displacees of Sehala. It encompasses some measures: shifting of family (100 per cent; n=140 of 140), shifting of assets (62.86 per cent; n=88 of 140) and shifting of livestock (18.57 per cent; n=26 of 140) (Table 1.1). The proportion of the displacees which formulated and undertook the strategy for a change in location is more than twice that of the displacees of Kazipur (Haque 1991). Haque (1991) found that 43.5 per cent of the displacees moved their families, 9.3 per cent livestock and 15.5 per cent shifted their belongings from erosion-affected areas to comparatively safer places.

Case #1

Idris Hossain Gharami is a 32 years old man. He has experienced the displacement status thrice in his lifetime. His homestead was at a distance of three metres which the erosion-attack engulfed in one night. As a matter of fact, he did not have any chance to dismantle his hut. Also, he lost all his land and became a landless labourer. During the last displacement, his family members were given shelter by his relatives in Sehala.

Means of Transportation

The displacees used different means of transportation in reducing their losses due to erosion. They possessed indigenous means of transportation which included bullock carts (28.57 per cent; n=40 of 140), bicycles (14.29 per cent; n=20 of 140) and *dingis* (country boats) (12.86 per cent; n=18 of 140) (Table 1.1). They used these means in carrying dismantled housing materials, wood, livestock and other tangible goods during the onslaught of riverbank erosion and high floods. This task of shifting goods is not manually manageable. The displacees with no means were helped and supported by their neighbours and relatives in carrying their goods from the erosion-affected homesteads to safer places.

It can be inferred from the field data that the displacees of Sehala undertook multiple strategies in accepting and reducing losses and in shifting their lives and properties. It is predicted that the dominant strategies would be land desertion (100 per cent) for loss-acceptance, salvaging housing structure (96.43 per cent) for loss-reduction and shifting family (100 per cent) for location change. The resilient strategies of loss-acceptance, loss-reduction, and shifting of lives and properties contribute crucially to the process of displacees' environmental adaptation to their vulnerable riparian habitat.

Place of Shelter

The displacees of Sehala took shelter on the flood-protecting embankments, on neighbour's land, on *khas* land, beside the roads and highways and under the shed of kin or neighbours. It is noticeable that 59.29 per cent (n = 83 of 140) displacees took shelter on the BWDB embankment (Figure 1.1). They rebuilt their small huts on both the sides of the embankment and took shelter in as in a squat dwelling.

Figure 1.1 Riverbank Erosion Displacees in the Places of Temporary Shelter

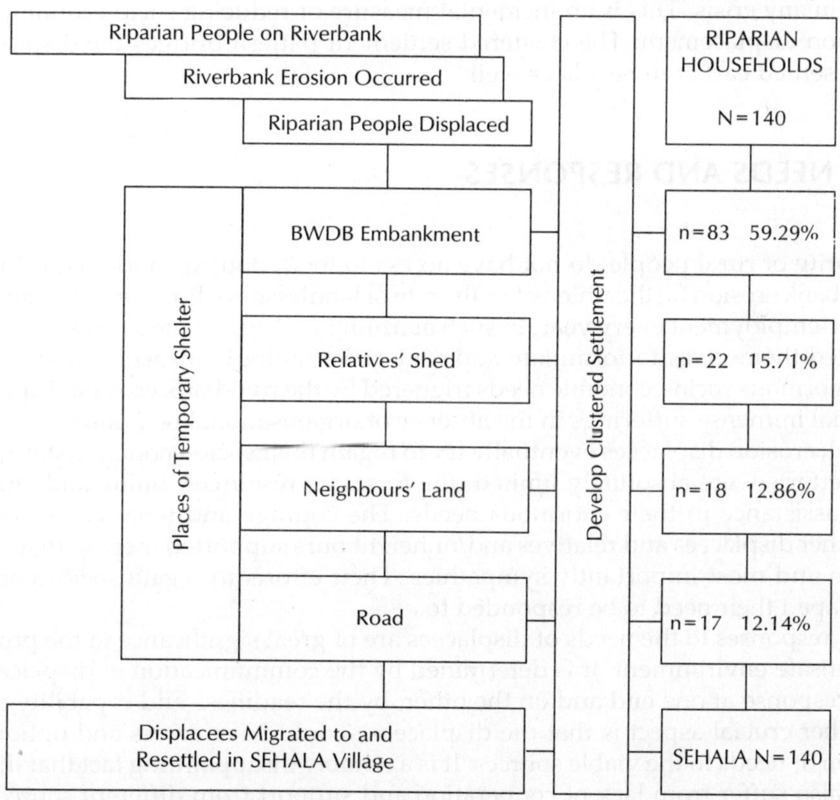

Note: Multiple responses computed.

They were found to suffer from the lack of drinking water, sanitation facilities and emergency health care services. Some displacees took shelter under their relatives' sheds (15.71 per cent; n=22 of 140) and on their neighbours' land (12.86 per cent; n=18 of 140). They were also supported by their relatives and/or neighbours in having access to drinking water and sanitation facilities to some extent. A few displacees (12.14 per cent; n=17 of 140) were found to take shelter beside the roads. It was reported that they were in such places of shelter for a period of three months at the lower limit, and of one year at the higher limit. Finally, they left these places and migrated to Sehala for developing their new settlement.

Clustered Settlement Pattern

The displacees lived in a clustered settlement on the original homestead plots and in their places of temporary shelter as well. They also started to settle themselves in Sehala staying closely together. They formed a cluster and/or contiguous settlement in their present residential locality. They were also clustered into a major squat on the flood-preventing embankment (Zaman 1986b).

The settlement of displacees in a cluster and contiguous pattern is a corrective type of strategy for adapting to a new social environment. One's homestead neighbours on another's helps them maintain their *samaj*[4] ties. It is observed that the *samaj* members extend assistance and cooperation to one another in any crisis. This is an incidental measure of reducing socio-economic loss due to riverbank erosion displacement. The clustered settlement pattern bridges the displacees and the non-displacees settled earlier in Sehala as well.

NATURE OF NEEDS AND RESPONSES

While the majority of rural people do not have access to food, housing and medical facilities, the disaster of riverbank erosion further intensifies the rate of landlessness, homelessness, and unemployment and under-employment every year. In such alarming and aggravated conditions as these, the displacees resiled themselves to formulate and undertake multiple measures and techniques for meeting their enormous socio-economic needs triggered by the riverbank erosion displacement and the consequential immense sufferings in the absence of organisational responses.

The riverbank erosion displacees eventually try to regain their socio-economic status. It is noticeable that their efforts were absolutely limited due to scarce resources, finite land and absence of organisational assistance to their enormous needs. The courage and tenacity of displacees was multiplied as other displacees and relatives and/or neighbours supported them with manual labour, accommodation and most importantly sympathies. Their efforts to regain socio-economic status significantly shaped their need to be responded to.

The levels of responses to the needs of displacees are of great significance in the process of their adaptation to unsafe environment. It is determined by the communication of displacees' needs to the sources of response at one end and on the other, by the readiness and capability of sources of response. Another crucial aspect is that the displacees should have access and option in relating their articulation of needs to the viable sources. It is a notably disappointing fact that the displacees in Bangladesh also suffer from lack of cooperation and support from different sources which are necessary for bridging the gap between them and the organisational responses.

It is the sources of response at the individual level that primarily provided the displacees with assistance in the immediacy of riverbank erosion displacement. These sources were usually confined to relatives, neighbours, *samaj* members, friends, etc. They also may be considered as the channels for mobilising the articulation of the nature and extent of the displacees' needs for their desperate state of livelihood. The organisational sources of response comprise non-government and government organisations. The governmental organisational sources are at the local and national levels. In addition to that, the non-government organisations at the international level may be available for responding to the victims of environmental disasters as well. Both the government and non-government organisations may work as viable sources in responding to the needs of displacees. They may also mobilise for spreading awareness about hazardous after-effects of riverbank erosion displacement among the people on the erosion-affected and erosion-prone areas. The local level sources can be pressed into service for assisting the displacees in the immediate aftermath of riverbank erosion. The effective efforts positively articulate the needs of displacees. Eventually, they highlight the significance for responding to the displacees at national and international levels.

The riverbank erosion displacees in Bangladesh do not receive any response from organisational sources to their prodigious needs for environmental adaptation in a significant way. Whatever responses they got were mostly by their relatives, neighbours, *samaj* members and friends; but they got no support from the organisations at the governmental level. A considerable number of studies (Elahi 1989, 1991; Elahi and Rogge 1990; Haque 1988, 1991; Haque and Zaman 1989; Hossain 1984; Hossain and Greendberg 1985; Rahman 1991; Rogge 1991; Wiest 1991; Zaman 1986a, 1988, 1989, 1991; Zaman and Wiest 1991) have made inferences that no governmental organisational support is provided to assist the displacees in getting shelter or employment. The adaptation strategies undertaken by them have received, to date, little attention at the organisational level (Mahbub and Islam 1991). It is also noticeable that the inadequate economic and social laws of a class-society like Bangladesh, are significantly worsening the socio-economic status of the displacees (Amin 1991).

The government needs to consider the size of erosion-affected population and the severity of catastrophic situation induced by riverbank erosion. Unfortunately, the government of Bangladesh has not yet formulated any long-term and ongoing response strategy for the riverbank erosion displacees.

Pattern of Expected Needs and Responses

The present study identifies seventeen principal needs for displacees' adaptation to their precarious and vulnerable riparian environment. It also mentions the respective sources of response received by the displacees of Sehala.

Immediate Needs

More than nineteen per cent (n = 27 of 140) of displacees expected that the local and/or national government should undertake measures for the prevention of erosion (Table 1.2). It is not possible for the individuals to undertake effective measures in this respect since it requires large-scale engineering works. It is reported that only two (1.43 per cent) displacee households received assistance in

Table 1.2 Differential View of Sources of Response Expected and Responded to Displacees' Immediate Needs

		Sehala				
	Households			*N=140*		
	Potential Responses			*Actual Responses*		
		Displacees Expected		*Displacees Received*		
Areas of Displacees' Needs	*Sources Expected*	*n*	*%*	*n*	*%*	*Sources Responded*
Immediate Needs: Erosion Prevention	Governmental Organisations	27	19.29	2	1.43	Relatives Neighbours *Samaj* Members
Physical Labour to Move	Relatives Neighbours *Samaj* Members Friends	72	51.43	68	48.57	Relatives Neighbours *Samaj* Members Friends
Shelter	Relatives Neighbours *Samaj* Members Friends	100	71.43	17	12.14	Relatives Neighbours *Samaj* Members Friends
Temporary Hut	Governmental Organisations	90	64.29	90	64.29	Governmental Organisations
Financial Assistance	Governmental Organisations NGOs	99	70.71	20	14.29	Relatives Neighbours *Samaj* Members Friends NGOs
Immediate Relief Supply	Governmental Organisations NGOs	117	83.57	102	72.86	Member of Parliament (Opposition)
Moral Support	Relatives Neighbours *Samaj* Members Friends	45	32.14	39	27.86	Relatives Neighbours *Samaj* Members Friends

preventing riverbank erosion with their indigenous technology during the onslaught on their original homestead plot. Their relatives, neighbours, and *samaj* members assisted them in doing the task. It is noteworthy that their technology was not effective and sustainable on the long-term.

After displacement, the displacees had to move from their original homesteads to safer places. They expected physical labour for salvaging houses and carrying housing materials and other tangible goods. It was a corrective type of need at the first phase of adaptation to hazardous situation. More than fifty-one per cent (n=72 of 140) displacees of Sehala expected this help from their relatives, neighbours, *samaj* members and/or friends. It is found that they were well responded to by their expected sources in this respect. More than forty-eight per cent (n=68 of 140) displacees of Sehala received physical labour from their relatives, neighbours, *samaj* members and/or friends to move their housing materials and other tangible goods from their erosion-affected homestead plot to a safer place.

In the immediate aftermath of erosion, the displacees need to be sheltered at any cost. While the displacees (71.43 per cent; n=100 of 140) expected to be sheltered by their relatives, neighbours, *samaj* members and/or friends, only 12.14 per cent (n=17 of 140) of them received such response from these expected sources. The displacees (64.29 per cent; n=90 of 140) expected that the government would provide them with adequate assistance with shelter. In this context, they were not responded to directly by the government but the displacees who built their huts beside the roadside (12.14 per cent; n=17 of 140) and on the BWDB embankment (59.29 per cent; n=83 of 140) were silently allowed by the government to stay temporarily in those places of shelter.

The financial assistance is crucial for the resettlement of displacees. Their (70.71 per cent; n=99 of 140) expected sources were governmental organisations and/or NGOs. It is a disappointing fact that only 14.29 per cent (n=20 of 140) of them received it, to whatever smaller extent, from their relatives, neighbours, *samaj* members and/or friends. A few of them were also responded to by two NGOs, as they were members of these NGOs. The displacees generally received no financial support from these sources (government and/or non-government organisations).

Nearly eighty-four per cent (n=117 of 140) displacees expected that the government and non-government organisations would provide housing materials, food, clothes and other necessary goods as immediate relief. The displacees also needed moral support in their desolate state of livelihood. Only seventy-three per cent (n=102 of 140) displacees were provided with food relief once doled out by the Member of Parliament (opposition) of that locality.

Subsistence Needs

The displacees expected that the food ration (77.14 per cent; n=108 of 140) and health care (26.43 per cent; n=37 of 140) should be provided for by the government and/or non-government organisations on a regular basis during the emergency situation (Table 1.2). They had to confront with food crisis whereas the emergency food ration may have lessened their sufferings. They were usually saddled with famine as they lost their standing crops due to the erosion. They were also subjected to health hazards with the ultimate result being an epidemic. They were immediately attacked by diarrhoeal diseases. In spite of such an unsafe and critical situation, the displacees did not receive any help and support from organisational sources. Their adaptation was further hindered by the lack of adequate housing and health care facilities (see Greenberg and Hossain 1987).

The drinking water and sanitation facilities in the places of their shelter were not adequately available to satisfy the displacees' real needs. They were compelled to carry drinking water from the neighbouring villages and/or to use the river and pond themselves as drinking water sources. They also did not have any sanitation facilities. To improve their living environment, building of low-cost housing with sanitary latrine and safe drinking water facilities needed to be provided for the displacees (Elahi and Rogge 1990). While they expected that the government and non-government organisations should provide the displacees with water (18.57 per cent; n=26 of 140) and sanitation facilities (20.71 per cent; n=29 of 140), only a few of them received safe drinking water (3.57 per cent; n=5 of 140) and sanitation facilities (5.00 per cent; n=7 of 140)—that too from their relatives, neighbours, *samaj* members and/or friends. In addition to that, the Members of Parliament (opposition) had installed two sanitary latrines and three tubewells for safe drinking water in the squatting settlement of the displacees.

More than seventy-nine per cent (n=111 of 140) displacees expected that they would be provided with income generating activities by the government and non-government organisations. But they (17.86 per cent; n=25 of 140) were only responded to informally and irregularly by their relatives,

Table 1.3 Differential View of Sources of Response Expected and Responded to Displacees' Subsistence Needs

	Areas of Displacees' Needs	*Sehala* Households – Potential Responses: *Sources Expected*	*Displacees Expected* *n*	*Displacees Expected* *%*	N=140 – Actual Responses: *Displacees Received* *n*	*Displacees Received* *%*	*Sources Responded*
Subsistence Needs	Emergency Food Ration	Governmental Organisations NGOs	108	77.14	–		
	Emergency Health Care	Governmental Organisations NGOs	37	26.43	–		
	Safe Drinking Water Supply	Relatives Neighbours *Samaj* Members Friends Governmental Organisations NGOs	26	18.57	5	3.57	Member of Parliament (Opposition) Relatives Neighbours *Samaj* Members Friends
	Sanitation Services	Governmental Organisations NGOs	29	20.71	7	5.00	Member of Parliament (Opposition) Relatives Neighbours *Samaj* Members Friends
	Income Generating Activities	Governmental Organisations NGOs	111	79.29	25	17.86	Relatives Neighbours *Samaj* Members Friends
	Employment	Governmental Organisations NGOs	126	90.00	27	19.29	Relatives Neighbours *Samaj* Members Friends

neighbours, *samaj* members and/or friends. In addition, their expectation of employment (90.00 per cent; n=126 of 140) was responded to in the same way by their (19.29 per cent; n=27 of 140) relatives, neighbours, *samaj* members and/or friends.

Resettlement Needs

The requirement of land for resettlement is most crucial to their adaptation. Adequate responses to this need help them in formulating and undertaking survival strategies. The government is considered as a viable source of response to this need. This expectation was made by 81.43 per cent (n=114 of 140) of the displacees (Table 1.2). Among these displacees, twenty-nine per cent of them were provided with land for the purpose of resettlement only without any ownership title at all by their relatives and neighbours and fifty-nine per cent of them by their friends. They were provided with a piece of land for building their housing structures and to some extent with opportunities for developing homestead agriculture on that plot. While they (85.00 per cent; n=119 of 140) expected the provision for housing materials for their resettlement from the government and NGOs, some of them (14.29 per cent; n=20 of 140) were provided with such materials by their relatives, neighbours, *samaj* members and/or friends.

The displacees had to enroll in a new *samaj* where they were resettled. They needed the cooperation of their *samaj* members. To secure their cooperation they (100 per cent; 140 of 140) wanted to establish good neighbourliness with their *samaj* members. Their need for enrollment in new *samaj* was satisfied by the relatives, neighbours, *samaj* members and/or friends. Also, they (7.14 per cent; n=10 of 140) developed fictitious kinship ties with their new neighbours and/or *samaj* members.

It is to be noted that more than nine per cent (n=13 of 140) of displacees clearly expected no assistance from any sources. It is because of their good economic standing though they lost their original homestead plot due to riverbank erosion attack. The fact is empirically supported by what Halli (1991) found in Kazipur. According to him, the displacees expected no assistance as they had skills and experiences in some non-agricultural occupations in addition to their agricultural skills.

Table 1.4 Differential View of Sources of Response Expected and Responded to Displacees' Resettlement Needs

		Sehala					
		Households			N=140		
		Potential Responses			Actual Responses		
		Sources Expected	Displacees Expected		Displacees Received		Sources Responded
	Areas of Displacees' Needs		*n*	%	*n*	%	
Resettlement Needs	Land for Resettlement	Governmental Organisations	114	81.43	24	17.14	Relatives Neighbours Friends
	Housing Materials	Governmental Organisations NGOs	119	85.00	20	14.29	Relatives Neighbours *Samaj* Members Friends
	Enrollment in New *Samaj*	Relatives Neighbours *Samaj* Members Friends	140	100	140	100	Relatives Neighbours *Samaj* Members Friends
	Developing Fictive Ties	Relatives Neighbours *Samaj* Members Friends	26	18.57	10	7.14	Relatives Neighbours *Samaj* Members Friends

Note: Multiple responses considered.

Case #2

Sabdar Ali Mondal of Kaloni is a man of forty-six. He was displaced three times due to riverbank erosion attack. He lost his homestead plot of 0.33 acre but not cultivable land. He is a poor peasant. His cultivable lands are in the Barind Tract—a safer area. In addition to that, he has been sharecropping. He is illiterate. His son has been doing wholesale business of fish.

The displacees were surrounded with family, kin groups, *samaj* ties, neighbours, friends, mosque, education, socio-economic status, health, livestock, tangible goods, market, employment, food and the like prior to their displacement (Rogge 1991). These surroundings were replaced by a plethora of needs due to riverbank erosion displacement. The cataclysm of riverbank erosion destroys their established settlement and living environment in the riverine Bangladesh (Nazem and Elahi 1990). It affects all the riparian people through eroding land and destroying employment facilities, which the land could provide for them (Romanowski 1988). It claims adequate responses from sources at both individual and organisational levels. More or less, all the displacees of Sehala considered the sources at organisational level, especially the national government, to be as viable and effective in responding to their needs. But unfortunately their expectation is yet to be fulfilled by the governmental organisational sources.

CONCLUDING REMARKS

The social resilience of riverbank erosion displacees of a northern Bangladeshi village shows the mere corrective type of measures they formulate and undertake for adaptation to unsafe riparian habitat in their own ways. It is because they receive no or meagre support from the governmental and non-governmental organisational sources. Moreover, their low level of socio-economic and technological capacity fails to prevent the attack of riverbank erosion and the consequent massive loss. This adverse and critical situation compels them to design and undertake corrective rather than preventive measures for reducing their loss. It can be concluded that in the absence of organisational support, the displacees of Sehala were compelled to become socially resilient and dependent on the assistance given by their relatives, neighbours, *samaj* members and friends in formulating and undertaking strategies for adapting to their precarious environment.

POLICY RECOMMENDATIONS

The policy implications suggested here prefigure that the government and non-government organisations should respond to the prodigious needs of riverbank erosion displacees. These responses should be made in the immediacy of erosion attack for aiding them in their environmental adaptation on the hazardous riparian tract. The recommendations include:

1. The government should undertake large-scale engineering works and allocate financial costs for preventing the riverbank erosion.
2. In the immediateness of erosion, the displacees need to be sheltered at any cost. The government should provide them with adequate assistance for shelter.
3. The displacees have to grapple with food crisis and thus the emergency food rations may alleviate their sufferings. In the event of an onslaught of riverbank erosion, the government must provide the food rations.

4. The displacees are subject to health hazards which could ultimately result in an epidemic. The government should provide them with health care and low-cost housing with sanitary latrine arrangements and arsenic-free drinking water facilities. This will aid them in adapting to their hazardous riverine environment.
5. The homestead plot, housing materials and financial support are crucial to their resettlement. They receive financial assistance from their neighbours and/or relatives but hardly any from any organisational sources. This job should be carried out by the government and non-government organisations.
6. The government and non-government organisations should provide the displacees with adequate income-generating activities and finally regular employment in both agricultural and non-agricultural sectors for their survival.

NOTES

1. A mid-channel islet in the riverbed. It is any accretion in the river which may be seasonal or may survive for several decades. *Char*lands are abundantly found in the large rivers of Bangladesh, such as the Padma. The intense competition among floodplain inhabitants to cultivate these *char* lands and/or to settle on it creates terrible social clash. These lands were regulated in British India by Bengal Regulations XI, 1825 (Wilson 1855).
2. A tract called *Varendra* in the Sanskrit literature, which means a part of Bengal, north of the Ganges/the Padma, designating especially one great division of the *Brahmans* of Bengal (Wilson 1855). This tract is one of the terrace areas of Pleistocene age within the Bengal Basin. It has two terrace levels—one at 39.7m and the other between 19.8m and 22.9m (Rashid 1977). It is divided into five sections, namely, Northeastern Outlier, Eastern Barind, East-Central Barind, West-Central Barind and West Barind. The district of Nawabganj includes parts of the West-Central Barind and of the Western Barind.
3. *Khas* is an Arabic term used to mean select, eminent, noble and also private, peculiar, etc. (Wilson 1855). *Khas* as a revenue term is applied to lands held by Zamindars and cultivated by themselves for their own benefit (Wilson 1855). The term of *khas*land is considered by the present study as unused land owned by the government.
4. It is an indigenous village social organisation. It is not organised by the government. It is a village council, which may be compared with *parea* of a Punjabi village in Pakistan (Eglar 1960). *Samaj* performs religious, ritual, ceremonial and adjudicative functions (for more, see Karim 1990).

REFERENCES

Amin, A.T.M.N. 1991. 'Settlement Strategy for Riverbank Erosion Displacees in Bangladesh: A Human Resource Development Approach', in K.M. Elahi, K.S. Ahmed and M. Mafizuddin (eds), *Riverbank Erosion, Flood and Population Displacement in Bangladesh*, Dhaka: Riverbank Erosion Impact Study (REIS), Jahangirnagar University (JU).

Currey, Bruce. 1979. 'Mapping Areas Liable to Famine in Bangladesh'. Ph.D. dissertation (mimeo), University of Hawaii, USA.

Eglar, Zekiye. 1960. *A Punjabi Village in Pakistan*. New York: Columbia University Press.

Elahi, K.M. 1989. 'Population Displacement Due to Riverbank of the Jamuna in Bangladesh', in Clarke et al. (ed), *Population and Disaster*, Oxford: Basil Blackwell.

———. 1991. 'Riverbank Erosion, Flood Hazard and Population Displacement in Bangladesh: An overview', in K.M. Elahi, K.S. Ahmed and M. Mafizuddin (eds), *Riverbank Erosion, Flood and Population Displacement in Bangladesh*, Dhaka: Riverbank Erosion Impact Study (REIS), Jahangirnagar University (JU).

Elahi, K.M. and John R. Rogge. 1990. *Riverbank Erosion, Flood and Population Displacement in Bangladesh: A Report on the Riverbank Erosion Impact Study*. Dhaka: Riverbank Erosion Impact Study (REIS), Jahangirnagar University (JU).

Greenberg, C. and M.Z. Hossain. 1987. 'Adjustment of Riverbank Erosion Displacees: A Case of Serajganj in Bangladesh', in V.S. Pendakur and O.P. Dwivedi (eds), *South Asian Horizons*, Montreal: Canadian Asian Studies Association, 5: 27–33.

Halli, S.S. 1991. 'Economic Impact of Riverbank Erosion in Kazipur', in K.M. Elahi, K.S. Ahmed and M. Mafizuddin (eds), *Riverbank Erosion, Flood and Population Displacement in Bangladesh*, Dhaka: REIS, JU.

Haque, C.E. 1988. 'Impacts of Riverbank Erosion Hazard in the Brahamaputra-Jamuna Floodplain: A Study of Population Displacement and Response Strategies'. Ph. D. thesis (mimeo), Department of Geography, University of Manitoba, Canada.

——. 1991. 'Human Responses to Riverbank Erosion Hazard in Bangladesh: Some Lessons from Indigenous Adjustment Strategies', in K.M. Elahi, K.S. Ahmed and M. Mafizuddin (eds), *Riverbank Erosion, Flood and Population Displacement in Bangladesh*, Dhaka: REIS, JU.

Haque, C.E. and M.Q. Zaman. 1989. 'Coping with Riverbank Erosion Hazard and Displacement in Bangladesh: Survival Strategies and Adjustments', *Disasters*, 13(4): September–December.

Hossain, M.Z. 1984. 'Riverbank Erosion and Population Displacement: A Case of Kazipur in Pabna'. M.Sc. Thesis (mimeo), Department of Geography, Jahangirnagar University, Dhaka.

——. 1991. 'Displacees of Riverbank Erosion in Urban Squatter Settlements in Serajganj: The Process of Impoverishment', in K.M. Elahi, K.S. Ahmed and M. Mafizuddin (eds), *Riverbank Erosion, Flood and Population Displacement in Bangladesh*, Dhaka: REIS, JU.

Hossain, M.Z. and C. Greenberg. 1985. 'Population Displacement Due to Riverbank Erosion and Urban Squatters: A Case of Serajganj'. Paper presented at the workshop on 'The Impact of Riverbank Erosion and Flood Hazard in Bangladesh', Dhaka: Jahangirnagar University.

Karim, A.H.M. Zehadul. 1990. *The Pattern of Rural Leadership in an Agrarian Society: A Case Study of the Changing Power Structure in Bangladesh*. New Delhi: Northern Book Centre.

Mahbub, A.Q.M. and Nazrul Islam. 1991. 'Urban Adjustment by Erosion Induced Migrants to Dhaka', in K.M. Ahmed and M. Mafizuddin (eds), *Riverbank Erosion, Flood and Population Displacement in Bangladesh*, Dhaka: REIS, JU.

Nazem, N.I. and K.M. Elahi. 1990. 'Impact of Riverbank Erosion on Land Settlements in Bangladesh', in A.Q.M. Mahbub (ed.), *Proceedings of the Seminar on People and Environment in Bangladesh*, Dhaka: UNDP and UNFPA.

Rahman, M. 1991. 'Vulnerability Syndrome and the Question of Peasants' Adjustment to Riverbank Erosion and Flood in Bangladesh', in K.M. Elahi, K.S. Ahmed and M. Mafizuddin (eds), *Riverbank Erosion, Flood and Population Displacement in Bangladesh*, Dhaka: REIS, JU.

Rashid, H. 1977. *Geography of Bangladesh* (2nd revised edition 1991). Dhaka: University Press Limited.

Rogge, John R. 1991. 'Individual and Institutional Responses to Riverbank Erosion Hazards', in K.M. Elahi, K.S. Ahmed and M. Mafizuddin (eds), *Riverbank Erosion, Flood and Population Displacement in Bangladesh*, Dhaka: REIS, JU.

Romanowski, Jacek I. 1988. 'Impact of Riverbank Erosion on Agricultural Productivity: Conclusions from the Field Surveys of 1985', in Abstract of Paper at the International Symposium on 'The Impact of Riverbank Erosion, Flood Hazard and the Problem of Population Displacement', 11–13 April, Dhaka.

Wiest, R.E. 1991. 'Domestic Group Dynamics in the Resettlement Process Related to Riverbank Erosion in Bangladesh', in K.M. Elahi, K.S. Ahmed and M. Mafizuddin (eds), *Riverbank Erosion, Flood and Population Displacement in Bangladesh*, Dhaka: REIS, JU.

Wilson, H.H. 1855. A Glossary of Judicial and Revenue Terms, and of Useful Words Occurring in Official Documents Relating to the Administration of the Government of British India, from the Arabic, Persian, Hindustani, Sanskrit, Hindi, Bengali, Uriya, Guzarathi, Telegu, Karnata, Tamil, Malayalam, and Other Languages. London: WM.H. Allen and Co.

Zaman, M.Q. 1986a. 'The Role of Social Relations in the Response to Riverbank Erosion Hazards and Population Resettlement in Bangladesh', *Studies in Third World Societies*, 36: 177–99 (Special Issue on Natural Disasters and Cultural Responses, Guest Editor: Anthony Oliver Smith), USA: College of William and Mary.

——. 1986b. 'Rural Bastees: A Socio-economic Profile of Squatters on the Brahmaputra Right Bank Flood Embankment'. Paper presented for the 20th Bengal Studies Conference, Lake Geneva, Wisconsin, 16–18 May, USA, p. 17.

——. 1987. 'Endemic Land Conflict and Violence in Char Villages of Bangladesh', *REIS Newsletter*, 3: 8–11. Also in V.S. Pendakur and U.P. Dwivedi (eds), *South Asia Horizons*, 5: 50–57. Montreal: Canadian Association of Asian Studies, University of Montreal.

——. 1988. 'The Socio-economic and Political Dynamics of Adjustment to Riverbank Erosion Hazard and Population Resettlement in the Brahmaputra-Jumuna Flood Plain'. Ph.D. dissertation (mimeo), Department of Anthropology, University of Manitoba, Canada.

Zaman, M.Q. 1989. 'The Social and Political Context of Adjustment to Riverbank Erosion Hazard and Population Resettlement in Bangladesh', *Human Organization*, 48(3): 196–205.
——. 1991. 'The Displaced Poor and Resettlement Policies in Bangladesh', *Disasters: The Journal of Disaster Studies and Management*, 15(2), June.
Zaman, M.Q. and R.E. Wiest. 1991. 'Riverbank Erosion and Population Resettlement in Bangladesh', *Practicing Anthropology*, 13(3), Summer.

2

Environmental Degradation and its Impact on Dryland Ecosystems: A Case Study of Anantapur, Andhra Pradesh

Ramakrishna Nallathiga

Abstract: Dry ecosystems are natural ecosystems on dry lands that are inherently vulnerable to desertification hazards. They are, however, becoming more vulnerable due to environmental resource degradation in dry lands, which is caused, at least partly, by the socio-economic pressures of human population. As most dry ecosystems have natural resources at margins and experience the recurrence of extreme weather related events, they have poor ecological resilience; hence, reducing their susceptibility to desertification hazard becomes critical for the sustenance of both ecosystems as well as the inhabiting population. Further, building social resilience is also important so as to have a buffer against ecological and climatic catastrophes. Anantapur is one such dry land district located in the south-central part of peninsular India that has been experiencing desertification conditions due to environmental resource degradation, which is conditioned by natural factors and exacerbated by anthropogenic factors. The current chapter analyses the vulnerability of dry ecosystems in Anantapur to desertification hazards and discusses the need for mitigating them through a range of institutional and policy interventions. The chapter, however, does not venture into establishing direct cause-effect relationships, as these interactions are complex and intertwined.

INTRODUCTION

Desertification is a phenomenon in the susceptible dry lands characterised by progressive decline and degradation of environmental resources, predominantly land, water and biomass. Dry lands are the areas characterised by hot climate that can be categorised based on the parameters like their mean temperature and rainfall. For example, one such classification based on precipitation categorises dry lands into hyper-arid, arid, semi-arid and dry sub-humid areas. Dry lands can also be delimited by the nature of ecosystems, known as dry ecosystems, which follow the same categorisation but include vegetation types. It is a term sometimes used to refer to variation in topography, climate and soils. The unifying factor is the mean precipitation and biological productivity (Beaumont 1989). However, desertification in dry lands and dry ecosystems are similar in several ways even as dry ecosystems prevail upon dry lands. The degradation of dry ecosystems on the dry lands resulting in desertification can be better understood through the definition given by Dregne (1983) and Be (1990) (cited in Grainger 1990):

> ...Desertification is a phenomenon of impoverishment of terrestrial ecosystem under the impact of man. It is the process of deterioration in this ecosystem that can be ascertained by reduced productivity of desired plants, undesirable alteration in the biomes and diversity of micro as well as macro flora and fauna, accelerated erosion and increased hazards for human occupation....

Desertification, in this context, is different from deserts in the sense that it leads to desert like conditions (but reversible) rather than converting into deserts (permanent) (Mainguet 1994). Yet, this transformation can have some significant implications on both ecosystems as well as human systems themselves. On one hand, an accentuation of desertification condition can lead to poor resilience of ecosystems, making them much more vulnerable to breakdown in climatic extremes. The malign effects of the persistence of desertification and drought conditions on the vegetation and ecosystems in the Sahel are well documented (Eriksen 2001a). On the other hand, desertification in dry ecosystems also makes human populations much more vulnerable to climatic failures, particularly when the social resilience of the human systems is very poor.

It is also important to understand that the desertification in dry lands/dry ecosystems is primarily caused by anthropogenic factors but also conditioned by natural factors, as defined by UNEP (1992): '...desertification is land degradation in arid, semi-arid and dry sub-humid areas resulting from various factors including climatic variations and human activities...'. Desertification, in this context, is predisposition of the ecosystem to natural vulnerability and an exacerbation in it caused by the above factors. How these natural and human factors interact to cause desertification is well explained by Hulme and Kelly (1993). They attribute the desertification to resource management failures in dry ecosystems, which is again due to poor mismanagement of human activities. Barrow (1985) describes the process of this degradation much more in detail. Large scale climatic failures causing the breakdown of ecosystems is somewhat rarely observed and it occurs on very long time horizons, which cannot be easily prevented because of the fact that it is determined by several factors and processes of complex climate systems. Hence, what can be managed is the human use of natural resources, which requires the building-up of the coping strengths or the resilience on the part of ecosystems as well as human systems. This is the central part of the mitigation and combating strategies of African countries affected by desertification (Eriksen 2001b).

Desertification of dry ecosystems also assumes importance because of its scale and extent—it affects as much as forty per cent of global land (Dregne 1986). The United Nations Conference on Desertification, 1977 was the first attempt to bring forth this issue faced by the African countries and it received a renewed interest at the United Nations Conference on Environment and Development, 1992 in course of the sustainability debate (Eriksen 2001a). It might be surprising to note that even in India, dry lands comprise as much as two-thirds of land and are under the threat of desertification (Nadkarni 1985). Although several studies documented the various aspects of dry lands (for example, Jodha 1991, 1995), desertification conditions in areas other than western Rajasthan did not receive much attention in literature; whereas the work done in case of western Rajasthan is well known (for instance, ICAR 1977; Jodha 1980; Sen and Kar 1989). The studies by Jodha (1980), Chakraborty (1990) and Valdiya (1987) are noteworthy for their contribution to the understanding of desertification in arid lands, plain areas and mountain areas respectively in the Indian context. This chapter provides a case analysis of dry ecosystem degradation in Anantapur and suggests the institutional as well as policy measures that need to be undertaken for enhancing the resilience of natural and human systems in order to mitigate the resource degradation and its impact. The study is perhaps the first of its kind in the peninsular south India, which focuses on analysing the vulnerability of the dry ecosystems to the desertification hazards due to the environmental resource degradation. Desertification, in this area of study has been analysed in terms of the degradation of environmental resources—land, water, bio-mass—and in terms of the anthropogenic pressures.

Anantapur is relatively a very large district with a total geographical area of 19,130 sq km. It is situated in the dry interior part of peninsular India, located between 13° 41'N and 15° 51'N latitudes and 76° 30' and 78° 30'W in longitudes. It experiences low rainfall and hot temperatures and has rocky terrain, poor soil and vegetation cover. Population growth, dependency on agriculture and natural resources, and prevailing illiteracy are some of the socio-economic factors that render high pressure on resources, creating conditions for the occurrence of desertification hazards. We shall now present an analysis of resource degradation.

ANALYSIS OF DESERTIFICATION DUE TO RESOURCE DEGRADATION IN ANANTAPUR

Desertification in dry ecosystem is caused by the interplay of human activities and natural resources in not a synergistic manner but an antagonistic manner. This process can only be well explained by analysing the causal interactions with the help of supporting data and interpreting the current patterns and trends, which is what has been attempted in the following sub-sections. However, it is also important to understand the concepts of vulnerability, thresholds and resilience, which are used frequently to explain the causal interactions in the following work. Vulnerability refers to the susceptibility of systems (humans or regions or ecosystems) to any external force or event which in this case is desertification. Thresholds or margins are extreme points or areas beyond which the breakdown is most likely to take place. Resilience refers to the capacity to withstand repetitive stresses and strains, of both natural and human systems. In socio-economic systems, it is better understood through the concepts of coping that is, resilience being the ability to cope with or withstand the catastrophic events.

CONDITIONING BY CLIMATE AND LOCATION OF THE STUDY AREA

The geographical location of the district in the rainfall shadow zone renders it with a low amount of rainfall and high temperature. With an average annual temperature of 38°C and mean annual rainfall of 520 mm (BSE, 1995), this area has more of semi-arid type of climate. However, if the evapo-transpiration is given consideration, it falls into arid zone climate. The temperature and rainfall vary widely over space and time. This is reflected in the temperature extremes ranging from 15°C to 45°C (BSE 1995) and rainfall extremes ranging from less than 200 mm to more than 900 mm in a year. High temperatures are recorded in the summers, when the atmospheric particulates are also high; and high rainfall is recorded in the monsoons over a short period with a high rainfall intensity (16.08 mm/day) and, correspondingly, high soil loss. Interestingly, the amount of rainfall varies over space with the higher amount observed in the south-eastern part and lower amount observed in the south-western part of the district, which can be explained by the differences in their vegetation cover. The prevalence of hot and dry climate conditions the region's vulnerability to desertification hazards due to resource degradation. The frequent recurrence of droughts (once in two and a half years) implies the fragility under which these ecosystems are functioning and their susceptibility to desertification hazards (Ramakrishna 2001).

SOCIO-ECONOMIC PRESSURES

In this sub-section, the nature and structure of socio-economic pressures is presented. Although, it is difficult to establish a direct correspondence between them and resource degradation, an understanding of these pressures on the resource degradation is provided through relevant arguments which lead to the arguments made subsequently.

Population Growth and Urbanisation

Population growth increases not only the demand for food, water and land, but also the demand for several economic goods, which are spurred by urbanisation. This, in turn, results in greater demand for these goods through exploitation of resources even in a subsistence economy. However, increasing access to markets and the marketisation of economies also lead to intensification of degradation in dry lands through greater resource consumption and increased market risks (Sen 1985), which was evident in the case of Rajasthan (Jodha 1980). Urbanisation increases the access as well as the pace of market expansion, thus increasing resource consumption. The district has been showing a trend of increasing population and urbanisation; this is despite a relatively low population density in the district. The population growth over the last century has been shown in Table 2.1. The urban population has shown a higher growth rate than that of rural population, taking a share of 38.36 per cent of total population (BSE 1995).

Unsustainable Agriculture Crops and Low Food Supportive Capacity

Agriculture plays a vital role in societal development by providing food for human beings and fodder for livestock but at the same time it can be linked to resource utilisation. Dry lands traditionally have subsistence agricultural practices that impose restraint on resource use; but they also have

Table 2.1 Population Growth in Anantapur

Year	*Population*	*Growth Rate*
1901	10,25,322	–
1911	10,53,449	2.74
1921	10,46,116	-1.26
1931	11,38,081	9.41
1941	12,73,060	11.86
1951	14,83,591	16.54
1961	17,67,464	19.13
1971	21,15,321	19.68
1981	25,48,012	20.48
1991	31,80,863	24.94
1996	47,00,000	47.76+
2001	57,20,000	21.7++

Source: Bureau of Statistics and Economics (1995).
Notes: + projection of trend; ++ trend projection for 5 years.

subsistence levels of food production and consumption. However, dry lands are different from others in that they are sensitive with respect to environmental resources due to the thresholds, the surpassing of which may lead to serious disruptions. Economic development through increasing access to markets and marketisation leads to the breakdown of subsistence conditions, but it also increases the risks as well as the costs of external damages (for example, Jodha 1995). From this view point, the management of dry lands, particularly the agricultural use of them assumes greater importance, since the failure to do so shall eventually lead to desertification (Ridley 1990). Although this can be understood conceptually, it is very difficult to prove the linkage between agriculture and environmental resource degradation in a straightforward manner. As mentioned earlier, unsuitable cropping practices, such as rice cultivation, resulted in increasing use of water that also affected the quality of water. However, this resulted not only with the advent of canal water availability but also due to the increased access to markets and marketisation which led to the abandonment of subsistence crops like pulses that are more suited to this climate. However, the impact of market forces is more evident in the case of groundnut cultivation, which is the major crop cultivated in the district (as high as fifty per cent of cropped area). With the market price of groundnut being high and the crop being easily cultivable in the dry lands, it has become the favourite crop of the farmers in the region. This risk reduction strategy, however, comes along with another risk of pests and diseases due to mono-cropping. Moreover, groundnut crop does not provide an effective protection because the field operations like deep ploughing would still affect soil moisture and increase soil erosion hazard. The advent of market forces also affected the food consumption and lifestyle leading towards an increase in resource consumption.

Livestock Growth Pressure on Pastures

Livestock breeding is not a major occupation in the district, but the livestock population is quite high. Livestock is rather seen as a toiling animal contributing to cultivation and an asset easily disposable in the times of severe drought. The livestock's demand for fodder is exceedingly high, whereas, as observed earlier, the land use pattern implies very low allocation of land to pastures suggesting very high incidence of grazing. As compared to a permissible grazing index of 0.55 cow units/ha, the actual grazing index is 4.7 cow units/ha, which is 7.5 times more than the optimal grazing index (NWDA 1995). This clearly indicates the amount of pressure from livestock grazing which results in the transfer of this pressure on forests through illegal browsing. It is observed that in one decade there is an increase in the number of illegal browsing cases as well as forest offences in general (figures shown in Table 2.5).

Occupational Structure and Land Holding

The occupational structure and land holding structure influence resource degradation if the pattern of the structure has a bearing on resource consumption and resource sharing arrangements. The occupational structure has a majority of the population as agricultural labourers and non-workers living on subsistence levels of income and a minority of it on agriculture and other service sectors which characterise a feudal structure. A similar picture emerges in case of land holding pattern, which has a peculiar structure: small and marginal farmers constitute 55 per cent of total population sharing less than 25 per cent of the land, while large farmers constitute 13.4 per cent of total population sharing half of the total cultivated land (BSE 1995). This is an inequitable pattern that suggests the lack of implementation of land reforms as well as fragmentation of land holdings, and also intensification of agricultural practices on them. The institutional arrangements for resource sharing and conservation are not easily discernible and go well beyond the scope of this study. However, the occupation and land holding patterns reflect the subsistence conditions and poverty, which lead to continued dependence on natural resources and result in degradation due to the pressure of human activities.

Degradation of Land Resource

Land is an important resource of ecosystems and that of dry ecosystems in particular. The degradation of land can take place due to the natural conditions of soils, land use/cover and land form, which can be greatly influenced by the human use. We shall examine how these factors would have affected the status of land resource degradation in Anantapur.

Poor Soils and their Erosion

Under this sub-section, the land degradation and the land use patterns are discussed in the context of the study of resource degradation leading to desertification. The soils of the district are poor red soils or black cotton soils with low fertility and are highly prone to erosion hazards, particularly the former, which is the dominant soil (see Figure 2.1 for the spatial distribution of the soils). Higher rate of soil erosion over the natural rate of soil formation results in the soil loss, which in turn results in the poor minerals and soil organic matter leading to increased desertification. However, besides the soil type, land use/cover, soil moisture (all of which determine the physical condition of soil) and

Figure 2.1 Soil and Forest Land Distribution in Anantapur

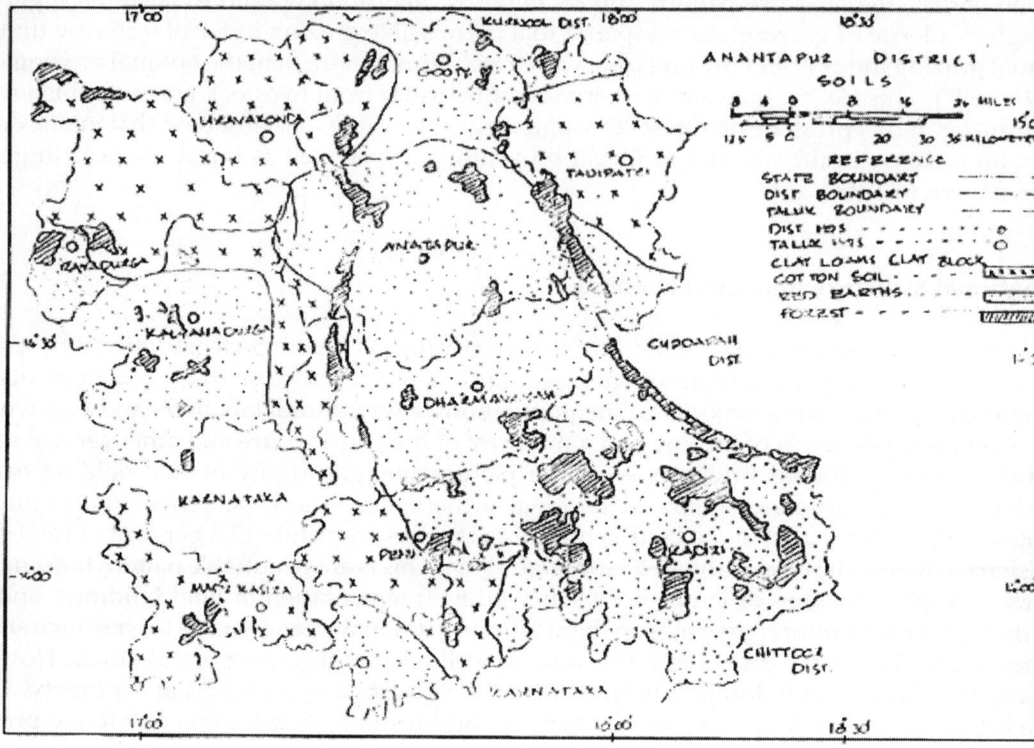

Source: Agriculture Department, Anantapur.

the rainfall intensity play a major role in the determination of soil erosion. In the previous section, it was well laid down that the intense rainfall over short duration results in low soil moisture retention and that the prevalence of hot and dry conditions in the district causes the soil to become loose, both of which condition easy erodibility of soils by the agents like wind and water. The role of land cover is, however, discussed below. The soil erosion rate in the district varies in the range of 11 to 17 t/ha/yr, which is well above the soil formation rate that varies between 2 to 11 t/ha/yr (Wischimeier and Smith 1978, cited in Gitay and Noble 1998).

Inappropriate Land Use/Cover

Land use, which primarily determines the land cover, is an important feature of land resource in the sense that the allocation of it also determines the status of the region or the ecosystem. Sustainable land use makes use of resilience of a resource, which varies with respect to time depending upon natural season, inter-annual variability, management practices and technologies (Middleton and Thomas 1997). Land use/cover pattern and its changes are often good indicators of desertification, particularly when they are observed over a larger spatial scale or time span. The land use pattern prevalent in the district (shown in Table 2.2) indicates a low amount of land resource allocated to

Table 2.2 Land Use Pattern in Anantapur

Type of Land Use	*Area (in ha)*	*Per cent of Geographical Area*
Forests	1,96,881	10.3
Non-agriculture	1,58,897	8.3
Barren and Uncultivable	1,75,750	9.2
Permanent pastures	23,352	1.22
Cultivable wastes	70,350	3.7
Current fallows	2,16,787	11.3
Other Fallows	1,07,032	5.6
Net area sown	9,76,775	51
Area Sown more than once	30,602	1.6
Miscellaneous trees and grooves	28,280	1.47
Gross area sown	10,07,377	52
Total Geographical area	**19,13,492**	

Source: Bureau of Statistics and Economics (1995).

the uses of more ecological value (e.g., forests and pastures) and a greater allocation to the uses of more economic value (e.g., agriculture). This might be expected, but a significant proportion of fallow land indicates the pressure on land resource from uses like agriculture and also a decline in soil fertility observed in the district. Besides this pattern depicting an intensification of the land resource use, the decreasing trend of barren land's share also suggests extensification of uses like agriculture, leading to the cultivation of marginal areas. The intensive use of land for agriculture and also its extensification result in an increase of soil erosion as well as salinity/alkalinity hazard. However, more stronger implications come from the forest cover, which is mentioned as 10.3 per cent of the total geographical area in the official records (BSE 1995); whereas, the actual forest cover based on the crown cover density observed through remote sensing (MOEF 1991) is only 2.6 per cent. Moreover, it is highly fragmented across the district as shown in Figure 2.1. The discrepancy is not an indication of differences in classification systems, but implies the degradation of forest cover to a great extent, which is not recorded in the official records.

Declining Water Resource

Water resources are critical resources for both natural and human sustenance. In particular, the availability of water in good quantity as well as quality will have a bearing on human life. In this subsection, criticality of water resource in the district is discussed to investigate the desertification due to water resource degradation in case of both surface and ground water.

Increasing Water Demand

The water resources availability and demands shown in Table 2.3 clearly indicate that, quantitatively, the district had adequate water resources to meet the demands of various uses only till mid-1980s. However, an increase in cultivation, discussed earlier, together with increasing population, to be

Table 2.3 Water Balance Status in Anantapur

Assessment Source	*Water Balance at 50% Dependability* 2001AD	2025AD	*Water Balance at 75% Dependability* 2001AD	2025AD
Irrigation Commission, 1972	-794.68	-1,268.64	-1,210.68	-1,702.64
National Commission on Agriculture, 1975	-2,214	-4,280.02	-2,630.8	-4,696.02
Master Plan	–	887.13	–	471.13

Source: National Water Development Agency (1995).

discussed later, exert a high demand on water. The canal water supply for irrigation provided water security to the farmers and enhanced their cropping options. But, it has also led to intensification and extensification of agriculture, accompanied by shifts in cropping pattern (a detailed discussion would be done later) towards cultivation of water intensive crops like rice. This has resulted in increased degradation hazards like salinisation and alkalinisation; besides it could have affected surface water quality through tail water discharges.

Declining Groundwater and its Quality

The intensification of agriculture exerts pressure on quantity of water not only from surface source but also from ground source, which brings forth additional complexities. The groundwater is increasingly used to irrigate the cropland, particularly during the non-monsoon period and in the off-canal areas. Technological improvements in deep well digging has led to an increase in the extraction of groundwater beyond the sustainable yields of aquifers (details shown in Table 2.4), while also posing water quality degradation problems. The geological formations of deep underlying rocks in the district are rich in fluoride. When they come into contact with water due to the increase in withdrawal, it results in groundwater quality degradation. The widespread prevalence of fluorosis in this district clearly indicates this problem.

Table 2.4 Unsustainable Withdrawal of Groundwater in Anantapur

Category of Groundwater Overdrawal	*Extent of the Problem* No. of Mandals	Percentage of Total
Dark or Black (85–100% drawal of yield)	3	5
Grey (65–85% drawal of yield)	5	9

Source: Central Ground Water Board, Hyderabad (1995).

Threatened Biological Resource

In this sub-section, the status of degradation of forest cover, which forms the major biological resource, is discussed. Although agriculture practices and pastures also produce biomass, it is not as high as that of forests and they also lack several other conditions like ecological services and biodiversity. Their status is discussed in the subsequent section.

Degradation and Fragmentation of Forests

The land use statistics indicating a constant area under forest cover vis-à-vis low forest cover of good crown density suggest the degradation of forest resources. According to the remote sensing classification criteria, forests are those with a crown cover density of more than forty per cent. The forests with crown cover density of less than forty per cent but more than ten per cent are called degraded forests, whereas the forests with crown cover density of less than ten per cent are called open forests (MOEF 1991). Historically, the district was endowed with a good forest cover with wild life but human pressures have resulted in their gradual decline as also reflected in the study of Rao and Rajeskar (1994). It is well documented that a decline in the vegetation cover would alter rainfall, producing convection circulation, leading to a decline in rainfall, which can lead to a decline in the run-off due to a greater loss of moisture in evaporation (Eriksen 2001a). The degradation of forest cover has led to its fragmentation into forest patches followed by non-forest land, which is also indicated by the spatial spread now concentrated in the hilly south-eastern part (see Figure 2.2). Incidentally, this is perhaps the residual forest cover, which still makes this part of the district receive a good amount of rainfall.

Threatened Wildlife Habitat

The degradation and fragmentation of forests also leads to the loss of habitat and wildlife, which results in the disruption of the ecological food chain. The increase in number of attacks by wild hyena on the nearby villages might imply this (figures shown in Table 2.6). This clearly indicates the degradation of the forest resource not only in terms of vegetation cover or biomass alone but also in terms of wildlife loss. Moreover, the human and livestock pressure on forests for fuel wood and fodder has also led to an increase in forest offences like illegal cutting and felling as well as browsing cases over time (figures shown in Table 2.5).

THE NEED FOR ENHANCING ECOLOGICAL AND SOCIAL RESILIENCE

Synthesis of Resource Degradation in Anantapur

The above sub-sections provide a theme wise analysis in terms of the conditioning factors, pressures and actual resource degradation (of land, water and bio-mass) in Anantapur. The discussion on land degradation suggests that poor soils and resource exploiting land use patterns and its trends, lead to increasing desertification from land resource degradation implied by greater rate of soil

Table 2.5 Forest Offences in Anantapur

Year	*Illegal Browsing Cases*	*Other Forest Offences*
1981–82	480	1,219
1982–83	419	1,171
1983–84	560	1,405
1984–85	394	4,422
1985–86	602	1,748
1987–88	756	1,481
1988–89	820	1,590
1989–90	808	1,650

Source: District Forest Office, Anantapur (1995).

Table 2.6 Wildlife Attacks in Anantapur

Year	*No. of Cases*	*No. of Deaths*	*No. of Wounded*
1985	19	14	5
1986	3	2	1
1987	12	9	3
1988	29	11	18
1989	16	13	3
1990	11	2	9

Source: District Forest Office, Anantapur (1995).

erosion and poor soil fertility. Also, the discussion on water resources implies that the scarce water resources are under pressure from demands for it (predominantly, agriculture and population); particularly, an increase in agricultural practices affects surface and ground water quantity as well as quality, while also leading to other degradation hazards. Essentially, water resource criticality is affected by the human uses to reach the margins of thresholds and also pose discernible degradation hazards that may lead to exacerbation of desertification conditions. Likewise, the discussion on biological resources implies that the vegetation cover is in a degraded state even in forest land and the fragmented nature of forest cover is accompanied by the degradation of habitat for wildlife/ biota, which, in turn, results in wildlife attacks and illegal browsing.

In summary, this process can be recaptured through the following flow diagram (Figure 2.2) which synthesises the above analysis and shows how various linkages lead to the formation of a vicious cycle. It then becomes imperative that institutional as well as policy measures for enhancing the resilience of natural and human systems are required in order to mitigate resource degradation, which is discussed in the following section.

THE NEED FOR ENHANCING ECOLOGICAL AND SOCIAL RESILIENCE

The above analysis of trends and patterns of resource degradation suggesting desertification of the dry ecosystems of Anantapur underline the need for mitigating these effects through appropriate

Figure 2.2 Vicious Cycle of Resource Degradation: The Need for Enhancing Resilience

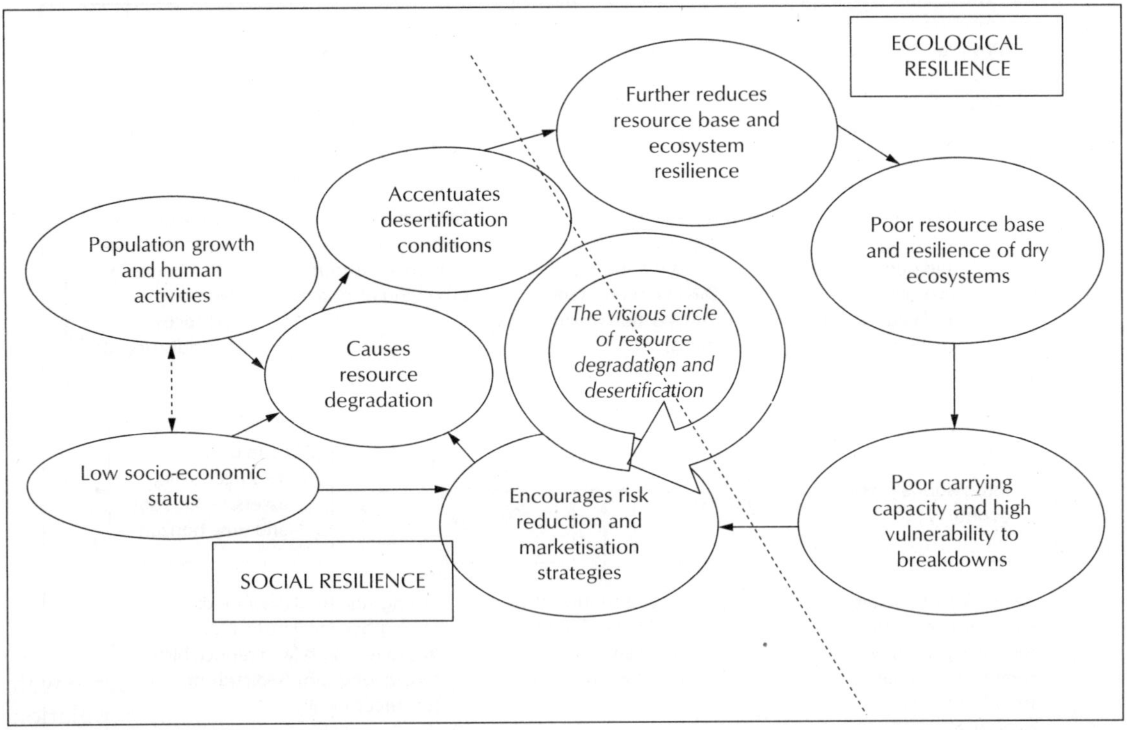

interventions/measures, that is, enhance the resilience of both ecological and social systems. These measures would reduce the vulnerability of the systems to the external forces, such as droughts, and enhance the ecological and social resilience (Subbaiah 1993). These measures shall be (*i*) strategic interventions at policy and decision-making and (*ii*) tactical interventions at local and block level. Besides identifying these measures, their implementation needs to make use of existing institutions, creating new institutions and designing an institutional framework for achieving the end results. The institutional framework suggested here is appropriate for management at a smaller spatial scale of a district or a region; a broader framework of institutional structure and interventions are discussed in Pachauri and Kanitkar (1997), which also looks into the possible assistance from research institutions in filling the gaps of ecological processes and human interventions. An example of such framework is shown below.

STRATEGIC INTERVENTIONS

The dry land ecosystem degradation and its exacerbation through human interventions can be ascribed primarily to the lack of appropriate resource management practices at the farm block

Figure 2.3 Institutional Interventions for Combating Desertification

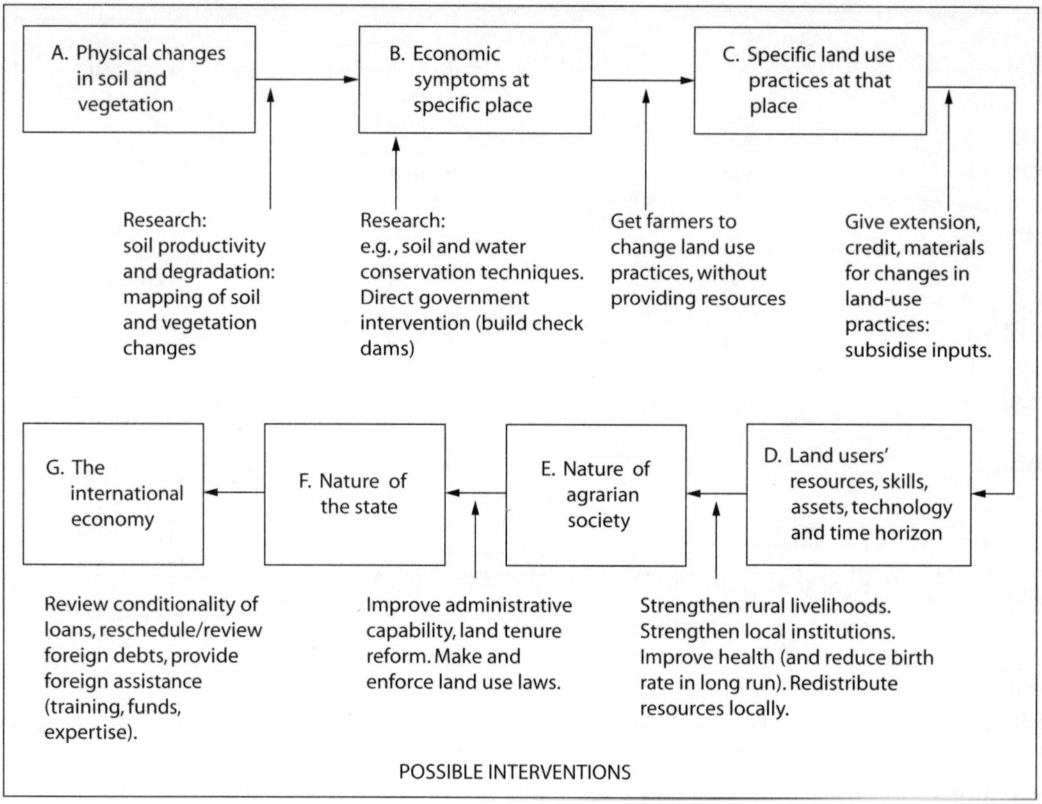

Source: Pachauri and Kanitkar (1997).

and regional level. Measures that enhance these practices and achieve resource conservation shall prove to be very useful in enhancing the ecological resilience (Figure 2.3). However, a lack of effective mechanisms for coping makes the people more vulnerable to crop failures due to climatic and ecological pressures. Interventions such as income generation through supplemental work, through public spending and designing crop insurance mechanisms will enhance the crop as well as human security (Ramakrishna 2001). The creation of such mechanisms of public support would greatly enhance the social as well as ecological resilience. Also, the provision of alternative livelihoods that provide economic opportunities is a critical element for enhancing social resilience and making policies that aim towards the creation/revamping of such industries is vital to achieve it. In fact, concentrating on pure agronomic measures like drought resistant crops cultivation by those African countries that had been affected by desertification did not yield the right results due to lack of demand for such crops and difficulty in persuading the farmers to cultivate them (Eriksen 2001b). While some efforts are needed on a sustained basis to do this, efforts also need to be made to enhance the agro-ecosystems and to set up agro-processing and cottage industries in order to create new livelihoods, thus, enhancing the coping capacities of humans.

Reducing Ecological Vulnerability

As dry ecosystems are threatened by human mismanagement, attention needs to be made to correct these actions through well-targeted large conservation programmes. Watershed based conservation of soil, water and vegetation would result in restoration of their losses through human activities like agriculture, livestock grazing and using biomass as an energy fuel. This strategy needs to have the following elements (*i*) conservation agriculture and horticulture, that is, water conserving crops, crop rotation, crop mulching and the use of genetic and irrigation engineering methods; (*ii*) efficient use of water through inter-row water harvesting in the form of ridge and furrow irrigation and deployment of water conserving technologies like drip irrigation; and (*iii*) enhancing soil water availability through rain water harvesting and enhancing tank storage and supplies (Swaminathan 2004). These activites can be planned on a big scale across the district by delineating the watersheds and prioritising them for phased targeting. Already, mapping of the watersheds has been done and prioritisation is in due course, then the design of programmes for resource conservation needs to be planned with the help of experts in agriculture, soil sciences and water resources. This integrated approach to resource conservation would yield better results than the separate conservation measures for one resource component each.

Enhancing the Economic Security or Coping

As the dependence on agriculture and allied activities for livelihood is predominant in the district, their vulnerability to crop failures becomes much stronger. This can be corrected through mechanisms of public intervention—both direct and indirect. In the direct intervention, support can be extended by way of distribution of chosen minimum (or subsistence) levels of food through public distribution system or by way of distribution of cash for purchasing either food grains or implements for cropping. While food distribution can be done through existing administrative channels and can be put to use during the times of drought, distribution of food can be done more efficiently through a creation of crop insurance system, with public payment of premium, which would step-in during the drought related crop failure. Another effective method of economic support would be by providing waged employment through rural infrastructure building programmes. These programmes can be targeted to create either physical infrastructure such as roads, irrigation canals, power lines, etc. or social infrastructure, such as schools, hospitals and community centres. This indirect suppport can have a long-term benign effect on the rural economy, society and its upliftment.

Creating Alternative Livelihoods

Agriculture is still the dominant occupation for both farmers and workers in the district. With the given distortionary land holding making it disadvantageous for small farmers, they shall not be able to enhance their incomes without resource degrading agricultural practices, which in turn shall affect their long-term sustenance. Likewise, farm workers also receive low wages due to poor returns on income to farmers. This situation can be only corrected by creating opportunities in non-agricultural sectors and providing incentives for setting up of industrial units. This need not focus on manufacturing alone, but several other industrial activities such as those based on rock-cutting, agro-processing and food-processing that extend the operations beyond farming. The cottage industries can also be revamped through cooperative institutions. But, most importantly, creation of effective education, training and credit facilities is vital for the revamping of the cottage industries as well as other basic industries.

TACTICAL INTERVENTIONS

While strategic interventions are required for bringing about the large-scale changes for mitigating the resource degradation, several micro-level interventions can be planned and executed as a priority at the village and block level administrative units. The Table 2.7 gives the range of interventions that can be planned with respect to the management of the resources. Implementing these measures would require greater amount of participation of local people and local institutions as well as a good amount of cooperation between public and local institutions. In fact, it was shown elsewhere that even small scale projects that enhance pastoral development can yield positive results (Pratt et al. 1997), but enough care needs to be taken to ensure that it would not mean overgrazing of public land.

Table 2.7 Matrix of Tactical Interventions

Resource	*Resource Component*	*Intervention(s)*
Forest/biological resource	Agro forestry	Encourage multiple cropping, crop mulching and relay cropping methods of farming
	Social forestry	Establish norms for management of commons like woods, pastures and plantations
	Forest conservation	Enhance the skills and methods of forest conservation through use of information technology and biotechnology
	Shelter belts	Implement tree plantations along the infrastructures like roads, railway lines, and water tanks for reducing the erosion
Land resource	Crop land	Impart training in soil conserving farming methods like multiple cropping, minimum tillage and crop mulching; encourage the complementary use of both organic and inorganic fertilisers
	Pastures/fallow land	Institute mechanisms based on community management to avoid overgrazing
	Barren/rocky land	Avoid land conversion for agriculture use; when it is used ensure the use of soil and water conservation methods like bunds, contours; use it for plantations
Water resource	Water availability	Promote water harvesting by means of infiltration wells, check dams and water tanks
	Water management	Ensure mechanisms in place to avoid poor water management impacts like soil salinity/alkalinity

INSTITUTIONAL FRAMEWORK FOR RESOURCE MANAGEMENT

The above interventions can only be implemented through the formation of an institutional framework, identifying the institutions responsible for implementation. This needs to make use of existing institutions, both formal and informal, and new institutions. As mentioned earlier, there is a need for creating a monitoring agency within the authority of the district collectorate which would coordinate the data generation and analysis on land, water and biological resources and execute management and improvement plans in liaison with the agencies such as departments of agriculture, land, water resources and forests. This would also require the inputs of spatial information acquired through satellite and other maps. When it comes to support programmes like giving cash, the direct support needs to be implemented through village panchayats and monitored at block level administration, wherein the public distribution system is already in place for the distribution of food grains. Creation of alternative livelihoods is possible through appropriate policies and incentives in the form of targeted subsidies for industrial growth as well as enhanced access to credit. The micro-interventions call for identifying the means of establishing cooperation between the public authorities and the local community so that they could implement the measures in a harmonised manner.

CONCLUSIONS

Desertification due to environmental resource degradation in dry ecosystems has different dimensions and it is driven by both natural and human factors. In this study, it is described in terms of the degradation of environmental resources that is conditioned by natural factors and exacerbated by anthropogenic factors. It was analysed by observing the trends and patterns in the degradation of resource components like land resource and water resource, in the light of socio-economic pressures. The continuation of these patterns and trends would aggravate the resource degradation further, and enhance the resources' vulnerability to the external forces like droughts, putting them on the brink of collapse in the event of crossing of thresholds. This calls for combating through public intervention by way of appropriate institutional and policy interventions for enchancing ecological and social security. This needs to be complemented by the cooperation and support at grassroots levels as well.

REFERENCES

Barrow, C.J. 1985. *Land Degradation: Development and Breakdown of Terrestrial Environments*. Cambridge: Cambridge University Press.

Beaumont, P. 1989. *Environmental Management and Development in Dry lands*. UK: Routledge Publishers Ltd.

BSE. 1995. *District Statistical Abstracts: Anantapur*. Hyderabad: Bureau of Statistics and Economics.

Bullock, P. and H. Le Houerou. 1996. 'Land Degradation and Desertification', in R.T. Watson, M.C. Zinyowera and R.H. Moss (eds), *Climate Change1995: Impacts, Adaptations and Mitigation of Climate Change*. Bonn: Working Group-II, Intergovernmental Panel on Climate Change.

Chakraborty, P.K. 1990. 'The Darjeeling Himalayas: A Study of Environmental Degradation', in B.N. Sinha (ed.), *Ecosystem Degradation in India*, New Delhi: Ashish Publishing House.

Central Ground Water Board. 1995. *Status of Ground Water in Southern India*, Hyderabad: Central Ground Water Board.

Dregne, H.E. 1986. 'Desertification of Arid Lands', in F. El Bazz and M.H.A. Hassan (eds), *Physics of Desertification*, Dodrecht, the Netherlands.

Eriksen, S. 2001a. 'Physical and Social Changes', in Linkages between Climate Change, Desertification in East Africa, *Arid Land News Letter*, 49, May/June.

——. 2001b. 'Policy Linkages Between Climate Change and Desertification', in Linkages Between Climate Change, Desertification in East Africa, *Arid Land News Letter*, 49, May/June.

Gitay, H. and I.R. Noble. 1998. 'Middle East and Arid Asia', in R.T. Watson, M.C. Zinyowera, and R.H. Moss (eds), *The Regional Impacts of Climate Change: An Assessment of Vulnerability*', pp. 231–52. Special Report of IPCC Working Group-II, Intergovernmental Panel on Climate Change, Cambridge, United Kingdom and New York: Cambridge University Press.

Grainger, A. 1990. *The Threatening Desert: Controlling Desertification*. London: Earth Scan Publishers Ltd.

Hulme, M. and M. Kelly 1993. 'Exploring the Links between Desertification and Climate Change', *Environment* 35(6): 5–11, 39–45.

ICAR 1977. *Desertification and its Control*. New Delhi: Indian Council of Agricultural Research.

Jodha, N. S. 1980. 'The Process of Desertification and the Choice of Interventions', *Economic and Political Weekly*, 25 June.

——. 1991. 'Drought Management: Farmers' Strategies and their Policy Implications', *Economic and Political Weekly*, 26 (39).

——. 1995. 'Common Property Resources and the Environmental Context: Role of Biophysical versus Social Stresses', *Economic and Political Weekly*, 30(51).

Johnson, D.L. and P.A. Lewis. 1989. Land Degradation: Creation and Destruction of Natural Environment. Oxford: Blackwell Publishers.

Mainguet, M. 1994. *Desertification: Natural Background and Human Mismanagement* (Spring Study Edition), Bonn: Springer Verlag Publishers.

MOEF. 1991. *The State of Forest Report*, New Delhi: Ministry of Environment and Forests.

Middleton, N. and D. Thomas. 1997. *World Atlas of Desertification*, UK: Arnold Publishers.

Miller. 1990. *Environmental Science: Living on the Earth*. USA: Wadsworth Publishers Ltd.

Nadkarni, M.V. 1985. *Socio-economic Conditions of Drought Prone Areas*, New Delhi: Concept Publishing Company Ltd.

Nallathiga, R. 2003. 'Production Systems Planning for Natural Resource Conservation in a Micro-watershed', *Electronic Green Journal*, 18(1) April.

NWDA. 1995. *Drought Conditions and Occurrence in Anantapur*, Project Report submitted to Central Water Commission, National Water Development Agency, Hyderabad, India.

Pachauri, R.K. and R.S. Kanitkar. 1997. 'Deforestation and Desertification in Developing Countries', in Yoichi Kaya and Keiichi Yokobari (eds), *Environment, Energy and Economy: Strategies for Sustainability*, Tokyo: United Nations University Press.

Pratt, D.J., F. Le Gall and L. de Haan. 1997. 'Investing in Pastoralism: Sustainable Natural Resource Use in Arid Africa and Middle East', Technical Paper No. 365, Washington DC: The World Bank.

Rao, G.N. and D. Rajeskar. 1994. 'Land Use Pattern and Agrarian Expansion in a Semi-arid Region: Case of Rayalaseema region', *Economic and Political Weekly*.

Ramakrishna, N. 2001. 'Drought Identification, Occurrence, Analysis and its Impacts: Policy and Planning for its Management', *Journal of the IPHE*, India, Vol 2001, No. 36.

Ridley, N. 1990. *Dry land Management: The Desertification Problem*, World Bank Technical Paper No. 116, Washington DC: World Bank.

Sen, A.K. 1985. *Poverty and Famines*. Oxford, UK: Oxford Book Publishers.

Sen, A.K. and A. Kar. 1989. *Land Degradation and Desertification in Asia and Pacific Region*, (eds), Jodhpur: Scientific Publishers.

Subbaiah, A.S. 1993. 'Indian Drought Management: From Vulnerability to Resilience', in D.A. Wilhite (ed.), *Drought Assessment, Management and Planning: Theory and Case Studies, Natural Resource Management and Policy Series*, pp. 157–79. Norwell (Mass) and Dordrecht: Kluwer Academic.

Swaminathan, M.S. 2004. 'Building Sustainable Food and Water Security Systems'. Presentation made in the Yashwant Rao Chavan Memorial Lecture on 25 November.

Tewari, A.K. 1975. *Desertification: Monitoring and Control*. Jodhpur: Scientific Publishers.
Times of India. 1994. 'Sands of Destruction', Sunday Times, 4, December, The Times of India Group, Bombay.
UNEP. 1992. *World Atlas of Desertification*. UK: Arnold Publishers.
Valdiya, K.S. 1990. Environmental Geology, New Delhi: Tata McGraw-Hill Publishers Ltd.
Wischmeier, W.H. and D. Smith. 1978. *Predicting Rainfall Erosion Losses: A Guide to Conservation Planning*. USDA-ARS Agriculture Handbook No. 537, Washington DC.

3

Economic Valuation of Ecosystem Changes: Salinity Ingress and Ground Water Depletion in Coastal Gujarat

Jyothis Sathyapalan[1] and *S. Iyengar*

Abstract: In this study, we try to estimate the economic value of ecosystem changes due to human action induced salinity ingress in the coastal areas of the state of Gujarat. In Gujarat, the salinity propagation is 0.5 to 1 km annually from the coastline as the ground water tables are receding by 4 to 5 metres. We found that in these areas, water quality parameters like total dissolved solid (TDS), hardness, electrical conductivity, etc. far exceed the permissible limits for drinking as conforming to Indian standards. Using cost-benefit analysis, we have found that the value of agricultural production in the sample area was sixty-two per cent less per acre of land in salinity affected areas as compared to a similar situation without salinity. The study noted that small farmers lose access to water resources as the water level goes down or the wells dry up. The rich farmers extract water by using new technology along with electricity which is available at a subsidised rate. In this context, the chapter presents a case for adopting measures to conserve water and land resources considering the social, economic and ecological implications of increased salinity and groundwater depletion in the coastal areas of the state of Gujarat.

INTRODUCTION

Salinity is the presence of soluble salts in or on soil, or in water. High levels of soluble salts may result in reduced plant productivity or the elimination of vegetation cover (QDPI 1987). It is well understood that salinity and water-logging play a crucial role in the decline of agriculture productivity, income and employment in irrigated agriculture (Joshi and Agnihothri 1984; Joshi 1987; Chopra 1989; Joshi and Jha 1992; Nayak 2002). Most of these studies noted that water-logging and salinity have affected the fertility of the soil which in turn leads to the decline of agricultural productivity. As a result, the investments made on the development of canal irrigation in the salinity affected areas become uneconomic to the society. The main source of this sort of salinity is due to the rise in the subsoil water table resulting from the perennial canal systems (GoI 1972). However, the salinity ingress in the coastal areas is due to the depletion of ground water sources which has received less attention in literature. It is a serious problem in many parts of coastal Gujarat (GoG 1978). The estimates show that the groundwater in the state of Gujarat, which has an area of 1,06,000 sq km with 1,659 km of coastline, has been adversely affected by the salinity ingress. It was reported that every year on an average 0.5 to 1.0 km inland areas in the coastline is affected. Thus, by now, 5 to 7.5 km wide strip of the inland area has been rendered saline. The groundwater quality has also deteriorated and more than 2,000 ppm of TDS is found in many places in coastal Gujarat (Barot 1996). In many areas of Gujarat, the groundwater draft is more than the groundwater recharge, which has resulted in lowering groundwater tables (Gujarat Ecological Commission 2001). This over exploitation of groundwater is mostly for irrigation needs. It was also found that more than ninety per cent of the total cultivated area in north Gujarat is being irrigated by groundwater sources. Not only this, the net area suitable for groundwater recharge is also on the decline across the state. The figures on the decline of net recharge area correlate with the large areas affected by salinity (Gujarat Ecological Commission 2001). In short, the ingress of salinity is not only a major environmental degradation or ecological problem facing the coastal belt of Gujarat, but also the resultant externalities impose severe economic strains on individual farmers, the state and the society. It should be remembered that the capital intensity in groundwater extraction makes it easier to exclude rival users especially in fragile resource regions where the high cost of groundwater extraction coupled with low and inequitable asset ownership makes the resource privy to a few well-to-do households. This give rise to 'free riding externalities which differ in nature and intensity depending on biophysical and climatic conditions' (Reddy 2003). Since the overexploitation of groundwater and the resultant degradation in terms of salinity ingress is largely the result of a lack of proper integrated water resource management practice, there is a real need to understand the magnitude of economic losses due to salinity ingress so that appropriate measures can be adopted for desalinization and groundwater replenishment.

OBJECTIVES, DATA AND METHODOLOGY

The main objective of this chapter is to report some findings of an ongoing study and estimate the forgone agricultural benefits due to salinity ingress (an external cost of salinity). The study also attempts to generate information on water quality in the region of the study and tries to explore policy options for desalinisation and groundwater replenishment. Data has been generated through a primary survey of 755 households that was carried out in thirteen villages located in the coastal areas of south Gujarat during the year 2004. These villages are located within seven kilometres of the seashore with an average elevation of six metres. The area of study spreads over two talukas of the Surat district in Gujarat state, that is Chorasi and Olpad. The former is a fast industrialising taluka of the district. The findings reported here are from the village located in taluka Olpad.

The sample represents ten per cent of the total households of the selected villages. The study has adopted a simple random sampling to select the households for the final sample survey. In order to assure good representation of the population, we have drawn samples from all the village hamlets. It is to be noted here that the selected sample households represent different caste groups since each hamlet represents a particular caste. The enumerators have chosen the households for collecting data by moving in different directions in the hamlet until the predetermined size of each sub-sample was eventually reached (the so called random walk technique). The distribution of sample households has been presented in Table 3.1. A structured interview schedule has been used to collect data from the households.

Table 3.1 Distribution of Sample Households in the Study Area

Villages	*Number of Sample Households*	*Total Number of Households in the Study Area*
Tunda	40	392
Tena	22	195
Pinjarat	39	356
Khosadiya	24	225
Bhandut	32	290
Chhini	26	252
Total Olpad Taluka	**183**	**1,710**
Vansva	70	650
Rajgari	41	360
Damka	117	1,108
Suvali	37	368
Junagam	62	575
Mora	79	683
Hazira	166	1,650
Total Chorasi Taluka	**572**	**5,394**
Grand Total	**755**	**7,104**

Data on Water Quality Parameters

In order to understand the water quality, we have adopted physical, chemical and micro biological tests of water. These include: colour, turbidity, conductivity, alkalinity, total hardness, total dissolved solids, salinity, calcium, magnesium, chloride, pH, nitrate and fluoride contents of the water. We have also examined the coli form of few samples from the most frequently used wells. The methods adopted for testing these parameters are given in Table 3.2.

Economic Value of Ecosystem Changes

The Opportunity Cost Approach has been adopted to estimate the economic value of ecosystem changes due to salinity ingress. The value has been estimated in terms of the forgone agricultural production due to increased salinity which has been resulted due to the depletion of groundwater. The data for this purpose has been generated from Pinjarat village located in the Olpad taluka of Surat district. The results of the cost benefit analysis have been presented for two categories of agriculture plots, that is, plots with and without salinity. The important crops grown in these areas are rice, *jowar*, wheat, *bajra* and vegetables. These crops are grown in three different seasons, *kharif* (June to December), *rabi* (December to March) and summer (March to June). The farmers who have access to sweet water from the government canal cultivate during all the three seasons. It is reported that those who depend on groundwater cultivate during both *kharif* and *rabi* seasons. In the highly salinity affected areas, farmers cultivate during one season only, that is *kharif*. To assess

Table 3.2 Methodologies Adopted for the Physical, Chemical and Micro Biological Analysis of Water

	Parameter (Units)	*Methodology Adopted*
1	Colour (colour units)	Visual Comparison
2	Turbidity (NTU)*	Nephelometric
3	Conductivity ms/m	Conductivity Metre
4	Alkalinity mg Ca CO_3 L at pH 4.5	Titration
5	Total Hardness mg/l	EDTA Titrimetric
6	Total dissolved solids mg/l	Dried at 180°C
7	Salinity 0/00	Argentometric
8	Calcium mg/l	EDTA Titrimetric
9	Magnesium mg/l	Calculation
10	Chloride mg/l	Argentometric
11	pH	pH metre
12	Nitrate mg/L	Colorimetric
13	Nitrate mg/l	Cadmium Reduction
14	Fluoride mg/L	SPADNS
15	Total coli form MPN Index/100 ml	Multiple tube fermentation technique.

Notes: *NTU (Nephelometric turbidity units; **0/00 parts per thousand; MPN means most probable number.

the external cost of salinity ingress in terms of the forgone agricultural benefits, we have computed the net benefits of agriculture in different scenarios. Two alternative viability measures adopted as part of the cost benefit analysis are Net Present Value (NPV) and Benefit Cost Ratios (BCR). The net present value is defined as the present value of benefits minus the present value of costs at 2004 prices where cash flows are summed for a period of twenty-five years. The period has been decided based on the farmers' opinion on the decline in the productivity of land once the land gets saline. It was opined that within twenty-five years, land would become highly unproductive once it became saline. However, the study assumed the flow of constant returns throughout the period. The net present value and benefit cost ratio are derived using the following formulae.

$$NPV = \sum_{t=1}^{t=n} \frac{Bt - Ct}{(1+i)^t} \tag{1}$$

$$BCR = \frac{\sum_{t=1}^{t=n} \frac{Bt}{(1+r)^t}}{\sum_{t=1}^{t=n} \frac{Ct}{(1+r)^t}} \tag{2}$$

Wherein B = benefit; C = cost ; t = 1, 2, .. n; n = number of years; i = discount rate.

Benefit-cost ratio is the present value of benefits expressed as a ratio to the present value of costs. The cost side of agricultural cultivation includes cost of preparing the plot, seeds, sowing, weeding, fertiliser, pesticides, labour, repair and maintenance of agricultural implements, supervision and cost of harvesting. In addition to this, there are fixed costs by way of agricultural implements and irrigation investments. The benefits include total production of crop and crop residuals. The NPV and BCR have been computed at three alternative discount rates: 8, 10, and 12.[2] The differences in the NPV for the two different scenarios, that is, a scenario without and with salinity, have been interpreted as the external costs of saline intrusion in the coastal area.

SOCIO-DEMOGRAPHIC PROFILE OF THE STUDY AREA

A brief overview of the socio-demographic and economic characteristics prevailing in the area of study has been presented in Table 3.3. The household size in the area of study is estimated to be 5.1 which is nearer to 2001 census estimate for Surat district which was 4.9. The overall literacy rate of the area of study is about 86.8 per cent. In Chorasi taluka, the literacy rate is 88.2 per cent which is comparatively higher than the literacy estimate of 82.6 per cent of Olpad taluka. The sex ratio estimated for the entire area of study is 860 females for 1,000 males, which is very close to the 2001 census estimate provided for the Surat district, that is, 872 females per thousand male population. It is also observed that the women-headed households are significantly higher in many of these villages. Nearly 18.4 percentages of the total households are headed by women. In Olpad taluka, 23.3 per cent of the total households are headed by women while it is only 13.9 per cent in Chorasi taluka.

Table 3.3 Demographic Characteristics of the Selected Sample Villages

Villages	*Household Size (No.)*	*Illiterates (No.)*	*Literacy Rates*	*Sex Ratio (Estimated from the Primary Data)*	*Percentage of Women Headed Households to Total Number of Households*	*Percentage of Households with Women Who are Widows, Separated or Divorced*
Tunda	5.1	31 (16.6)	156 (83.4)	860	25.0	25.0
Tena	5.4	8 (7.6)	97 (92.4)	1,018	13.6	18.2
Pinjarat*	5.1	32 (18.2)	144 (81.8)	989	35.9	38.5
Khosadiya	4.5	26 (26.3)	73 (73.7)	1,020	16.7	16.7
Bhandut	5.6	26 (15.4)	143 (84.6)	830	21.9	31.3
Chhini	5.5	27 (21.4)	99 (78.6)	844	26.9	23.1
Olpad	5.2	150 (17.4)	712 (82.6)	913	23.3	25.4
Vansva	5.2	35 (10.4)	303 (89.6)	849	20.0	20.0
Rajgari	5.1	22 (11.5)	169 (88.5)	809	7.3	7.3
Damka	5.3	75 (13.4)	483 (86.6)	823	20.5	23.9
Suvali	5.1	34 (18.8)	147 (81.2)	889	5.4	18.9
Junagam	5.4	43 (13.9)	267 (86.1)	931	8.1	12.9
Mora	5.1	45 (12.2)	325 (87.8)	906	15.2	19.0
Hazira	4.8	65 (8.6)	688 (91.4)	909	20.5	24.7
Chorasi	5.1	319 (11.8)	2,382 (88.2)	876	13.9	18.1
Total	**5.1**	**469 (13.2)**	**3,094 (86.8)**	**860**	**18.4**	**21.9**

Source: Primary Survey Data.
Notes: Figures in parenthesis are percentages; *case study village.

Most of the women heading the households belong to the category of divorced, widow or separated. Nearly 70.3 per cent of these women in Olpad earn their livelihood from general labour and household domestic work while it is 56.2 per cent in Chorasi taluka. Occupation in agriculture and allied sectors is the main source of livelihood for these households. This is an interesting paradox. With declining male to female ratio, the families are also losing men as heads of the households. This calls for a separate study.

A caste-wise classification of the occupational category indicates that Koli–Patel, Ahir, Rabari, Kumbhar are the dominant caste groups engaged in agriculture and animal husbandry. The caste groups like Halpatis, Rathods, Mahyavansees and Harijans are mostly daily wage employees working either as general labourers or agricultural labourers. It is also observed that members from the Halpati and Harijan groups go for fishing activities. These are the poor people of the region. The survey data shows that nearly sixty-four per cent of the total population are in labour force, while dependents constitute about thirty-six per cent. It is also found that the dependent population, that is, 37.5 per cent in Olpad taluka, is considerably higher as compared to the estimate for the Chorasi taluka where it comes around 35.6 per cent of the total population.

LAND, WATER AND SALINITY

In the selected villages, although the most important productive asset is land, nearly 40.1 per cent of the total population is landless. Out of the total landholding households (59.86 per cent), we found that 28.7 per cent of the households were affected by salinity intrusion. The distribution of households owning salinity affected land by their landholding class is presented in Table 3.4. Although the distribution of salinity affected land is uniform across all landholding class, its absolute impact is more on marginal and small farmers.

Table 3.4 Salinity Affected Households by Landholding Class

		Landholding Households (No.)		
Landholding Class (Area in Acres)		*Salinity Affected*	*Not Affected*	*Total Households (No)*
Landless	303 (100.00)	–	–	303 (100)
Marginal (above 0 to 2.5)		188 (48.45)	200 (51.55)	388 (100)
Small (2.5 to 5)		16 (37.20)	27 (62.80)	43 (100)
Medium (5 to 10)		9 (60.00)	6 (40.00)	15 (100)
Large farmers (above 10)		5 (83.33)	1 (16.66)	6 (100)
Total	**303 (40.13)**	**218 (28.87)**	**234 (16.66)**	**755 (100)**

Source: Primary Survey Data.
Notes: Figures in parenthesis are percentages; 1 Acre = 0.4 hectares).

In the area under study, people generally depend on open wells, canals and village tanks to meet their water requirements. We found that open wells were still an important source of drinking water in the villages even in salinity ingressed areas. Nearly, 63.7 per cent of the people use open wells, 30.5 per cent use open wells and public taps, 2.3 per cent depend on only public taps and the remaining 3.6 rely on tanker supply for drinking water. For irrigation and other domestic uses, people fetch water from the village tanks or canals. Most of the tanks which get water from government canals are found to be in bad conditions with high silt and ravaged by eutrophication. In order to understand the water quality of the region, we have collected water samples from ten open wells situated in the village Pinjarat. The characteristics of the sample wells are provided in Table 3.5.

Table 3.5 Characteristics of the Wells Examined in and around Pinjarat Village

Well No.	*Depth of the Well (ft)*	*Distance from Sea (km)*	*Quality as per Villagers Opinion*	*Availability of Irrigation Water nearer to the Well*	*Age of the Wells in (years)*
1	30	6	Drinking	Available	100
2	40	6	Domestic Use	Not available	100
3	22	1	Domestic Use	Not available	54
4	20	2	Domestic Use	Not available	100
5	30	6	Domestic Use	Not available	50
6	25	7	Abandoned	Not available	15
7	25	3	Domestic Use	Not available	50
8	40	7	Drinking	Not available	100
9	15	4	Drinking	Available	10
10	35	4	Domestic Use	Available	50

Note: Micro biological test has also been conducted for water sample from well number 1, Domestic use includes all other uses of water other than drinking that is, water for cattle, irrigation, washing cloths etc.

SOURCE: PRIMARY DATA

It is clear from Table 3.5 that villagers fetch water from three wells for drinking purpose. These wells are located approximately six kilometres inside from the sea shore. Another important character of these wells (two wells) is that they are proximate to irrigation canals. They are getting constantly recharged due to their proximity to canal water. The well which is away from the canal water is situated seven kilometres inside the sea shore. It is opined that one well which is located one kilometre inside from the sea shore is not usable (abandoned) since it received highly saline water. The water from all other seven wells is used for bathing, washing cloths, livestock and agriculture. The water from these wells is not used for irrigation purposes during summers since the salinity is more during that period. As a result, the farmers from the non-irrigated area of the village have to forgo some amount of agricultural output. The forgone agricultural benefits of these farmers as compared to the benefits of similar agricultural plots from the irrigated area has been interpreted as the external cost of salinity ingress. We shall be discussing this aspect in the latter part of the chapter.

Table 3.6 provides the physical and aggregate properties of water collected from these sample wells. It is clear from this table that certain properties of drinking water exceed the permissible limit as per the Indian and WHO standards for drinking water. We found high TDS in all the wells which are used for drinking water purpose although other parameters are in the upper limit by Indian standards. The dissolved solids for the three drinking water wells are in the range of 560–1,800 mg/l which are in excess of permissible limit. It is reported that the water with high TDS is inferior in

Table 3.6 Physical and Aggregate Properties of Water Samples Collected from the Pinjarat Village

Well No.	*Conductivity (Ms/m)*	*Colour Units*	*Turbidity (NTU)*	*Alkalinity (mg Ca CO3/L at pH 4)*	*Hardness (mg Ca CO3/L)*	*TDS (Mg/L)*	*Salinity*
1	1.788	10.00	2	105	370	1,400	0.41
2	1.748	15.00	2.5	108	182	1,800	0.33
3	16.216	75.00	2.5	165	2,470	11,600	8.77
4	6.170	40.00	22	170	568	3,900	2.59
5	1.320	10.00	10	146	330	1,220	0.16
6	5.320	5.00	2	140	850	3,200	2.34
7	9.790	<5	<1	76	3,100	7,600	4.72
8	1.110	10.00	<1	98	3,000	600	0.14
9	0.930	10.00	2	76	280	560	0.11
10	2.350	25.00	3	118	1,000	1,860	0.75
Indian Std	6			150	200	500	
WHO Std	6			150	150	500	

Source: Primary data.
Note: Std (standard).

palatability and may induce an unfavourable physiological reaction in the transient consumer (Srinivas and Ravi 2000).

The total hardness is also an important parameter for the quality of water. Depending on the interaction of other factors such as pH and alkalinity, water with hardness above 200 may cause deposition of scales in the distribution system which may result in excessive soap consumption and subsequent scum formation (Srinivas and Ravi 2000). Hardness of drinking water wells in the area under study ranges from 280 to 3,000 mg Ca CO3/l. The reason for such high solidity and hardness is that these are unprotected wells. In these drinking water sources, calcium and magnesium are found within the permissible limit, that is, 75 and 50 mg/l respectively. Chlorides, nitrates and fluorides are also found within the permissible limit (Table 3.7). It is important to mention here that all these parameters are kept within the permissible limit just because of the wells' proximity to the sweet water canal. However, the micro biological test conducted for a water sample of the most frequently used well for drinking water showed an MPN index 130 (Most Probable Number) which is quite high. It suggests that water is not of high-quality for drinking. The well water used for agriculture and other purposes was also found to be highly contaminated.

Low salinity water can be used for irrigation with most crops on most soils with little likelihood of developing any salinity problem (Elango et al. 1992). Excellent irrigation water is found with TDS less than 200 mg/l. The permissible limit of TDS in the irrigation water ranges between 500 to 3,000 mg/l (Elango et al. 1992).

The water samples from all the wells, except one, (well no. 5) show the value of TDS that exceeds 1,500 mg/l. This indicates that water is highly saline and not suitable for irrigation under ordinary conditions. It makes farmers forgo their agricultural benefits due to high salinity. It was reported that water is more saline during *rabi* and summer seasons. Many farmers are not able to cultivate during this period, especially the small farmers who cultivate in unirrigated areas.

Table 3.7 Inorganic Non-metallic Constituents of Water Samples Collected from the Pinjarat Village

Well No.	*pH*	*Calcium (Mg/l)*	*Magnesium (m Mg/L)*	*Chloride (Mg/l)*	*Nitrate (Mg/L)*	*Phosphate (Mg/l)*	*Fluoride (e Mg/L)*
1	7.9	53.7	57.35	211.90	0.670	0.050	0.15
2	8.5	36.07	22.35	167.90	0.300	0.070	0.15
3	8.4	561.12	260	4,838.40	0.700	0.175	ND
4	8	120.24	65.12	1,419.50	0.250	0.240	.20
5	7.8	84.17	29.16	71.97	0.200	0.095	.15
6	8.2	64.13	167.67	1,279.60	1.100	0.130	ND
7	7.7	480.96	461.7	2,599.10	0.950	0.040	ND
8	8	68.14	31.59	59.98	0.900	0.040	ND
9	8.2	40.08	43.74	43.98	0.950	0.070	.16
10	7.8	112.22	174.96	399.80	1.000	0.090	.13
Indian Std	7.50–8.50	75	50	250	50		1.5
WHO Std	7.50	75	50	250	50		1.5

Source: Primary data.
Note: ND (not detectable).

COST OF RESOURCE DEGRADATION

The Pinjarat village has a total area of 3.382 sq km and a total population of 5,183 (976 households).[3] Majority of the households depend on the open wells for drinking water. As per the village records, the total land under cultivation is 1,029 acres out of which 1,000 acres of land is irrigated by the government canal. The remaining 15,72.5 is unirrigated land. Total non-cultivable area of the village is 5,462.5 acres.

Some of the important crops grown in the village are rice (750 acres), jowar (125 acres), wheat (250 acres), bajra (150 acres) and vegetables (100). Since it was reported that water is more saline during *rabi* and summer seasons, many farmers, especially the small farmers who cultivate in unirrigated areas, are unable to cultivate during this period. The average value of output and the cost of agriculture for a period of one year at 2004 prices have been given in Table 3.8. The data has been generated from six plots located in the village among which three are salinity affected.

An estimate of the forgone benefits of agriculture for an acre of land in two scenarios is provided in Table 3.9. Scenario one represents the situation where the farmers cultivate during all the three seasons (since they have access to irrigation water). Scenario two represents the situation where the farmers do not have access to irrigation water and hence cultivate during one or two seasons. It also represents rain-fed agriculture supplemented with well water irrigation during *rabi* season. During summers, farmers hardly cultivate their land due to high salinity ingress.

Table 3.9 shows that for a period of 25 years, the net benefits at 12 per cent discount rate in scenario one is les than the net benefits in scenario two. The benefit cost ratio of the first scenario is 1.44 which is far less than the benefit cost ratio of the second scenario that is 1.96 at 12 per cent discount rate. The difference in the net present value of the two scenarios at 12 per cent discount rate is Rs 72,221

Table 3.8 Average Cost and Output (in Rupees) of Agricultural Cultivation per year for One Acre Plots with and without Salinity Ingress

Costs/Output	*Average Cost per Acre (Rice, Wheat and Vegetables) Without Salinity*		*Average Cost per Acre (Rice, Wheat) With Salinity*	
	Average Cost (Rs)	*Percentage to Total Cost*	*Average Cost (Rs)*	*Percentage to Total Cost*
Land preparation	500	2.96	480	3.29
Seeds	2,262.03	13.38	450	3.08
Chemical fertiliser	2,412.5	14.27	1,322	9.05
Pesticides	2,825	16.70	120	0.82
FYM	444	2.63	400	2.74
Irrigation	2,000	11.83	2,000	13.70
Equipments (rental)	0	0.00	3,600	24.66
Transport	748	4.42	1,484	10.16
Sowing and Harvesting	3,631	21.47	2,015	13.80
Agriculture Implements	3,750	22.17	4,270	29.25
Total Cost	18,572	109.82	16,141	110.55
Total output Value	**16,912**	**100.00**	**14,600**	**100.00**

Source: Estimated from Primary data.

Table 3.9 Sensitivity Analysis of Net Benefits from Agriculture under Alternative Assumptions (for Cash Flows summed up over Twenty-Five Years at 2004 Prices)

Scenarios	*Discount Rate in Percentages*	*Discounted Benefits in Rupees*	*Discounted Cost in Rupees*	*Net Present Value in Rupees*	*Benefit Cost Ratio*	*Differences in NPV of Two Scenarios*	*Percentage Difference in NPV of Two Scenarios*
Scenario I (With Salinity)	12	1,42,745	98,516	44,228	1.44	72,221	62.01
	10	1,65,202	1,13,488	51,713	1.45	83,518	61.76
	08	1,94,280	1,32,856	61,424	1.46	98,145	61.51
Scenario II (Without Salinity)	12	2,37,647	1,21,197	1,16,449	1.96		
	10	2,75,034	1,39,801	1,35,232	1.96		
	08	3,23,445	1,63,876	1,59,569	1.97		

Source: Estimated from Primary data.

per acre, which is to be considered as an external cost of salinity ingress. Therefore, the total external cost due to salinity ingress is approximately Rs 11,35,68,041 in the case of the village of Pinjarat. This justifies all initiatives of investments that aim at desalinisation of ground water. It is an externality created by farmers themselves with excessive extraction of ground water. Therefore, there are many policy options available with us to control this external cost which ranges from developing new water institutions among farmers, controlling water markets, direct state intervention, etc.

Apart from these costs, the study has made a few field level observations on how the incidence of these costs are shared among various strata of farmers such as small, medium and large ones. We found that small and marginal farmers usually face severe salinity problems because of their lack of access to water from alternative sources. As a result, farmers disproportionately face the consequences of salinity. This type of externality is referred to as vertical externality (Reddy 2003). It was also observed that groundwater markets can take care of such vertical externalities (equity problems) to a large extent (Shah 1993). But the evolution of water markets is possible only in regions where groundwater is available in sufficient quantities, which is not in the present case. Markets do not evolve where there is not enough water to share or to sell (Reddy 2001). Therefore, it warrants an integrated water resources management approach by considering social, economic and ecological aspects. The present study proposes to work on these policy matters, especially towards policies with respect to groundwater development that include the role of subsidised credits/power/diesel in groundwater, development pricing of groundwater, drafting land rights and groundwater rights.

CONCLUSION

The chapter tries to value the changes in the ecosystem due to human action-induced increase in salinity and groundwater depletion and their impact on the lives of people living in the coastal areas of Gujarat. The water samples collected from the dug wells at different distances in the coastal areas have been studied on the parameters of water quality. While it is difficult to generalise the total impact of salinity ingress on the coastal areas over a period of time, it is possible to raise some important issues relating to the impact of salinity ingress. The dominant discourse of development in the country has been that of developing agriculture and allied sectors in the rural areas. The strategy adopted and encouraged by the state has been by way of green revolution technologies. We can see that in Pinjarat, the area of groundwater is consistent with this strategy. However, we have also brought out the fact that overexploitation of groundwater by private farmers has imposed environmental externality on the society. Although, we have pointed out that efforts for ground water recharge and arresting of salinity ingress have social benefits over the public investment in it, the cost is to be borne by the government and not the private farmers who created this externality. Another important dimension of the problem brought out is the deteriorated water quality unfit for human consumption. This impact is hardly foreseen even when ecological assessment of the coastal region has been undertaken with respect to watershed development. The local source for drinking water, which is more sustainable, has been lost in achieving agricultural development. Since drinking water is a basic amenity, the burden on the state increases further. If our results from the larger study of thirteen sample villages in the two talukas of Surat district further confirm the same trends that we have observed in Pinjarat, then, the costal areas of Gujarat are likely to be in serious trouble. It is necessary, therefore, to initiate macro level assessment of the entire region in the state.

NOTES

1. Corresponding author.
2. The discount rate represents the time value or time preference of money. One hundred rupees received today will be preferred over Rs 100 (even if at real prices) received a year later due to time preference. (For more details regarding cash flow analysis and selection of discount rate, see Ninan K.N. and Jyothis S. 2004)
3. Village data have been collected from village records.

REFERENCES

Barot, J.M. 1996. *Reaching the Unreached: Challenges for 21st century*. 22nd *WEDC Conference* held in New Delhi.

Chopra, K. 1989. 'Dimensions and Causalities', *Indian Journal of Agriculture Economics*, 44(1).

Elango, L.S. Ramachandran and Y.S.N. Chowdhary. 1992 'Groundwater Quality in Coastal Regions of South Madras', *Indian Journal of Environmental health*, 34(4): 318–25.

Government of Gujarat (GOG). 1978. *Report of the High Level Committee to Examine the Problems of Salinity Ingress Along Coastal Areas of Saurashtra*. Gandhinagar: Public Works Department.

Government of India (GOI). 1972. *Report of the Irrigation Commission*, Vol 1. New Delhi: Ministry of Irrigation and Power.

Gujarat Ecological Commission. 2001. *State of Environment Gujarat*. Gujarat: Gujarat Ecological Commission.

Joshi, P.K. and A.K. Agnihotri. 1984. 'An Assessment of the Adverse Effects on Canal Irrigation in India', *Indian Journal of Agriculture Economics*, 39(3).

Joshi, P.K. and Dayananda Jha. 1992. 'An Economic Enquiry into the Impact of Soil Alkalinity and Waterlogging', *Indian Journal of Agriculture Economics*, 47(2).

Joshi, P.K. 1987. 'Effects of Surface Irrigation on Land Degradation: Problems and Strategies', *Indian Journal of Agriculture Economics*, 42(3).

Joshi, Vidyut and S. Jyothis 2004. *Socio Economic Baseline Study, Suvali Hazira Area*. Surat: Centre for Social Studies (Mimeo).

Nayak, Sanatan. 2002. 'Extent and Costs of Environmental Degradation in Sriramsagar Project Command Area: Integration with Cost Benefit Analysis', *Artha Vijnana*, 44(2–4): 291–304.

Ninan, K.N. and Jyothis S. 2004. 'Social Forestry: A Case Study from Karnataka', in Gopal K. Kadekodi (ed.), *Environmental Economics in Practice: Case Studies from India*, New Delhi: Oxford University Press.

Queensland Department of Primary Industries (QDPI). 1987. *Landscape Soil and Water Quality*. Proceedings of Brisbane Regional salinity Workshop, Brisbane, Australia.

Reddy, Ratna V. 2001. 'User Valuation of Renewable Natural Resources: A Study of Arid Zone', in V. Ratna Reddy (ed.), *User Valuation of Renewable Natural Resources: The User Perspective*, New York: Nova Science Publishers.

———. 2003. *Costs of Resource Degradation Externalities: A Study of Groundwater Depletion in Andhra Pradesh*. Working Paper No. 49. Hyderabad: Centre for Economic and Social Studies.

Shah, T. 1993. *Groundwater Markets and Irrigation Development: Political Economy and Practical Policy*, New Delhi: Oxford University Press.

Srinivas, C.H. and S.P. Ravi. 2000. 'Studies on Ground Water Quality of Hyderabad', *Poll Res*, 19(2): 285–89.

4

Reconciling Weak and Strong Sustainability

B.S. Sri[1] and *M.S.V. Prasad*

Abstract: A conceptual and analytical approach is presented to reconcile weak and strong sustainability. It involves a reconsideration of the conception of total capital from an ecological economic system perspective. In particular, natural capital is classified into non-renewable resources, renewable resources that are harvested, and those that are not used in production. Strong sustainability is defined in terms of constant environmental quality. Weak sustainability is characterised by non-decreasing value of aggregate income and environmental quality.

INTRODUCTION

The idea of sustainability as a management rule and policy principle is by no means new. However, interpretations and management objectives have changed over time. As a fundamental principle of resource management, it has a long tradition in forestry, and has logically been extended to other concerns of natural resource use (Hediger 1997). In addition, the notion of sustainability has been adapted to the context of various objectives of economy, society and the environment (Tisdell 1991), including the context of economic growth (OECD 1960) and nature conservation (IUCN 1980). As a consequence, the interpretation has become ambiguous and the operational content increasingly unclear. It was the contribution of the 'Brundtland Commission' (WCED 1987) to provide a still vague but nonetheless meaningful definition of sustainable development which involves a shift of the predominant development paradigm (Munasinghe 1993), and 'a subtle but extremely important transformation of the ecologically-based concept of sustainable development, by leading beyond concepts of physical sustainability to the socio-economic context of development' (Adams 1990: 59).

Correspondingly, sustainable development is an important concept of integrating social, economic and ecological dimensions of development and jointly addressing the objectives of conservation and change. Since these objectives cannot be achieved simultaneously as a rule, trade-offs across the various objectives are inescapable. They must be made explicit to have a clear idea of what is meant by 'sustainability' and 'sustainable development', respectively (cf. Barbier 1987; Tisdell 1988, 1990; Hediger 1997). Nonetheless, much of the present literature does neglect these trade-offs, by either concentrating on issues of environmental preservation or economic development. This has culminated in mutually exclusive concepts of 'weak' and 'strong' sustainability that are either grounded in an ethical premise of keeping the general production capacity of the economy constant, or maintaining essential functions and capacities of the environment intact over time. It is dividing economists and environmentalists, rather than reinforcing and integrating different perspectives and comprehensively addressing the overall challenge of sustainable development.

An integrated approach is required that goes beyond traditional debates on economic development versus environmental conservation, and monodisciplinary conceptions of sustainability. Hence, an important task is to bridge the gap between the value principle of weak sustainability and the physical principle of strong sustainability. This is the aim of this chapter which provides a conceptual and analytical approach to reconciling different concepts of sustainability. In section two, we briefly present basic principles of 'weak' and 'strong' sustainability, and discuss the limitations of either conception. In section three, we reconsider the understanding of total capital, that is, an economy's generalised productive capacity and the aggregate of natural capital as a reference base for defining terms of sustainability in a consistent framework. In this background, we elaborate an integrated approach of sustainability in an ecological-economic context of development and the long-term preservation of the environment as our life-support system. This problem of integrating the objectives of economic growth and environmental preservation is formally analysed in section four, where a 'sustainability-based social value function' is proposed. This shall contribute to the elaboration of an integrated framework of sustainable development which is required for the definition of operational and consistent terms of sustainability and the evaluation of environmental and developmental policies from a systemic perspective.

BASIC INTERPRETATIONS OF SUSTAINABLE DEVELOPMENT AND SUSTAINABILITY

Sustainable Development

In correspondence with the WCED (1987), sustainable development is generally defined as 'development that meets the needs of the present without compromising the ability of future generations to meet their own needs'. It has evolved from different development paradigms, encompasses economic, social and ecological perspectives, and contains within it the key concepts of equity, needs and limitations.

First, the concept of sustainable development is generally motivated by an ethical imperative of intergenerational equity, a concern that must logically be extended to requirements of intragenerational equity. This is not only fundamental to the idea of sustainable development but also corresponds to the development paradigm of equitable growth that evolved in the 1970s, when distributional objectives were recognised as distinct from and as important as economic efficiency (Munasinghe 1993). Correspondingly, priority should be given to meet the essential needs of the world's poor; that is, the alleviation of poverty and inequity (WCED 1987). Yet, this is not sufficient for sustainable development. Rather, the satisfaction of both human needs and aspirations (needs and wants) is seen as the major objective of development—'Sustainable development requires meeting the basic needs of all and extending to all the opportunity to satisfy their aspirations for a better life' (WCED 1987: 44).

Second, the concept of sustainable development implies limitations that need to be imposed on economic development. These are not absolute limits, but rather limitations that are determined by:

- institutional arrangements and technologies that are under control of social and individual decision makers; and
- current capacities of the environment as our global life-support system.

The recognition of such limitations reflects a paradigm shift that occurred in the 1980s, when environmental protection had become the third major objective of development. This does not imply preservation of ecosystem everywhere. Rather, it requires that 'sustainable development must not endanger the natural systems that support life on Earth' (WCED 1987: 45), and that the adverse impact on the quality of the environment (air, water and other natural elements) are minimised so as 'to sustain the ecosystem's overall integrity' (WCED 1987: 46). Correspondingly, sustainable development is broader in conception than the traditional conservation versus development problems. It requires that the goals of social and economic development should be defined in terms of sustainability. However, this is not as straightforward.

Sustainability

Divergent interpretations of the objectives and notional definitions of terms of sustainability are sources of confusion, rather than contributions that could help to reinforce the basic idea of sustainable development. As a consequence, there is disagreement about the conceptual and operational content of the terms.

Among other reasons, this is caused by:

- differences in disciplinary perspectives, including different paradigms and axiomatic foundations of the dynamic models within which the concepts have been explored; and
- differences in the philosophical and ethical interpretation of sustainable development. This has resulted in different paradigms of economic and ecological, or 'weak' and 'strong' sustainability. These can also be referred to as economic value principles and ecologically-based physical principles, respectively.

First, 'weak' sustainability is an economic principle which is founded within the body of neoclassical capital theory. It is a value principle with the necessary condition that some suitably defined value of aggregate capital including human-made capital and the initial endowment of natural resources must be maintained intact over time. However, this definition is subject to different interpretations with respect to the objectives of sustainable development. In narrow terms, it requires that the generalised production capacity of an economy is maintained intact, such as to enable constant consumption per capita through time (Solow 1974, 1986). This is also referred to as 'Solow sustainability'

(Common and Perrings 1992), or 'very weak sustainability' (Turner et al. 1994). In broader terms, 'weak sustainability' requires that the welfare potential of the overall capital base remains intact (Opschoor 1996). This is not restricted to sustaining a material standard of living, or consumption, but also includes values that are related to non-consumptive uses (existence and bequest values) and the public good character (amenity and recreational values) of the environment (cf. Freeman 1986; Siebert 1987; Munasinghe 1993).

Second, the conception of 'strong' sustainability emerged from the basic paradigm of ecological economics that the economy is an open subsystem of the finite and non-growing global ecosystem, that is, the environment (Costanza et al. 1991; Daly 1991a, 1991b). It is a physical principle which is founded upon the laws of thermodynamics, and requires that certain properties of the physical environment must be sustained. As a minimum necessary condition, 'strong sustainability' requires that the total stock of natural capital remain constant over time (Costanza et al. 1991; Costanza 1991; Daly 1991a; Pearce et al. 1994). Apparently, this implies an 'ecological value principle' which measures the total 'value' of the heterogeneous stock of natural capital from an ecosystem perspective (Hediger 1998). In contrast, to make it an operational principle, the above criterion of constant natural capital has been translated into principles of 'safe, minimum sustainability standards' (Costanza 1991). These imply a set of ecological criteria which every project should meet (Costanza 1991; Daly 1991a, 1991c). This is a stationary-state principle which is also referred to as 'very strong sustainability' (Turner et al. 1994).

Differences among these concepts are a result of different visions about what a sustainable world can and should look like. This cannot be reduced to differences in the representation of economic transformation processes (production functions) and assumptions about the substitutability of manmade and natural capital. It is also a consequence of notional differences, and particularly involves differences in the way the environment and its functions are integrated and valued. This implies different objectives of what it is that should be sustained, and different conceptions of capital. It is therefore important to investigate these differences.

CONCEPTIONS OF CAPITAL AND SUSTAINABILITY RECONSIDERED

Conceptions of Capital

Conventionally, total capital is defined as consisting of human-made capital and natural resources. Yet, this is too narrow a conception for defining operational terms of sustainability. Rather, as illustrated in Figure 4.1, we may usefully distinguish four main categories of physical assets: human-made capital, non-renewable resources, renewable resources that are used (harvested) and those that are non-used (not harvested) for economic production. These are the key compartments for defining the main aggregates of total capital that can be referred to as economic capital, natural capital and ecological capital. To provide an overall framework, this must be extended to also include immaterial assets such as human capital, social organisation, institutions and the level of technology as well as land and other determinants of the ecosystem's carrying capacity.

The generalised productive capacity, or total stock of economic capital (production capital) is an aggregate of human-made, non-renewable and harvested stocks, as well as the above mentioned

Figure 4.1 Aggregate of Economic and Environmental Capital

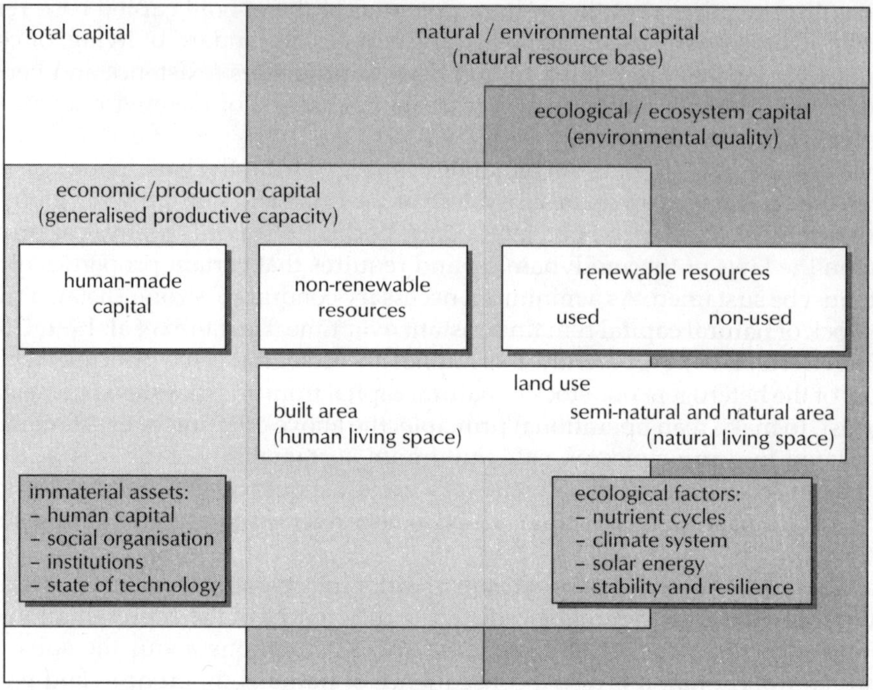

immaterial assets. Non-used renewable resources are excluded, unless an option value is associated to them for potential future uses. However, there may always be natural areas secluded/precluded from development and biological populations that have no instrumental value in present and future production processes. Nonetheless, these 'non-used' ecological assets are essential and therefore valuable from a total system perspective. They are essential components of the ecological, or ecosystem capital. The latter can also be referred to as overall quality of the environment and be defined as an increasing function of used and non-used biological resources and the total carrying capacity of the ecosystem which depends upon the space available (that is, the land area of the ecosystem), the flow of nutrients and other factors that are essential for the integrity and productivity of the ecosystem (Hediger 1998). Furthermore, we may define natural, or environmental capital as the overall natural resource base; that is, all forms of non-renewable and renewable resources, as well as the total land area (built through natural landscapes), and other ecological factors.

From this perspective, and under consideration of the WCED's overall conception of sustainable development (WCED 1987: 43–46), we can judge the various principles of sustainability as follows.

Strong Sustainability

Strong sustainability is a principle of environmental conservation. It is generally formulated in terms of keeping some aggregate of environmental assets, or natural capital constant over time. However, these aggregates can be variously defined. Therefore, without additional specification, the concept of strong sustainability remains unclear, even from an ecological perspective. In general terms, it can either be referred to as a physical criterion of maintaining the economy's material resource base intact for production through time, or an ecosystem principle for protecting the natural environment as our life-support system, or both.

As a physical principle of production, it implies the strong requirement of balancing the depletion of non-renewable resources with enhancing the stock of renewable resources. In other words, it implies a need for enhancing the ecological capital base (see Figure 4.1). This is only feasible, if the rate of regeneration of used resources would be increased, or if additional resources would be developed to enlarge the base of resources that can be used in economic transformation processes. This may be questionable from an ecological point of view. Hence, any viable alternative implies an improvement of the generalised productive capacity of the economy without degrading the overall quality of the environment. This will require an intensified use of waste materials through recycling processes, investment in technological progress and human capital and improvement of institutions and social organisations. Correspondingly, sustainability criteria that refer to the physical resource base of an economy should be analysed from an economic capital perspective, and thus integrated in the framework of an economic value principle. Similar caveat applies if the maintenance of the total stock of natural capital is at issue (see Figure 4.1).

As a consequence, a feasible and viable criterion of strong sustainability is most suitably defined in terms of maintaining ecological capital intact over time. This is an ecosystem principle which corresponds to the request for protecting our global life-support system (Costanza et al. 1991), and to minimise adverse impact on the quality of the environment so as to sustain the overall integrity of the ecosystem (WCED 1987). To this end, the aggregate structure of the ecosystem as the life-support system for all species, including humanity needs to be taken into account. Critical issues that should be addressed in this context are the stability and resilience of ecosystems, interdependencies of ecological processes, and the non-substitutability of some components of ecological capital (cf. Common and Perrings 1992; Perrings 1991, 1996). We shall take up this matter later.

Very Strong Sustainability

Very strong sustainability is a stationary-state principle which requires limiting human scale (zero population growth and zero economic growth), enforcing technological progress that is reducing throughput of matter and energy, and complying with a set of 'safe, minimum sustainability standards' (Costanza et al. 1991; Costanza 1991; Daly 1991a, 1991b). These criteria may be sufficient for a steady-state economy, but not viable for sustainable development as a process of change. In particular, such stationary-state criteria are not sensitive to essential properties and dynamics of ecosystems, non to the existence of natural resources, or species, that are not harvested and therefore out of direct control. Correspondingly, the strict application of safe, minimum sustainability standards as a decision-rule imposed upon each asset will not in general be sufficient to meet total system requirements, such as the maintenance of natural, or ecological capital (Hediger 1998). However, it may be useful to selectively integrate stationary-state criteria into an ecosystem principle of strong sustainability.

Very Weak Sustainability

The principle of very weak sustainability, or 'Solow sustainability', is the economic counterpart to the environmental principle of strong sustainability. It requires an economy's generalised productive capacity ('economic capital', production capital of the economy) to be transmitted across generations to maintain a constant level of per capita consumption (cf. Solow 1986). This is an ethical principle based on an application of Rawls' (1971) maximin principle to a problem of intergenerational equity and intertemporal capital accumulation of non-renewable resources (Solow 1974). Moreover, it is a value principle which contains within it Hicks' (1939) definition of income as the maximum amount that may be spent on consumption in one period without reducing real consumption expenditures in future periods and corresponds to the standard welfare interpretation of Net National Product (NNP) as the largest permanently maintainable value of consumption (Weitzman 1976).

An important feature is that growth in Hicksian income is by definition sustainable, otherwise it could not be counted as income (Daly 1991a: 249). But, this essential property of development is excluded from the Solow model for reasons of intergenerational equity. Rather, Solow sustainability requires an initial stock of total capital big enough to support a decent standard of living, or else it perpetuates poverty (Solow 1974). Correspondingly, Solow sustainability is 'less a criterion for sustainable development than a condition for the efficiency of inter-temporal resource allocation' (Common and Perrings 1992: 11). It is blind to the dynamic properties of ecosystems, and to working in a framework of stationary-state conditions that apart from non-renewable resource depletion and accumulation of reproducible capital provide a rationale for constant consumption per capita through time, given the initial stock of renewable resources (cf. Hartwick 1978).

Yet, if we would relax Solow's inter-temporal maximin criterion, and enable growth in per capita income, then we may obtain a principle of economic development defined in narrow terms of consumption growth that would be sustainable by definition. This would not provide a criterion for sustainability, but rather introduce economic efficiency as an additional criterion for the evaluation of sustainable development boundaries; that is, the benchmarks of a 'sustainable development corridor' within which aggregate economic growth would not be in conflict with any sustainability criterion. Correspondingly, the maximin criterion for constant consumption per capita provides a minimum necessary condition for economic development. In this sense, it can be interpreted as 'sustainable development frontier' on the economic side of the overall system, rather than as a criterion for sustainability.

Weak Sustainability

Finally, weak sustainability is defined here as an integrative value principle, which requires that the total value of aggregate economic activity and environmental quality should be intact over time. In essence, this is an integrated framework including the ecosystem principle of strong sustainability and the above principle of economic development. It does not need that either stock of ecological capital or economic capital should be maintained over time. Rather, weak sustainability requires that some suitably defined value of services of these stocks should be maintained over time.

The rationale of this value principle is that changes in environmental quality can be evaluated and traded-off against changes in aggregate income, and vice versa. This involves the distinction between renewable resources that are harvested and those that are not directly used. This distinction corresponds to the above conceptions of ecological capital and economic capital, that are integrated into this welfare economic principle.

Correspondingly, we can take into account that, apart from instrumental values in physical transformational processes in an economy (production and consumption), the environment provides non-consumptive services. The latter include:

- functional benefits that are provided through ecological processes, like the regeneration of natural resources and assimilation of waste and pollutants; and
- ecosystem values that are generally associated with the public-good attribute of the environment in providing social benefits to present and future generations, such as recreation and amenity services, or existence and bequest values (cf. Freeman 1986; Munasinghe 1993).

Altogether, these non-instrumental values are integrated with the concept of environmental quality which is defined as an increasing function of the totality of environmental assets. They determine the interactions and relationships which are essential for the integrity of an ecosystem ('ecological capital'), and provide public-good functions to society (Hediger 1998). Hence, the concept of weak sustainability integrates the various environmental benefits (value of ecosystem capital) with those of economic development (value of production capacity).

Integrating Weak and Strong Sustainability

In sum, the principle of weak sustainability is crucial for making sustainable development a meaningful and operational concept. It brings about an evaluation of trade-offs among different systemic goals. As a minimum criterion, it requires that the total value of aggregate economic activity and environmental quality should be non-decreasing from one generation to the next. This can be compatible with environmental degradation if suitably compensated by growth in Hicksian income. However, weak sustainability is not sufficient for sustainable development, which also requires that the ecosystem's overall integrity should be sustained (WCED 1987). This is generally conceptualised as an ecosystem principle of strong sustainability which is consistent with ecosystem resilience, and thus with the maintenance of the overall quality of an ecosystem. This principle of maintaining the physical stock of ecological capital intact over time can also be justified on economic grounds. In the presence of irreversibility and uncertainty, loss aversion felt by many individuals, and the criticality (non-substitutability) of some components of natural capital, strong sustainability applies as a valid decision criterion. However, this does not necessarily imply a stationary-state economy, and conservation of every ecosystem and environmental asset. Rather sustainable development requires a process of change that is in harmony with:

- both ecological and economic minimum conditions, such as ecosystem resilience and basic human needs; and
- the satisfaction of preferences beyond these limits.

Altogether, this calls for integrating the objectives of weak and strong sustainability, and changing resource allocation over time (Turner et al. 1994) which is usefully formalised in terms of an intertemporal allocation model (cf. Common and Perrings 1992; Hediger 1998).

This gives rise to two key questions that are related to the vision of sustainable development. The first question is concerned with the positive dimensions of development. It involves the understanding of the functioning and interdependence of the different subsystems which is fundamental for determining the feasibility and viability of any development path. The second question deals with the normative foundations of sustainable development. It is concerned with the underlying value system and the desirability of optional states and development paths; that is, the integrated objective of sustainable development. This involves further questions about what we want to sustain—consumption per capita, environmental quality, or some function of social welfare and the evaluation of trade-offs across different systemic goals of sustainability and development. Apparently, this is an issue for a formal analysis.

FORMAL ANALYSIS FROM AN ECOLOGICAL ECONOMIC PERSPECTIVE

The Opportunity Space of Weak and Strong Sustainable Development

From an overall ecological-economic system perspective, an important task is to integrate the objectives of weak and strong sustainability with the goals of positive socio-economic development. This chapter elaborates an analytical framework of environment versus development trade-offs that are implicit to any conception of either weak or strong sustainability, or an integrated principle of both. This framework is based on the above conceptions of capital and sustainability. It allows the formal representation of the relative values that are associated with economic and environmental assets. The corresponding analysis inevitably involves the fundamental question about how ecosystems are valued with reference to the economy. Apparently, this is different for different sustainability principles.

If the principle of strong sustainability applies, then environmental quality can be said to be an astronomical, or asymptotically infinite value with reference to that of aggregate economic activity. In other words, the environment or an ecosystem can be said to be considered as sacred capital (Taylor 1996). In contrast, under the weak sustainability criterion, there exist possible trade-offs between economic activities and the overall quality of the environment. This implies an economic value principle which is based on a total system perspective. It is usefully conceptualised in terms of a social value, or social preference function.

For this analysis, we assume constant population, and that a social preference measure U exists, which is represented by a strictly concave function of current consumption, or income Y and environmental quality or ecological capital Q:

$$U = U(Y, Q)$$
$$U_Y, U_Q > 0, Y_{YY}, U_{QQ} < 0, U_{YQ}, U_{QY} > 0,$$
$$lim\ y \to 0\ UY\ (Y, Q) = \infty,\ lim_Q \to 0\ U_Q\ (Y, Q) = \infty$$

In principle, this can be thought of as reflecting some aggregate of individual preferences, or those of a benevolent dictator. It can be referred to as an individualistic, preference-based aggregate value function. For the formal analysis, this can be introduced into different frameworks. In any case, the value function U provides the normative representation of the analytical framework which constitutes the basis for evaluating trade-offs between economic development and environmental quality.

Preference-based social value function: $U = U(Y, Q)$.
Weak sustainability $U = UY\ Y = UQ\ Q \geq 0$
Strong sustainability $Q \geq 0$

Integrated weak and strong sustainability

$$U = UY\ Y + UQ\ Q \geq 0 \text{ and } Q \geq 0$$

Economic growth ('development')

$$Y \geq 0$$

Weak sustainable development

$$U \geq 0 \text{ and } Y \geq 0$$

Strong sustainable development

$$Q \geq 0 \text{ and } Y \geq 0$$

Weak sustainability requires that at least the aggregate value U should remain constant over time. That is, the weighted sum of changes in aggregate income and environmental quality should not inter-temporally decrease. The weights are determined by the marginal value U_Y and U_Q for aggregate income and environmental quality, respectively. These values are not constant. Rather, they reflect the characteristics of the underlying value function at the current stage of economic development and quality of the environment.

The minimum requirement of strong sustainability is the maintenance of the current stock of ecosystem capital. This is a physical principle that can also be included in a value concept of weak sustainability. The result is an integrated principle of weak and strong sustainability that requires both compliance with non-decreasing aggregate values and a constant stock of ecological capital. This may be compatible with a decrease in aggregate income and therefore be in conflict with the requirement of income growth. The latter is an essential means for enabling development, but in itself is a highly imperfect proxy for development (World Bank 1992). Nonetheless, it is crucial for sustainable development. Indeed, since growth in Hicksian income is sustainable by definition, growth of income per capita should be considered as a minimum requirement for sustainable development. Otherwise, we may refer to a concept of sustainability, rather than development. Correspondingly, an integrated framework is required which adequately matches elements of weak and strong sustainability with those of economic development.

The above conditions for weak and strong sustainability and economic development are illustrated in Figure 4.2. For any given situation, or state of development (Y_0; Q_0), the frontiers of weak sustainability, strong sustainability and economic development are determined by U_0, Q_0, and Y_0, respectively. Apparently, these frontiers are state dependent. They are also historically determined. In other words, sustainable development frontiers are the consequences of past development. Together, they determine the boundaries of the opportunity space for sustainable development.

Figure 4.2 Illustration of Reconciling Weak and Strong Sustainability

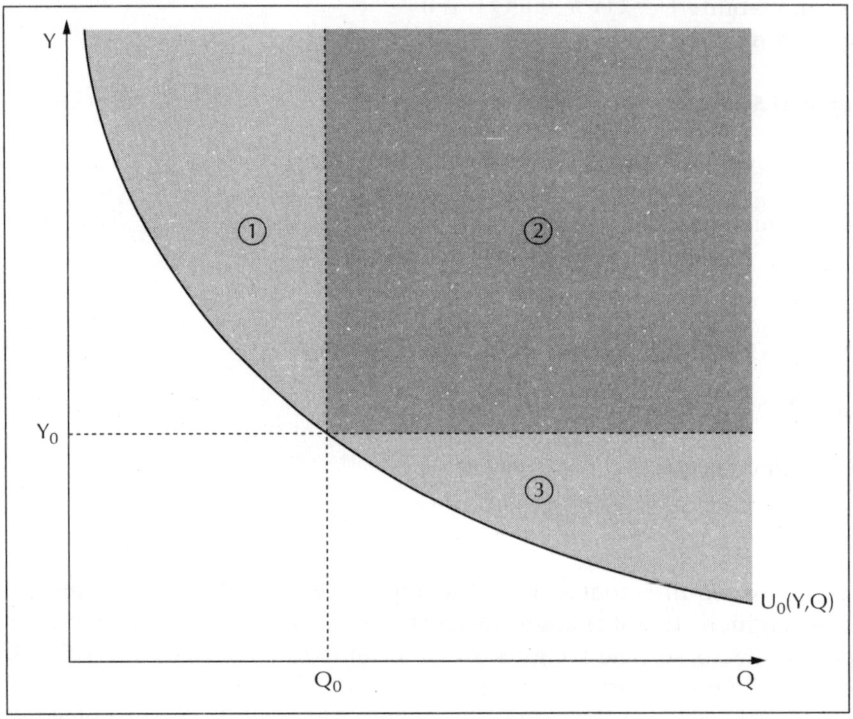

As formally represented above, and graphically illustrated in Figure 4.2, the opportunity space for sustainable development can be categorised into two domains–weak and strong sustainable development. From the Figure 4.2, the weak form would enable development into areas 1 and 2, whereas strong sustainable development is characterised by an opportunity space that is restricted to area 2 . In contrast, any transition into area 3 would be both consistent with weak and strong sustainability, but not compatible with the requirement of growth in Hicksian income, and thus with a minimum requirement for economic development. Correspondingly, development into area 2 is the sole opportunity of change that is consistent with the conjoint requirements of weak and strong sustainability and economic development. Development into area 2 is not only sufficient for weak sustainable development, but also for strong sustainable development (see equations above). However, it can only be realised:

1. if investments are made into technological progress or change that is sufficient for reducing adverse impact upon the quality of the environment; or
2. if it would be possible to improve the stock of ecological capital without reducing the level of aggregate human activity.

The first, technology option seems very promising to various groups of people. It involves engineering solutions and induces investments within the economic system. Correspondingly, it has gained much attention in the economic literature. Potential and limitations of this option have been largely discussed in the broad context of economy and the environment (cf. Faber et al. 1987; Tisdell and Maitra 1988; Victor 1991; Giampietro 1996; Hourcade et al. 1996; Gowdy and O'Hara 1997). In contrast, the second option is by far less attractive to engineers and economists. It requires investments in environmental assets, so as to improve the ecological capital base. Yet, this is only feasible if beneficiary ecological effects could either result from increasing the assimilative or regenerative capacity of the ecosystem, or from making land use changes reversible (cf. Hediger 1998). Otherwise, any feasible and viable option of sustainable development must be formulated in terms of a value principle which enables a trading off of environmental assets against economic benefits.

This accentuates a dilemma about sustainable development that on the one side, aggregate income growth would generally go along with a decrease in ecological capital (development into area 1 of Figure 4.2), and, on the other side, ecosystem improvement would only be possible at the cost (sacrifice) of income (development into area 3).

These alternatives bear the risk of alterations of the ecosystem or socio-economic system that may be undesirable from an overall systemic perspective. To reduce this threat, we need to integrate some minimum requirements of strong sustainability along with that of weak sustainability and economic development.

We can formally define the terms of sustainability and sustainable development from an ecological-economic system perspective and reconcile the above concepts of weak and strong sustainability. The minimum conditions for integrating weak and strong sustainability are:

Ecological-Economic Sustainability (EES):

$$W = W_Y Y + W_Q Q \geq 0$$
$$Y > Y_\# \text{ and } Q > Q_\#$$

Ecological-Economic Sustainable Development (EESD):

$$\text{EES plus } Y \geq 0$$

This says that from an ecological-economic system perspective, sustainability requires that the social value W, should not decrease through time. However, the underlying value function W, and therefore the above condition for ecological-economic sustainability (EES) is only defined for $Y > Y_\#$ and $Q > Q_\#$; that is, the region of compliance with aggregate economic and ecological minimum standards for sustainable development.

CONCLUSION

Sustainable development is a global challenge, which calls for envisioning both conservation and change. It is a normative concept, which involves conflict across various system goals that may be optimally achieved through an adaptive process of trade-offs. This cannot exclusively be addressed with either an economic value principle of 'weak' sustainability, or an ecologically-based physical principle of 'strong' sustainability. Rather, an integrated approach is required that is based on an ecological-economic system perspective of development and preservation of the environment as our life–support system. This needs to be suitably extended to the social context of development and the environment, which remains subject to further research. The focus of this contribution is to reconcile the conceptions of weak and strong sustainability in the context of economy-environment interactions.

One major problem of sustainable development analyses is that traditional principles of weak and strong sustainability are based on different conceptions of capital and different objectives of what it is that should be sustained. As a consequence, they are mutually exclusive, rather than being as reinforcing the root idea of sustainable development. For this reason, the conceptions of capital aggregates and sustainability principles have been reconsidered. The illustration in Figure 4.1 underlines that it would not be sufficient for sustainable development if some suitably defined aggregate of capital, natural or total, would be kept intact over time. First, the root of the problem is the co-existence of environmental assets that are used in economic processes and those that are non-used. The former have both an instrumental value in the production of goods and services, and a functional value within the ecosystem. They are part of the economy's generalised productive capacity (economic/production capital), as well as part of the ecosystem's capital base and determinants of environmental quality. The latter includes elements of the natural resource base that have no instrumental use, but may be essential for the functioning of the ecosystem. As a consequence, the economic and ecological capital bases are overlapping. Second, the physical criterion of keeping the material resource base of the overall system intact through time will not in general be sufficient for sustainable development. Rather, this should be extended to protecting the natural environment (ecosystem capital) as our life-support system. In order to comprehensively address the challenge of sustainable development, the ecosystem principle of strong sustainability may be usefully integrated with the value principle of weak sustainability. The latter requires that the total value of aggregate economic activity and environmental quality should be non-decreasing from one generation to its successors.

Having accepted such integrated framework of weak and strong sustainability, we need to identify the requirements for sustainable development. These can be referred to as minimum conditions of sustainability and development, respectively. First, assuming that there is a constant population, one minimum requirement for development is that, if suitably defined, aggregate income should not decrease. This is consistent with the fact that growth in Hicksian income is sustainable by definition. But, it gives rise to a dilemma of sustainable development. The problem is that:

- either technological progress or opportunities for environmental improvement must be sufficient to enable a strong sustainable development; or

- economic development results in continuous degradation of the ecosystem capital as it is not generally possible to realise economic growth without physical expansion of the economy which involves land use and change in cover.

Under these circumstances, sustainable development requires stabilisation of ecosystem capital at a level that is compatible with the minimum standard of ecosystem resilience.

NOTE

1. Corresponding author.

REFERENCES

Adams, W.M. 1990. *Green Development: Environment and Sustainability in the Third World.* London and New York, NY: Routledge.

Arrow, K.J. and A.C. Fisher. 1974. 'Environmental Preservation, Uncertainty and Irreversibility', *Quarterly Journal of Economics*, 88: 312–19.

Barbier, E.B. 1987. 'The Concept of Sustainable Economic Development', *Environmental Conservation*, 14(2): 101–10.

Barbier, E.B. and A. Markandya. 1990. 'The Conditions for Achieving Environmentally Sustainable Development', *European Economic Review*, 34: 659–69.

Chichilnisky, G. 1977a. 'Development Patterns and the International Order', *Journal of International Affairs*, 31(2): 275–304.

———. 1977b 'Economic Development and Efficiency Criteria in the Satisfaction of Basic Needs', *Applied Mathematical Modelling*, 1: 290–98.

Common, M. and Perrings, Ch. 1992. 'Towards an Ecological Economics of Sustainability', *Ecological Economics*, 6: 7–34.

Costanza, R. 1991. 'The Ecological Economics of Sustainability', in R.Goodland, H. Daly, S. El Serafy, and B.von Droste (eds), *Environmentally Sustainable Economic Development: Building on Brundtland*, UNESCO, Paris: 83–90.

Costanza, R., H.E. Daly and J.A. Bartholomew. 1991. 'Goals, Agenda and Policy Recommendations for Ecological Economics', in R. Costanza (ed.), *Ecological Economics: The Science and Management of Sustainability*, pp. 1–20. New York: Columbia University Press.

Daly, H.E. 1991a. *Steady-State Economics*. 2nd edn with new essays. Washington DC and Covelo, CA: Island Press.

———. 1991b. 'Elements of environmental macroeconomics', in R. Costanza (ed.), *Ecological Economics: The Science and Management of Sustainability*, pp. 32–46. New York, NY: Columbia University Press.

———. 1991c. 'Ecological Economics and Sustainable Development', in C. Rossi and E. Tiezzi, (eds), *Ecological Physical Chemistry*, pp. 185–201. Proceedings of an International Workshop, 8-12 November 1990, Siena, Italy; Elsevier, Amsterdam.

Faber, M., H. Niemes and G. Stephan. 1987. *Entropy, Environment and Resources: An Essay in Physico-Economics*. 2nd edn Berlin: Springer.

Freeman III, A.M. 1986. 'Estimating the Benefits of Environmental Regulations', in J.D. Bentkover, V.T. Covello and J. Mumpower, (eds), pp. 211–29. *Benefits Assessment: The State of the Art*, Dordrecht: D. Reidel.

Giampietro, M. 1996. 'Fossil Energy Resources as a Prerequisite for Intensive Agriculture: Problems and Perspectives for the Future'. Paper presented at the SAGUF Symposium, *How Will the Future World Population Feed Itself*, 9–10 October, Zurich, Switzerland.

Gowdy, J. and S. O'Hara. 1997. 'Weak Sustainability and Viable Technologies', *Ecological Economics*, 22: 239–47.

Hartwick, J.M. 1978. 'Investing Returns Form Depleting Renewable Resource Stocks and Intergenerational Equity', *Economics Letters*, 1: 85–88.

Hediger, W. 1991. *OpportunitaÈ tskosten der Umweltverschmutzung: Eine dynamische oÈ kologischo È konomische Analyse*. Churund ZuÈ rich: Verlag RuÈegger.
——. 1997. 'Towards the Ecological Economics of Sustainable Development', *Sustainable Development*, 5(3): 101–09.
——. 1998. 'Ecosystem Management and Sustainability: An Ecological-Economic Model', in S. Dwyer, M. O'Connor and U. Ganslosser (eds), *Ecology Society Economy: Life Science Dimensions*, Germany: Filander Press.
Henry, C. 1974. 'Investment Decisions Under Uncertainty: The "Irreversibility Effect"', *American Economic Review*, 64(4): 1006–12.
Hicks, J.R. 1939. *Value and Capital*. Oxford: Oxford University Press.
Holling, C.S. 1973. 'Resilience and Stability of Ecological Systems', *Annual Review of Ecology and Systematics*, 4: 1–24.
——. 1986. 'The Resilience of Terrestrial Ecosystems: Local Surprise and Global Change', in W.C. Clark and R.E. Munn (eds), *Sustainable Development of the Biosphere*, Cambridge: Cambridge University Press.
Hourcade, J.C. et al. 1996. 'A Review of Mitigation Cost Studies', in J.P. Bruce, H. Lee and E.F. Haites (eds), *Climate Change 1995: Economic and Social Dimensions of Climate Change*, pp. 297–366. Contribution of Working Group III to the Second Assessment Report of the Intergovernmental Panel on Climate Change, Cambridge: Cambridge University Press.
IGBP/HDP. 1995. *Land-Use and Land-Cover Change: Science/Research Plan*. IGBP Report No. 35/HDP Report No. 7. Stockholm and Geneva.
IUCN. 1980. *The World Conservation Strategy: Living Resource Conservation for Sustainable Development*. Geneva: International Union for Conservation of Nature and Natural Resources, United Nations Environment Programme and World Wildlife Fund.
Moon, B.E. 1991. *The Political Economy of Basic Human Needs*, Ithaca and London: Cornell University Press.
Munasinghe, M. 1993. *Environmental Economics and Sustainable Development*, World Bank Environment Paper, No. 3. Washington, DC: The World Bank.
OECD. 1960. Convention signed in Paris on 14 December 1960.
Opschoor, J.B. 1996. 'Institutional Change and Development Towards Sustainability', in R. Costanza, O. Segura and J. Martinez-Alier (eds), *Getting Down to Earth: Practical Applications of Ecological Economics*, pp. 237–350. Washington, DC and Covelo, CA: Island Press.
Pearce, D.W. and, R.K. Turner. 1990. *Economics of Natural Resources and the Environment*. Baltimore, MD: John Hopkins University Press.
Pearce, D.W., G.D. Atkinson. and W.R. Dubourg. 1994. 'The Economics of Sustainable Development', *Annual Review of Energy and the Environment*, 19: 457–74.
Perrings, C.H. 1991. 'Ecological Sustainability and Environmental Control', *Structural Change and Economic Dynamics*, 2(2): 275–95.
——. 1996. 'Ecological Resilience in the Sustainability of Economic Development', in S. Faucheux, D. Pearce and J. Proops (eds), *Models of Sustainable Development*, pp. 231–52. Cheltenham: Edward Elgar, Aldershot.
Pillet, G., W. Hediger, S. Kypreos and C. Corbaz. 1993. *The Economics of Global Warming*. PSIBericht Nr. 93–02, Paul Scherrer Institute, Villigen PSI, Switzerland.
Rawls, J. 1971. *A Theory of Justice*. Cambridge, MA: Harvard University Press.
Siebert, H. 1987. *Economics of the Environment: Theory and Policy*. 2nd revised and enlarged edition. Berlin, Heidelberg and New York, NY: Springer.
Solow, R.M. 1974. 'Intergenerational Equity and Exhaustible Resources', *Review of Economic Studies*, pp. 29–45. Symposium on the Economics of Exhaustible Resources.
——. 1986. 'On the Intergenerational Allocation of Natural Resources', *Scandinavian Journal of Economics*, 88(1): 141–49.
Taylor, R. 1996. 'Forms of Capital and Intrinsic Values', *Chemosphere*, 33(9): 1801–11.
Tisdell, C.A. 1988. 'Sustainable Development: Perspectives of Ecologists and Economists, and Relevance to LDCs', *World Development*, 16(3): 373–84.
——. 1990. *Natural Resources, Growth, and Development: Economics, Ecology, and Resource-Scarcity*. New York, NY, Westport, CT, and London: Praeger.
——. 1991. *Economics of Environmental Conservation, Development in Environmental Economics 1*. Amsterdam, London, New York, NY, Tokyo: Elsevier.
Tisdell, C.A. and A. Maitra (eds). 1988. *Technological Change, Development and the Environment: Socio-Economic Perspectives*. London: Routledge.

Turner, R.K., P. Doktor and N. Adger. 1994. 'Sea-level Rise and Coastal Wetlands in the UK: Mitigation Strategies for Sustainable Management', in A.M. Jansson, M. Hammer, C. Folke and R. Costanza (eds), *Investing in Natural Capital: The Ecological Economics Approach to Sustainability*, pp. 267–90. Washington, DC, and Covelo, CA: Island Press.

Victor, P. 1991. 'Indicators of Sustainable Development: Some Lessons from Capital Theory', *Ecological Economics*, 4: 191–213.

WCED. 1987. *Our Common Future, The World Commission on Environment and Development*. Oxford and New York, NY: Oxford University Press.

Weitzman, M.L. 1976. 'On the Welfare Significance of National Product in a Dynamic Economy', *Quarterly Journal of Economics*, 90: 156–62.

World Bank. 1992. World Development Report 1992: 'Development and the Environment'. New York, NY: Oxford University Press.

Section 2

Ecosystem Services and Quality of Life

5

Fish Biodiversity of Digha Fishery in Eastern India: An Empirical Analysis[1]

Anita Chattopadhyay Gupta[2] and *Rabindra N. Bhattacharya*

Abstract: The problem of water pollution is linked with loss of biodiversity and the resultant impact on the fish harvested. These twin problems have been addressed simultaneously and modelled in an aggregated Gordon-Schaefer model for the Digha fishery. An economic biodiversity index and a variable for environmental quality have been included modifying the aggregated Gordon-Schaefer model. For estimating the parameters of the model, the Schnute method has been used. Since the Digha Fishermen and Trader's Association regulates fishing activities and acts as a profit-maximising unit within a larger competitive fish market, this chapter focuses on the dynamics of the profit-maximising regime and explores the dynamic maximum economic yield and net present value of profit by fishery that are maximised here. Small variations in discount rates and intrinsic growth rates have been done as part of sensitivity analysis and their impact on optimal profit has been examined. This has been done under different biodiversity scenarios. It is found that in Digha fishery, there exists a trade-off between economic biodiversity conservation and profit maximisation. Policy measures have to be so designed as to minimise the level of conflict between them.

INTRODUCTION

Fisheries are a major world industry exploiting natural resources for food. Globally, fishery products directly provide approximately fourteen kg of food per person and approximately 28 per cent of global fishery products are used for animal feed and other products that do not contribute directly to human food (FAO 1997). Yet, the condition of coastal ecosystems, from the standpoint of fish production, is poor. Many marine fisheries are on decline and globally their production has reached a plateau of about eighty-four million metric tons. Yields of 35 per cent of the most important commercial fish stocks went down between 1950 and 1994[3] (Grainger and Garcia 1996). As of 1999, the Food and Agricultural Organisation (FAO) reported that 75 per cent of all fish stocks for which information is available are in urgent need of better management—with 28 per cent being currently overharvested and 47 per cent being fished at their biological limit and therefore vulnerable to depletion if fishing intensity increased any further (Garcia and DeLeiva 2000). The improved management of fisheries, which are either already depleted from past overfishing or are in imminent danger of depletion, could realise substantial long-term benefits. Robinson (1980) estimated that possibly some 10–15 million tones of additional fish could be landed as a result of improved management.

Declining marine fishery catches has been the subject of much recent attention in the technical literature (Botsford et al. 1997). The concept of sustainable yield has long dominated the analysis of renewable resources (Schaefer 1954; Beverton and Holt 1957). For many years, the objective of fishery management was to maintain this maximum biologically sustainable yield from fishery, with limited concern for social, economic and environmental factors. However, in recent times, optimum yield has come to be defined as the amount of fish prescribed on the basis of the maximum sustainable yield from such fishery as modified by any relevant economic, social or ecological factors.

Globally, the number of people living within 100 km of the coast increased from roughly two billion in 1990 to 2.2 billion in 1995—39 per cent of the world's population. However, the number of people whose activities affect coastal ecosystems is much larger than the actual coastal population because rivers deliver pollutants from inland watersheds and human settlements into estuaries and surrounding coastal waters. As coastal and inland populations continue to grow, their impact—in terms of pollutant loads and the development and conversion of coastal habitats—can be expected to grow as well. These have a marked influence on fish catch and as such offer arguments for changing fishery management paradigms towards a more coherent ecosystem approach. An ecosystem-restoration plan is needed which should work on water discharges and water pollution prevention and other causes of estuarine habitat degradation, which affect the fish catch.

The role of fisheries management as a whole is that of managing the harvesting of this renewable resource and maintaining stocks at levels which permit their rational and sustainable exploitation. But often poor estuarine management has led to concerns related to the misuse and overuse of resources through excessive harvesting of fish stocks. A pattern of 'sequential exploitation' of fish resources has occurred, whereby fish stocks have been gradually drawn down from accessible to less accessible areas and from valuable to less valuable species.

A long-held view of the development of fisheries is that there is initial exploitation of more abundant, more easily caught species and then, switching in time to increasingly less abundant, less easily caught species. More recently, a study of FAO global fishery statistics has shown this to be true. It is seen that harvesting does lead to a shift from long-lived, high-trophic level fish towards short-lived, low-trophic level fish species. This has been termed 'fishing down the food web' (Pauly et al. 1998). The economic explanation is that fisheries are market-driven and this influences fish harvesters to focus on high economically valued species gradually moving to low economically valued species as the first category gets fished out. Kasulo and Perrings (2001) have called this phenomenon 'fishing down the value chain'. Such extraction has been made possible through the increased use and development of new technologies like improved gears and vessels. However, such practices have meant considerable losses of biological diversity.

Several measures like imposition of closed seasons, minimum sizes of fishes caught, regulation of mesh size and limitation on gear types have been advocated to address this problem. Here the methods involve mostly imposing restrictions on the types and specifications of equipment used during fishing. At the core of such measures, lies the specification for minimum mesh sizes, as a means of enhancing the ability of the fishing gear to select larger, more mature fish, leaving behind more juvenile stock. But conventional fishery management is reluctantly coming to accept that such technological measures alone (for example, mesh size of fishing gear) would be sufficient to maintain or revitalise the depleted fish stock. Rather mesh regulations must be coupled with biodiversity concerns in a multi-species fishery environment. This is because the efficiency of mesh regulations depends upon the mechanical selection of different sizes of fish by the gears being used. In reality, each species requires a different mesh size. Excessive fishing effort with any mesh size not only reduces potential catches and profitability, it may in the long run reduce the size of stock to the point where the stock can no longer reproduce satisfactorily, causing it to collapse completely. So one must incorporate a fishing effort that is adjusted to the needs of multi-species fishery environment through the conservation of biodiversity. This is an important consideration in fishery management decisions relating to protection of biodiversity and which ultimately leads to measures that also sustain fisheries.

The study of biodiversity includes ecological and economic considerations. The ecological aspect relates to human actions that affect the numbers and persistence of species. The economic aspect looks at the driving forces that affect biodiversity as a result of human intervention and are a cause for their loss (Holling et al. 1995). One widely used ecological index, the Simpson Index is expressed as:

$$D_t = \sum_{i=1}^{s} (Y_{it} / Y_t)^2$$

where Y_{it} is the catch of the ith species harvested in period t, Y_t is the total catch in period t and s_t is the number of species harvested in period t as a popular measure of species' dominance. To capture the economic value of species, the Simpson's biodiversity index is modified so that it uses market values of species caught rather than the total amount of species caught. Now, the actual amount of the species will be weighted by price. Therefore, the Simpson's Index becomes:

$$B_t = \sum_{i=1}^{s} (P_{it} Y_{it} / TR)^2$$

where B_t is the economic biodiversity index, P_i is the per unit price of species i, and TR is the market value of the total fish catch. When all the species have the same market value, the solution for economic biodiversity index is the same as the ecological biodiversity index. When the community is dominated by species of high market value, economic biodiversity index will be greater than an ecological biodiversity index of the same community and vice versa (See Kasulo and Perrings, 2001).

Another related concern deals with long enduring perturbations in coastal environments through the continuous influx of various wastes, which is a severe threat to the ability of these systems to generate ecological services. For many nearshore species, land-based and coastal threats are harmful, as these fish species depend on coastal areas for spawning, growth and stock replenishment. Changes that occur in larvae and juveniles here can greatly affect populations and distribution of adult fish. Estuaries have long been the focal point for much human activity. As the meeting place of sea and river, they have provided quiet and sheltered water for fishing. But the pouring of effluents into the estuaries have badly polluted the lower reaches of many of them. Fish and fisheries must be considered in relation to all these uses and abuses[4].

THE PRESENT STUDY

Our chapter examines the problem of fishing when water is polluted and this pollution leading to loss of biodiversity. The twin problems have been addressed simultaneously and modelled in an aggregated Gordon-Schaefer model. It has then been applied to the Digha fishery. While Digha is located in the eastern part of Midnapore district of the state of West Bengal in eastern India, it also lies in the southern most part of the state in the Bay of Bengal. It is situated nearly mid-way along the relatively straight shoreline between the huge Ganga-Brahmaputra delta in the east and the joint Mahanadi-Brahmani-Boitaroni delta in the west. From the geo-morphological point of view, Digha is located on the eastern fringes of the Subarnarekha delta along the south west shoreline of West Bengal and on the eastern borders of Orissa. The landing centers in this area are: (see Figure 5.1)

1. *Paschim Gadadharpur*: Situated about six km west from Digha, it is a landing centre for drag net hauls.
2. *Udaypur*: A place on the western side about five km from Digha, it is a landing centre for drag net hauls.
3. *Ongaria ghat*: Situated about four km on the western side from Digha, it is also a landing centre for drag net hauls.
4. *Jatranala ghat*: About six km from Digha, it is a landing centre for drag net hauls.
5. *New Digha*: Situated about two km west from Digha, it is a landing centre for drag net hauls.
6. *Hospital ghat*: Situated about one km west from Digha, it is a landing centre for drag net hauls.
7. *Sea Hawk ghat or Digha 'Ghole'*: A place on the eastern side about one km from Digha, it is a landing centre for drag net hauls.
8. *Digha Mohona*: More or less an estuarine zone, about four km on the east from Digha, where two irrigation canals, that is, Khadalgobra canal and Ramnagar canal fall into the sea. It is a landing centre for both fishing vessels and drag haul nets.
9. *Sankarpur Harbour*: A minor fishing harbour, about six km away from Choudamile (near Ramnagar) in the Champa canal which falls into the sea across Digha Mohona (Estuary).

The area of our case study is Digha *Mohona* (estuary) of West Bengal, India, where the Khadalgobra and the Ramnagar canals meet the Bay of Bengal. It is a key breeding area mostly for *hilsa*, a popular and traditional species of fish from Bengal. They live in the sea for most of their lives, but migrate at least 1,200 km up any river system according to the trends in their spawning behaviour.

Digha has a sprawling fishing economy. The catch generally includes fishes like *hilsa*, pomfret, mackerel, brawns, sharks, sea urchins etc. In fact this whole belt is endowed with rich biotic diversity (Meenbarta 1998). Some villages are completely dominated by fishermen. With the introduction of diesel-run powerboats, deep fishing and mechanisation in fishing is taking an upturn[5]. Tourism is a phenomenon that has come into prominence much later in the occupational profile of the resident population of Digha (Chattopadhyay 1995). But this area of the sea has lost its purity. Waste disposal from its burgeoning tourism industry causing pollution is affecting fisheries adversely (Ghosh et al. 2004). In 1996, the public health engineering department of the Government of West Bengal had

Figure 5.1 Fish Landing Centres of the Digha Coast

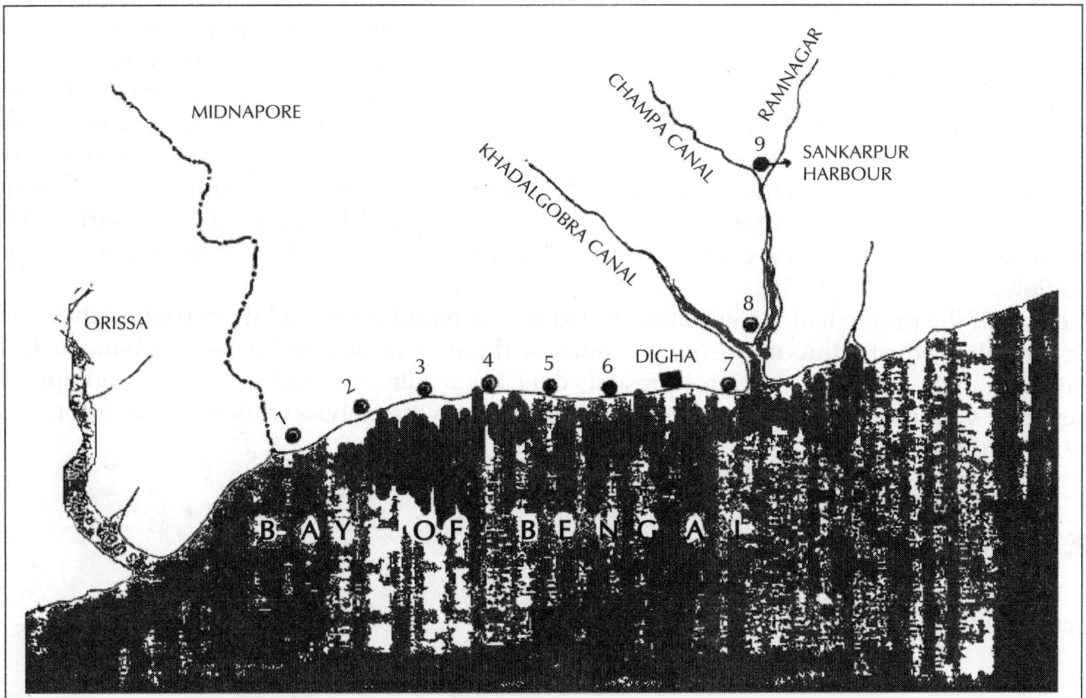

1. Paschim Gadadharpur 2. Udaypur 3. Ongaria ghat 4. Jatranala ghat 5. New Digha 6. Hospital ghat 7. Sea Hawk ghat or Digha 'Ghole' 8. Digha Mohona 9. Sankarpur Harbour

undertaken a project to collect all the wastewater discharged by the hotels in Digha and treat it before discharging into the sea. This sewage treatment system set up by the public health engineering department to prevent marine pollution and to stop the sewage from flowing into the sea from this bustling tourist resort has proved to be completely ineffective (Jana 2003).

Digha, being quite near to many urban centres and also being very easily accessible, is thus a major tourist hub and is experiencing high coastal pollution associated with the rapid growth of the town itself. Combination of such influences act against the backdrop of long-term environmental changes that have an effect on fisheries resources and their exploitation and so must be considered in any integrated assessment.

The estuaries in West Bengal have seen both freshwater fishes and migratory marine species spawning during the monsoon. In the last three decades (1960–90), there have been catastrophic changes. *Hilsa*, an anadromous fish, which used to constitute about 70 per cent to 80 per cent of total fish landings, is disappearing (Rao 2000). Changes in salinity and hydrology cause large shoals of *hilsa* to congregate near the head of the Bay but they do not enter the estuary. Instead, marine catfish have become dominant here. The last few decades have witnessed anthropogenic changes in the coastal waters of the Bay of Bengal. Biodiversity has been adversely affected and multi-species communities are changing to single-species' dominance. The pollution of coastal areas, which serve as nursery grounds for commercially valuable species of *hilsa* and prawns, might ultimately affect their stocks in the Bay. Along the coast, the catch of *hilsa* is on the decline because of pollution and conservation measures are now necessary. The major rivers of the Bay of Bengal drain 200 km^3 of water and 12.0×10^9 tons of silt during the monsoon season which influence and govern the dynamics of the ecosystem (BOBP 1994). These enormous quantities of silt discharged by the river systems in the Bay of Bengal have the potential to act as carriers of pollutants discharged by industries at the mouth of the estuaries. The effects are reflected in decreased growth rates and reduced potential for fishing. Digha, with its vibrant tourism industry, has attained significant urbanisation. Thus, the general picture indicates a decline in the estuarine fishery caused by pollution and reduced water flow.

A study of the problem of water pollution-led loss of biodiversity and its impact on the level of catch per unit of effort is thus necessary. We address these two major problems simultaneously and model them in the Section 3. Section 4 presents data, the results of model estimation and optimum values of the variables. Section 5 reports the results of sensitivity analysis while Section 6 incorporates some concluding remarks.

THE MODEL

In a standard fishery model, we have the expression for net rate of growth of fish biomass as

$$\dot{X} = rX(1 - X/K) - qXE \tag{1}$$

where r is the intrinsic growth rate of fish stock, X is the fish stock, K is the carrying capacity of fish stock, q is the catchability coefficient and E is the effort used.

At steady state, $\dot{X} = 0$ and this implies that

$$X = K\,(1 - q/rE) \tag{1a}$$

The Gordon-Schaefer fish production function is

$$Y = qXE \tag{2}$$

where Y is the catch rate or harvest rate of fish stock. Substituting X from (1a) in equation (2), we have

$$(Y/E) = qK - q^2(K/r)E \tag{3}$$

In the Gordon-Schaefer model, environmental factors can affect fish biomass through its growth function. Hence, in this study, the environmental variable is incorporated into the model through its growth function. Introduction of an environmental factor implies a parametric shift of the logistic curve and we can rewrite equation (1) as

$$\dot{X} = rX(1 - X/K - eW) - qEX \tag{4}$$

where W is the environmental quality variable, and e is a parameter that relates how much one unit increase of the environment variable reduces the relative growth of the fish biomass X. The need for such an inclusion stems from the fact that one needs to study whether besides their level of effort the total catch of fishermen is significantly affected by the weather variable, which here represents the level of pollutants washed into the estuary. The amount of rainfall in Midnapore district is used as a rough indicator of the level of water pollution. Intensive rainfall can lead to larger deposits of sediments and nutrients in the estuarine area. The Bay of Bengal receives an annual precipitation of 11,000 km^3 which it washes off sediments brought by the rivers amounting to 4×10^9 m^3 (Rao, 2000). So, the use of rainfall figures captures the impact on water quality.

The most marked effect of biodiversity loss occurs on the productivity of the resource. Changes in fish diversity affect fisheries through their impact on the wider aquatic ecosystems that support fish production (Barbier et al. 1994). Therefore fish production depends not only on the level of effort and fish stock but also on the level of fish biodiversity. The biodiversity variable can be introduced in the Gordon-Schaefer model through the production function that specifies a relationship between fish biodiversity as an input and fish catch as an output. The effect of species diversity on fish productivity can be captured through an additional term in the fishieries production function (Kasulo and Perrings 2001). Equation (2) now becomes:

$$Y = qAEX \tag{5}$$

where A is the biodiversity index. Here AE is the biodiversity-adjusted effort applied in the fishery. Here, A may represent either ecological or economic biodiversity index, that is, A = D or B as defined earlier. The growth function becomes $\dot{X} = rX\,(1-[X/K]) - qAEX$.

When both biodiversity-adjusted effort and an environmental quality variable is introduced in the standard model, we have the expression for growth of fish biomass as

$$\dot{X} = rX[1 - (X/K) - eW] - qAXE \tag{6}$$

where eW focuses on the impact of water pollution, A is the biodiversity index and AE is the biodiversity-adjusted effort applied in the fishery. In the steady–state equilibrium, when $\dot{X} = 0$

$$X = K\,[1 - (qKA/r) - eW] \tag{7}$$

The Gordon-Schaefer production function then is

$$Y = qAXE \tag{8}$$

Substituting (7) in (8), we have

$$Y = qKAE\,[1 - (qAE/r) - eW] \tag{9}$$

The catch per unit of adjusted effort can then be expressed as

$$(Y/AE) = qK\,[1 - (qAE/r) - eW] \tag{10}$$

For estimation purposes, we have followed the approach of Schnute (1977). The Schnute equation can be modified through the simultaneous introduction of biodiversity index and an environmental quality variable. It defines a population growth function in terms of U, defined as the catch per unit of adjusted effort (Y/AE). Then,

$$\dot{U} = rU(1 - U/qK - eW) - qAEU \tag{11}$$

where A can be the ecological or economic biodiversity index, as the case may be.

The Gordon–Schaefer production function Y = qAEX implies that X = U/q. Dividing both sides of Equation (11) by U, we have

$$(\dot{U}/U) = r - qAE - (r/qK)U - reW.$$

This implies that

$$1/U(dU/dt) = r - qAE - (r/qK)U - reW.$$

After time averaging and thereby smoothing out the data, this equation can be framed as

$$\ln X_t^* = r - qE_t^* - (r/qK)U_t^* - reW_t^* \tag{12}$$

where

$$X_t^* = U_t^*/U_{t-1}^*;$$
$$E_t^* = (E'_{t-1} + E'_t)/2,$$
$$E'_t = E_t A_t$$

Also

$$A_t = D_t \text{ or } B_t;$$
$$W_t^* = (W_{t-1} + W_t)/2.$$

The parameter estimates obtained from the regression analysis of equation (12) show the effect of the inclusion of both biodiversity and environmental quality variables over time. Here, the main intention is to capture the effect of loss of biodiversity and that of water pollution on relative catch per unit of adjusted effort that may affect the productivity of a fishery adversely.

DATA, REGRESSION RESULTS AND OPTIMUM VALUES OF THE VARIABLES

Data for this study has has been collected from the Digha Fishermen and Fish Traders' Association[6] in Digha covering the period 1993–94 to 2002–03. Catch is measured as kg of fish landed and the effort is represented by fishing months[7]. It has been observed in Digha coastal areas that total marine fish landing mainly consists of sardine, *hilsa*, coila, pomfret, croakers, Bombay duck, catfish, ribbon fish, shark, *shankar*, prawn[8] etc. Thus total thirty-seven varieties are found here. These varieties have been divided into five groups considering their importance from the viewpoint of their demand and price (Das et al. 2000). Among them, the contribution of *hilsa* in total catch per trip was found to be maximum in Digha estuary. It was followed by two types of pomfrets: Chinese and Silver pomfret being one variety and the other being Black pomfret. So we see from the collected data that there are mainly four varieties of marine fish that dominate the Digha fishing industry in terms of both prices and quantity. They are *hilsa*, Chinese and Silver pomfret, Black pomfret and prawn. More than 50 per cent of the total value of catch was contributed by these four species. Individual contributions of the other thirty-three species in terms of value are not very significant. Also these other thirty-three varieties of fish such as (in local vocabulary) *sardine, mackerel, chela, para* and *American bhetki* have a very low price range in the market. These have been grouped under the heading 'others'. Digha Estuary, in recent years, has seen a shift in fish species harvested towards catches of fish species of very low local value (ranging between Indian Rs 4–Rs 35 per kg) consisting of *sardine, chela* and *kaante* which we have clubbed under the heading 'others'. This transition in fish catch from high valued to low valued species points to the role of the market and the effects of economic forces towards loss of biodiversity. The decline in the dominance of *hilsa* in total catch reflects not only a decline in the trophic level of fishes but can also be associated with its economic value.

A comparative analysis of the unweighted and weighted Simpson indices is carried out by using the data on catch per species for the fisheries of Digha estuary. The Simpson economic biodiversity is constructed by weighting the simple ecological Simpson's index by average prices so as to capture the fluctuations in value. It will capture any shift that may occur in fish value resulting from the over-exploitation of high-valued species. Figure 5.2 shows a comparison between the Simpson

Figure 5.2 Comparison between Ecological and Economic Biodiversity Indices in Digha Mohona (Estuary) Fishery

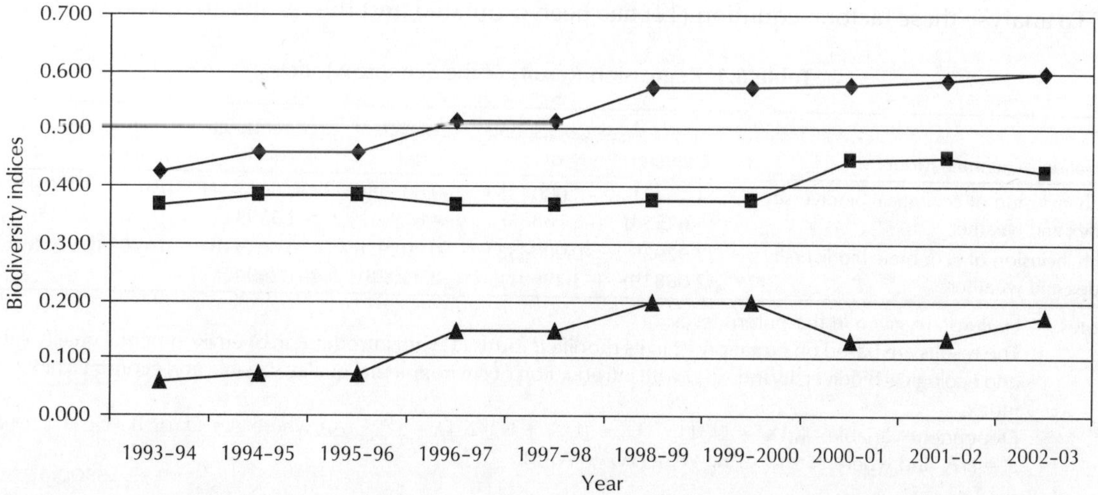

unweighted and weighted indices, and it is seen that the value of the weighted index is lower than that of the unweighted index. It is because of the differences in the value of the species caught that the differences in the two indices occur. The lower values of the weighted indices in comparison with the unweighted index reflects that on an average, catches are dominated by less valuable species. If catches had been dominated by valuable species, price weighting would increase their dominance even further and the weighted indices would have higher values than unweighted indices (Kasulo and Perrings 2001).

This is seen in Figure 5.2 where it suggests a decline in economic biodiversity associated with a shift in fish catch from high-valued to low-valued species. Curve D reflects the ecological biodiversity index over the years and Curve B the economic biodiversity index over the period 1993–2003. The third curve D-B plots the difference between the ecological and economic biodiversity indices over the same period. The curve D-B takes on the shape of a somewhat inverted U-shape with the difference peaking in the year 1998–99. In the early years of our study like 1993–94, we find that the D-B curve is more or less flat in shape at a low level, implying that the difference between the D and the B curves is fixed at a low level.

The implication is simple. At the initial stages of the liberalisation regime, we find that fishing was not much mechanised in Digha *Mohona* (Estuary) due to which fishermen there used to catch fish by using country boats or non-mechanised fishing boats. Naturally the cost of fishing at that time was not as high as we find in present situation. Due to lack of mechanised trawlers, the fishermen at that time were unable to enter the deep sea as a result of which they used to sell more or less similar type of fish species in the market. This situation is more specific for the local markets near Digha *Mohona* (Estuary). Apart from this, fish marketing in Digha *Mohona* (Estuary) was not as much developed as we find in the present situation. All these factors did not lead to much fish price per species by variation along with its low price. Hence it can be argued that during the initial years of liberalisation, the difference between ecological and economic biodiversity indices was not much prominent. The gradual fall-off of the difference in the next few years show the loss of economic biodiversity that fishery is facing with the more expensive fishes being fished out.

To analyse these factors, equation (12) has been estimated and the results are given in Table 5.1.

Table 5.1 Regression Results of the Schnute Models

Equation of Schnute Model	*Constant*	*Coefficient of E_t^**	*Coefficient of U_t^**	*Coefficient of W_t^**	*R^2 Statistic*	*$\bar{R}^2$ Statistic*
with inclusion of ecological biodiversity index and weather	4.16421 (3.67554)	–0.0004353 (–5.78885)	–0.0029904 (–4.8797)	–0.0015041 (–2.55941)	0.861737	0.802481
with inclusion of economic biodiversity index and weather	2.92939 (2.00819)	–0.0004551 (–3.29691)	–0.0029269 (–2.94028)	–0.0009481 (–1.26947)	0.896241	0.851772

Notes: t-values are given in the parentheses.

1. The results are based on equation 12 in its modified forms (1) with introduction of environmental quality variable and ecological biodiversity index (2) with introduction of environmental quality variable and economic biodiversity index.
2. Dependent variable: X_t^*; $X_t^* = U_t^*/U_{t-1}^*$; $U_t^* = (U_{t-1} + U_t)/2$, $U_t = (Y_t/E_t\, A_t)$ where $A_t = D_t$ or B_t; $E_t^* = (E'_{t-1} + E'_t)/2$, $E'_t = E_t A_t$ and where $A_t = D_t$ or B_t; $W_t^* = (W_{t-1} + W_t)/2$.

When the ecological biodiversity index is introduced along with the environmental quality variable in the model and regressed, R^2 is 86 per cent, with all parameter estimates having expected signs and are statistically significant at 5 per cent level of significance. The model including the environmental factor and the economic biodiversity index registered a better performance. It can be seen that an improvement occured after the introduction of a weighted economic biodiversity index and now the model explains about 89 per cent of the variation in fish biomass, which is a good fit for the regression line. Since it helps to explain a large proportion of the variation in fish biomass, this equation is instrumental in using the estimated values of the parameters for the purpose of sensitivity analysis.

One can now estimate the optimum values of the variables under different biodiversity scenarios and in the context of the alternative regimes. The economic biodiversity index that has been considered in our regression procedure is actually the average of the economic biodiversity indices constructed over the time horizon of our study. Generally, the value of B ranges from 1 for the lowest diversity (where the fishery is dominated by species of the same value) to 1/s, where s is the number of species, giving a high level of fish diversity (where fishes have a wide range of differentiated market values). In this context, we have compared three different biodiversity scenarios, reflecting fishing down the value chain that is occurring in the fishery in Digha: one, a scenario where the fishery has a wide range of differentially valued fish species, that is, high level of economic diversity (past situation), another being the average level of economic biodiversity in the fishery (current situation) and the third, a scenario where the fishery has fish species mostly of similar values, that is, low level of economic diversity (projected future situation).

The optimal values of the variables have been calculated under different regimes: maximum sustainable (biologically ideal regime), the open access (hypothetical regime) and the profit-maximising regime (actual regime prevalent in Digha fishery). Both the static and dynamic values of the variables have been calculated for the profit-maximising regime. The need for deducing the dynamic values become important because in order to judge the sustenance of the fishing industry, static rent maximisation is not optimal if the objective is to maximise the present value of profit. The results have been obtained by using the parametric values given in Table 5.2.

These values are tabulated under three alternative biodiversity scenarios of the fishery. Since in Digha fishery, the Digha Fishermen and Traders' Association regulates the local fishing activities and acts as a competitive profit-maximising unit in the larger regional fish market, we have ultimately focused on the dynamics of the profit-maximising regime.[9] We have introduced here two ecosystem-factors, environmental quality variable and economic biodiversity under different regimes and estimated the values of the variables therein. The different property rights solutions are necessary to look into the institutional structure of the fisheries and how they influence the level of fishing effort and through it the mix of exploited species. Here we have looked into three regimes: (1) biologically maximum sustainable regime as the base-line case; (2) open access regime; and (3) competitive profit-maximising regime as the actual case of the Digha fishery. Here we have considered a discrete time framework instead of a continuous time framework as it helps to analyse the sensitivity analysis part of our chapter (Perrings 2000; Gupta 2004, 2005).

Table 5.2 Values of the Parameters of the Digha Mohona (Estuary) Fishery Model

Parameter	*Notation*	*Value*	*Unit*
Intrinsic growth rate[a]	R	1.5	dmnl./year
Catchability coefficient[b]	Q	0.000019	1/fishing hours
Environmental carrying capacity of fish stock[c]	K	2725631.8	Kg.
Coefficient of environmental quality variable[d]	E	0.0003236	dmnl./mm.
Average weighted economic biodiversity index[e]	B	0.3946	Dmnl
Environmental quality variable[f]	W	1382.5	mm.
Average price[g]	P	43.637616	Rs/kg./year
Average cost[h]	C	46.65569	Rs/fishing hour
Discount rate[i]	Δ	0.11	dmnl./year

Notes: [a] the intercept value of the regression of the modified equation 12 (with A=B)
[b] the value of the coefficient of the effort function of the regression of the modified equation 12 (with A=B)
[c] calculated by using the value of the coefficient of the catch per unit effort function and then plugging in the values of r and q
[d] calculated by using the value of the coefficient of the environmental quality variable and then plugging in the value of r
[e] the average of the economic biodiversity indices constructed for the period under study
[f] here rainfall has been taken as a proxy to the environmental quality variable and so for W we use the average level of rainfall of the period under study
[g] it is the aggregative average of the prices of all fish species under our consideration during the time period of our study
[h] it is the total cost incurred by fishermen calculated on the basis of their wages both for labour in boats and

Biological Maximum Sustainable Regime Solution

The Maximum Sustainable Yield (MSY) level of effort is derived by modifying the sustainable-yield function. The logistic growth function is

$$X_{t+1} - X_t = rX_t(1 - (X_t/K) - eW) \tag{13.1}$$

and the Gordon-Schaefer production function is

$$Y = qBE_tX_t \tag{13.2}$$

In equilibrium,

$$X_{t+1} - X_t = rX_t(1 - (X_t/K) - eW) - qBE_tX_t = 0 \tag{13.3}$$

and so,

$$X_t = K(1 - qKB/r - eW) \tag{13.4}$$

We get the sustainable yield function as

$$Y_t = qKBE_t(1 - qBE_t/r - eW) \tag{13.5}$$

Differentiating equation (13.5) with respect to effort, setting the derivative to zero, and solving for effort, we have

$$E_{msy} = r(1 - eW)/2qB \tag{13.6}$$

Setting the derivative of the logistic growth function with respect to X to zero, the associated level of stock is

$$X_{msy} = [K(1 - eW)]/2 \tag{13.7}$$

Substituting (13.7) in the sustainable yield equation (13.5) gives:

$$Y_{msy} = [rK(1 - eW)^2]/4 \tag{13.8}$$

The above equation gives the maximum sustainable catch that occurs at $X_{msy} = K(1 - eW)/2$ corresponding to effort level $E_{msy} = [r(1 - eW)]/2qB$.

The maximum sustainable values of the variables in the modified framework have been calculated using the values of the parameters given in Table 5.2. The results are shown in Table 5.3.

Table 5.3 Optimum Values of the Variables under Different Regimes and Alternative Economic Biodiversity Scenarios

	Maximum Sustainable Solution			*Open Access Solution*			*Profit Maximising Solution (Static Framework)*		
	Stock (kg.)	*Harvest (kg./year)*	*Effort (Fishing Hours/ Year)*	*Stock (kg.)*	*Harvest (kg./year)*	*Effort (Fishing Hours/ Year)*	*Stock (kg.)*	*Harvest (kg./year)*	*Effort (Fishing Hours/ Year)*
Situation 1	7,53,123	3,12,145	1,09,070	2,81,397	37,932	1,77,393	5,24,223	1,51,242	75,923
Situation 2	7,53,123	3,14,131	55,633	1,42,634	42,785	1,01,413	5,93,616	1,95,161	43,851
Situation 3	7,53,123	3,12,145	21,814	56,273	44,903	41,998	6,36,786	2,23,157	18,444

The Open Access Regime Solution

Under open access, as

$$\Pi_t = pY_t - cE_t = 0, \tag{14.1}$$

so,

$$pY_t = cE_t$$

where

Π_t = net revenue
p = price of fish and
c = unit cost of effort

In terms of the logistic growth function, the equilibrium-effort level is determined by the equations $X_t = K(1 - qKB/r - eW)$ and $Y_t = qBE_tX_t$. Thus,

$$\Pi_t = pY_t - cE_t = pqBE_tK(1 - qKB/r - eW) - cE_t = 0 \quad (14.2)$$

Solving this equation for the equilibrium-effort level gives

$$E_{oa} = [r(1 - c/pqBK - eW)]/Bq \quad (14.3)$$

The corresponding catch and stock levels are

$$Y_{oa} = cBE_{oa}/p \quad (14.4)$$

$$X_{oa} = c/pqB \quad (14.5)$$

The open access values of the variables in the modified framework have been calculated using the values of the parameters given in Table 5.2. The results are shown in Table 5.3.

The Optimal or Profit-maximising Regime Solution

Static Framework

Here we intend to estimate the values of the fish biomass, level of harvest and associated effort under a static framework for the Digha estuarine fishery. Under steady-state,

$$X_t = K[1 - (q/r)BE_t - eW] \quad (15.1)$$

and

$$Y_t = qBE_tX_t \quad (15.2)$$

Profit-maximisation occurs at $d\Pi_t/dE_t = 0$ and implies that

$$E^*_{static} = r(1 - eW)/2qB - rc/2pq^2KB^2 \quad (15.3)$$

Similarly, by substituting equation (15.3) in (15.1), we get

$$X^*_{static} = K/2(1 - eW) - c/2pqB \quad (15.4)$$

Substituting equations (15.3) and (15.4) in equation (15.2), we get the value of Y^*_{static} by

$$Y^*_{static} = qBE^*_{static}X^*_{static} \quad (15.5)$$

Finally, optimal profit can be estimated as

$$\pi^*_{static} = pY^*_{static} - cE^*_{static} \quad (15.6)$$

The optimal values of the variables in the static framework have been calculated using, again the values of the parameters given in Table 5.2. The results are shown in Table 5.3.

Table 5.3 compares between the different values of the variables computed under the three regimes at different levels of diversity present in the fishery. In Situation 1, with high economic diversity in the fishery, the maximum sustainable yield (3,12,145 kg./year) is the highest when compared with the other two regimes. Following this is the catch level under profit-maximising solution (1,51,242 kg/year) followed by that under open-access regime (37,932 kg/year).

The stock level is the highest under the maximum sustainable solution (7,53,123 kg) followed by that of the profit maximising level (5,24,223 kg) and is the lowest under the open access situation (2,81,397 kg). Effort level under open access (1,77,393 fishing hours/year) is the highest while that under profit-maximisation, level of effort (75,923 fishing hours/year) is the smallest. So, we find that the open access solution registers the smallest catch level associated with the highest level of effort.

In Situation 2, with average level of economic biodiversity prevalent in the fishery, we find that the harvest level is still the highest in case of maximum sustainable regime solution (3,14,131 kg/year) and its value is larger than what it was in Situation 1(3,12,145 kg/year). The harvest levels of the other two regimes (open access solution: 42,785 kg/year; profit-maximising solution: 1,95,161 kg/year) also register an increase compared to the previous case of Situation 1. The open access level of effort (1,01,413 fishing hours/year) is the highest combined with the smallest catch (42,785 kg/year), even as effort levels under all the three regimes increase with increase in economic diversity level. The open access effort level (1,01,413 fishing hours/year) remains the largest effort level under all three types of regimes.

Under Situation 3, we have B = 1 representing a fishery where all the species are equally valued and can thus be treated as a 'single-species' fishery. Effort level for all the three types of regimes decreases as one fishes down from a high economic diversity to a low economic diversity situation of the fishery. The optimal effort level is the smallest at 18,444 fishing hours/year (maximum sustainable regime solution: 21,814 fishing hours/year, open access solution: 41,998 fishing hours/year).

Dynamic Framework

In the dynamic framework, fishers seek to maximise the present value of profits over a time horizon 0 to T subject to the constraint of net growth of fish stock. The problem can be stated as

$$\text{Max. } \pi = \sum_{t=0}^{T}(pY_t - C_t)\rho^t,$$

where

$\rho = 1/1+\delta$, and

δ is the rate of discount, subject to

$$X_{t+1} - X_t = rX_t(1 - X_t/K - eW) - Y_t,$$

where

$$Y_t = qX_t\, B\, E_t \text{ and } C_t = cE_t \tag{16.1}$$

We can rewrite the problem as

$$\text{Max.} \sum_{t=0}^{T} (pqX_t\, BE_t - cE_t)$$

subject to

$$X_{t+1} - X_t = rX_t\,(1 - X_t/K - eW) - qX_tBE_t \tag{16.2}$$

The current value Hamiltonian, H_c, for this problem is

$$H_c = (pqX_t\, BE_t - cE_t) + \rho v_{t+1}\,(rX_t\,(1 - X_t/K - eW) - qX_t\, BE_t) \tag{16.3}$$

where ρ (the co-state variable) is the current value shadow price associated with a change in the fish stock, E_t is the control variable and X_t is the state variable. The first-order necessary conditions for a maximum are

$$\partial H_c/\partial E_t = 0, \tag{16.4}$$

$$\rho v_{t+1} - v_t = -\partial H_c/\partial X_t \tag{16.5}$$

and

$$X_{t+1} - X_t = \partial H_c\,/\partial \rho v_{t+1}. \tag{16.6}$$

Equations (16.4) and (16.5) give

$$(pqX_t\, B - c) - \rho v_{t+1} q\, X_t B = 0$$
$$\rho v_{t+1} = p - (c/qX_t), \text{ and}$$
$$v_{t+1} = (1+\delta)[\, p - (c/qX_tB)] \tag{16.7}$$

Equation (16.5) gives

$$\rho v_{t+1} - v_t = -pqE_t - \rho v_{t+1} r\{1 - (2X_t/K) - eW) + \rho v_{t+1} qBE_t\} \tag{16.8}$$

Steady-state implies that

$$\nu_{t+1} = \nu_t = \nu^* \text{ and } X_{t+1} = X_t = X^*.$$

So, equation (16.8) becomes

$$\nu^*(\rho - 1) = qBE_t(\rho\nu^* - p) - \pi\nu^* r\{1 - (2X_t/K) - eW\}. \quad (16.9)$$

Putting $\rho = 1/(1 + \delta)$ and using equation (16.7) (after putting $\nu_{t+1} = \nu^*$), we get

$$BE_t = (1/cq)(pqBX_t - c)[\delta - r\{1 - (Xt/K) - eW\}]. \quad (16.10)$$

Again, equation (16.8) at steady-state gives

$$BE_t = (r/q)\{1 - (X_t/K) - eW\}. \quad (16.11)$$

Comparing (16.10) and (16.11) and by letting $\Omega = c/Bpq$, we get

$$X^*_{dyn.} = \tfrac{1}{4}[\{\Omega + K(1 - \delta/r - eW)\} + \sqrt{[\{\Omega + K(1 - \delta/r - eW)\}^2 + 8K\Omega(\delta/r)]}]. \quad (16.12)$$

Once $X^*_{dyn.}$ is known, we can determine the optimum levels of effort and catch as

$$E^*_{dyn.} = (r/q)\{1 - (X^*_{dyn.}/K) - eW\}/B \quad (16.13)$$

$$Y^*_{dyn.} = qX^*_{dyn.}\, BE^*_{dyn.} \quad (16.14)$$

Hence, the optimum level of NPV of profit is

$$NPV^*_{dyn} = \sum_{t=0}^{T}(pY^*_{dyn.} - cE^*_{dyn.})(1/1+\delta)^t \quad (16.15)$$

Here $X^*_{dyn.}$, $Y^*_{dyn.}$, $E^*_{dyn.}$ and $NPV^*_{dyn.}$ are respectively the optimal values of fish stock, harvest, effort and net present value of profit.

The optimal values of the variables in the dynamic framework have been calculated, as before, using the values of the parameters given in Table 5.2. The results are shown in Table 5.4.

Table 5.4 Dynamic Profit-maximising Values of the Variables under Three Alternative Economic Biodiversity Scenarios

Alternative Scenarios of Economic Biodiversity of the Fishery	*Profit-Maximising Solution (Dynamic Framework)*			
	Stock (kg)	*Harvest (kg/year)*	*Effort (Fishing Hours/Year)*	*NPV of Profit (Rs)*
Situation 1	14,03,568	79,312	14,870	9,74,558
Situation 2	13,00,020	1,47,547	12,630	17,33,954
Situation 3	12,91,008	1,52,924	6,234	22,47,745

From Table 5.4, we find that with progressively lower levels of biodiversity, stock size and effort decrease while fish harvest rises. This is evident when we look at Situation 1 (catch: 79,312 kg./year; effort: 14,870 fishing hours/year) and compare it with Situation 2 (catch: 1,26,330 kg./year; effort: 12,630 fishing hours/year) and Situation 3 (catch: 1,52,924 kg./year; effort: 6,234 fishing hours/year). We also observe NPV of profit to increase with lower diversity (an increase from Rs. 9,74,558/year in Situation 1 to Rs. 17,33,954/year in Situation 2 and a further rise to Rs. 22,47,745/year).

So economic diversity reduction is associated with a rising level of NPV of profit masking the existence of the potential threat of a loss of the valuable fish species in the fishery.

For all the three types of regimes, we have derived results that are typical of single species Gordon-Schaefer models (Kasulo and Perrings 2001; Conrad and Adu-Asamoah 1986; Gallastegui 1983). The maximum sustainable solution gives the highest level of catch while the open access solution gives the lowest level of catch among the three. The optimal profit-maximising solution gives the lowest level of effort. These results are true for all three Situations 1, 2 and 3, that is, at different levels of falling economic biodiversity. We find here that for the maximum sustainable regime, a fall in economic diversity does not affect the stock level but decreases the effort. In case of catch level, it first increases and then falls back to its original value with falling biodiversity. In case of open access and optimal profit-maximising regimes, stock size and effort falls while harvest rises as there is a decrease in economic diversity of the fishery.

SENSITIVITY ANALYSIS

Change in Discount Rate on Dynamic Profit-maximising Optimal Solutions

We consider first, *ceteris paribus*, the impact of perturbations on the discount rate δ, (base $\delta = 0.11$) on the optimal values of the variables. The optimal values of the variables given in Table 5.3 and 5.4 are considered as the base values in our sensitivity analysis. The discount rate here approximated by the market rate of interest represents the opportunity cost of investing in the fishery vis-à-vis other assets or allied industries like tourism. Here, we have considered a range from 0.9 to 0.13 between which the market rate of interest has varied over the last 10 years as obtained from Reports on Currency and Finance of the Reserve Bank of India.

One can infer from Table 5.5 that NPV of profit situation 3 > NPV of profit situation 2 > NPV of profit situation 1. Another important observation is that the gain in NPV of profit associated with decreasing levels of biodiversity is highest when we contrast between Situations 1 and 3 than between Situations 1 and 2. So paradoxically, it can be seen that greater endeavour to capture the most expensive species leads to greater losses associated with the fishery. This occurs as demand in the markets triggers off greater exploitation of the expensive species and in this process large amounts of cheaper by-catches are discarded. So the current trend towards the exploitation of only valuable fish species raises doubts about the profitability and hence sustainability of the fishery.

One can hence conclude that maximisation of profit and economic biodiversity considerations are ultimately in conflict with each other in the context of sustenance of a fishery. This underlines the importance of economic scarcity and the role of market demand for fishes in a fishery. Changes in the discount rate have been found to have effect on the fish catch and the NPV of profit and hence

Table 5.5 Impact of Perturbations of the Discount Rate δ

Value of δ	*NPV of Profit in Situation 1*	*NPV of Profit in Situation 2*	*NPV of Profit in Situation 3*	*Gain in NPV of Profit*	*Gain in NPV of Profit*
	(1)	*(2)*	*(3)*	*(4) = (1) – (2)*	*(5) = (2) – (3)*
0.09	9,77,194.24	18,87,813.70	25,03,757.30	6,15,943.60	9,10,619.46
0.10	9,79,283.75	18,10,318.90	23,71,692.00	5,61,373.10	8,31,035.15
0.11	9,74,557.84	17,33,954.10	22,47,745.50	5,13,791.36	7,59,396.26
0.12	9,61,369.91	16,55,341.10	21,11,416.70	4,56,075.60	6,93,971.19
0.13	9,43,252.23	15,78,135.60	20,07,586.00	4,29,450.40	6,34,883.37

Notes: Reference/Base value: δ = 0.11 on optimal value of NPV of profit under three alternative scenarios: (1) high economic diversity; (2) average economic diversity; and (3) low economic diversity in the fishery model.

loss of economic biodiversity. The higher the discount rate or rate of interest on a project, the smaller the present value of a given payment in the future. It is important to consider the factors that lead to changes in the discount rate such as excess demand for money or decrease in government spending. Policies that lead to reduction in interest and discount rates would also lead to losses in biodiversity.

Change in Intrinsic Growth Rate on Dynamic Profit-maximising Optimal Solutions

The Digha estuarine fishery is already experiencing problems related to environmental pollution which can have a distinct impact on intrinsic growth rate of fishes. So an analysis with regard to changing intrinsic growth rates becomes more important in this connection. We consider, *ceteris paribus*, the impact of perturbations on the intrinsic growth rate r, (initially, r = 1.5) and on the optimal values of the variables.

From Table 5.6, we find that in comparison with the base value NPV of profit, the percentage change in level of NPV of profit is most dramatic for Situation 2 followed by Situation 3 and finally by Situation 1. Also with declining values of r (from 1.5 to 1.0 or further down to 0.5), levels of fish stock, catch and effort decrease.

Table 5.6 Impact of Perturbations in the Intrinsic Growth Rate, r

Value of r	*NPV of Profit Situation 1*	*NPV of Profit Situation 2*	*NPV of Profit Situation 3*	*Gain in NPV of Profit*	*Gain in NPV of Profit*
	(1)	*(2)*	*(3)*	*(4) = (1) – (2)*	*(5) = (2) – (3)*
0.5	7,86,758.63	10,40,402.00	12,11,162.00	1,70,760.00	2,53,643.37
1.0	9,22,927.95	16,03,978.30	17,71,744.50	1,67,766.20	6,81,050.35
1.5	9,74,557.84	17,33,954.10	22,47,745.50	5,13,791.36	7,59,396.26
2.0	10,05,012.10	20,17,655.80	27,02,611.40	6,84,955.60	10,12,643.70
2.5	10,27,096.70	22,92,898.90	31,49,107.10	8,56,208.20	12,65,802.20

Notes: Initially r = 1.5 on the optimal value of NPV of profit under three alternative scenarios: (1) high economic diversity; (2) average economic diversity; and (3) low economic diversity in the fishery model.

Not only that, the NPV of profit falls from Rs.17,33,954/year (for r = 1.5, that is, the base case) to Rs. 16,03,978/year (for r = 1.0) and drastically to Rs.10, 40,402/year (for r = 0.5). So, it can be observed that at very low levels of intrinsic growth rate, there is a greater slide in the value of total profit than otherwise. This is because fish harvest is affected adversely by pollution resulting from sewage disposals, a by-product of the tourism industry of Digha affecting the mortality of fish species and hence its productivity. Any further reductions could be detrimental for this fishery, as it would have a direct impact on the NPV of profit which would get drastically reduced. This is very important from the perspective of fishery management and the policymakers have to be careful in treading the narrow line between the development of the sister industries of fishing and tourism thriving in Digha.

Again, the Digha fishery has its effort mostly targeted at the most expensive fish species, *hilsa* and reducing it ultimately puts the sustainability of this valuable fish species at stake. So it seems that as r changes, the threat to economic biodiversity gets associated with increasing levels of profit. So just by regulating the level of pollution and increasing fish productivity, the fishery cannot become profitable and sustainable. Policymakers and fishermen must also take into account the fact that any reduction of economic biodiversity can actually put the sustainability of the very fishery at stake. So economic biodiversity conservation is essential for a high profit-maximising regime.

CONCLUDING REMARKS

The bioeconomic model initially presented in this chapter considers the important aspects of economic exploitation of a renewable resource as well as associated fundamental biological processes. Any fishery model needs at the aggregated level to incorporate important factors that affect fish harvest. One such component of biodiversity occurs through the depletion of a number of fish species. Incorporation of biodiversity into a fisheries model is a difficult task. The main difficulty lies in choosing the relevant biodiversity measure to be used for our study. Following the work of Kasulo and Perrings (2001), we have used the economic biodiversity index which has been formulated on the basis of Simpson's biodiversity index. This has been done by using the market prices of the fish species as weights. The indices have been calculated for the Digha *Mohona* (Estuary) fishery. It has been found that the economic (weighted) biodiversity index has lower values than the ecological (unweighted) biodiversity index. This suggests that the fish catch in Digha *Mohona* (Estuary) fishery are dominated by less valuable species.

Our work started with the standard Gordon-Schaefer model and it has been modified to include environmental quality and biodiversity variables. The environmental variable captured the effect of discharge of sewage causing water pollution. The amount of rainfall has been used as a proxy for water pollution. The biodiversity variable captured the impact of changes in species mix on fish productivity. The parameters of the model are estimated using Schnute's (1977) method, as it reduces the bias in the estimates resulting from errors in the measurement of variables. Application of the model has been done on the fishery of Digha *Mohona* (Estuary) in West Bengal. Introduction of the environmental variable has improved the fitness of the model. The coefficient of the environmental variable is negatively related to changes in fish biomass. The model fit and the significance of the parameter further improved after introduction of the ecological biodiversity index. The improvement of the model fit increased even further when the weighted economic biodiversity index was introduced in the model. This clearly shows that biodiversity plays an important role in the value of the fishery. This is specially true if the fishery is subjected to market forces where the market value of the species is taken into account.

The Gordon-Schaefer model was modified by introducing economic biodiversity and environmental quality variables and estimated by using Schnute method. It adds realism in exploring sustainability of fish catch in Digha estuary. This model is used to analyse the economic impact on the fishery in Digha due to a reduction in the intrinsic growth rate and an increase in the discount rate—two parameters that are of fundamental importance to the population biodiversity and dynamics of the fish species. The change in population dynamics results in an adjustment of the sizes of the fish stocks, influencing the harvest of fish and subsequently the returns from fishing.

We have also done sensitivity exercises by perturbing the discount rate and intrinsic growth rate and examined their impact on net present value of profit. Interestingly, it was found that biodiversity conservation and profit-maximisation are in conflict with each other. The conflict between profit-maximisation and biodiversity conservation underlines the importance of economic scarcity. The fishery is exploited to meet the market demand and any signals of scarcity as reflected by market prices induces further exploitation of the fishery. So a reduction in the over-exploitation of the fishery will increase not only the biodiversity but also the value of the fish catch. It is here that institutional factors reflecting alternative fishery management regimes play a major role as they are a principal cause of this over-exploitation.

We have compared our results both under the cases of maximum sustainable regime (base-line case) and profit maximisation regime (actual case) with that of the open access regime (hypothetical case). We have already observed that under the hypothetical situation of open access, fish catch falls with falling biodiversity while in case of optimal exploitation, it actually increases with falling biodiversity. Generally open access equilibrium occurs at points beyond the maximum sustainable levels. This leads to over-exploitation of the fishery. Over-exploitation not only means a loss of biodiversity but also threatens the sustainability of the fishery. Fishery management must be such that it should ensure that the fishery is not over-exploited as it can happen under open access and so the role of institutional factors to put in place appropriate fishery regimes becomes important.

Traditional fishery management strategies mostly involve gear restrictions, closed seasons and licensing. Gear restrictions involve prohibition of certain methods of fishing like restrictions on the size of the mesh. The enforcement of mesh size regulation is very difficult. Closed seasons aim at protecting fish stocks during critical stages like breeding. These policies have not been very effective in case of Digha *Mohona* (Estuary) fishery. Licensing is aimed at limiting entry into the fisheries. It seeks to control the amount of effort by directly regulating the number of fishermen. This method has been partially effective in Digha *Mohona* (Estuary) fishery.

The traditional fishery management strategies have largely focused on the biological aspect of the resource. Biological analysis relates sustainable catch with the amount of fishing effort but importantly, this effort level is itself driven by economic forces. If this aspect is not considered, it has a negative impact on the biodiversity of the resource. Economic incentives and disincentives through price and tax policies can help the stakeholders to conserve biodiversity. If fishing costs are sufficiently high relative to the price of fish, the fishery will not be exploited.[10] A user tax may help to reduce fishing effort as the fishery owners will try to control the costs to maintain their level of profit. Costs also include the opportunity cost of fishing. Low opportunity cost in the fishing industry means over-exploitation of the fishery. The opportunity cost of fishing may be enhanced by creating better employment alternatives, raising minimum wages and improving the availability of credit for small scale businesses. Employment opportunities outside the fishing industry have to be created because controlling of fishing effort without creating alternative employment opportunities will mean increasing poverty.

Our analysis finds that in the Digha fishery, there exists a trade-off between economic biodiversity conservation and profit maximisation. Policy measures have to be devised to minimise this level of conflict between them. One can increase the opportunity cost of fishing by providing alternative employment in nature-preserving tourism industry in Digha as employment generation can be fostered through tourism at a low cost. Such measures can restrain over-exploitation of the fishery and can help to conserve biodiversity. It is also essential that an enabling environment with a greater emphasis on institutional support involving stakeholders be developed. For proper development of the fishery, the possibility of stakeholders' participation and community-based participatory approach should be considered and explored.

ACKNOWLEDGEMENTS

The authors acknowledge with gratitude the comments from participants and discussants at all the conferences. They are highly indebted to Prof. Charles Perrings of the University of York, UK for his help in framing this research problem. The authors are solely responsible for any error that may exist in this chapter.

NOTES

1. The present chapter is a part of the doctoral dissertation of the first author, which is in progress at the Department of Economics, University of Calcutta. Earlier versions of this chapter was presented by her at the 8th Biennial Scientific Conference of the International Society for Ecological Economics (ISEE) held in Montreal, Canada from 11 to 14 July 2004, 4th Biennial Conference for Ecological Economics (INSEE) on 'Ecology and Human Well Being' held in Mumbai from 3 to 4 June, 2005 and 14th International Economic Association World Congress held in Marrakech, Morocco from 29 August to 2 September, 2005.
2. Corresponding author.
3. For example, Atlantic halibut once commonly found off New England are now rare. Some fisheries have been subject to severe curtailment and closure in North America, most notably cod off Newfoundland, groundfish off New England, and some salmon species in the Pacific Northwest. Atlantic salmon and American shad have largely disappeared from many rivers of the eastern United States (Merrett and Haedrich 1997).
4. GESAMP (1990) states that eutrophication caused by excess nutrients from sewage discharged into coastal waters is an expanding problem. The initial effects are of altered species compositions leading to local changes in biodiversity gradually moving towards more severe effects like mass mortalities of fish. This is because the pressures from a wide array of human activities show no signs of diminishing and the maintenance of economically healthy estuaries and their fishes depends very much on well-defined management policies.
5. The trawlers are in operation for about 10 to 11 months in a year and generally they are rested in the months of April-May. Usually a fishing trip lasts for about five days in sea and after the catching, the catch is landed in the harbour from where they are sold in the wholesale market at Digha. According to the Digha Fishermen and Fish Traders' Association, the wholesale fish market at Digha Estuary accounts for Rs 70 crore a year (Ganguly 2004). Also a large portion of the

catch is ice packed and sent to the wholesale market in Sealdah of Kolkata and Howrah for auctioning by the local agent auctioneers. Besides sending the catch to Kolkata, sometimes fishes like pomfret and sea urchins are sent directly to Chennai or Viasakhapatnam for marketing.

6. Annual Reports of the Digha Fish Traders' Association, Various issues.
7. Fishing effort has been calculated on the basis of a composite index constructed by us. It has been taken to be the weighted average of number of fishing hours involved in catching fish through fishing boats and trawling boats where the weights are the number of trips by fishing boats and trawling boats in an year.
8. The biological names of some of the fishes are given in the parantheses: sardine (*Sardinella gibbosa*), hilsa (*Hilsa Tenualosa ilisha)*, coila, silver pomphret (*Pampus argenteus*), black pomfret (*Parastromateus niger*), croakers (*Johnius belangerii*), Bombay duck (*Harpadon nehereus*), catfish (*Arius jella Day*), ribbon fish (*Eupleurogrammus muticus*), shark (*Carcharhinus limbatus)*, mackerels (*Rastrelliger kanagurta*), prawn.
9. This is utilised further in sections 5 and 6 in our exercise in sensitivity analysis and resultant policy implications.
10. It is quite reasonable to assume that the fishermen are not able to influence the fish price. Fishery in Digha Estuary and in the surrounding region suggests that the fishermen there are mainly price takers.

REFERENCES

Barbier, E.B., J.G. Burgess and C. Folke. 1994. *Paradise Lost*. London: Earthscan Publications Limited.

Beverton, R. and S .Holt. 1957. 'On the Dynamics of Exploited Fish Populations', *Fisheries Investigation (Series 2)*, vol. 19. UK: Ministry of Agriculture, Fisheries and Food.

BOBP (Bay of Bengal Programme). 1994. 'Bay of Bengal'. http://na.nefsc.noaa.gov/lme/text/lme34.htm.

Botsford, L.W., J.C. Castilla and C.H. Peterson. 1997. 'The Management of Fisheries and Marine Ecosystems', *Science*, 277: 509–15.

Chatterjee, T.K., R.S. Talukdar and A.K. Mukherjee. 2000. *Fish and Fisheries of Digha Coast of West Bengal*. Records of the Zoological Survey of India.

Chattopadhyay, K. 1995. *Economic Impact of Tourism Development – An Indian Experience*. Delhi: Kanishka Publishers Distribution. pp. 143–51.

Conrad, J.M. and R. Adu-Asamoah. 1986. 'Single and Multispecies Systems: The Case of Tuna in the Eastern Tropic Atlantic', *Journal of Environmental Economics and Management*, 13: 50–68.

Das, T.K., R. Neogy and D. Chakraborty. 2000 'Sustainability of Marine Fishing: A Case Study of West Bengal', *Applied Economic Letters*, 7: 707–10.

FAO. 1997. *The State of World Fisheries and Aquaculture*. Rome: Food and Agricultural Organisation of the United Nations.

Folke, C., S. Holling and B.O. Jansson (eds). 1995. *Biodiversity Loss: Ecological and Economic Issues*, pp. 44–83. Cambridge: Cambridge University Press.

Gallastegui, C. 1983. 'An Economic Analysis of the Sardine Fishing in the Gulf of Valentia (Spain)', *Journal of Environmental Economics and Management*, 10: 138–50.

Ganguly, T. 2004. 'Unscientific Policies Lead to Decline of Marine Fish Production and Jobs, *The Week*.

Garcia, S.M., and J. Deleiva. 2000. 'Trends in World Fisheries and their Resources: 1974–99'. *In the State of Fisheries and Aquaculture 2000*. Rome: FAO.

GESAMP (International Joint Group of Experts on the Scientific Aspects of Marine Pollution). 1990. *The State of the Marine Environment*. Oxford: Blackwell Scientific Publications.

Ghosh, S., A.K. Hazra, S. Banerjee and B. Mukherjee. 2004. 'The Seasonal Toxicological Profile of Four Puffer Fish Species Collected along Bengal Coast, India', *Indian Journal of Marine Sciences*, 33(3): 276–80.

Grainger, R.J.R. and S.M. Garcia. 1996. 'Chronicles of Marine Fishery landings: 1950–94', *Trend, Analysis and Fisheries Potential*. Fisheries Technical Paper No. 359. Rome: FAO.

Gupta, K. 2004. 'Issues in Ecological Economic Modelling with special reference to India', (mimeo), Rabindra Bharati Universty, Kolkata, accepted for publication in an edited volume (editor: R.K.Sen) to be published by Deep and Deep Publications, New Delhi.

Gupta, K. 2005. 'Mangrove-Fishery Linkages Under Alternative Policy Regimes', in N. Sengupta and J. Bandyopadhyay (eds), pp. 272–93, *Biodiversity and Quality of Life*. Macmillan India Ltd.

Holling, C.S., D.W. Schindler, W. Brian and J. Roughgarden. 1995. 'Biodiversity in the Functioning of Ecosystems: An Ecological Synthesis', in C. Perrings, K. Mäler, C. Folke, C.S. Holling and B-O. Jansson (eds), *Biodiversity Loss*, Cambridge: Cambridge University Press.

Jana, N. 2003. 'Digha Move to Plug Sewage Flow into Sea, *The Telegraph*.

Kasulo,V. and C. Perrings. 2001. 'Fishing Down the Value Chain: Modelling the Impact of Biodiversity Loss in Fresh Water Fisheries—the Case of Malawi', mimeographed, UK: Environment Department, University of York.

Meenbarta. 1998. *Special Issue on Wetlands*, Government of West Bengal.

Merrett, N.R. and R.L. Haedrich. 1997. *Deep Sea Demersal Fish and Fisheries*, London: Chapman and Hall.

Pauly, D., V. Christensen, J. Dalsgaard, R. Froese and F. Torres Jr. 1998. 'Fishing Down Marine Webs', *Science*, 279: 860–63.

Perrings, C. 2000. 'The Biodiversity Convention and Biodiversity Loss in Sub-Saharan Africa', in C. Perrings (ed.), *The Economics of Biodiversity Conservation in Sub-Saharan Africa*, pp. 1–45. Cheltenham: Edward Elgar Publishing Limited.

Rao, D.V. 2000. 'The Bay of Bengal', in C. Sheppard (ed.) *Seas at the Millennium: An Environmental Evaluation*, Elsevier Science Ltd.

Robinson, M.A. 1980. 'World Fisheries to 2000: Demand and Management', *Marine Policy*, 4: 1.

Schaefer, M.B. 1954. 'Some Aspects of the Dynamics of Populations Important to the Management of Commercial Marine Fisheries', *Bulletin of the Inter-American Tropical Tuna Commission*, 1: 25–56.

Schaefer, M.B. 1957. 'Some Considerations of Population Dynamics and Economics in Relation to the Management of Commercial Marine Fisheries', *Journal of Fisheries Research Board of Canada*, 14: 669–81.

Schnute, J. 1977. 'Improved Estimates from the Schaefer Production Model: Theoretical Considerations', *Journal of Fisheries Research Board of Canada*, 34(5): 583–603.

6

Aquaculture vs Wild Shrimp Fishery: A Bio-economic Analysis of West Bengal and Orissa[1]

Kausik Gupta[2], *J. Chowdhury* and *Raman Khaddaria*

Abstract: The chapter attempts to examine the impact of an increase in aquaculture shrimp production on wild shrimp fishery. In a major departure from the earlier works in this area, we have tried to establish the link between the aquaculture industry and wild shrimp fishery. This has been done through a structure that analyses the impact of an expansion of aquaculture shrimp industry, as reflected through an increase in the proportion of shrimp fry consumption. This issue has been captured by incorporating technological improvement of the shrimp industry via the increase in stocking density. On the basis of our structure, we can thus conclude that technological change leads to an expansion of aquaculture industry and the contraction of the wild fishery. This result is important from the point of view of policy makers and emphasises on the need for defining more socially and ecologically responsible aquaculture industries that enhance traditional fishery and reduce current user conflicts that are in existence now.

INTRODUCTION

There is a growing per capita demand for fish products since the 1950s. The United Nations Food and Agricultural Organisation predicts that in this century, world consumption of aquatic proteins will increase to 150–160 million tons (FAO 2000). Traditional Fisheries can provide no more than 100 million tons, so the bulk of the increase will need to come from aquaculture. Over the past ten years, India has become the third largest shrimp producer in the world. Organisations like the World Bank and the Asian Development Bank have invested in shrimp aquaculture in developing countries[3] in the expectation that this industry would create jobs which would lead to new avenues for earnings. This would also help the policy makers to uplift the condition of the impoverished sections of the society living mainly in rural areas. However, the experience of developing countries is something different. A Report published by National Environmental and Engineering Research Institute (NEERI 1995) has concluded that resource destruction caused by shrimp aquaculture is actually greater than the amount of income generated from the exporting of shrimp in case of India. However, due to the increased demand for shrimp by people in countries like the United States and

Japan, shrimp aquaculture has become a very big business in India, since traditional fisheries (wild fisheries) can supply only a small portion of the total demand. Against such a backdrop, it is found that there is a rapid expansion of shrimp aquaculture in India.

FAO has defined aquaculture as the farming of aquatic organisms, including fish, shrimp and aquatic plants. It has collectively termed the land-based and water-based brackish and marine aquaculture practices as 'coastal aquaculture'. When we consider farming of shrimps, we refer to it as aquaculture shrimp farming. Wild shrimp fishery on the other hand, refers to ocean capture shrimp fisheries. As traditional fisheries, that is, wild fisheries and aquaculture—both are used for meeting the increased food demand, the competition between both cannot be ignored. The rapid expansion of shrimp aquaculture industry in India creates ecological conflicts and hence this expansion must be accomplished by promoting an alternative aquaculture development model, that is, an 'Ecological Aquaculture Model' (EAM). EAM considers not only the technical aspects of ecosystems' design and ecological principles pertaining to aquaculture, but also incorporates comprehensive planning for the wider social, economic and environmental contexts of aquaculture (Costa-Pierce 2002; Costa-Pierce and Bridger 2002).

The expansion of the aquaculture industry is at the expense of growth of wild fishery since both these sectors are facing the same demand situations. Moreover, the expansion of aquaculture industry means a scarcity of wild shrimp fry. This is because the aquaculture industry uses wild shrimp fry as an input. This very issue raises the question of sustainability of shrimp farming. So the competition between aquaculture and fisheries should be so managed that they benefit both wild fishery and aquaculture operations as well as maintain ecological balance.

The literature on the trade-off between cultured fisheries and captured fisheries is not much developed. Some studies have considered the economic viability of aquaculture shrimp production vis-à-vis natural land uses (Srinath et al. 2000; Bhatta and Bhat 1998). Some other studies have considered the adverse impact of commercial shrimp production on coastal environment (Barbier E.B. and Strand I 1998; Boyd 1997). The aim of the present chapter is different from the above. It attempts to examine the impact of an increase in aquaculture shrimp production (through an increase in the absorption of wild shrimp fry) on wild shrimp fishery. The issue is addressed in terms of an aquaculture shrimp farm/industry model. Such a structure is necessary for making policies in determining the optimal allocation of shrimp fry between aquaculture and wild shrimp fishery.

The structure of the chapter is as follows. Section 2 deals with the theoretical model. Results and discussions have been given in Section 3. Finally, the concluding remarks are provided in Section 4.

THE THEORETICAL MODEL

We assume that the aquaculture industry consists of 'n' identical farms[4] under a competitive set up. The aquaculture industry is actually competing with the wild fishery industry. In our model, we assume that the product of aquaculture and traditional fisheries are perfect substitutes. It implies that the prices are same for both the products. We also assume that the competing farms within the aquaculture industry are adopting the same technology. The producers of aquaculture farms use

Shrimp Fry (SF) as an input along with other input effort (E). Producers buy their input SF, from local fry collectors. This way, we ignore the role of hatcheries in shrimp production.[5] Though it is a simplifying assumption that the introduction of hatchery will not bring any significant qualitative difference.

The aquaculture industry uses a fixed proportion of total SF and the remaining amount of SF goes to wild fishery. The aquaculture shrimp farms need shrimp fry as an input along with other inputs like chemicals, fertilisers, machines etc. Take for example, let ν proportion of SF to aquaculture industry and (1 – ν) proportion go to wild shrimp fishery. The expansion of wild shrimp fishery is actually dependent on the shrimp fries that are not used for matured shrimp production in aquaculture industry. The shrimp fries left for wild shrimp fishery is ultimately converted into matured shrimp in the sea.

Therefore,

$$SF = SF_A + SF_W \tag{1}$$

where,

SF_A: Amount of shrimp fry which goes to aquaculture industry
SF_W: Amount of shrimp fry which goes to wild shrimp fishery

In this model,

$$SF_A = \nu SF \text{ and } SF_W = (1-\nu)\, SF \tag{1a}$$

We next consider the production function of a representative aquaculture farm. We assume it to be a Cobb-Douglas production function[6] and it is given by

$$Y_A = ASF_A^{b1}\, E^{b2} \tag{2}$$

where,

Y_A = Total yield of a farm.

The cost equation of the farm is given by

$$C = F + P_f.\, SF_A + W.E \tag{3}$$

where,

P_f: Price of SF
W: Cost/unit of E

We have expressed effort 'E' in terms of labour days.[7] Hence we take the price of effort as wage given for per unit of labour days. We denote it by 'W'.

We now focus on the technology part of the production function. For the aquaculture farm owner, technology for shrimp production is a function of its stocking density, which we can specify as

$$A = A(SF_A/\Omega) = A(\nu SF/\Omega),\ A^{/} > 0,\ A^{//} < 0 \tag{4}$$

where Ω is the given pond area.

Under the assumption that pond area and total stock of shrimp fry are given (for a given time period),[8] we find that an increase in ν, that is, an increase in proportion of total SF that goes to aquaculture, leads to an increase in the level of the technology of the aquaculture farm through an improvement in its stocking density.[9]

Then we minimise cost, given by equation (3), subject to the production function, given by equation (2), and set up the Lagrangian to derive the input demand functions for 'SF' and 'E' by a particular farm.[10] They are derived as follows

$$E_d = (b_2/b_1)^{b_1/(b_1+b_2)} (Y_A/A)^{1/(b_1+b_2)} (P_f/W)^{b_1/(b_1+b_2)} \tag{5}$$

$$SF_{A\,d} = (b_1/b_2)^{b_2/(b_1+b_2)} (Y_A/A)^{1/(b_1+b_2)} (W/P_f)^{b_2/(b_1+b_2)} \tag{6}$$

where,

E_d: Effort demand
$(SF_A)_d$: Demand for shrimp fry for a particular farm

The total cost function of a particular farm can be expressed as

$$C = F + [\{P_f (b_1.W / b_2.P_f)^{b_2/(b_1+b_2)}\} + \{W.(b_2.P_f / b_1.W)^{b_1/(b_1+b_2)}\}]^{1/(b_1+b_2)}. (Y_A/A)^{1/(b_1+b_2)} \tag{7}$$

Hence, marginal cost (MC) function of the farm is

$$MC = 1/(b_1 + b_2).(1/A)^{1/(b1+b2)} [\{P_f (b_1.W / b_2.P_f)^{b2/(b1+b2)}\} + \{W.(b_2.P_f / b_1.W)^{b1/(b1+b2)}\}]^{1/(b1+b2)}.(Y_A)^{\{1-(b_1+b_2)\}/(b_1+b_2)} \tag{8}$$

that is, MC $\rangle$ 0 as long as $(b_1 + b_2) \leq 1$.

Hence, under the conditions of CRS/DRS the supply curve is positively sloped. The possibility of IRS does not arise since it is not compatible with the assumption of perfect competition.

As all the farms are price-takers, the demand curve of a particular farm for the final product must be a horizontal straight line having an intercept. Let the price of shrimp be 'P'. Thus, from the equilibrium condition P = MC, we can derive the equilibrium output of a particular farm and we denote it as Y^*.

The supply curve of the aquaculture industry can be derived by simple horizontal summation of individual supply curves of all the farms.

The supply curve of the aquaculture industry is nothing but

$$SS_A = \Sigma MC_i. \tag{9}$$

Let the supply curve for the wild shrimp fishery be given by SS_W.

And the total supply of the shrimp fishery sector is SS_F.

Hence,

$$SS_F = SS_A + SS_W \tag{10}$$

The equilibrium output of the shrimp fishery sector can be determined by the intersection of SS_F and the demand curve (DD) for the shrimp fishery sector as shown in Figure 6.1.

Figure 6.1 Allocation of Total Supply between Aquaculture and Wild Shrimp Fishery

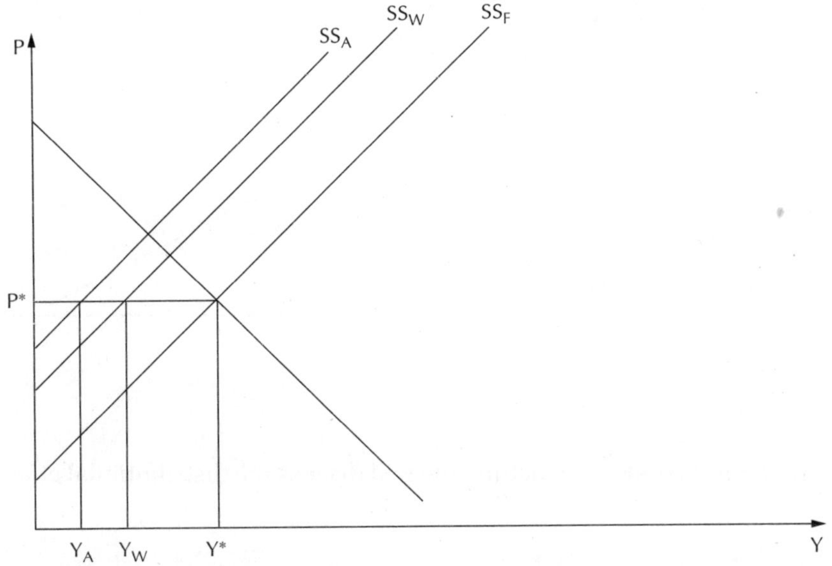

According to Figure 6.1, the equilibrium values of price (P) and output (Y) are P^* and Y^* respectively. In this situation, the aquaculture industry supplies Y_A whereas the wild sector fishery supplies Y_W. Thus though our theoretical model, we can establish the allocation of total supply between aquaculture industry and wild shrimp fishery.

Next, we now want to analyse with the help of Figure 6.2, the case when aquaculture industry expands for a given total supply of shrimp fry in the region for both aquaculture sector and wild shrimp fishery sector. In Figure 6.2 we find that with the expansion of aquaculture industry, the requirement of 'SF' increases which means 'v' increases and with this increment, SS_A also increases and hence 'SS_A' curve shifts parallel and rightward as SS_{A1}. It is to be noted that an increase in v actually implies an expansion of aquaculture farm and hence of aquaculture industry through technological upgradation (see equation [4]). This technological upgradation leads to a rightward shift of 'SS_A' curve. However, with the increase in 'v', '(1 – v)' falls. As a result, SS_W shifts to the left and becomes SS_{W1}. Thus, supply from wild shrimp fishery decreases and that of aquaculture increases. This is true for a given supply of total shrimp fry.[11] From Figure 6.2 we find that aquaculture supply increases to Y_{A1} whereas supply of wild shrimp fishery decreases to Y_{W1}.

Figure 6.2 Expansion of Aquaculture Industry and Contraction of Wild Shrimp Fishery

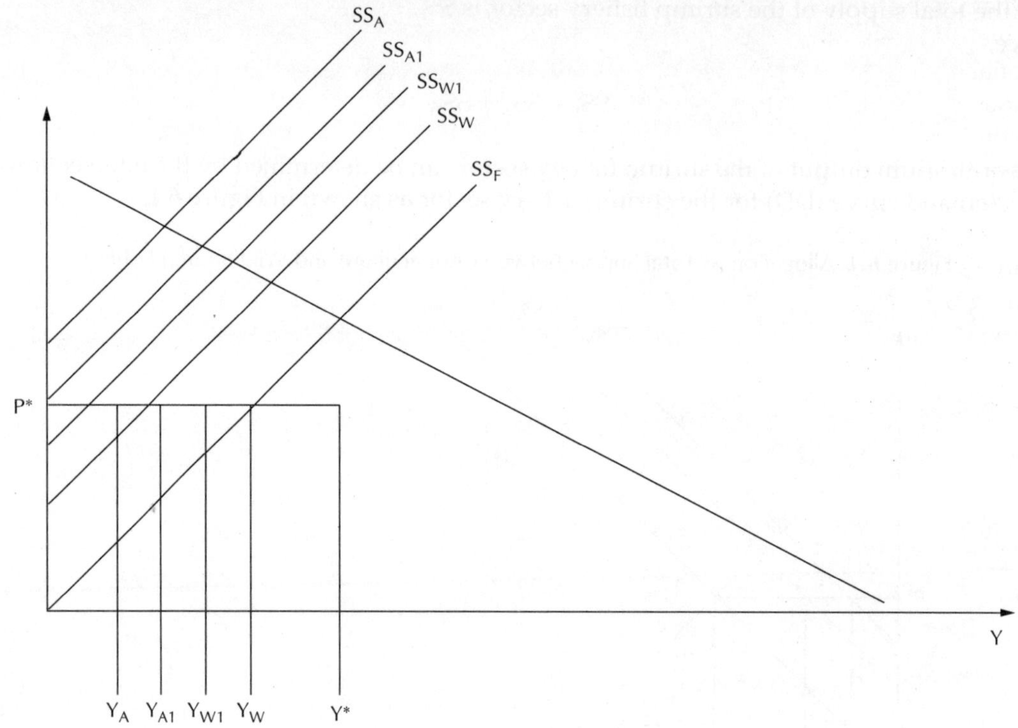

In the next part of our theoretical structure, we will discuss the distributional effects of the above problem.

RESULTS AND DISCUSSION

In this section, our attempt is to test our theoretical structure empirically. For this purpose, we first estimate the Marginal Cost (MC) function to derive the supply function of an individual farm. Since we have dealt with cross-section data, we do not have marginal cost data. So, we have considered price (P) of shrimp in place of MC.[12] The level of price varies from farm to farm due to spatial differences. They vary around the equilibrium market price so that the average of all prices can be assumed to be the equilibrium market price. To derive the supply function, we equate MC with various levels of P as we find from our data and regress P on Y_A. We take natural logarithm of equation (8) of the theoretical model and find that lnMC is a function of lnY_A. Finally we equate lnMC with lnP and find that lnP can be expressed in terms of lnY_A.

For our estimation, field surveys are conducted in various blocks of the Sundarban (in West Bengal) and Dhamara (in Orissa). Here, in total we have considered ninety-five aquaculture shrimp farms for our study. Out of the ninety-five farms, eighty-one are from the Sundarban region of West Bengal and fourteen are from the Dhamara region of Orissa. At first sight, the selection of such few farms from Dhamara may seem puzzling. However, if we go through the type of technology used in the farms, we find that in the Sundarbans varieties of technologies are used. Most of the farms use traditional varieties (Santhakumar et al. 2003). There are a few farms that use extensive and semi-extensive types of technologies. The number of shrimp farms in Sundarbans is also much higher when compared to the number of shrimp farms in any other mangrove of India. The farms in Dhamara (near Bhitarkanika mangroves) are mainly homogeneous type from the point of view of technology used. Most of them use semi-intensive type of technology. Again as the total number of farms in Dhamara is much less than the Sundarbans, for our study we have selected only a few farms in Dhamara compared to the Sundarbans. Selection of the site Dhamara not only helps us to compare the relative performance of shrimp farms near the two big mangrove swamps of India, but it also helps us to exhaust the list of various types of shrimp farms in the country when the mode of classification is on the basis of technology.[13] The following supply function has been estimated for two sets of data, that is,. Sundarban region and Dhamara region.[14]

$$Z = \alpha + \beta X$$

where, $Z = \ln(P)$, $X = \ln(Y_A)$

$$\alpha = [1/(b_1 + b_2)] \ln[1/(b_1 + b_2).\,(1/A)^{1/(b1+b2)}\,[\{P_f\,(b_1.W\,/\,b_2.P_f)^{b2/(b1+b2)}\} + \{W.\,(b_2.P_f\,/\,b_1.W)^{b1/(b1+b2)}\}]]$$

and $\beta = \{1/(b_1 + b_2)\} - 1$

The results are given in Table 6.1.

Table 6.1 Regression Results

	No. of Observations(n)	R^2	$\bar{R}^2$	$\hat{\alpha}$	t Values (Intercept)	$\hat{\beta}$	t Values (Slope)
Dhamara	14	.53	.49	4.67	26.12	.2	3.68
Sundarban	81	.73	.72	4.49	62.2	.17	14.52

From the Table 6.1, on the basis of the values of R^2, we find that the regression fit of Z on X is quite good given the fact that we have considered cross-section data for our analysis. This is true for both Dhamara and Sundarban. Moreover, all the estimated values of parameters are significant along with expected signs.

Using the average estimated value of Z, we can find the average value of X for all the data series. Taking the antilog of X, we can find the average yield that is $\bar{Y}_A$. This $\bar{Y}_A$ can be considered as the equilibrium output of a representative aquaculture shrimp farm. Multiplying this with the number of farms[15] (given by the number of observations), we can find the total yield of aquaculture industry for each of the above mentioned two data series. The equilibrium price is obtained by taking the average of Z, that is $\ln \bar{P}$, and then taking the antilog of $\ln \bar{P}$. As aquaculture industry supplies only 33 per cent of total requirements, we can conclude that wild fisheries supply the rest 67 per cent of total requirements that is not supplied by aquaculture industry (see Costa-Pierce 2002).[16] So we can calculate the supply of wild shrimp fishery and hence the total supply of shrimps.[17] Table 6.2 provides all such details.

Table 6.2 Average Price, Average Yield and Total Supply of Different Sectors for Different Regions

	$\bar{P}$ (Rs)	$\bar{Y}_A$ (Kg)	Y_A (Kg)	Y_W (Kg)	Y^* (Kg)
Dhamara	202.35	2,405	33,670	68,360.3	1,02,030.30
Sundarban	254.68	428.38	34,698.78	70,449.04	1,05,147.82

In Table 6.2 we find that though the average yield of an aquaculture farm in Dhamara is greater than that of Sundarbans, the total supply of aquaculture and wild sector fishery. Hence the composite fishery sector in Sundarbans is quite greater than the Dhamara region. Next we will examine the impact of changes in technology on wild fishery and aquaculture sectors.

In our model, changes in technology can be captured by changes in stocking density. As pond area is given to an owner for a particular time period, changes in stocking density is nothing but changes in the stocking of shrimp fry for a farm. Moreover, we have not considered the role of hatcheries in our model. Thus, changes in the stocking of shrimp fry of an aquaculture farm will definitely affect the availability of shrimp fry for the wild fisheries.[18]

We now consider the case of increase in the stocking density of an aquaculture farm, that is, an increase in the proportion of shrimp fry that goes to aquaculture industry. This means that an increase in the value of 'v' will definitely increase the value of 'A', since $A^/ > 0$. An increase in the value of $A^/$ means an improvement of technology which ultimately reduces the value of '$\hat{\alpha}$'. For the time being, we will consider the case of reduction in the value of '$\hat{\alpha}$' by 1 per cent. The impact of a reduction in the value of '$\hat{\alpha}$' on aquaculture industry and wild fishery has been summarised in Table 6.3.

Table 6.3 Impact of Changes in Technology on ψ_A and ψ_W in Sundarban and Dhamara

	% Changes (Increase) in ψ_A	% Changes (Decrease) in ψ_W
Dhamara	30.98	15.26
Sundarban	41.91	20.64

From Table 6.3, it can be seen that an improvement in technology for the aquaculture farm will cause a reduction in the supply of wild fishery and an increment in the supply of aquaculture industry products for both the regions.

CONCLUDING REMARKS

In India, shrimp industry plays an important role not only in bringing foreign exchange but also in catering to domestic needs. This industry consists of two parts: wild fishery and aquaculture industry. Aquaculture industry has expanded due to the fact that wild shrimp fishery alone cannot meet the increased demand for shrimp in the world market. However, there exists a trade-off between wild fishery and aquaculture shrimp industry and so, they should be better managed for the benefit of both traditional fishing and aquaculture farming communities. In a major departure from the earlier works in this area, we have tried to establish the link between aquaculture industry and wild shrimp fishery. This has been done through a structure that analyses the impact of an expansion of aquaculture shrimp industry, as reflected through an increase in the proportion of shrimp fry consumption. This issue has been captured by incorporating technological improvement in the shrimp industry via the increase in stocking density.

On the basis of our structure, we can thus conclude that technological change leads to an expansion of aquaculture industry and contraction of the wild fishery. This result emphasises on the need for defining more socially and ecologically responsible aquaculture industries that enhance traditional fishery and reduce current user conflicts. It is important from the point of view of policy-making. Researchers can explore this area for getting more insights on shrimp aquaculture management practices.

ACKNOWLEDGEMENTS

The authors greatly acknowledge the funding provided by Shastri Indo-Canadian Institute, New Delhi (INDIA) and Calgary (CANADA) to carry out this research. The authors also acknowledge with thanks the comments received from the team of Indian researchers of the project, especially Anita Chattopadhyay Gupta, in preparing this chapter. Any remaining errors that may be, however, are the sole responsibility of the authors.

NOTES

1. The present chapter is related to the collaborative ongoing Shastri Indo-Canadian Applied Research Project, entitled 'Assessing Environmental Management Options to Achieve Sustainability in the Shrimp-Mangrove System in the Indian Coastal Zone of Bay of Bengal', between Simon Fraser University (Canada) and Jadavpur University (India).
2. Corresponding author.
3. See the website http://www.worldtrek.org
4. At first sight it seems to be a restrictive assumption. However, if we consider the technology of the farms, then it seems very reasonable since all the farms in a region adopt more or less the same technology. Moreover, from the point of view of the product (shrimp) which they are producing, it is essentially homogeneous in nature. Thus, we can conclude that as they are adopting the same technology to produce a homogeneous product, they are identical in nature.
5. Our study area is mainly the Sundarbans, where the role of hatcheries in shrimp fry production is not at all significant. However in Dhamara, shrimp fries are supplied both by local fry collectors and hatcheries. So as a starting point, the exclusion of the role of hatcheries is quite a reasonable assumption when our purpose is to compare the role of aquaculture vs wild fishery from the point of view of wild shrimp fry collection.
6. It is a common and a simplifying assumption.
7. Effort is expressed in terms of labour days by using the equation E = (total variable cost of all inputs)/(average wage rate). For this, we have converted variable cost of all inputs in terms of labour days by dividing it by the wage rate. In this way, we have established the equivalence between variable cost of inputs other than labour in terms of labour days. For example, cost of chemicals is expressed in terms of labour days by using the expression (cost of chemicals)/(average wage rate). It is to be noted that labour input is already expressed in terms of labour days.

8. Pond area of a farm is given for a particular period (short run). The owner of the farm cannot change his pond area within that period. The decision to change its pond area is a planning decision, which can be implemented only in the long run. It is also true for stocking of shrimp fry. Moreover we are using cross sectional data. Hence we can assume that pond area and stocking of shrimp fry are given for a particular time period that is in short run.
9. As pond area is given in the short-run, which is mentioned in footnote 7, we can say that improvement in the stocking of shrimp fry implies improvement in the stocking density, which ultimately reflects the technology of a representative farm. Since producers generally operate in the economic zone of production function that is, second zone of production function, increase in stocking density will definitely increase the yield of a particular farm which will ultimately increase the total supply of the aquaculture industry.
10. Since profit maximisation and cost minimisation are dual to each other, in this case we have adopted the technique of cost minimisation. See Sathirathai (1998) in this context.
11. See footnote 14 for an explanation.
12. Under the assumption of perfect competition, P = MC condition holds in equilibrium. So we can take P instead of MC.
13. It is to be noted that intensive shrimp farming is not practiced in India. So when we say that the classification of shrimp farm is on the basis of technology, we consider all types of classification starting from purely traditional to semi-intensive type.
14. For estimation purposes we have used E. VIEWS package.
15. Here we assume that the total number of farms forming the aquaculture industry is equivalent to the number of surveyed farms in a particular region.
16. We have obtained this proportion from the literature.
17. As we assume that the aquaculture industry is nothing but the sum of total surveyed farms we can say that total supply of aquaculture industry is nothing but the requirements that are not supplied by wild shrimp fishery. This is true for the total population of matured shrimps. Here, however, we have considered a representative sample of aquaculture shrimp farms. Naturally we have focused only on a representative part of total population of matured wild shrimp. It has been implicitly assumed here that the ratio between the supply of aquaculture industry and wild sector remains the same, whether it is population ratio or sample ratio.
18. For a given time period total amount of shrimp fry is given for a region, since changes in shrimp fry will only occur when there is a new recruit of shrimp fry or there is a migration of shrimp fry from other regions which is not possible in case of a shorter time period.

REFERENCES

Barbier, E.B. and I. Strand. 1998. 'Valuing Mangrove-Fishery Linkages: A Case Study of Campeche, Mexico', *Environmental and Resource Economics*, 12: 151–66.

Bhatta, R. and M. Bhat. 1998. 'Impacts of Aquaculture on the Management of Estuaries in India', *Environmental Conservation*, 25(2): 109–21.

Boyd, C.E. 1997. 'Shrimp Farming and the Environment', *MPEDA Newsletter*, 11(6): 14–22.

Costa-Pierce, B.A. (ed.). 2002. *Ecological Aquaculture: The Evolution of the Blue Revolution*. Blackwell Science, 30–43.

Costa-Pierce, B.A. and C.J. Bridger. 2002. 'The Role of Marine Aquaculture Facilities as Habitats and Ecosystems', in R. Stickney and J. McVey (eds), *Responsible Marine Aquaculture*, pp. 105–44. CABI Publishing.

FAO. 2000. *The State of World Fisheries and Aquaculture*. Rome, Italy.

India and China Stage: Team Dispatch. 2000. '*How a Little Shrimp Can Destroy a Whole Farm*'. http://www.worldtrek.org, May 27.

National Environmental Engineering Research Institute. 1995. 'Inspection Report on Impacts of Aquaculture Farming and Remedial Measures in Ecologically Fragile Coastal Areas in the States of Andhra Pradesh and Tamil Nadu', Nagpur, India.

Santhakumar, V., E. Haque and R. Bhattacharya. 2003. *Economics of Coastal Wetlands: A Study of India and Bangladesh*. Summary Report on the Project sponsored by South Asian Network for Economic Initiatives (SANEI), Global Network. http://www.saneinetwork.org.

Sathirathai, S. 1998. *Economic Valuation of Mangroves and the Roles of Local Communities in the Conservation of the Resources: Case Study of Surat Thani, South of Thailand*. Final Report submitted to the Economy and Environment Programme for South East Asia (EEPSEA), Singapore: EEPSEA.

Srinath, K., M. Sridhar, P.N.R. Kartha and A.N. Mohanan. 2000. 'Group Farming for Sustainable Aquaculture', *Ocean and Coastal Management*, 43(7): 557–71.

7

Linking Conservation with Livelihood: Lessons from Management of Gir-Protected Area in Western India

Amita Shah

Abstract: Recent discourse on the management of Protected Area (PA) speaks eloquently about linking the conservation efforts with development of periphery, especially the pastures and community forests. The approach is particularly important in the case of the PAs having large human population in the periphery with direct stakes in resources within the PA. It has been increasingly realised that conservation of the core is contingent upon development of the periphery of the PA—the philosophy being echoed by a number of Eco-Development Projects (EDPs) across various parts of the world. India has also gone a long way in implementing EDPs for conservation of major PAs in the country. The central focus of the EDPs has been to enhance livelihood support system through regeneration of CPLRs and other resources in the peripheral villages, thereby reducing people's dependence on the PA. This is to be achieved by involving local communities in planning as well as implementation so as to obtain reciprocal commitment for conservation and protection of the PA. The experience from the various EDPs at best is mixed; a lot more is desired to be done in order to achieve the laudable goals of regeneration and conservation. One of the major constraints in the design of EDPs, notwithstanding the faulty implementation, is that the efforts for regeneration of CPLRs in the periphery are seen in isolation of the efforts, or management plan for regeneration of resources inside the PA. This kind of disjointed approach for resource management not only affects the planning exercise, but it also hampers people's involvement owing to the limited stakes and inadequate incentives for protection. It is thus plausible that linking-up of regeneration efforts within and outside the PA by treating them as an integrated ecological system may help better management and also protection of PA through people's commitment for conservation. The chapter demonstrates this by exploring alternative approaches for management in the context of Gir National Park and Sanctuary in Western India.

INTRODUCTION

The policy discourse on management of protected areas (PAs) has come a long way from purely conservationist strategies to participatory approaches. In between these two, there is a wide range of options that combine different elements of resource sharing, market regulation and privatisation. Experience from a large number of developing economies suggests that none of the pre-conceived, 'blue-print' solutions may work across different PAs, though it might have worked in the situations of wilderness without much of human activities around (Chopra 1998). This implies that analysis of the cost of bio-diversity loss and the development of appropriate institutions and incentives should primarily be a local exercise (Perrings 2000). The choice of PA-management approach therefore, has to be in tune with the location specific situation—ecological, socio-economic, political and financial. Also, the choice is time specific; it may undergo changes along with different stages of PA-management. Exploring options and evolving new approaches therefore are important aspects of policy formulation on PAs.

Located in the western part of India, Gir is surrounded by a substantially large human as well as livestock population having direct stakes in the ecology of the PA. The region had faced severe risk of extinction of its core wild life specie, that is, the lion, before it was notified as a sanctuary in 1965. Subsequently a number of conservation measures were initiated, leading to successful revival of wild life within the PA (Singh and Khamboj 1995). By the turn of the century, the wildlife population had overshot what was earlier considered as carrying capacity of the PA. To a large extent, the success could be attributed to effective protection and habitat development practices, featuring the PA-management plan. The next stage therefore, is to evolve sustainable strategies for regeneration and conservation of vegetation and biodiversity in Gir. It is envisaged that evolving appropriate institutional arrangement for sharing of the regenerated resources, especially from pastures within and outside the PA, might help both—conservation as well as people's livelihood in a sustainable manner.

MARKET LINKED APPROACH

One of the possible strategies is to adopt a market-linked approach, which seeks to combine important elements of the two alternative approaches noted above (Shah 2003). Essentially, the approach involves regeneration of ecology under the existing conservationist management system while incorporating people's livelihood needs/stakes as a legitimate component of the regeneration strategy. Conceptually, the approach offers a fairly practical solution for reducing people's pressure on the ecosystem by making adequate provisions for the supply of resources like fodder, fuel wood, non timber forest produce (NTFPs), water, and silt on a sustainable basis. It envisages multi-stakeholder professional organisations to look after the resource management and sharing of responsibilities; these aspects are generally missing in the other two approaches. The approach therefore renders some kind of a supply management system with technological interventions through resource regeneration and market development. Another important feature of the approach would be to define a specific timeframe of say, twenty years within which the results should be achieved. If properly executed, the strategy may turn out to be cost-effective (that is, requiring relatively lower amount of subsidies) and at the same time, ecologically more effective (that is, reducing degradation within a 'reasonable' time frame).

Given this backdrop, the chapter seeks to explore alternative management strategy for Gir, which consists of large tracts of common pool resources both within and in the periphery of the PA. This is being explored in the light of a detailed mapping as well as valuation of the existing resources and the use thereof.

OBJECTIVES

The specific objectives are to examine:

- the present status and benefits flowing from Gir-PA;
- people's dependence on the PA; and
- alternative approaches for PA-management with a special focus on regeneration of pastures within and outside the PA.

The chapter is divided into five sections including this introduction. The next section examines the status of Gir-ecology and the major benefits flowing from the resources along with estimates of the cost of PA-management. This is followed in Section 3 by a detailed account of the extent and nature of people's dependence on the PA. Section 4 discusses implications of the present patterns of resource – use as well as management, and explores alternative approaches. The last section discusses policy recommendations. The study is based on secondary as well as primary data collected from a sample of villages in the periphery and also from selected hamlets and forest settlements inside the sanctuary area.[1]

GIR-PA AND THE BENEFITS FLOWING FROM ITS RESOURCES

Gir ecosystem is the last surviving habitat for Asiatic lions. Spread over an area of 1,412.1 sq km, Gir is one of the largest compact tracts of dry deciduous forest in semi-arid regions in the country. Apart from being the only home of the Asiatic Lion, the ecosystem assumes special significance because of its tremendous regenerating, self-supporting and sustaining capacity for the rich and diverse fauna and flora (Singh and Kamboj 1995). Recognising the special ecological features of the region, it was first notified as a sanctuary in 1965 and subsequently as National Park in 1975 under the Wild Life (Protection) Act (1972).

Gir forest represents an important ecological formation in western parts of India. Apart from being the only home of the surviving Asiatic Lion, it constitutes catchments of the seven major rivers thus, providing ecological security to the surrounding drought prone region. Conserving this ecosystem therefore would serve some important functions (Singh, M. 1995) as shown in Figure 7.1.

It is important to note that the region including the PA has been viewed as a fodder bank especially during drought years, attracting livestock from a number of places like the rest of Saurashtra, Kachchh, North Gujarat and even Rajasthan. While there is no systematic estimate of intrusion of

Figure 7.1 Gir-Ecology

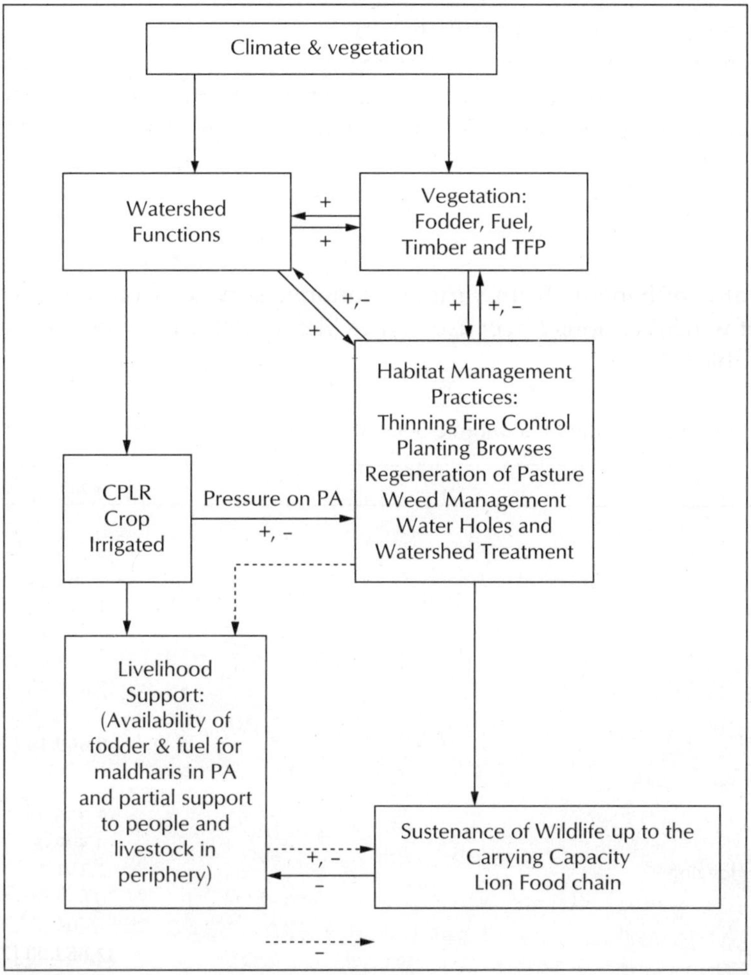

Notes: + Positive Impact; – Negative Impact.

people from other regions—seasonal, occasional and permanent, there are evidences which suggest that the region has been performing an important drought proofing function both formally as well as informally (Sinha 1967). Essentially, regeneration of ecology should be based on development of watersheds covering the seven rivers flowing from the PA. In that case, the regeneration plan should also cover those areas of the watersheds, which lie outside the PA. Linking up these areas in the periphery would amount to incorporating people and their economies as integral parts of the ecology. The peripheral region and people therein thus, become important stakeholders though, their stakes may assume a relatively lower priority in management of the PA (see Figure 7.1).

Benefits from Gir-PA: Select Estimates

The close interactions and continued conflicts between people and the PA suggest the need for a major shift in the management strategy of Gir region. Before discussing that, it would be useful to have a brief account of economic and ecological services rendered by the PA. This section presents a summary of the valuation exercise conducted in a larger study undertaken by the author (Shah 2003).

Direct Benefits

The annualised value of benefits from various economic services from the PA is estimated to be Rs 47,705.1 lakh of which, various direct use-values like fodder, fuel wood, irrigation etc. comprise 20 per cent (See Table 7.1).

Table 7.1 Value of Economic Services from Gir-PA (at 1994–95 prices)

Economic Benefits	*Value (Rs lakh)*	*Percentage*
Direct Use		
Fodder	4,114.23	42.55
Fuel wood	1,406.25	14.54
NTFP	1,319.02	13.64
FYM (from Neses)	392.44	4.06
Tourism	19.64	0.20
Irrigation	2,411.40	24.94
Medicinal plants	6.16	00.06
Sub-total (a)	**9,669.14 (20.27)**	**100**
Indirect Use		
Timber Teak	7,250.00	19.14
Non-teak	4,499.00	
Fuel wood from timber logging	2,751.00	
	30,633.00 (64.2)	80.86
Sub-total (b)	**37,883.00 (79.41)**	**100**
Drought proofing	153.00	
Grass collection (Tonne) Supporting about 2,040 livestock for 4 months)	(0.32)	
Total (a + b)	**47,705.14 (100)**	

Note: Based on the estimates presented in Tables 1 to 6 in Shah (2003).
These values are significantly under estimated due to non-valuation of the ecological diversity.

However, if we consider the value of fuel wood that might be realised through logging or maturation as well as damage due to natural factors like cyclones etc., as having direct use value, the share of direct use value increases to about eighty-five per cent. Two issues are important in this context. First, the estimated fodder value is based on the national average of 3,000 kg/hectare for the Indian forest (Tewari 1994). Local prices have been used for converting the estimated fodder production into monetary value. This was essential because the existing studies on Gir do not provide any estimates of fodder production from the PA. The second aspect relates to the estimates of NTFPs. Since there are no systematic estimates of the production of a large number of NTFPs available from the PA, we have once again, resorted to using the national average to estimate the market values. It may however, be noted that according to the official norms (adopted by the Central Statistical Organisation), the actual production is generally ten times the value realized in the market. We have however, not incorporated these projected values of NTFPs in our estimates. Together, these estimates lead to a downward bias in valuation of the benefits resulting from direct use of PA-resources. This is to ensure that the estimates do not become unrealistic when compared with the estimated cost of investment, necessary for regeneration of the PA.

Valuation of Non-Use Benefits

More than direct as well as indirect use-values, non-use benefits have special relevance in the context of a protected area. These include benefits like existence value, rarity and aesthetic value, option value, cultural value and ecological value. Assessing the monetary value of these benefits however, is difficult. Alternatively we have tried to capture people's perceptions on relative importance of the major attributes of and also on desirability of conservation of the PA. This was obtained by asking the respondents to rank the five major attributes, which can be broadly classified as Watershed Functions, Rarity of Lion, Bequest Value, Religious-Aesthetic Value and Consumptive Value (grazing + fodder). The exercise is based on qualitative information collected from 162 households from four villages in the periphery of Gir-PA (Shah 2003).

It is interesting to note that apart from consumptive use, people in the peripheral villages attach significant importance to religious-aesthetic aspects of the PA, which is closely followed by watershed services, rarity and bequest value. It may be noted that the religious aspect has a close link with the overall ambience of the forest ecology and its aesthetic value. It is largely perceived that the religious spots may also lose their cultural-aesthetic importance if the forest cover in the PA gets deteriorated. To a large extent, these perceptions are in conformity with the observations made in an earlier study by Debnath et al. (2001). The estimated benefits could be compared with the cost of PA-management. A summary of the major benefits and costs has been made/presented in Table 7.2.

Table 7.2 Summary Benefits and Costs (Rs Lakh at 1995–96 Prices)

Value of Benefit		*Value of Cost*	
Details	*Value*	*Details*	*Value*
Direct Use	9,669.14	Average Budget for Management per year	1,191.40
Indirect Use	37,883.00	Crop Damage	419.80
Opportunity Cost*	39,524.98	Loss of livestock	143.16
Loss of crops to replace the fodder	2,592.00		
Potential loss of fodder	1,170.33		
Soil Loss	9,793.25		

Note: * Refers to the value of land under alternative use that is for crops and plantation. This is the cost of lost opportunity, which may be treated as minimum value of conservation.

PROTECTED AREAS AND THE PEOPLE

There are three sets of human settlements within and in the periphery of Gir. These include neses (clusters of cattle herders, that is, Maldharis), forest settlements and revenue villages. The first two are located in the sanctuary whereas revenue villages are located on the periphery of the PA. It may be noted that people in neses have greater access to resources within the PA and are therefore considered to be the most crucial category from the viewpoint of the PA management. The forest settlements are next in terms of people's access to the PA resources. The revenue villages, as such do not have any 'legal' access or rights to obtain any direct use value from resources within the PA. Obviously therefore, the analysis of people's dependence on the forest will have to keep in mind the two separate categories of people—those living within and outside the PA. In what follows, we discuss the nature and extent of forest-dependence among these two categories of people and identify issues that emerge from the interface between the two.

People Inside the Protected Areas

At present, there are about fifty-four hamlets (neses) and fourteen forest settlements in Gir-PA. Together they inhabit a population of about 8,000 persons and 11,000 livestock. While these people living inside the sanctuary draw upon the various resources such as fodder, fuel, land, water, NTFP, timber etc., for satisfying their livelihood needs, they also seem to be contributing towards sustenance of the ecology. Two important aspects are often noted in this context. First, grazing of livestock with a well laid out seasonal rotation helps sustaining biodiversity of grasses; this may also help reducing the incidence of forest fire, which has a high probability of occurrence under the dry-hot weather in the region. Another ecological function that seems to have been performed by the people is that of keeping up the chain of herbivorous species, in the absence of which, damage to the peripheral agro-economic system and crops might have been more severe.

Recognising the conflicts between wild life and people inside the PA has led to a policy approach, which seeks to relocate these people outside the PA as noted in the special scheme prepared for Gir-Sanctuary way back in the early seventies, and subsequently in the management plan prepared during the mid-nineties. In the same vein, the management approach focuses mainly on protection measures and vigilance against interference by the local communities as well as other vested interests from industry, mining and developmental activities. Together this has led to the usual scenario of conflicts between people and the PA or between conservation and livelihood. The conflicts become more severe during droughts. What has aggravated the situation is 'inappropriate' use of land as well as water resources in the peripheral villages. This is reflected by the fact that thirty-three per cent of the forest area in and around the PA are degraded and/or highly degraded (Singh and Kamboj 1995). Apart from these, the PA has a network of about 600 km of road length and 15 km of railway tracks. More than two lakh vehicles pass through Gir every year causing problems of noise as well as air pollution on the one hand and damages to wildlife on the other. The presence of a number of religious places adds to these problems.

Table 7.3 Budget Estimates for PA-Management in Gir (at 1994–95 Prices)

Budget Head	*Estimated Expenditure for 1995–2000 (Rs Lakh)*	*Average per Year (Rs Lakh)*	*Percentage*
A.			
Demarcation	20.80	4.16	0.51
Habitat Improvement	131.00	26.20**	3.20
Development of Peripheral Coastal Forest	716.90	143.38**	17.55
Protection	161.00	32.20*	3.94
Research, Education and Training	82.80	16.56	2.02
Vehicle and Equipments	151.40	30.28	3.71
Quarters and Buildings	135.10	27.02	3.31
Tourism	170.00	34.00*	4.16
Socio-Economic (including Resettlements)	309.00	61.80*	7.57
Recurrent Expenditure	2,205.00	441.00	54.00
Total	**4,083.00** (68.54)	**816.60**	**100.00**
B.			
Eco-Development Project			
Village Eco-Development	1,239.00	247.8*	66.11
Improvement of Protected Area	396.00	79.2**	21.13
Education and Awareness	49.00	9.8	2.61
Research	190.00	38.0	10.14
Total	**1,874.00** (31.45)	**374.8**	**100.00**
Grand Total	**5,957.00** (100.00)	**1,191.4**	**0.0**

Source: Singh and Kamboj (1995).

Notes: * Indicates allocation for activities that might indirectly contribute to regeneration; ** Indicates allocation for measures directly relevant for regeneration.

Cattle Herders in Neses: Resource Use and Damages

Cattle herders (Maldharis), residing inside the PA, face maximum resistance from the forest management team, as they have relatively higher dependence on forest resources especially, fodder, as compared to other communities residing in forest settlements. Unlike Maldharis, the residents of forest settlements are involved in crop cultivation and hence, their dependence on livestock is very limited. Maldharis, on the other hand, keep a large size of cattle herds, for which free grazing is the common norm. As a result, way back in the early seventies, the Maldharis had to face involuntary resettlement outside the PA. Unfortunately, the resettlement process was so unfavourable that many of them refused to move out, and some of those who did get resettled, eventually returned to their hamlets inside the PA (Singh; Choudhary 2000). We have tried to assess benefit-cost for these nearly 3,000–5,000 people who continue to stay within the PA.

Table 7.4 presents estimates of benefits and costs accruing to Maldharis living in neses within the PA. The benefits are mainly in terms of greater access to forest resources as compared to those living outside the PA. Against these, the costs are mainly in terms of lack of physical infrastructure, social and economic alienation and conflicts with the forest department. It may be noted that the loss of livestock forms is only a marginal proportion, that is, about 4–5 per cent of their total stock every year. Since a substantial part of the livestock-loss is likely to be consisting of less productive cattle (as the more productive cattle are better protected and taken care of), the actual loss could be treated as a 'rent' for occupying the housing space within the PA.

Table 7.4 Benefits and Costs to Maldharis in Gir

Economic Benefits and Costs	*Rs Lakh (1994–95)*	*Other Gains and Losses*
A. Benefits		Gains
Fodder	784.48	Clean air and water
Fuel wood	5.93	Less risk of droughts
Timber	4.30	Free housing
FYM	156.98	Grazing outside livestock
MTFP	NA	Natural ambiance
Grazing of outside animals	196.12	Losses
Water, Housing, Other Amenities	NA	Absence of schools
		Absence of electricity
		Lack of health facilities
		Limited scope for occupational diversification
Total economic benefits	**1,147.81**	**Limited links to market**
B. Loss of livestock (750/Year)	112.50	Problems of mobility
C. Total net benefits (A–B)	1,035.31	Conflicts with FD staff
D. Cash Income		
Milk	750.00	
FYM	156.98	
Total	**906.98**	

Source: Calculated on the basis of the information collected from various studies. For details see Chapter 3 in Shah (2003).

Notes: Based on the information obtained from Maldharis about average consumption of fodder per livestock and fuel wood per households. The norms used for fodder consumption by cow and buffalo are 20 and 25 kg per day per animal respectively. Fuel consumption per household was estimated at 6kg per day. For timber the norm used is 10 cubic metres per household for 20 years. The FYM production per livestock is 8 tones per year and the net price received is Rs 0.2 though, the market price is Rs 0.5. The prices used for fodder, fuel wood and FYM are Rs 1,1.25 and 0.75 per kg respectively.

It is observed that the estimated value of the benefits in terms of direct use of forest resources is Rs 1,147.81 lakh per annum. Against this, the cost borne by the Maldharis works out to be Rs 112.5 lakh. The net benefit is Rs 1,035.31 lakhs. Alternatively, we worked out the net returns from selling of milk and farmyard manure. This worked out to be Rs 906.98 lakh per annum. These estimates clearly suggest that the Maldharis derive significant economic benefits from the PA. Conversely, it implies that shifting them out would require a fairly attractive compensation package that may ensure at least similar if not the same level of livelihood support after the resettlement. Or else, these

Maldharis should be convinced to cooperate with the conservation objectives through participatory processes for awareness generation and improved compliance, given the existing norms of 'rights and restrictions.'[2]

Peripheral Economy and Interface with Protected Area

The periphery of Gir consists of ninety-nine villages in the radius of 5–7 kms. Table 7.5 provides information about these villages with varying distances from the PA-boundary. In 1991, these villages had 26,397 households with a population of over 1.52 lakhs. By now, the human population would have increased to about 1.8 lakhs using the average growth of 2 per cent per annum. According to the official estimates, these households own about 95,000 livestock.

Table 7.5 Population and Employment among Villages in the Periphery of Gir

All (99) Villages on Periphery	*Total HHs*	*Total Population*	*Total Main Workers*	*Cult. as % to Main Worker*	*Agril. Labourers as % to Main Workers*	*Livestock etc as % to Main Workers*	*Non-Workers as % to Main Workers*
1971	18,386	1,06,620	28,200	68.38	35.27	2.80	248.41
1991	26,397	1,52,032	41,513	51.27	32.17	2.68	166.53
Difference	8,011	45,412	13,313	-17.11	-3.1	-0.10	-81.89
% change	43.57	42.59	47.20	–	–	–	–

Sources: Census of India, 1971 and 1991.

To a large extent, this population (human + livestock) depends upon various ecological as well as economic services provided by the PA. The most important among these are water (that is, the seven rivers originating from Gir) and fodder (with high degree of biodiversity and quality) which sustain a large number of faunal diversity species including milch animals. Together these resources have claimed a special agro-ecological significance to the region, which is the only green fertile patch of land in the dry/semi-arid region in the western part of Gujarat. In turn, this has been reflected by the relatively higher productivity of land as well as livestock, cultivation of high valued crops like sugarcane as well as mango (and other horticulture) plantation and the presence of scenic beauty with a number of religious places in the region.

However, agriculture in the peripheral villages is facing certain challenges. For instance, better availability of ground water and soil moisture in the region has led to increased extent of water intensive crops like sugarcane and cotton. This has resulted in depletion of groundwater and drying up of streams. This, in turn, has increased the risk for the wildlife since a large number of wells and water holes have to be created in order to provide adequate drinking water for them inside the PA (Ramachandran et al. 2001).

Finally, natural disasters like cyclones and droughts have also affected the balance between ecology (including wildlife) and human requirements. For instance, a devastating cyclone during 1982–83 had destroyed about twenty-eight lakh timber trees besides other shrubs and plants. Similarly, frequent droughts and the resultant water scarcity in the region have led to stunted growth and sparse vegetation in the large tracts of degraded (345.5 sq km) and highly degraded areas within the PA (122.2 sq km). Besides this, there are other factors such as natural calamities, causing degradation within and in the periphery of the PA.[3]

Overall therefore, Gir forest has undergone significant changes over the past two centuries leading to drastic reduction in the forest area as well as its resources (Singh 1997). It is noted that 'encroachment and destruction of natural surroundings of the PA, increasing population of carnivore and herbivore and increasing disturbance to wild animals force them (that is, lions) to move outside and to cause crop damages and killing of livestock. Hence the man-animal conflicts are increasing, threatening the wildlife in turn' [Singh and Kamboj (1995); also see Sinha (2001)].

Population Growth, Changing Land Use and CPLRs

As noted above, the rich ecological resources of the region are surrounded by densely populated human settlements. Between 1971 and 1991, the population increased at the rate of 2.19 per cent per annum. This is slightly lower than the district average of 2.23 percent and the state average of 2.74 per cent per annum. Prima facie, this observation supports the generally held view that the PA-ecology attracts more human as well as livestock population in the immediate periphery, that is, in the radius of less than 3 km. Prima facie, the low level of population growth in these talukas could be attributed to two important changes that have taken place since the mid-seventies. First, due to protection measures, people's access to forest resources has declined (though, not stopped) over time. Second, decline in the quality and quantity of Common Property Land Resources (CPLRs) has led to out-migration to the nearby urban centres. While these propositions are difficult to ascertain in the absence of a detailed investigation, we have tried to find plausible explanations by examining some of the important changes during 1971–91 in the peripheral villages (Table 7.6).

These are:

1. Proportion of forest to the total area has increased by eighteen and fifteen per cent in the nearby and distant villages respectively.
2. Against this, there has been a decline in the area not available for cultivation. However, a major part of the increase in forest area seems to have come from conversion of village pastures into forest vidis at the time of demarcation of the PA and subsequently while redefining the boundary.
3. Irrigation has also increased substantially but, more so in the distant villages. In 1991, the total area under irrigation was 8,088 ha among seventy-seven nearby villages vis-à-vis 6,237 ha in twenty distant villages.
4. As a result, the area under crop also increased in fifty-two out of the ninety-nine villages. In the remaining forty-seven villages, Net-Cropped Area (NCA) had declined by 20, 646 hectares. To a large extent, this decline is mainly due to demarcation of the PA boundary. It is observed that as many as thirty-two villages in the periphery had lost more than 100 hectares of privately cultivated land in each village, besides several villages having lost their CPLRs under Section 4 of the Land Acquisition Act.

Table 7.6 Changes in Land Use among Peripheral Villages

Distance from PA	*Year*	*Total Area (ha)*	*Forest Area as % of Total Area*	*Irrigated Area (ha)*	*Cultivable Wasteland (ha)*	*Not Available for Cultivation 95% to Total*	*NCA*
<3	1971	79,494	15.59	4,339	14,340	10.44	43,380
<3	1991	7,8,685	33.67	8,088	7,994	7.33	47,002
Difference	–	–809	18.08	3,749	–6,346	–3.11	3,622
>3	1971	32,715	3.93	2,171	7,012	9.53	23,001
>3	1991	34,386	19.32	6,237	2,823	3.34	15,183
Difference	–	1,671	15.39	4,066	–4,189	–6.19	–7,818
Both	1971	1,12,209	19.52	6,510	21,352	19.97	66,381
Both	1991	1,13,071	52.99	14,325	10,817	10.57	62,185
Difference	–	862	33.47	7,815	–10,535	–9.40	–4,196

Source: Primary Survey.

Increased irrigation in the region along with regeneration of the PA should imply higher rate of population growth vis-à-vis the district or the state average. But this, as we noted earlier, is not the case. If so, the lower growth rates in peripheral villages suggest two possibilities in terms of population movements. That is, people in the distant villages are either pushed into the nearby villages and/or have been pushed out of the region probably due to declining size and quality of CPLRs in these villages. Given the fact that a large number of villages have also lost a part of the cropped land, out-migration from the peripheral villages appears to be a more predominant phenomenon than the movement nearer to the PA.

The issue of CPLRs has been probed further by obtaining information from twenty-nine villages in the periphery. The information has been collected through repeated visits and informal discussions with individuals as well as groups of people in these villages. This was essential because the issue of CPLRs is very sensitive and highly politicised. In turn, this makes it difficult to get accurate information on the size and status of CPLRs.

The situation becomes more complex as many of these villages have continued conflicts with Forest Department due to inclusion of CPLRs within the PA-boundary. Notwithstanding these limitations, we have tried to capture some of the basic information pertaining to CPLRs in these twenty-nine villages (Table 7.7). It is observed that the size of CPLRs has declined substantially in eighteen out of twenty-nine villages. To a large extent, this has happened due to notification of village pastures as forest area within the PA. Moreover, there is a significant problem of encroachment of CPLRs by the village communities. As a result, seven out of the twenty-nine villages have no or very small (that is <10 hectares) area left as gaucher (or pasture) land. Another fourteen villages have about 10–50 hectares of pastureland. It is therefore crucial that these pastures are properly regenerated and managed so that people in these villages do not have to depend much on the PA.

How far do people in the periphery actually depend on the forest resources? What is the extent of their dependence on these resources? What is the nature of conflict over these resources? And what is their perception about future plan for regeneration of pastures and vidis within and outside the PA? These issues have been examined through a sample survey of four villages, four neses and

Table 7.7 Status of CPLRs in Selected Villages

	Status of Gauchar						Other Grazing	
					Current (in Ha)			
Village	*Earlier (in ha)*	*Encroached (in ha)*	*Donated (in ha)*	*Notified Forest (in ha)*	*Available*	*Condition*	*Vidis (in ha)*	*Private (in ha)*
Kamdadi	34.89	5.87	0.00	23.16	5.87	A	0.00	0.00
Hirava	111.15	0.00	0.00	111.15	0.00		111.15	185.24
Paniya	0.00	0.00	0.00	0.00	0.00		0.00	15.44
Gigasan	30.87	0.00	0.00	0.00	30.87	A	0.00	0.00
Shivad	15.44	0.00	0.00	0.00	15.44	A	0.00	0.00
Jhankia	10.03	0.00	0.00	0.00	10.03	C	0.00	0.00
Fareda	77.18	30.87	0.00	0.00	46.31	C	78.11	15.44
Dron	385.92	46.31	169.81	0.00	169.81	A	0.00	0.00
Nitli	293.30	15.44	0.00	0.00	277.86	A	0.00	0.00
Juna Ugla	30.87	6.17	15.44	0.00	9.26	A	0.00	0.00
Itvaya	92.62	46.31	0.00	0.00	46.31	A	0.00	0.00
Khilvad	77.18	46.31	0.00	0.00	30.87	A	0.00	15.44
Bhalchel	231.55	0.00	0.00	231.55	0.00		0.00	0.00
Kenedipur	571.16	308.74	77.18	108.06	23.16	A	38.59	0.00
Ambala	61.75	30.87	0.00	0.00	30.87	B	0.00	52.02
Amrapur	120.41	0.00	0.00	108.06	12.35	A	0.00	0.00
Jalandhar	648.35	324.17	0.00	0.00	15.44	A	287.13	0.00
Khodiyar	154.37	0.00	7.72	0.00	146.65	A	0.00	77.18
Ratang	385.92	30.87	108.06	0.00	246.99	A	0.00	0.00
Limadra	231.55	0.00	0.00	0.00	231.55	A	0.00	0.00
Monpari	77.18	0.00	0.00	30.87	46.31	A	0.00	0.00
Laduli	185.24	0.00	100.34	0.00	84.90	A	154.37	0.00
Jepur	277.86	30.87	38.59	154.37	54.03	A	23.16	12.35
Jambur	308.74	46.31	77.18	154.37	30.87	A	0.00	0.00
Rasulpara	15.44	0.00	0.00	0.00	15.44	A	0.00	0.00
Bhojde	540.29	0.00	0.00	540.29	540.29	C	0.00	0.00
Borvav	277.86	46.31	77.18	0.00	77.18	A	77.18	0.00
Surajgadh	12.35	0.00	0.00	0.00	12.35	A	0.00	0.00
Chitrod	123.49	0.00	0.00	123.49	0.00		0.00	0.00

Source: Primary Data.

Notes: A indicates that land supports livestock of the village for 2 or 3 season for grazing and frequent harvesting of grass is possible.

B indicates that land supports livestock of the village for monsoon season and harvesting of grass is not possible every year.

C indicates that land partially supports village livestock during monsoon.

two forest settlements in Gir-PA.[4] In what follows, we present a summary of the major observations based on the secondary as well as the primary data.

We have tried to examine these aspects by conducting a house listing in eight revenue villages, four neses and three forest settlements (FS). The exercise was conducted by combining a survey method with informal discussions by forming groups of the homogeneous categories of households. The information is also supplemented by functionaries of outside agencies having close familiarity with the village communities over a long period of time. The main observations emerging from this exercise have been discussed in the subsequent analysis.

Livestock and Fodder

It is observed that as large as forty-five per cent of the households in the peripheral villages do not own land. Similarly, thirty-two per cent households in these villages do not own any livestock. This is quite significant. The households in neses and FSs are not permitted to own land, though land is made available to households in the FSs for cultivation on lease. The large proportion of landlessness in revenue villages however, reflects dynamic changes in the land market where many of the traditionally cultivating communities like Kolis are coming from other (less irrigated) regions to till the land of other households in the Gir region. Thus, it is possible that a part of these landed households in the villages under study are owners of land in their own villages.

Notwithstanding this specific feature of the landless households, what we have generally observed in the region under study is a fairly close relationship between those without land and those without livestock. This of course, leaves out the traditional herder community, which owns substantial number of livestock, at times without much of a land base. These communities traditionally depend on the village pasture and/or the PA for sustaining their livestock. Among the remaining households, average number of milch animals is found to be fairly small that is 2.4, 11.8 and 23.3 in revenue villages, FSs, and neses respectively. These estimates are worked out by considering only those households, which had some livestock. The gross average would be even lower than this.

Prima facie, the limited ownership of livestock in the peripheral villages would suggest lower dependence on the PA for fodder. While it is difficult to get a realistic estimate of people's dependence on the PA, findings from our primary survey suggest that nearly thirty-five per cent of the households in the peripheral villages obtain up to fifty per cent of their fodder requirement from the forest vidis. Only thirteen per cent obtain more than fifty per cent of the fodder requirement from these resources. The remaining forty-eight per cent did not report accessing fodder from the forest. As noted earlier, a part of these forty-eight per cent households may not have any livestock; the proportion of households without any livestock was found to be thirty-two per cent. This implies that only sixteen per cent of the households owning livestock did not depend on forest for their fodder requirements. These estimates seem to be fairly reasonable.

Overall the findings, notwithstanding the lower livestock population per household, suggest substantial dependence on the PA for meeting at least a part of the fodder requirement even in the peripheral villages. Obtaining a realistic estimate of the total livestock population thus, becomes crucial for assessing the total dependence on fodder among the peripheral villages. In its absence, the micro level estimates based on the households' reported access to the PA, may not help in working out the aggregate estimates of the actual availability of fodder from the PA and people's dependence on that.

Fuel Wood

Compared to fodder, people's dependence on PA for fuel wood is much higher as already shown by the IIFM study and also our house listing. However, with a closer investigation and the information obtained through informal discussions with the people, it is learnt that nearly eighty per cent of the households in peripheral villages depend on PA for the fuel wood requirements. This excludes households belonging to socially as well as economically better-off communities, that is, Patels, Brahmins, Luhanas, Ismailis, and Mahajans. The above phenomenon has been further confirmed by the available estimates suggesting that as large as seventy-four per cent of the fuel wood requirement of households in the peripheral villages is being met by fodder collection from the forest or, through market purchase, a large part of which is likely to have come from the forest.

Of course, fuel wood collection varies significantly across households as observed during our survey in the sample villages. Basically, the dependence on forest would depend on the households' capacity to shift to alternative sources like kerosene, cooking gas (LPG) and bio-gas. While most of the households in the peripheral villages use kerosene, it constitutes only a part of their requirements for fuel. To a large extent, these households obtain a fixed quota of kerosene, that is, ten litres per month at a subsidised rate. This might be sufficient at the most for one-third of their requirement. For the rest, these households depend on fuel wood either through direct collection from the forest or through purchase from market/other households.

According to recent estimates, fuel wood requirement per household is 6 kg per day. For the 26,397 households in 1991, the total requirement would work out to be 57,809 tonnes per year. Assuming that fuel wood constitutes half of the total requirement of these households, the demand for fuel wood in the periphery would be 28,904 tonnes per year. This is based on the assumption that the remaining fuel requirement is met by kerosene, dung cake, LPG etc.

Thus the total requirement and the estimates demand by the peripheral villages (subtracting the kerosene, dung, LPG) are 57,809 tonnes and 28,904 tonnes per year. These estimates are fairly lower than the estimated availability of fuel wood (of the tune 1.87 lakh tones per year) from the PA (that is, sanctuary area). This kind of vast difference between the total requirement and the estimated availability, notwithstanding the limitations in estimation of the latter, would suggest substantial amount of fuel wood extraction for commercial purposes. This corroborates the estimated requirement by the people from a larger periphery covering 150 villages. According to this, the required fuel wood is 1.17 lakh tonnes per year. It appears reasonable to argue that a large part of the fuel wood requirement of these 150 villages is met by Gir-PA through collection and/or market purchase.

Timber

Extraction of timber is strictly prohibited. However there are occasional evidences where people from the periphery indulge in illegal felling either directly or indirectly. Such instances often surface during informal discussions with people where it is reported that about 5–7 per cent of the village community in the immediate periphery of Gir (that is <3 km radius) are involved in such activities. These households/individuals often belong to economically and socially very vulnerable groups of the society. However what is really of concern is that their involvement in such activities, at times, is triggered by some of the resourceful households in the villages often having political patronage. The economically vulnerable individuals fall prey to the 'greed' of the resourceful persons in the time of extreme distress when they need immediate cash. In other instances, they do undertake this risky activity because of their sheer need and the ability to maneuver the protection system. It is thus essential to distinguish the circumstances that lead and make it possible to extract timber from the PA.

Differential Pattern of Dependence among Households

The above observations along with our informal interactions with the village communities suggest a broad pattern of interface between people and PA across different categories of households in the periphery of the PA (See Table 7.8). Prima facie, we have categorised these households into three: The first consists of the resource-poor households with no or small piece of land and limited livestock. The next category consists of middle level agriculturalists with medium size of land and livestock ownership. The third category represents households with large land holdings and/or livestock and also socio-political power. It is postulated that the households in the first and the third categories 'depend' significantly on the PA—the former does that out of the 'need' to meet their subsistence requirements, and the latter out of the 'greed' to maximise their earnings. Apart from the economic base, the actual dependence is also determined by the household's capacity to manipulate 'rules and rulers' of the PA.

Table 7.8 Differential Interface across Households

Type of Households	*Asset Base*	*Potential/Actual Benefits from the PA*	*Losses due to the PA*	*Likely Response to the EDP*
Poor	Landless or marginal farmers with no or limited livestock	Fuel wood, NTFP, Illegal grazing for small ruminants	Limited	Good response if (a) alternative grazing space is provided; and (b) alternative fuel is affordable
Middle range of Farmers	Moderate land and livestock	Moderate use for fuel wood	Moderate to high (depending on the location of the farmers)	Good response if, effective protection to farms is provided
Better-off	Large land holdings and livestock	Fodder	High	Limited response because the loss of fodder benefit might exceed the limited protection which could be provided under the project

Understanding these dynamics is very crucial for evolving the right kind of incentives as well as compensation packages for different categories of households so as to reduce their dependence on the PA. It is crucial to note that whereas all the households have similar access to the forest resources, in practice, the access varies significantly depending upon the socio-economic and political base.

Negative Externalities

Despite the direct benefits from the PA, people in the periphery face severe problems with the wildlife damaging the crops. This aspect has already been discussed while estimating the economic cost arising out of the conservation of the PA. However, apart from the actual damage to the crops, people have to face a lot of problems in protecting their crops especially during night hours. Majority of people reported that they had to continue guarding their crops from various herbivores such as blue bulls, chital and wild boars. The problem actually starts right from the time when the seeds for the crop are sown. Farmers have to keep awake throughout the night for protecting the fields as the herbivores cover as much as 20–25 km of area both ways while going as well as returning early morning.

To a large extent, the phenomenon of herbivores going out to the field is an outcome of the degraded as well as improper vegetation within the PA. Availability of irrigation might have aggravated the situation. The result therefore, is migration of lions in search of the herbivores. While it has been argued that lions have always been moving out in the radius of 20–25 km, the frequency has increased due to the frequent droughts. It may be noted at this stage that the increased frequency of droughts is more a manifestation of the high rate of soil and water erosion rather than a result of the declined rainfall in the region. Hence, in the absence of proper measures for watershed management inside the PA, the vegetation is likely to remain thin, which in turn, pushes the herbivores outside the PA. Lions happen to follow this food chain and in the process get into conflicts with the people or the livestock. Interestingly, people in the sample villages reported that they would rather have lions on their fields so that the herbivores keep away! Breaking this cycle would therefore require appropriate management of vegetation inside the PA, which in turn, necessitates proper measures for soil and water conservation.

The recent debate among the team management however, views increased vegetation as non-conducive for lion-habitat. But, this argument needs further qualification. It appears that increased vegetation density has taken place mainly due to plantation activities in the National Park Area. This kind of vegetation is preferred only by *sambar*. Other herbivores prefer more of open grassland with shrubs which is found in the sanctuary area in the western part of the PA. Given the degradation (rather than increased density) of vegetation, the western region may not be able to sustain more herbivores so as to be able to increase the lion population beyond 150 or 160. This is perhaps why one observes that the increase in lion population in the past few decades has taken place mainly in the eastern region. This however, still does not imply that improving the density of vegetation especially, grass and shrubs in the sanctuary area is unsuitable for habitation of lions. Resolving this issue is very crucial, for increased vegetation and its proper management (including 'cut and carry' operations for collection of grass, weed-management etc.) has a significant bearing on economic benefits derived by people in the periphery. These issues have been discussed in the subsequent sections.

ALTERNATIVE APPROACHES FOR PA-MANAGEMENT

The Present Status

The foregoing analysis of various economic and ecological services derived from PA and people's interface with resources therein has highlighted some important issues that need special attention while exploring alternative approaches for its future management. The issues pertain to (*a*) habitat management which is conducive for the 'core' wildlife specie; (*b*) regeneration of vegetation that could sustain wildlife and also people's needs subject to the carrying capacity of the ecosystem; (*c*) sustainability of resource-use; (*d*) institutional mechanism for sharing of resources; and (*e*) effectiveness of the protection measures. In fact, all these issues are closely interrelated, hence should be seen in a comprehensive manner rather than as isolated entities while designing a management plan for the PA.

The forest department of Government of Gujarat has already worked out the second phase of the management plan, envisaging a special focus on regeneration of pastures, and significant expansion of the home range in order to sustain a population of about 500 lions (Singh and Pathak 2000). This of course, involves a detailed planning for resource management, people's livelihood and an implementation strategy. Given the need for regeneration of vegetation within and outside the PA and the critical role of soil-moisture and water thereof, we have tried to explore alternative land +(and) water use planning for the region. This is based on three basic principles—First, soil-water conservation assuming a top priority; Second, a more balanced allocation of water-resource within and outside the PA; and third, using a part of regenerated resources from the PA as incentives for reducing the pressure by checking haphazard and 'illegal' use of forest-resources on the one hand, and over-exploitation of ground water on the other.

We have identified alternative approaches for land-water use and the requisite resource sharing mechanism as well as other subsidies/support to compensate the loss of income in the short/medium term. Subsequently, implications of each of these alternatives have been mapped out for three sets of stakeholders, that is, farmers with irrigation, farmers without irrigation and landless, and the cattle herders (Maldharis). This, of course, should be treated as indicative planning for regeneration, conservation and sharing of resources in the region.

Two considerations are important while exploring alternative strategy for PA-management—First, Gir-ecology has a vast tract of degraded and highly degraded areas, hence vegetative regeneration is crucial. The second aspect pertains to the involvement of people in the periphery for effective conservation or protection of the ecology. Together these considerations bring to the core the issue of land regeneration and land use planning. Assessment of benefits and costs presented earlier may provide a basis for exploring an alternative strategy that could address these issues.

Evolving an alternative land use (and vegetation) plan, essentially requires setting up of a suitable mechanism for accessing (or sharing) these resources by the people whose livelihood needs are closely linked with the ecological status of the PA. At present, the existing legal structure does not recognise the stakes of the people especially, in the periphery. This, as we have seen in the previous section, is not in tandem with the ground realities that have come to be over a period of time. Non-recognition of people's rights thus, leads to a situation of a legal status quo where people continue to access the forest resources, but without the formal system taking note of that. What is worse is the existing legal framework of 'command and control' that gives way to the usual scenario where protectors themselves turn out as appropriators; in a non-transparent system such as this, the chances of being caught are fairly low. The formal perception therefore, treats this as 'stray incidences' of illegal activities rather than as a regular practice as a part of the people's livelihood base. What makes this worse is the fact that such extractions take place not only at the instance of those who 'need' them for their survival but, also by those who have economic-social-political power to get into faulty alliances without being caught for their illegal activities. While exploring alternatives for a more effective management in future, one should therefore try to look into the changing pattern of people's resource base within and outside the PA, people's livelihood requirements and the illegal alliances for extracting resources from the PA. This section tries to look into these issues with a view to identify alternative approaches for PA-management. The analysis has been carried out with the help of primary data collected from a sample of households in the region under study.

PEOPLE'S LIVELIHOOD BASE: THE PRESENT SCENARIO

Land, Irrigation and Livestock

The analysis in the previous section had indicated certain patterns in terms of population movements, changing land use pattern, and people's dependence on forest. We propose to take this analysis further by looking at the livelihood base among five major categories of households covered by the primary survey. The following observations depict important features of the livelihood patterns and implications for resource-management within the PA.

1. A large proportion of the farmers (that is about eighty-one per cent) with irrigation pursue live-stock as a supplementary source of income, whereas many of those without irrigation and the landless cannot afford to have livestock. The proportion of households having income from livestock is sixty-three per cent among farmers without irrigation and twenty-seven per cent among landless. Thus, livestock as a source of income is associated more closely with access to irrigation rather than land.
2. Landless households depend more on the prospects from agriculture by seeking employment on farm. This in turn, is influenced more by access to irrigation rather than by access to fodder and livestock. What is however, surprising is that nine per cent of the landless households reported collection of forest produce as the source of income (among others) and another thirteen per cent reported trading, which is also likely to be related to the various forest produce. Thus landlessness, as expected, is closely associated with dependence on forest.
3. Similarly a large proportion of households from traditional herder communities have to depend upon agriculture. This might imply that livestock alone is no more an adequate source of employment and/or income even among these communities. This observation is substantiated by relatively smaller size of livestock owned by these households. Declining access to CPLRs as well as fodder from the PA might be an important factor responsible for this phenomenon.

Together, these observations substantiate the earlier findings that households at the two ends of the spectrum in terms of access to land and irrigation tend to depend more on forest resources whereas, those with land and irrigation tend to access fodder for their livestock, the landless (excluding herders) may depend on forest mainly for NTFP, illegal extraction of timber, collection of fodder etc.

Status of Ground Water and Shift in Cropping Pattern

The decline in ground water table has been fairly widespread as reported in Table 7.9. In fact, those in the nearby villages recognised the problem more clearly than in the distant villages that are likely to be in the proximity of the command area of irrigation dams in the region. Obviously therefore, the extent of irrigation is higher in the distant villages (41 per cent) vis-à-vis the nearby villages (17 per cent). While we do not have details of cropping pattern in all the ninety-nine villages in the periphery, the observation about relatively better access to irrigation in the region suggests predominance of some of the more water intensive crops like sugarcane, cotton, castor, groundnut, wheat etc. Since

Table 7.9 Distribution of Households Reporting Changes in Water Table

(% of Households)

	Before Ten Years				*At Present*			
Water Table in Ft.	*<50*	*51–100*	*100–151*	*>151*	*51–100*	*101–300*	*301–500*	*>501*
Kendipur	97	3	–	–	73*	–	–	–
Madhupur	86	11	3	–	69	31	–	–
Govindpur	63	30	5	2	48	30	12	–
Dadli	85	10	3	2	65	32	–	3
All	**82**	**14**	**3**	**1**	**70**	**23**	**5**	**2**

Source: Primary Survey.
Note: * 24 per cent households reported <50 feet.

the nearby villages constitute a large proportion, that is, about sixty-eight per cent of the net-cropped area within the region, the pressure for using ground water is likely to be much more stronger than in the distant villages. If so, it may exert a negative impact on ground water resources within the PA. An important way out is to change the cropping pattern from more water intensive to less water intensive crops, especially in nearby villages.

We have tried to explore this option by obtaining perceptions of the sample farmers. While a large number of farmers agreed that the present cropping pattern is not conducive for ground water situation in the region, they were not willing to accept the proposed changes in cropping pattern. For, most of them felt that shifting to mango plantation in place of sugarcane or, groundnut instead of cotton will adversely affect their net returns. Nevertheless, a large number of farmers did recognise the fact that there has been a significant overuse of water and that, there is a scope for improving water-use efficiency. Table 7.10 depicts people's perceptions about the measures that could help checking ground water depletion in the region. It is interesting that farmers, though unwilling to change their cropping pattern, recognise alternative crop mix as an important mechanism for mitigating the problem of depletion of ground water. Incidentally, water-harvesting measures turned out as the most important aspect in this context.

Table 7.10 Farmers' Responses for Adoption of Measures to Improve Efficient Use of Water

Measures	*Kendipur*	*Madhupur*	*Govindpur*	*Dadly*	*All*
Changing Crop mix	35	22	24	27	108
Less Use of water	37	35	34	22	128
Use of Drip Irrigation	21	19	27	17	84
Control of High Power Electric Motor	17	19	28	20	84
Water Storage and Management	40	39	38	38	155
Well recharging	16	15	12	29	72

Source: Primary Survey.

A central point, which has emerged out of the above discussion, is that PA-management needs land plus water use planning where management of water (rather than land) should take a lead. However, before we discuss this issue in further details, we take a brief account of people's perceptions about the preferences for regenerating CPLRs and pastures outside as well as within the PA.

Use of CPLRs and Perceptions about their Regeneration

Table 7.11 presents information about use of Common Property Resources (CPRs) in the villages under study. It is observed that a large proportion, that is, sixty-two per cent of households access fodder/fuel from the village pastures, whereas forty-six per cent also access forest vidis. This is substantially high considering the fact that about twenty-two per cent of the households do not have milch animals and fourteen per cent of the households do not have any livestock. Moreover, it is likely that the actual use of forest vidis is under reported. This kind of extensive use of CPLRs and forest vidis, when seen in conjunction with limited number of livestock per household, reinforces the need for better management of these resources, especially when an alternative strategy for cropping pattern and land + water-use is being explored.

Table 7.11 Use of CPRs among Sample Households

Use of CPRs	*Kendipur*	*Madhupur*	*Govindpur*	*Dadly*	*All*
Gaucher	37	9	21	34	101
Forest vidi	19	16	7	32	74
Check dams/pond	6	12	10	13	41
Other colio	–	–	–	–	–
All	**43**	**39**	**40**	**40**	**162**

Source: Primary Survey.

Regeneration of Village Pastures

We have tried to obtain people's perceptions about their preferences for regeneration of CPLRs and also for reducing pressure on the PA. This is based on discussions with the households covered under the survey and also with the village communities. It was noted that whereas a majority of people in villages in the western zone preferred development of fodder alone, those in the eastern zone, felt that fodder + plantation might be a good strategy for regeneration of village pastures. This apparently suggests importance of livestock in the former vis-à-vis the latter, suggesting mutually reinforcing impact of irrigation and preference for availability of fodder in Gir-West. Those having relatively low access to irrigation as in the case of Gir-East, may like to access NTFPs from the CPLRs since their livelihood base is fairly low. Prima facie, this kind of preferences, viewed in the light of a proposed water-use planning, would imply increased allocation of water for fodder in Gir-West and for plantation in the case of Gir-East. It is encouraging that reducing the pressure on PA has turned out to be the most important reason for the increased development of the CPLRs/forest vidis. This is followed by an increased income from livestock, and then by drought relief.

Regeneration of Pastures on Forest Land

We tried to understand people's perceptions about improving the status of the forest and especially by reducing the pressure thereof. Among the various measures suggested, management of fodder collection and distribution, development of village pastures, providing alternative source of livelihood and protection were reported as important steps.

While these are some of the usual responses with respect to PA-management, what is important to note that a large proportion of people (that is sixty per cent) perceived economic plus ecological services from the PA as unsustainable given the present scenario of PA-management and people's pressure on resources thereof. Evidently, large farmers with irrigation and households from herders' community do not share this perception. This kind of divergence in perceptions indicates differential stakes across households with different socio-economic characteristics. It is however, encouraging that there is almost a consensus on the desirability of conservation measures for sustenance of the ecology, especially because the present management system is viewed as highly satisfactory. We have tried to ascertain what kind of support people would expect in case the restrictions on resource-use from the PA are further tightened in order to achieve better protection of the PA. The responses, in a way, reflect people's willingness to accept complete ban on accessing the PA-resources. The responses in their relative importance are availability of alternative employment and income, setting up of a system ensuring smooth supply of fodder and fuel, access to land (private as well as common), provision for alternative sources of fuel and development of agriculture (see Table 7.12).

Table 7.12 People's Expectations from Management of Gir-PA

Expectations	*Revenue Villages %*	*FSs %*	*Neses %*
Adequate employment + self-employment schemes	40	6	14
Access to fodder and fuel	22	38	22
Pasture development on degraded vidis	3	26	36
Measures of agricultural development	4	–	–
Allocation of land to landless	8	–	–
Settling down the issue of land lost of the PA	7	–	–
Distribution of gobar gas	12	–	–
Other amenities	4	30	28
All responses	**100**	**100**	**100**

Source: Primary Survey.

The above responses indicate two important aspects. First, people attach significant value to conservation of the PA and seek alternative arrangements for its effective management. And second, in the absence of an adequate livelihood base as well as development of CPLRs, they continue to depend on the PA despite their realisation that the use is unsustainable.

It is in this backdrop that we have tried to recapitulate main features of the status of various resources within the PA and the problems faced in management thereof. This has been presented in Table 7.13.

EXPLORING ALTERNATIVE MANAGEMENT SCENARIOS

The above description of resources, status and issues for management of Gir-PA, highlighted the critical importance of improving vegetation in a manner that can simultaneously address the twin

Table 7.13 Status and Issues Pertaining to PA-Resources: A Recapitulation

PA-Resources (Present Stock)	*Status*	*Issues*
Wild life (No.) Lion 300 to 320 Ungulates 36,555	• Increased number	• Possibility of exceeding the carrying capacity • Increased damages to crops/livestock • Need to develop coastal corridors • Problem of water for drinking
Timber (Teak + Non-teak): (No. in lakh) Teak 27,192 Non-teak 63,448	• Low density and slow regeneration after the cyclones in the mid-eighties	• Teak not suitable for the ecology • Appropriate mix of trees and browsing species so as to maintain medium density
NTFPs + Medicinal plants: Ambala, Harde, Jamun, Gum, Timru etc.	• Substantial diversity	• Need for regeneration and regulated management • Support livelihood among landless
Fodder: Estimated productivity: 3,000 kg/Ha or 1,500 kg/Ha Total production (T/Year) 4,11,423	• Large tracts of degraded and highly degraded areas • Balancing of vegetation for habitation of wildlife and livestock.	• Degradation due to: Natural conditions (drought) Increased pressure Ineffective protection • Declining size of CPLRs in peripheral villages due to: Loss of CPLRs to PA Encroachment Continued degradation • Limited intervention in terms of collection and distribution thereby leaving a large proportion of the fodder resources to be exploited by the people
Fuel wood: Total availability (T/Year) 1,87,500	• Substantial supply and heavy dependence by the people even through market channels	• Need to regulate supply through appropriate channels so that people can find some employment income without over-exploiting the resources • Promoting alternative sources of fuel through proper incentives
FYM: Production (T/Year) 78,488	• Large quantity of supply	• Selling out by Maldharis for very low revenue-realization • Selling of fertile soil by Maldharis • Scope for restricting the sales and retaining a part of it within PA • Scope for composting and value addition
River streams and seven dams: Total cultivable command area 39,010 (Ha)	• Major source of irrigation and income from agriculture as well as livestock outside the PA	• Limited measures for SWC • High level of soil-moisture erosion in the catchments, aggravating the problems of low regeneration of vegetation • Depletion of groundwater due to overuse by farmers • Imbalance between availability of water within and in periphery of the PA that is between the upstream and the downstream • Private control of groundwater and lopsided incentive structures against the measures for efficient use of water

(*Table 7.13 continued*)

(Table 7.13 continued)

PA-Resources (Present Stock)	*Status*	*Issues*
Livestock:13–14,000 within PA 95,000 in the periphery	• Declining livestock population though systematic estimates are not available	• Livestock population inside the PA is well within the carrying capacity • Infiltration of livestock from outside PA perhaps consisting of less productive livestock • Grazing vs stall feeding • Landless and small farmer without irrigation not being able to afford livestock
People: Population Periphery 1.8 lakh Maldharis in PA 35,000 FSs	• Shifting of population to the nearby villages partly due to declining NCA and CPLRs in distant villages	• Recognise the value of conservation but continue to exploit resources due to: Prevalence of the 'Tragedy of the Commons' Conflicts with the FD-staff Need + Greed of the people
The PA-Management: The next plan is under preparation	• Fairly good understanding of the problems and significant achievement in the first phase of conservation	• Problems of second generation, policy formulation • Absence of proper database on resources, stakeholders and dependents • Faulty alliance between people and protectors • Water scarcity as critical constraint for regeneration efforts • Budgetary constraints
Funders: National + global	• Support through eco-development project	• Inadequate consultation with stakeholders and managers
Researchers & global communities interested in bio-diversity: Various disciplines	• High level of awareness and large number of quality research	• Need for synthesis • Projection for fund raising and tourism • Absence of a policy dialogue

objectives of ecological regeneration and livelihood support. While the PA-management realises this critical need, there is perhaps inadequate recognition of people's stakes in the resources, especially fodder and fuel. As a result, it tends to maintain an artificial boundary between the pastures within and outside the PA while preparing a regeneration plan. The alternative approaches may therefore focus on conservation, allocation and utilisation of water resources within and outside the PA, that is, in the upstream and downstream of the watersheds in an integrated manner.

Prima facie, the objective function of a watershed-based planning in the region should be to maximise surplus resources to support the livelihood of the people in a sustainable manner. Here, 'surplus resources' is to be defined with respect to the requirement of an optimum size of the core specie, that is, lion and the ecological chain thereof. This kind of coexistence of wildlife and people (+ their livestock) is increasingly being accepted in the ongoing debate especially in the context of developing countries with sizeable population dependent on PAs (Parker 1983). This has given way to a wide range of alternative arrangements for PA-management by evolving collaborations between the statutory conservation bodies and private landowners (Biglake 2000).[5]

The recent literature on PA-management highlights a wide range of management approaches to deal with the issues of the functional relationship between parks and agriculture on the one hand, and competition between wildlife and livestock on the other. Also there has been an increasing emphasis on privatisation and/or people's participation in PA-management. What has however, remained relatively less explored is identification of an appropriate combination of public-private partnership where the former retains the overall responsibility and regulatory role of protection within which specific functions have to be carved out for private initiatives through development of markets as well as institutions. This is important because depending on regulation and restrictions alone may lead to conflicts, corruption and over-exploitation. And, too much of emphasis on people's participation may also result in neglect of some of the basic functions of conservation, habitat management, and long-term sustainability.

In what follows, we present alternative approaches for PA-management with specific focus on three sets of communities, that is, farmers with irrigation, resource-poor households, and Maldharis.

Farmers with Irrigation (and Livestock)

There has been a significant increase in irrigated area in the periphery of Gir. The present use of irrigation has two major problems. First, in the absence of proper SWC-measures in the upstream region, increasing irrigation in the downstream is often at the cost of its availability within the PA. And second, water-use is quite inefficient in terms of both—for irrigation as well as selection of crops. Thus, the issue of water availability centres around its allocation between PA and the periphery and across households within the periphery. Two alternatives can be explored with respect to the allocation of water arising from a watershed based planning wherein soil-water conservation within PA is considered to be the first step and the top priority. As an immediate impact of increased soil-water conservation measures, availability of water (surface + ground) resources might decline in the periphery. This could be compensated through two alternative approaches as described in Tables 7.14 to 7.16:

POLICY IMPLICATIONS AND RECOMMENDATIONS

While the present Management Plan has already recognised critical importance of regeneration of pastures within and outside the PA, interdependence between the two and its implications for mobilising people's commitment towards protection of the PA need to be clearly spelt out. As of now, the management plan (including Eco-Development Project) does not adequately focus on the fact that feasibility as well as effectiveness of regeneration of village pastures in the periphery is essentially dependent on efficacy of soil-water conservation in the upper catchments of watersheds, that is, inside the PA. Similarly, the plan does not seem to visualise that sorting out the issues pertaining to people's stakes might help significantly in mobilising cooperation or participation of people in protection of the PA. This is reflected by the fact that apart from fodder supply during droughts, people in the periphery do not have any direct claims on the PA-resources. This suggests a rather conservationist approach where people especially in the periphery, do not have any legal rights, hence their involvement in PA-management. But as argued earlier, not recognising people's stakes (if not the legal rights) leads to greater exploitation because of the tendency of overlooking illegal extraction not only by people, but also by the protectors. The next phase thus, needs to go beyond this strictly legal framework pertaining to people's stakes and involvement in PA-management. The basic

Table 7.14 Approaches for Alternative Water Use

	Alternative Water-Use Approaches	
Components	*I*	*II*
Crop-mix	Same crops with predominance of cotton, sugarcane, mango plantation, groundnut and wheat	Change to less water intensive crops like. Groundnut → Castor Sugarcane → plantation/groundnut Cotton → Castor Groundnut → Bajri + Fodder Wheat → Bajri + Jiru
Water-use	Improve the field channels to reduce waste, Adoption of modern methods of irrigation (like drip, sprinkler), Agronomic practice	Reduce number of watering Improved efficiency of irrigation Reduced demand for farm labour Reduced availability of crop residue Fencing to reduce crop-damage
Livestock	Reduced number and/or improved quality of livestock	Reduced quality/number of livestock
Increased fodder-supply from PA	Improved quality of livestock	Improved quality of livestock and reduced no. of livestock
Income and compensation/subsidy	More or less same from crops Subsidies on modern methods of irrigation Reduced income from livestock Employment on SWC	Same/reduced income from crops Subsidies on modern irrigation methods More or less same income with reduced number of livestock Compensation for the loss of income through supply of plantation material, compost from PA, fencing on farms, bio-gas/LPG etc. at a 'reasonable price'.
Cost to PA-management in the short-run	Increased cost of SWC-measures Increased subsidy on irri. methods Support for bio-gas/LPG etc.	SWC-measures Increased subsidies on irrigation methods Supply of fodder and other material at 'reasonable price'. Cost of compensation against net loss in income crop
Benefits to PA in the long-term	Moderate increase in vegetation Pressure for grazing may continue at moderate level Crop-damage may continue	Significant increase in vegetation, Pressure for growing may reduce Crop damage reduces due to fencing etc.

proposition is that—if people's stakes for subsistence needs are taken care of on a sustainable basis, rather than merely as drought relief measures, it can help reorganising the livelihood system and also improve compliance of protection measures by the people. The specific policy recommendations have been discussed as follows:

1. While regeneration of vegetation should primarily look into the requirements of wildlife, it should at least for next 10–15 years also provide a stable supply of fodder, fuel, and NTFP through a regulated management system adopting 'cut and carry' method. Improved vegetation and habitat management should thus ensure that incidence of attacks on crops and wildlife is reduced. Essentially, management of pastures within and outside the PA should be undertaken as an integrated activity with people's participation and reciprocal commitment for protection.

Table 7.15 Farmers with Unirrigated Small Holdings and Landless with Limited/No. of Livestock

	Alternative Water and Land-Use Approaches	
Components	*Fodder + Fuel*	*Plantation + Fodder + Fuel + NTFP*
Crop-mix on Private Land	*Same Crops*	*Shift to Plantation and/or Fodder*
Regeneration of village pastures	Fodder + fuel wood	Plantation + fodder + fuel wood
Livestock	Increased from the present size	Increased from the present size
Increased availability of water	SWC-measures on private and public land	SWC-measures + increased allocation of water from irrigation dams as well as other structures within the villages
Protection of CPLRs	Incentives through supply of fodder	Supply of fodder + fencing/watchman etc.
Employment & income	On SWC, forest vidis + CPLRs (for collection of grass and MTFP)	On SWC, forest vidis, CPLRs, NTFP collection and SWC-work
Sources of fuel	Fuel wood from CPLRs and forest through regulated markets	Fuel wood from regulated markets

Table 7.16 Maldharis within PA

	Alternative Locations for Settlement	
Components	*Outside PA*	*Within PA*
Livestock	*Reduced*	*Same*
Grazing practices	Grazing in specially developed plots	Seasonally regulated pattern + cut & carry method
Supply of fodder from PA	On regular basis through cut and carry method + droughts	During droughts
Outside livestock	Stopped completely	Only in limited number during normal years
FYM	Compost for the development of the fodder plot	FYM selling restricted to half
Availability of water	Irrigation for fodder plot + water for livestock	Water for livestock
Compensation	To ensure development of fodder plot + rights to access (not graze) fodder & fuel + cash compensation through term deposits and institutional backing + package of amenities	Incentives for improving quality of livestock without increasing their number

The latter should also involve defining carrying capacity of PA in terms of livestock population. This can be done if, access to fodder is ensured on a sustainable basis. Soil-water conservation measures should take a lead in the process of regeneration of ecology in Gir-PA.

2. While the management plan has recognised the need for developing irrigated fodder plots in the periphery, its actual implementation is found to be difficult. The experience of Eco-Development Project is also not so encouraging with respect to regeneration of pastures in the peripheral villages. It may therefore be important to explore alternative institutional mechanisms to help

developing pastures in the periphery and also organising fodder supply system by pooling resources from pastures both—within and in the periphery of the PA. A professional agency preferably, a non-profit making organisation, may be involved in managing these tasks.

3. A reliable fodder supply system may also help stabilising the livelihood base of Maldharis relocated outside the PA. A comprehensive plan for their effective rehabilitation on various land-based activities should be worked out. This is essential not only for checking further deterioration of their livelihood base, but also for mitigating the problem of 'illegal' re-entry of human as well as livestock population into the PA.

Given the large area covered by the PA, and also considering expansion of the home range for accommodating an increased population of about 500 lions, it is essential that the next stage of PA-management is much more interactive and inclusive rather than exclusive of the people living in thc periphery of the expanded boundary. Management of pastures within and outside the PA, holds the key to operationalisation of an approach such as this.

NOTES

1. Conservation Values: Largest compact tract of dry deciduous forests in the semi-arid western part of the country; Last home of 'Asiatic lions', Panthera leo persica, last surviving gene pool' in nature on earth; Rich biodiversity area supporting large number of species including several endangered species; Highest concentration of top carnivores—lions and leopards (over 500), and possibly the single largest population of marsh crocodiles in the country; Catchment area of seven major rivers which sustains economic prosperity of this drought prone region; Ecological security and environmental amelioration for the region, climate, water, salinity prevention and pollution absorption; Important biological research area with considerable scientific, educational, aesthetic and recreational values; Mother of cultural and religious evolution in Saurashtra.
2. Of course, both these are highly contested issues. While some ecologists as well as social activists perceive these people and their domestic livestock as parts of the ecology of Gir, there are however, some differences of opinion among the PA-managers. For, it is often argued that the people (especially, Maldharis that is cattle herders) living within the PA are recent settlers and, are largely responsible for degradation of floral bio-diversity as well as for forest fire. It is also felt that the domestic livestock providing easy prey for the lion, has led to distortion of the genetic characteristics of this core wildlife of species. In turn, this forces lions to go out of the PA in search of the domestic animals and thus results in increased damages to the property and people in the peripheral region.
3. The compensation package prepared in the early seventies, consisted of 3 hectares of cultivated land with proper treatment, access to CPLRs at 16 hectares per 100 livestock, a plot of 600 sq metres for housing and cash subsidy for construction cost, seed and agricultural equipments, and other amenities. The cost of the package works out to be about 2 to 3 lakh (at 1994–95 prices) per household. This is fairly small compared to the annual flow of benefits derived from the PA.
4. The primary survey consisted of sample households selected from five categories viz; large farmers with irrigation (LI); small farmers with irrigation (LI); farmers without irrigation (UI); landless (LL); and traditional herder communities (LH). The sample households were selected by adopting a stratified random sampling procedure. Table 7.2 presents distribution of the sample households in different categories.
5. (i) About 33 per cent of the forest area is degraded or highly degraded and above 44 per cent of the area with trees has a density of less than 0.2. (ii) Proportion of teak in the total timber tree has declined from 45 to 38 per cent. (iii) A large part of the PA belongs to the category of moderate to severe soil erosion. (iv) Water table in peripheral region has declined. (v) Fodder collection though, increased over time, is subject to very high year-to-year fluctuations.

REFERENCES

Biglake, R.C. 2000. 'Functional Relationships between Parks and Agriculture Areas in South Africa and Namibia', in H.H.T. Prins, J.G. Grootenhuis and T.T. Dolan (eds), *Conservation of Wildlife by Sustainable Use*, pp. 169–202. London: Kluwer Academic Publishers.

Chopra, K. 1998. 'Valuation of Bio-Diversity in Protected Area: Alternative Approaches and Case Study', Discussion Paper No. 3, Delhi: Institute of Economic Growth.

Choudhry, K. 2000. 'Development Dilemma: Resettlement of Gir-Maldharis', *Economic and Political Weekly*, 35(30): 2662–68.

Debnath, D., S. M. Hasan and P. Rishi. 2001. Assessment of People's Attitudes towards Park Resources and Park Management: A Study of Gir National Park, Bhopal, India: Indian Institute of Forest Management.

Parker, I.S.C. 1983. 'Conservation, Realism and the Future', in R.N. Owen-Smith (ed.), *Management of Large Mammals in African Conservation Areas*, pp. 281–90. Pretoria: Haum.

Perrings, C. 2000. 'Modelling Sustainable Ecological Economic Development', in K. Chopra, C. Perrings, U.R. Rao, K. Parikh and C.H. Hanumantha Rao (eds), *Ecological Economics for Sustainable Development*, pp. 45–72. Delhi: Indian Society for Ecological Economics.

Ramachandran, A.V., Mankodi, P.C. and Vachhrajani, K.D. 2001. 'Assessment of Water Quality in Major Streams and Rivers in Gir Protected Area and Siltation Rate in Four Reservoirs', Baroda, India: The M.S. University of Baroda.

Shah, Amita. 2003. 'Conservation of Gir Ecosystem; Assessment of Benefits and Costs under Alternative Management Sys-tems', Working Paper-WB 7, Mumbai: Environmental Economics and Resource Centre, Indira Gandhi Institute of Development Research.

Singh, H.S. 1997. 'Gir Lion-Present Scenario and Future Conservation Strategy'. Paper presented at the National Workshop on Regional Planning for Conservation and Development, Organised by Wild Life Institute of India and Government of Gujarat, 7–9 November 2000, Ahmedabad: Indian Institute of Development.

Singh, H.S. and Kamboj, R.D. 1995. 'Biodiversity Conservation Plan: A Management Plan for Gir National Park and Sanctuary', Forest Department, Gandhinagar: Government of Gujarat.

Singh, M. 1995. 'Issues for Integrated Management Plan for Gir'. Paper presented at the National Workshop on Regional Planning for Conservation and Development, Organised by Wild Life Institute of India and Government of Gujarat, 7–9 November 2000, Ahmedabad: Indian Institute of Development.

Singh, M. and J. Pathak. 2000. 'Ecodevelopment in Gir—Initiative and Future', Forest Department, Gandhinagar: Government of Gujarat.

Sinha, S.K. 1967. 'A Draft Scheme for Settlement of Maldharis in the Gir Forest Division', Unpublished Report, Government of Gujarat.

Sinha, S.P. 2001. 'Recommendations on the Man-Animal Conflict around Gir Protected Area', Submitted to Forest Department, Government of Gujarat.

Tewari, D.N. 1994. *Tropical Forest Produce*. Dehradun: International Book Distributors.

8

Environmental Implications of Integrated Pest Management: Farming of Paddy and Vegetables

Alka Singh[1], *Ranjit Kumar, P.K. Jain* and *A.K. Mangal*

INTRODUCTION

Pesticides coupled with other modern inputs have undoubtedly enabled the Indian farmers to achieve unparallelled increase in agricultural productivity over the last five decades. Evidences indicate that pests cause twenty-five per cent of the losses in paddy, fifty per cent in cotton, thirty per cent in pulses and twenty per cent in sugarcane (Dhaliwal and Arora 1996). Until recently, chemical pesticides were increasingly relied upon to limit the production losses, but now there is a growing concern about the health hazards associated with pesticide usage. Environmental contamination from the use of pesticides ranges from water, air and soil pollution to alteration of the ecosystem resulting in detrimental effects on non-target organisms. Evidences of pesticide threats to human health and the trade-off between health and economic effects have been documented in several studies in the past (Rola and Pingali 1993; Antle and Pingali 1994). Although the pesticide consumption in India is low (0.57 kg/ha) as compared to countries like Japan (12 kg/ha), Taiwan (17 kg/ha) and West Germany (3 kg/ha), pesticide residues in food products in India, especially vegetables, are the highest in the world. This is mainly due to unregulated use of pesticides. India accounts for one-third of the total pesticide poisoning cases in the world (Puri 1998).

Experience has shown that such methods of plant protection have proven to be increasingly unsustainable and cost-ineffective due to development of pest resistance, rising pesticide costs, pesticide-induced outbreaks of pests and the negative effects of pesticide use on human health and the environment. The synthetic organic insecticides widely used in agriculture are general biocides having innate ability to cause injury to all living organisms as well as to the quality of environment. The presence of residues of these pesticides in food commodities and other components of the environment has proved toxic to human beings, domestic animals, birds, fish and other non-targeted fauna of the agro-ecosystem.

Despite the fact that the consequences of injudicious use of pesticides in Asia are well documented, crop protection continues to be dominated by the dependence on chemicals. The practice of calendar spraying is common among Asian farmers and pesticide subsidies remain a major aspect of plant protection policies in many countries (Gopalan 1998). Though negative externalities cannot be eliminated altogether, their intensity can be minimised through development, dissemination and promotion of environment friendly technologies such as biopesticides and bio-agents as well as better agro-economic practices commonly known as Integrated Pest Management (IPM) rather than solely relying on chemical pesticides.

Therefore, a major challenge to plant protection specialists worldwide is the ability to integrate effectively different pest control measures, which must be selected on the basis of their cost effectiveness, sustainability and eco-friendly nature. In India, the paradigm shift in pest management policy in favor of IPM during the nineties has helped a lot in reducing pesticide consumption in the country. A number of direct and indirect regulatory and policy measures were taken, including import restriction on hazardous chemicals used in agriculture, reduction of subsidies for chemical pesticides, promotion of biopesticides, development of IPM packages, training of extension workers and farmers in IPM by establishing Farmers' Field Schools (FFS) and through IPM demonstrations. Yet, there is an urgent need for newer incentives to encourage farmers to reduce pesticide use and ensure safe food supply.

Despite these concerns, little empirical work has been done to estimate the aggregate environmental benefits of IPM, even in developed countries. Such estimation is difficult because assessing the physical or biological effects of pesticide use on different components of the environment is a cumbersome and sometimes uncertain process. To explore these issues further, the aim of this chapter is to quantify the effects of IPM and the risks posed by pesticides to different categories of environment including human beings and to assess the farmers' willingness to pay for safer options.[2]

DATABASE AND METHODOLOGY

Sampling Framework

The State Department of Agriculture, Directorate of Plant Protection, Quarantine and Storage, Ministry of Agriculture, Government of India, State Agricultural Universities and other agencies like Uttar Pradesh Diversified Agricultural Project (UPDASP) have been successfully conducting Farmer's Field Schools (FFS) to sensitise and train the farmers on IPM in several crops. The present study was conducted in Karnal and Kaithal districts of Haryana for paddy IPM and Ghaziabad district of Western Uttar Pradesh for vegetable IPM. The area of study represents one of the most progressive regions in terms of productivity and input usage and is also characterised by highly commercialised agriculture. Paddy-wheat rotation is most common, and cropping intensity has recently increased with the introduction of summer paddy in some parts of Karnal and Kaithal districts. The mono-cropping and high cropping intensity have accentuated the pest problems, depleted ground water resources and worsened soil quality. Tomato and cabbage are commercially grown in Ghaziabad district and consume high quantity of insecticides and fungicides. Farmers prefer to cultivate vegetables commercially in the area under study because of its vicinity to Delhi and possessing a good supporting infrastructure which allows for quick and easy transportation to Delhi and nearby markets, and hence enables them to generate high profits.

For selection of sample farmers, two top-ranking blocks in terms of area under paddy (Karnal and Kaithal districts) and vegetables (Ghaziabad) were chosen, and from each selected block, two villages were selected, one where Farmers' Field School on IPM had already been conducted and the other where no such programme was ever organised. Finally, ten farmers were chosen from each village to make the total sample size of 160. Hence, the study is based on primary data, collected for the year 2003–04 from a sample of forty IPM trained farmers (received formal training regarding IPM in FFS) and forty NIPM trained farmers (not attended IPM training in FFS) growing paddy in Haryana and forty IPM and forty NIPM farmers growing vegetables (tomato and cabbage only) in western Uttar Pradesh . The primary data on the socio-economic characteristics of sample farmers, cultivation practices with particular emphasis on plant protection, adoption of IPM practices and farmers' willingness to pay for safer pesticides in crop production was collected through personal interview method. Besides, secondary data related to toxicity level for different environmental categories (human beings, animals, beneficial insects, birds, and aquatic species) were also collected from published sources for each pesticide used by sample farmers in paddy and vegetables cultivation.

ANALYTICAL APPROACH

Pesticide risk to the environment is often related to the amount of active ingredient applied or expenditure incurred on pesticides. However, both these measures are not the best indicators of risk because pesticides differ with respect to their toxicity, mobility and persistence and thus pose different levels of risk to different components of the environment. Analysis of the environmental benefits of reduced pesticide use must examine the toxicity, mobility and persistence characteristics of the pesticides being used. When farmers reduce the total quantity of pesticidal active ingredient applied but simultaneously substitute highly toxic, mobile and persistent chemicals for relatively lower quantities, it is difficult to argue that environment has gained (Mullen et al. 1997).

Most of the studies have focused on the human health effects of pesticides (Rola and Pingali 1993) and little attention has been paid to other environmental categories. A few studies have suggested possible approaches for measuring the aggregate environmental costs of pesticides and the benefits of IPM (Kovach et al. 1992; Highly and Wintersteen 1992; Owens et al. 1997; Mullen et al.1997; Cuyno et al. 2001). These studies considered the effects of pesticides on different components of environment namely surface water, ground water, aquatic organisms, birds, mammals, beneficial insects and humans (acute and chronic toxicity).

The present study identifies five environmental categories which include human health (acute and chronic effects), animals, birds, aquatic species and beneficial insects. Active ingredient of each pesticide was assigned three levels of risk, that is, high, moderate and low for each of the five environmental categories. These risk levels were rated on a scale from one to five with one having a minimal impact on environment or low toxicity and five considered to be highly toxic or having a major negative effect on the environment. Information regarding hazard rating as well as toxicity database for each pesticide was obtained from databases such as EXTOXNET, Pesticide Manual and the previous studies. Both toxicity and exposure potential criteria were considered in arriving at the assigned risk for each pesticide used in paddy and vegetable production in the area of study. A brief summary of these criteria is presented in Appendix I. These criteria make use of the current state of knowledge with respect to data that indicate pesticide risk to individual environment category. Recognising the limitations of available data and information, the criteria and hazard categories (as given in Appendix I) make the *ceteris paribus* assumption that highly toxic and persistent chemicals pose a greater risk to different environment categories than pesticides that are less toxic and deteriorate quickly. The detailed description of different criteria and hazard ratings are as follows:

ACTIVE AND CHRONIC HUMAN HEALTH CRITERIA

The assignment of acute human health risk level is based on the WHO criteria or EPA criteria. As these criteria require all pesticides to be labelled with 'Danger', 'Warning', or 'Caution' tag based on toxicity, every pesticide has a corresponding signal word which can be correlated to a high, moderate or low rating. Chronic toxicity of a specific pesticide is calculated as the average of the ratings from various long term laboratory tests conducted on small mammals. These tests are designed to determine potential reproductive effects, teratogenic effects, mutagenic effects and oncogenic effects. Criteria for assigning chronic health risk levels are based on the results of the above mentioned tests.

AQUATIC SPECIES CRITERIA

A given pesticide does not affect all aquatic species to the same degree. In this study, the highest level of risk a pesticide poses to any aquatic species is the risk level assigned to that pesticide.

Birds: Assignment of risk to pesticide with respect to avians is based on the highest level of risk the pesticide poses to any species within the category, high if LD_{50} is <100 ppm, moderate if LD_{50} is between 50–500 ppm, and low if LD_{50} is >500 ppm. There are some pesticides for which toxicological tests have not been conducted. A pesticide is assumed to pose a moderate level of risk to any category where gaps in data exist.

Animals: Pesticide toxicity towards mammalian farm animals was assumed to be same as that for human beings.

Beneficial Insects: The assignment of 'beneficial insect risk' to an active ingredient is based on insect toxicity ratings and is characterised as high, medium or low. However, in the case of some pesticides, toxicity of pesticidal compounds to beneficial insects has not been formally assessed. Hence, in such cases a low level of risk was assigned to that pesticide.

After the data on individuals' risk level associated with each environment category was collected, pesticides were grouped by classes (insecticide, fungicide and herbicide) and scores assigned to each pesticide active ingredient were combined with usage data to arrive at an overall eco-rating for each pesticide. An overall eco-rating score was then calculated separately for IPM and NIPM categories of farmers. The difference between the two represents the amount of risk avoided due to adoption of IPM practices. The formula for eco-rating can be expressed as

$$ES_{ij} = (IS_j) \times (AI_i) \times (Rate_i)$$

Where

ES_{ij} = eco-rating score for active ingredient i and environmental category j
IS_j = pesticide risk score for environmental category j
AI_i = per cent active ingredient in the formulation
$Rate_i$ = application rate per hectare of ith active ingredient

The present analysis covers only a single year and pesticide use may vary considerably depending on weather conditions; this holds true for both IPM adopters as well as non-IPM adopters.

To examine the farmers' preferences for safer pesticides, the values of willingness to pay (WTP) were obtained through contingent valuation (CV) method using a survey of forty farmers practising vegetable IPM and forty farmers practising paddy IPM in the surveyed area. The respondents were asked to provide WTP values for different formulations of their favorite pesticides. Five formulations were offered, one that avoids risk to each of the five environmental categories. The farmers were asked to rank the five categories in the order of their preference. They were then asked about the minimum amount that they were willing to pay per kg of active ingredient for their most preferred category and their least preferred category. The other categories were valued between the upper and lower bounds of these values. The respondents were given the chance of rearranging their ranks until they were completely satisfied that the rankings and WTP values were representative of their preferences.

RESULTS AND DISCUSSION

Table 8.1 shows that pesticide use on paddy sample farms was estimated to be 2.07 and 2.42 kg active ingredient per ha respectively, on IPM adopted and non-IPM adopted farms. On an average, paddy crop was treated four times with pesticide, one application each of herbicide and fungicide and two applications of insecticide. Among major insecticides used in paddy farms included Endosulfan, Monocrotophos, Chlorpyriphos, Phorate, Diclorvos and Pyrethroids such as lamdacyhalothrim etc. All these insecticides are classified as highly hazardous to moderately hazardous (Category I and II) according to the WHO classification (Table 8.2). In case of vegetables, on an average, pesticide consumption in tomato farms was found to be much higher than that in cabbage farms on both types of farms. On an average, tomato and cabbage crops were treated four times with pesticides

Table 8.1 Pesticide Use Pattern on Sample Farms

			Average No. of Sprays		*Quantity (Active Ingredient/per ha)*	
State	*Districts*	*Crops*	*IPM*	*NIPM*	*IPM*	*NIPM*
Haryana	Karnal and Kaithal	Paddy	4	4	2.07	2.42
Western Uttar Pradesh	Ghaziabad	Tomato	4	9	2.02	3.71
Western Uttar Pradesh	Ghaziabad	Cabbage	4	7	1.61	2.63

Table 8.2 Types of Chemical Pesticide used in Paddy and Vegetable Cultivation

Pesticides	*Safety Hazard Level**
Organochlorines	
Endosulfan 35% EC	Class II
Organophosphates	
Monocrotophos 35% SL	Class I b
Dichlorvos (DDVP) 76% EC	Class I b
Phorate 10% G	Class I a
Quinalphos 25% EC	Class II
Dimethoate 30% EC	Class II
Triazophos 40% EC	Class I b
Profenofos 50% EC	Class II
Chlorpyrifos 20% EC	Class II
Methyl parathion 50% EC	Class I a
Synthetic pyrethroids	
Lambda-cyhalothrim 5% EC	Class II
Alpha-cypermethrin 10% EC	Class II
Cypermethirin 25% EC	Class II
Others	
Cartap hydrochloride 50% SP	Class II
Butachlor 50% EC	Class U
Streptocycline	Class U
Anilofos 30% EC	Class II
Carbandazim 25% DS	Class U
Pretilachlor	Class U
Propiconazole 25% EC	Class II
Tricyclazole 75% WP	Class II
Imidacloprid	Class II
Fipronil	Class II
Copper oxichloride 50% WP	Class III
Mancozeb 75% WP	Class U
Hexaconazole 5% SC	Class U

Notes: * WHO classification: Class I a – Extremely Hazardous, Class I b – Highly Hazardous, Class II – Moderately Hazardous, Class III – Slightly Hazardous, Class U – Unlikely to present acute hazard in normal use.
EC: Emulsifiable concentrates, SC: Suspension concentrates, WP: Wettable powders, SL: Soluble concentrates, G: Granules, SP: Soluble powders, DS: Powder for dry seed

on IPM farms, whereas, the frequency increased to nine and seven times respectively on non-IPM farms. Among pesticides, only insecticides and fungicides were used and that too, almost in equal proportions. The major insecticides used are cypermethrin, chlorpyriphos, monocrotophos and quinalophos whereas, mancozeb and copper oxichloride are the major fungicides used in tomato and cabbage cultivation in the area of study.

A majority of respondents opined that the use of pesticides reduced the pest population, and thereby increased crop yield. However, many of the non-IPM adopters were of the opinion that the prescribed doses in the package of practices were not effective in controlling insects and diseases. On the other hand, some IPM farmers were also found to spray either no chemical pesticides or used lower concentrations. The source of pesticide supply was mainly private pesticide dealers and the distance of the place from where pesticides were bought was within one to three km of the area of study. Since pesticides have adverse effects on non-target organisms, many of the respondents reported a significant decline in population of beneficial organisms, especially birds, earthworms etc. The respondents in the area under study were concerned about increasing crop losses due to insects and diseases. Some of the major pests and diseases causing economic loss are shown in Table 8.3.

Table 8.3 Major Insects and Diseases

Crops	*Insects*	*Diseases*
Tomato	Fruit borer	Tomato leaf curl Virus
	White fly	Damping off
	Aphids	Stem and Fruit cracker
Cabbage	Cabbage borer	Club rot
	Diamond back moth	Ring rot
	Aphids	Cabbage yellow
	Termite	Soft rot
		Root rot
Paddy	Stem borer	Blast
	Leaf folder	False smut
	White-backed plant hopper	Sheath blight
	Rice Hispa	Zinc deficiency
	Gundhi bug	

In paddy crop, insects such as yellow stem borer and leaf folders are causing major damage whereas among diseases, blast and false smut are gaining prominence in the area of study. Indiscriminate usage of pesticide coupled with mono-cropping and high fertiliser usage has further compounded the issue. In vegetables such as tomato, diseases caused by fruit borer, white fly and tomato leaf curl virus have assumed worrisome proportions whereas cabbage borer and diamond back moth are the major insects in the cabbage crop in the area.

WILLINGNESS TO PAY

Table 8.4 shows that in case of paddy cultivation, forty-one percent of the sample farmers ranked first the pesticides considered safe for human beings. They were willing to pay up to thirty per cent price premium for those formulations that were certified to have either zero or the least harmful effects on human health. The average willingness to pay for these pesticides was estimated as ten per cent over the present value. However, more than fifty per cent of the respondents rated pesticides safer for beneficial insects as the most preferred one. For pesticides with this characteristic, they were ready to pay a maximum of thirty-three per cent premium on the current price.

Similarly, in case of vegetable cultivation in western Uttar Pradesh, more than two-third sample farmers expressed their first choice towards safer pesticides formulations for beneficial insects. For that, farmers were ready to pay an on an average a thirty per cent premium on the current price.

Table 8.4 Rank and Willingness to Pay for Risk Avoidance in Each Environment Category

Environmental Category	*% of Sample Opting First Number Choice*	*Average WTP (%)*	*Average WTP (%) Opting First Number Choice*	*Max WTP (%)*
Paddy				
Human	41.38	10.00	17.91	30.00
Animal	3.45	2.24	0.00	12.00
Birds	0.00	0.17	0.00	5.00
Beneficial insect	55.17	13.96	20.00	33.00
Aquatic species	0.00	0.00	0.00	0.00
Vegetables				
Human	30.00	16.38	28.33	40.00
Animal	0.00	3.75	0.00	14.00
Birds	2.50	2.38	0.00	12.00
Beneficial insect	67.50	21.75	29.07	38.00
Aquatic species	0.00	0.00	0.00	0.00

Those who had ranked human health as the first category were found to be willing to pay a maximum of forty per cent. Aquatic species, land animals and birds are the least preferred environment category with respect to the willingness to pay in both the crop regimes. These results confirm that a market exists for safer or environment friendly pesticides in the area of study.

ENVIRONMENTAL IMPACT OF IPM

The risk scores for most commonly used pesticides in the region under study for each environment category, that is, human beings, animals, birds, aquatic species and beneficial insects are presented in Table 8.5. As mentioned earlier, the information regarding hazard rating as well as toxicity database for each pesticide was obtained from databases such as EXTOXNET, Pesticide Manual and the previous studies. Higher values indicate higher risk associated with respective pesticide.

The scores assigned to each pesticide active ingredient were combined with usage data to arrive at an overall ecological rating for each pesticide. These estimates are presented in Table 8.6 by each category of pesticide. These results show higher aggregate eco-ratings for each environment category on NIPM farms as compared to IPM farms demonstrating higher environmental concerns. The estimates also show that eco-ratings were reduced from twenty to thirty per cent as a result of adoption of IPM practices by IPM adopters in each paddy growing season. Similar results were also reported in western Uttar Pradesh where IPM is being practiced in vegetable cultivation. The estimates show that eco-ratings were reduced by up to thirty-nine to forty-six per cent as a result of adoption of IPM practices in tomato and vegetable cultivation in each season (Tables 8.6 and 8.7). These reductions represent the per cent of pesticide risk avoided due to reduced pesticide application as well as judicious selection of environment friendly pesticides on IPM farms in crop cultivation in the area under study.

Table 8.5 Risk Score for Paddy and Vegetables Pesticide Applied in the Study Area

	Risk Scores				
Name of Pesticides	*Human*	*Animals*	*Birds*	*Aquatic*	*Beneficial*
Cartap	3	3	1	3	1
Phorate	5	5	3	5	3
Endosulfan	5	5	3	5	1
Monocrotophos	5	5	5	3	5
Diclorvos	5	5	5	3	3
Cholorpyriphos	3	3	5	5	5
Lindane	5	5	5	5	5
Lambdacyhalothrim	3	3	1	5	5
Carbandazim	3	3	5	5	1
Propiconazole	3	3	1	3	1
Tricyclazole	3	3	3	1	1
Butaclore	1	1	1	5	3
Anilofos	3	3	1	3	3
Pretilachlor	1	1	1	3	1
Quinalphos	3	3	3	3	3
Cypermethrin	3	3	1	1	3
Dimethoate	3	3	3	3	3
Alpha-cypermethrin	3	3	1	1	1
Copper oxichloride	3	3	1	3	1
Mancozeb	1	1	1	3	1
Hexaconazole	1	1	1	3	3

Table 8.6 Environmental Risk Associated with Pesticide Use in Paddy by NFFS and FFS Farmers

		Eco-ratings		
Category	*Types of Pesticide*	*NIPM Farmers*	*IPM Farmers*	*Aggregate % Risk Avoided for Each Environment Category*
Human beings	Herbicide	42.89	40.64	30.08
	Insecticide	214.90	138.32	
	Fungicide	6.86	6.06	
Animals	Herbicide	42.89	40.64	30.08
	Insecticide	214.90	138.32	
	Fungicide	6.86	6.06	
Birds	Herbicide	39.35	35.62	26.10
	Insecticide	143.11	100.46	
	Fungicide	9.44	5.74	
Aquatic species	Herbicide	187.62	171.01	20.72
	Insecticide	208.54	144.00	
	Fungicide	9.73	6.78	
Beneficial insects	Herbicide	112.47	105.78	19.51
	Insecticide	131.75	90.62	
	Fungicide	2.29	2.02	

Table 8.7 Environmental Risk Associated with Pesticide Use in Vegetables by Non-IPM and IPM Farmers

Category	*Types of Pesticide*	*Eco-ratings*		*Aggregate % Risk Avoided to Each Environment Category*
		Non-IPM	*IPM*	
Human beings	Insecticide	238.38	137.10	39.16
	Fungicide	96.36	66.57	
Animals	Insecticide	238.38	137.10	39.16
	Fungicide	96.36	66.57	
Birds	Insecticide	196.70	117.52	40.64
	Fungicide	96.67	56.62	
Aquatic species	Insecticide	219.69	133.58	40.35
	Fungicide	191.21	111.52	
Beneficial insects	Insecticide	230.83	121.67	46.13
	Fungicide	57.40	33.59	

Note: No herbicide use was reported by sample farmers in vegetable cultivation.

CONCLUSIONS

The study estimated the farmers' willingness to pay for pesticides hazard reduction for five environmental categories. These results show that a market exists for environment friendly pesticides in the area of study and farmers are willing to pay a price premium. Databases were also compiled for assessing risk levels to eight environment categories, for more than twenty pesticides used in the area. These risk values may be used by researchers and farmers while recommending or using different pesticides in the field. It has also been estimated that current use of IPM technology has the potential of avoiding pesticide risk hazards to different environment categories by twenty to thirty per cent in paddy cultivation and thirty-nine to forty-six per cent in vegetable cultivation. Hence, developing the farmer's capacity by imparting information, knowledge and skill through in-depth and intensive training as well as awareness programmes about pesticide hazards would go a long way in enhancing environmental benefits due to IPM adoption.

NOTES

1. Corresponding author.
2. The data used in the present study was drawn for AP CESS project, 'Pesticide Use and Sustainability of Agriculture—Emerging Issues and Policy Options'.

REFERENCES

Antle, J.M. and P.L. Pingali. 1994. Pesticides, Productivity, and Farmer Health: A Filipino Case Study', *American Journal of Agricultural* Economics, 76: 418–30.

Birthal, P.S., O.P. Sharma and Sant Kumar. 2000. 'Economics of Integrated Pest Management: Evidences and Issues', *Indian Journal of Agricultural Economics*, 55(4): 644–58.

Cuyno, Leach C.M., George W. Norton and Agenes Rola. 2001. 'Economic Analysis of Environmental Benefits of Integrated Pest Management: A Philippine Case Study', *Agricultural Economics*, 25(2001): 227–33.

Dhaliwal, G.S. and R. Arora. 1996. *Principles of Insect Pest Management Concept and Approach*. New Delhi, India: Kalyani Publishers.

EXTOXNET, [Multi-university computer network providing toxicity-related electronic data; data base available on Oregon state server.] online available at http://extoxnet.orst.edu/pips/

Gandhi, P. Vasant and N.T. Patel. 1997. 'Pesticide and the Environment: A Comparative Study of Farmer Awareness and Behaviour in Andhra Pradesh, Panjab and Gujarat', *Indian Journal of Agricultural Economics*, 52(3): 519–28.

Gopalan, H.N.B. 1998. IPM in the Tropics-Current Status and Future Prospects. UNEP.

Higley, L.G. and W.K. Wintersteen. 1992. 'A Novel Approach to Environmental Risk Assessment of Pesticides as a Basis for Incorporating Environmental Costs into Economic Injury Level', *American Entomologist*, 38: 34–39.

Kovach, J., C. Petzoldt, J. Degni, and J. Tette. 1992. 'A Method to Measure the Environmental Impact of Pesticides', in *New York's Food and Life Science Bulletin*, No. 139, New York State Agricultural Experiment Station, Ithaca: Cornell University.

Mullen, Jeffery D., George W. Norton and Dixie W. Reaves. 1997. 'Economic Analysis of Environmental Benefits of Integrated Pest Management', *Journal of Agricultural and Applied Economics*, 29(2): 243–53.

Owens, N.N., S.M. Swinton and E.O. Van Ravenswaay. 1997. 'Farmer Demand for Safer Corn Herbicides: Survey Methods and Descriptive Results'. Michigan Agricultural Experiment Station, Michigan State University. Research Report 547, East Lansing, MI.

Passalu, I.C., B. Mishra, N.V. Krishnaiah and Gururaj Katti. 2003. Integrated Pest Management in Indian Agriculture. Proceeding No. 11, National Centre for Agricultural Economics and Policy Research (NCAP), New Delhi: National Centre for Integrated Pest Management (NCIPM).

Puri, S.N. 1998. 'Present Status of Integrated Pest Management in India' Paper Presented at a Seminar on IPM, Bangkok: Asian Productivity Organization at Thailand Productivity Institute.

Rola, A.C. and P.L. Pingali. 1993. 'Pesticides, Rice Productivity, and Farmers' Health: An Economic Assessment', The Philippines and World Resources Institute, Washington DC: International Rice Research Institute.

Tomlin, Clive. *Pesticide Manual*. Tenth edn. Farnhan: Crop Protection Publications.

APPENDIX I Pesticide Impact Scoring System

		Score		
Environmental Categories	*Indicators*	*High Risk = 5*	*Moderate Risk = 3*	*Low Risk = 1*
Human Health				
1. Toxicity				
Acute Toxicity	Pesticide Class (WHO Criteria) Signal Word (EPA Criteria)	Ia; Ib Danger/Poison	II Warning	III Caution
Chronic toxicity	Weight of Evidence of chronic effects	Conclusive Evidence	Probable Evidence	Inconclusive Evidence
Aquatic Species				
1. Toxicity	95 hr LC50 (fish) mg/L Fish/ other aquatic Species Toxicity	<1 ppm	1–10 ppm	> 10 ppm
2. Exposure	Runoff Potential Score	High	Moderate	Low
Beneficial Insects				
1. Toxicity	Insect Toxicity Ratings	Extreme/High	Moderate	Low
2. Exposure	Plant Surface Residue Half life	>4 weeks	2–4 weeks	1–2 weeks
Mammalian Farm Animals	For animals and human beings, same level of risk has been assumed			
Birds				
1. Toxicity	Birds Toxicity Ratings	High/Extreme	Moderate	Low
	8 days LC 50	1–100 ppm	100–1,000 ppm	> 1,000 ppm

Section 3

Valuation of Ecosystem

9

Social Perceptions and Valuation of Urban Wetlands in Kolkata

Sumana Bandyopadhyay[1], *K. Narayanan* and *A. Ramanathan*

Abstract: The process of development has affected many natural and sensitive environmental entities—the focus of this chapter is the wetland ecosystem. Of wetlands across the world, those located near urban regions face more critical problems. Urban wetlands need careful prioritisation as the process of urbanisation generates the greatest volume of wastes and pollutants as also the most large-scale conversion of land uses—reclaiming wetlands for urbanisation is the most common concern at the global level. This study focuses on the wetlands of Kolkata, one of the largest metropolitan regions of the world. An intricate water-based ecosystem consisting of a network of distributaries, canals, and natural waterbodies affected by tidal influences constitute a complex and unique attribute of its deltaic location. An attempt has been made to explore the nature of preferences of people (who are either direct or indirect stakeholders in the system) and the attributes associated with these preferences. The contingent valuation method has been used for this study to assess people's preferences. It is concluded that user group preferences differ across rural and urban respondents and these may be explained by different socio-economic and environmental variables.

INTRODUCTION

Planning for urban environments has been faced with innumerable challenges through time. One of the most dynamic of these challenges pertains to the issue of conservation of natural ecosystems within and adjacent to growing metropolitan regions. Sustainable urban planning is not only about providing adequate amenities and infrastructure and reducing pollution levels, but is also primarily concerned with protection of sensitive ecosystems. This chapter begins by placing the unique nature of the Kolkata wetlands in proper perspective and subsequently proceeds to identify the conservation attitudes and preferences of rural and urban populations towards the wetlands of Kolkata. In examining social perceptions of the use of wetlands, the contingent valuation method has been used in a broad and inclusive sense for the analysis of willingness to pay in favour of preservation of wetlands.

APPROACH AND RATIONALE

Wetlands and their multifarious functions have been extensively researched upon, from the point of view of hydrological, botanical, zoological, micribiological, limnological and biodiversity related parameters. These studies, almost unanimously lead to one question—about the importance of conserving these ecosystems. This necessitates weighing of losses and gains related to retention or conversion of those lands, which gives way to an interesting dimension of research—the benefit-cost approach. This approach is imminent in determining the future of the sensitive natural ecosystems of the world. It is also necessary to mention that the twin processes of industrialisation and urbanisation have created, by far, the largest negative impact upon natural ecosystems. Urban wetlands therefore may be considered to be in far greater danger of degradation than remotely located ones.

A large volume of literature has been compiled by environmental economists, as a series of methodological variants with supporting empirical examples have been evolved to study the social processes of environmental valuation. Institutional research like that of the World Bank studies on lakes and reservoirs (1995) and the studies on resource valuation under the European Union Environmental Action Plan has incorporated natural resource evaluation into the mainstream of institutional policy framework. Individual studies conducted on the basis of several case studies across the world also lends support to the fact that valuation may be used a method of understanding the values for natural ecosystems, hitherto treated as a 'free good' for generations. As far as the use of valuation techniques are concerned, literature abounds. Studies that have been referred to, include some that have used simple benefit-cost ratios (Adger et al. 2001) and contingent valuation (Brouwer et al. 2001, Hanley and Spash 1998). Valuation as a methodology has been developed in the context of Indian studies by Murty et al. (1999), Chopra (2002) and Parikh (2003). Many studies have been conducted upon urban water-based ecosystems in India—such as an assessment of the tangible and intangible benefits and costs of cleaning up of the Ganges by Markandya and Murty (2000), the values of ecological functions of the Yamuna floodplains by Prasad and Babu (2003), the Bhoj wetlands of Bhopal by Verma et al. (2003), the functions and values of the East Kolkata Waste Recycling Region by Chattopadhyay (2003) and that of water resource management of the Nainital lake and its watershed by Singh et al. (2003), to name a few.

All these studies have gone into estimating the values of benefits of the ecosystem. The conclusion of all such important studies is that preservation as a long-term benefit should be prioritised.

METHODOLOGY

Environmental phenomena have absolute and non-negotiable values, thus making it difficult to attempt any direct estimation of non-use values. Studies on wetlands from the point of view of the benefit-cost analysis approach have been conducted with different methodologies. By and large, the studies may be classified according to methods of valuation of benefits and functions of wetlands, namely, revealed preference and stated preference methods. Studies on tourism and recreation benefits largely use the travel cost approach; those on variation of real estate prices have taken up the hedonic pricing approach; while contingent valuation has been used for developing the people's preferences arguments. It has also been generally concluded that such studies can adequately identify the willingness to pay in a broad and inclusive sense. Contingent Valuation represents the general techniques or procedures used to elicit focus on stakeholders. It is a direct method for estimating values based on behavioural models for measuring environmental benefits, implying that the demand for an environmental change is measured by means of a constructed or hypothetical market.

Assessing people's preference, by eliciting responses on their willingness to pay for conservation of a certain environmental good has been relied upon in this study although environmental preference is a value-loaded concept and goes much beyond monetary measures. In this case, the CVM survey was designed after referring to the NOAA guidelines (1993) and studies by Murty et al. (1999) and Verma et al. (2003).

The household survey was conducted with the help of a structured questionnaire, which covered the socio-economic aspects. The CVM survey was included with the help of a bid format and some open-ended questions to gather qualitative information. Sample selection was guided by stakeholder identification (direct and indirect users of wetlands) and 184 rural households and eighty urban households were selected from an original survey of 200 and 100 respectively.

The area under study lies between 22° 35' to 22° 40' N and 88° 25' to 88° 30' E and spreads across two districts, namely, Kolkata Metropolitan District and the North Twenty Four Paraganas District. It may be important to mention that it lies to the northeast of Salt Lake City which is the first example of wetland reclamation for urban expansion. The survey consisted of both rural and urban households. The rural sample was collected across seven villages around Nowai and Haroa Khal, east of the National Highway (NH 34) and the airport. For the urban sample, newly developed housing societies were selected from Narayanpur, Kaikhali and VIP Road, located in the South Dum Dum and Rajarhat-Gopalpur municipalities.

THE WETLANDS OF THE EAST OF KOLKATA AND ITS SURROUNDINGS

Kolkata is located on the levee of the Hugli-Bhagirathi, a distributary of the Ganges. An extensive tract of wetlands is located to the east of the city, which has always been the city's natural drainage retention spill basin. Urban sprawl is gradually having its impact upon the extensive wetland tracts, which includes an inter-connected system of distributaries, streams, canals, natural waterbodies as well as man-made fisheries. Apart from this factor, its location upon the 'fringes of the active delta, hardly a few feet above sea level' and 'along the track of tropical cyclones coming up from the Bay of Bengal' (Munshi 1991) gives it a unique environmental characteristic not found in any other populous metropolis of the world.

Kolkata, although never a walled city, had imposing physiographic boundaries in the form of the Hugli distributary to the west and the extensive Salt Water Lakes to the East. The growth of the city has been limited by these physical boundaries for centuries. Over the last decade, however, the city has been witnessing the conversion of the wetlands lying to the east. Though many geographers and hydrologists have documented that the wetlands' loss would have negative impact on the city's ecological balance, the spiralling rate of constructions and conversions continue and are a cause for concern.

The Location of the City and the Form of the Wetlands

The present metropolis started as the nucleus of three villages (Sutanati, Kolikata and Gobindapur) along the eastern bank of the River Hugli, previously known as Bhagirathi. To the east of this nucleus was a very large area of salt water lakes which is now partially taken up by the satellite township of Salt Lake (or Bidhannagar). North of Sutanati, a creek joined the river to the eastern marshes. To the south of Gobindapur, another creek joined the river to the marshes. Between the two villages of Kalikata and Gobindapur, yet another creek flowed west-east joining the river and the marshes, the Adi Ganga. The water system that was generated by this network consisted of an interconnected system of wetlands with the creeks joining the main river to the eastern salt water lakes. Thus, much before the British factory settlement was established, this region had a typical deltaic topography consisting of active and moribund channels, creeks and swamps. It is documented that by the time Job Charnock landed at Sutanati, the Adi Ganga had decayed into a stream joining Bidyadhari river.

The Kolkata Metropolitan Area is located within the lower deltaic alluvial plain of the Ganges river system. The entire region is built up with sediments deposited by the river system. The river Hugli divides the metropolitan area almost equally, but it is the eastern portion that slopes away into a saucer shaped landform wherein lies an extensive wetland area comprising the natural receptacle. The east Kolkata wetlands, currently covering an area of about 12,000 ha, has recently been declared as a wetland of national importance and subsequently declared a 'Ramsar Site'. Out of this total area, waterbodies cover 8,000 ha. The average depth of water of the area is estimated to be less than one metre in the lean season, which rises to nearly two metres during the monsoons. From the hydrogeomorphic point of view, Rudra (2001) documents that the origin of these swamps can be attributed to:

(*a*) the decay of channels;
(*b*) drainage congestion;
(*c*) estuarine sedimentation;
(*d*) tidal intrusion;
(*e*) auto compaction of deltaic sediments; and
(*f*) anthropogenic intervention into the fluvial regime.

The eastern wetlands mark the junction between the fluvial and tidal regimes of the northern and southern parts of the Southern Deltaic West Bengal respectively. According to him, the swamps can also be treated as the areas of incomplete morphogenesis and even the possibility of slow subsidence of these areas lying to the south of stable shelf zone cannot be ruled out. These swamps are now being filled up by intra-basin transfer of sediments, augmented by the dumping of 2,600 tonnes of solid wastes and deposition of 2,076 tonnes of suspended solids per day carried by east-flowing canals of Calcutta (Nath 1995).

Tracing the pattern of development that has been long affecting these wetlands, one may find that the present situation has made it evident that costs and benefits be weighed right away and the critical ecosystem protected from intrusion.

Urban Expansion in the Context of the Wetlands: Its Pattern and Functions

The first major developmental pocket to affect the wetlands directly by way of conversion of land use was the Salt Lake City—targeted to be a satellite town aimed at depopulating the city centre. It was planned primarily as a residential township and as a centre for administrative functioning. The location was the marshy region comprising the salt water lake, which was 'formed by the tidal action of the streams flowing through the district of 24 Parganas where '… the Bay waters found an easy passage through numerous water courses and estuaries of Sundarbans and could freely spill over the extensive region….' (Chattopadhyay 1990). The intricate river system, which facilitated the spillover, comprised the Hugli, Bidyadhari, Piali, Kalindi, Jamuna, Ichhamati, Padma, Kalyan Khal, Banstala and ten other major channels. According to Chattopadhyay, the extent of the swamp area has been variously recorded. According to the 1951 Census, the Salt Lakes which lie to the east and southeast of Calcutta were vast tracts of water-logged swamps covering nearly an area of thirty square miles. Reclamation resulted in dwindling of this area to a great extent. The Calcutta Municipal Gazette of 1940 recorded that the salt lake area contracted from over forty square miles towards the beginning of the 19th Century to about eleven square miles in 1940. These statistics apart, it is definite that the salt water lakes at one time stretched northward up to Dum Dum and southward up to Sonarpur but are no longer in existence at present. Beyond this area lies a vast stretch of wetlands, which are gradually being confronted by the demand of planning authorities to provide space for the growing population and of the developers to convert the land into urban uses without prioritising conservation methods.

The wetlands of Kolkata have a multifaceted role in supporting the urban economy as well as the ecosystem. It acts as a reservoir for excess rain water during monsoons and helps mitigate adverse impact of floods; it also ensures microclimatic stability and contains numerous species of flora and fauna. It helps facilitate waste treatment through its vast stretches of sewage-fed fisheries and natural waste recycling. It ensures food security and livelihood of the rural and urban poor around the city and renders an 'invisible subsidy to the urban economy' (Ghosh and Sen 1987).

Studies on Kolkata Wetlands

Many studies have been conducted upon the general nature and characteristics of the Kolkata wetlands, its judicious use and issues of its sustainability.

The initial idea about the marshes (or wetlands) of Kolkata (then Calcutta) was that it was a difficult and uninhabitable terrain. The first plan for reclamation of the salt lakes was made by the British in 1945 with the 'Salt Lakes Reclamation Plan' by Government of Bengal and since then developmental trends have followed the path of reclamation. The most significant contributions regarding the importance of these wetlands have come from authorities involved in the planning and development mechanism, which in itself shows that the positive feedback process would have been successful, if given an adequate opportunity.

An account of the gradual changes in the marshes in east Kolkata to extensions of the urban sprawl has been well documented by Chattopadhyay (1990), a historian, in his book *From Marsh to Township*. He analyses the region from its hydrological and geographical point of view, documenting the transformation and the changes in the drainage patterns.

Ghosh and Sen (1987) document the history of the wetland conversions and conclude that there is a possibility of the threats of conversions being graver in the future. Rudra (2001) studies the hydrogeomorphological conditions and analyses the morphogenesis of the wetlands thereby highlighting its functional entity as an important part of the natural drainage system that supports the urban environment. His study gives a comprehensive account of the origin and changes of this unique geohydrological region. His account provides the basis for a holistic understanding of the wetland ecosystem, which is essential for determining policy measures for conservation. Nath (1995) highlights the extent of pollution and the containing of urban waste by this wetland system.

Two recent studies assume importance in the context of the present study. The first is by Chattopadhyay (2003), which is a study of valuation of the productive functions of a part of the wetlands, namely the Waste Recycling Zone, which is an important subset of the East Kolkata Wetlands. The other important study is by Sikdar et al. (2002), which attempts to highlight the problems of conversion through an environmental impact assessment of a proposed info-tech complex in the East Calcutta Wetlands. This study concludes that there is a disjointed complex of water bodies which will be adversely affected by the project and highlights the need to identify restoration policies.

ANALYSIS OF WILLINGNESS TO PAY FOR CONSERVATION OF URBAN WETLANDS

Willingness to pay (WTP) is the amount of payment that a respondent is willing to make for the protection of the resource or willing to contribute if the resource is conserved instead of being converted for other uses (such uses as may be considered ideal for generation of benefits while conserving the resource). This section analyses both willingness and unwillingness to pay and attempts to justify preservation preferences in terms of the socio-economic parameters related to the sample population. It then goes on to conclude which group among rural and urban areas has better user preferences.

For those willing to pay, the payment ranges from Rs 60 per year to Rs 1,800 per year. The highest payment in rural sample is Rs 1,200 per year, while that in the urban sample is Rs 1,800. Considering the high-income levels in urban areas, the payment as a proportion of their income is very low. The overall mean for 264 households is Rs 300.72, with a standard deviation of Rs 400.60, which indicates that there is high overall variation in the amount people are willing to pay for conservation. If rural and urban willingness to pay are considered separately, it appears that while the rural mean is Rs 143.04 per annum, with a standard deviation of Rs 222.73, the urban mean is Rs 663.38 with a standard deviation of Rs 477.75. This clearly demarcates the difference in the nature of payment possibilities across rural and urban populations. While urban payments are higher, there is lesser variation. On the other hand the situation is reverse in case of rural population, where the payments are lesser but show high variation. This implies that rural population is more forthcoming in payment whenever their conditions permit them to do so. The WTP values were grouped into seven categories

(Figure 9.1) to analyse the nature of distribution of households. The distribution shows concentration of households in different payment groups.

Figure 9.1 Willingness to Pay across Sample Households

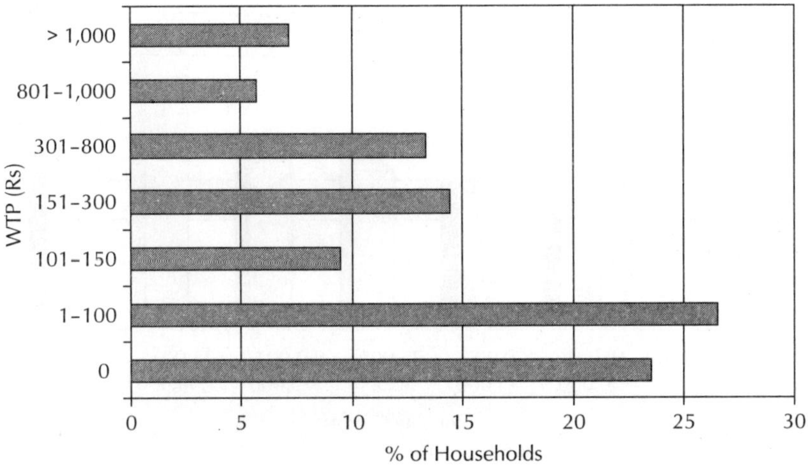

It is evident from Figure 9.1 that the frequencies are highest for the lowest range of WTP of Rs 1–100 per year, that is, 70 per cent of the population is willing to make the least payment. In the groups with higher payments, number of households decline considerably. The lowest number of households is in the second highest payment group—only 5.7 per cent are willing to pay anything between Rs 801–1,000 and 7.2 per cent are willing to pay more than Rs 1,000. It was verified that the households belonging in this group are either direct stakeholders like owners of fisheries or feel strongly about conservation benefits.

Rural Willingness to Pay

In the rural area, the WTP values range from Rs 60 to Rs 1,200 per annum in the seven villages taken together. It may be mentioned here that high WTP of Rs 1,000–1,200 is strictly restricted to the *bheri* (fish farm) owners, who are comparatively financially better off than other rural respondents. The total amount of payment expected from 184 rural households amounts to Rs 26,320.

Importantly, it is to be observed that the payment for conservation is proportionate to increases in income in case of rural households, which is evident from Figure 9.2. This reiterates our assumption that rural willingness to pay is likely to be influenced primarily by income.

However, as a whole, the proportion of income that households are willing to sacrifice is very low, ranging from 0.08 to 0.87. Understandably, while the higher income groups can afford to pay for protection of their sources of livelihood, the poor cannot secure their livelihoods even though they may want to. It may be mentioned here that higher income groups in the sample were all *bheri* (fish farm) owners and they are willing to contribute higher amounts as the wetlands are their source of profit and livelihood.

Figure 9.2 Rural Willingness to Pay as a Proportion of Income

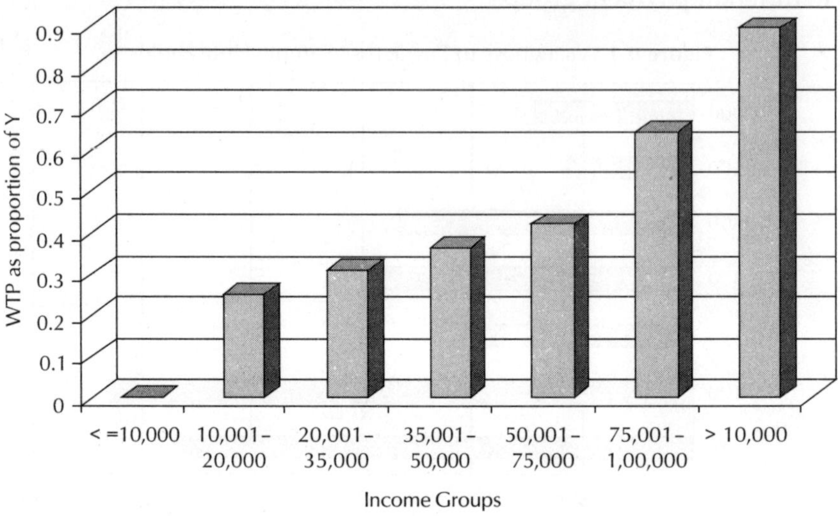

Urban Willingness to Pay

As a whole, the urban willingness to pay is higher than the rural because of higher levels of income and greater security in terms of employment and savings. Total willingness to pay by urban households amounts to Rs 53,070, more than double the payment expected from the rural sample population.

However, its variation across income groups does not follow the same rationale as in the rural sample. As shown in Figure 9.3, the maximum share of payment comes from the middle-income group, that is, Rs 1,50,001–3,00,000.

Comparative Aspects of User Preferences

Higher level of incomes earned by the urban households is not reflected in their willingness to pay as a whole. Though the total value of WTP is higher, the proportion of income that they are willing to forgo is lower than rural households. Table 9.1 summarises the relationship between income and willingness to pay.

In spite of lower incomes and poverty, rural households were likely to contribute 0.42 per cent of their income while urban households are likely to pay 0.34 per cent, lower by 0.08 percentage points. Though a very marginal difference, it shows better user group preference for rural population. Presumably, the reason is that a greater proportion of the rural population is directly dependent upon the wetlands for agriculture and pisciculture for their livelihood and, therefore, comprises of direct stakeholders. On the other hand, there are no direct stakeholders at present among urban households except for being consumers of fish and vegetables produced in the wetland areas. The

Figure 9.3 Willingness to Pay across Urban Income Groups

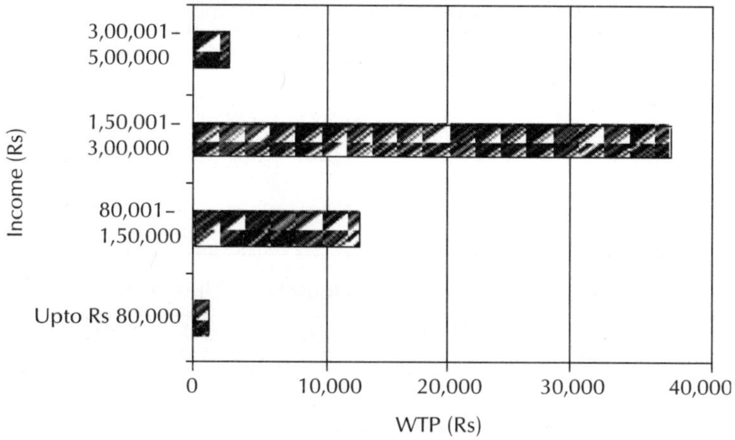

Table 9.1 Willingness to Pay as Proportion of Income

Samples	*Total Income (Rs/yr)*	*Total WTP (Rs/yr)*	*WTP as a % of Income*
Rural	63,13,400	26,320	0.42
Urban	1,54,83,000	53,070	0.34

urban population, however, comprises indirect beneficiaries of the positive functions of the wetland ecosystem but is not well aware or informed about these functions.

The above discussion of conservation attitudes among rural and urban population revealed that positive conservation attitudes were shown by 95 per cent of urban respondents and 80 per cent of rural respondents. However, it is evident that the willingness to pay as a proportion of income does not necessarily match with the perception of urban population towards conservation. While 95 per cent respondents are willing to conserve, they are willing to forgo only a very small proportion of their income for the purpose. Rural perceptions are also strongly in favour of conservation and, in spite of low levels of income, are willing to forgo 0.42 per cent of their income, which amounts to Rs 26,320.

Assessment of People's Unwillingness to Pay

There are respondents in both rural and urban samples that are not willing to make a payment, whatever the payment vehicle may be. Figure 9.1 in the above section shows that 23.5 per cent of the total population is unwilling to pay. Most of these households cannot pay because of their financial instability. But some also feel that they would not pay because they are unsure about how their hard-earned money would be utilised by the institution to which (or the person to whom) they will have to pay. There is another section who feel that they should not pay as it is the government's responsibility to pay for resource conservation. Tables 9.2 and 9.3 show the distribution of rural and

Table 9.2 Distribution of Rural Households Unwilling to Pay

Income Groups	*No. of Household*	*Households Unwilling to Pay*	*Groupwise%*	*% of Total Household*
< =10,000	11	11	100.00	6
10,001–20,000	64	32	50.00	17.5
20,001–35,000	43	3	7.00	1.5
35,001–50,000	28	1	3.60	0.5
50,001–75,000	26	0	0.00	0
75,001–1,00,000	6	3	50.00	1.5
> 1,00,000	6	0	0.00	0
Total	**184**	**50**		**27**

Table 9.3 Distribution of Urban Households Unwilling to Pay

Income Groups (Rs)	*No. of Household*	*Household Unwilling to Pay*	*Groupwise % of Household*	*% of Total Household*
< = 80,000	3	1	30	1.25
80,001–1,50,000	27	2	8	2.5
1,50,001–3,00,000	42	3	7	3.75
3,00,001–4,50,000	8	6	75	7.5
Total	**80**	**12**		**15**

urban households unwilling to pay across different income groups—27 per cent of rural households are unwilling to pay, whereas 15 per cent of urban households are unwilling.

In case of rural population, it is noted that 27 per cent of the households are unwilling to make any payment. Table 9.2 and corresponding Figure 9.4 indicate the income-wise distribution of rural households not willing to pay. Out of the 27 per cent, none of the households belonging to the very low-income group are willing to make any payment. Two reasons are very clear regarding non-payment by the lower income group. First, their hand-to-mouth existence does not permit them to spare anything for conservation payments, which is a luxury for them.

Figure 9.4 Income-wise Distribution of Rural Households Not Willing to Pay

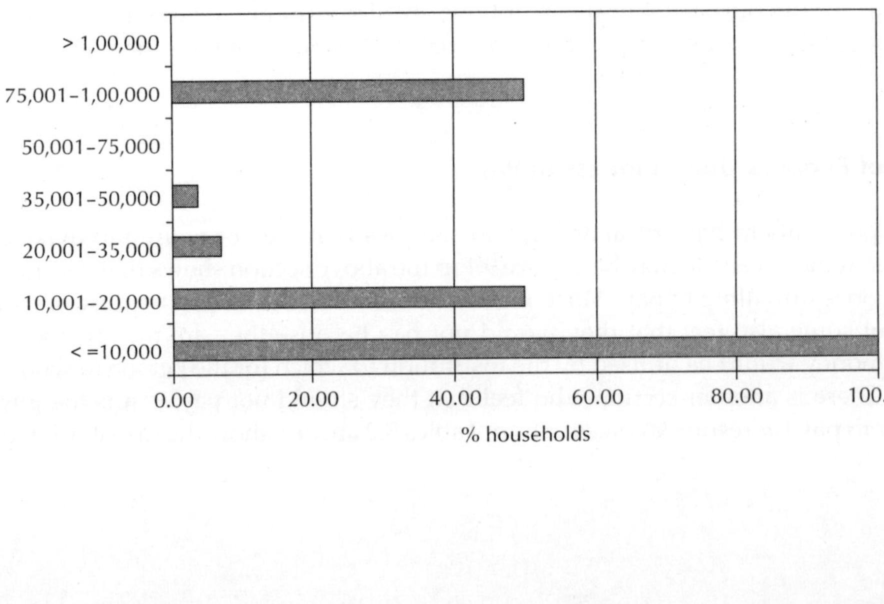

Second, some feel that conversion of wetlands to residential and commercial uses is likely to bring greater employment opportunities in the form of construction work, need for daily labourers, domestic helps, shops for daily needs and a range of other opportunities. This possibility of employment is a far greater need for the poor and unemployed than wetland conservation. In higher income groups, payment possibilities are noted to be increasing. The reasons were determined to be their greater stake in the wetlands as most of these respondents are fish farm owners or shareholders.

On the other hand, urban households unwilling to pay, arranged by income groups, has been shown in Table 9.3. The scenario is reversed in this case. The highest income group of Rs 3,00,001–4,50,000 is largely unwilling to pay. Of the lowest income group in urban population 30 per cent is not willing to pay.

The unwillingness to pay by members of high-income groups may be explained by the fact that their expenditure for improvement of standard of living does not spare much for conservation. Another assumption may be made, that the members of high-income groups are alienated from the ideas of ecosystem benefits and conservation needs for the wetlands. It is an important assumption that implies that higher income does not have any association with higher level of awareness of the environmental resources.

The descriptive statistics based on the field observations show the difference in variables between the urban and rural samples and it is likely to have differential impact on the perceptions of the ecosystem between the two groups. Understanding of the difference between these two groups—rural and urban, is important as a basis for identification of the strategies to be developed for conservation of the wetlands. The following variables have been selected for analysis:

Size of the Family (Household Size): While mean rural family size is 5.99, that of urban is 3.21. Variances are more divergent, that of rural is 4.279 and urban is 1.005. This implies that number of family members in rural areas is varying to a much greater extent than in the urban area. This is essentially because of the fact that average family sizes in rural areas are much larger: first, because of lack of awareness and education; and second, because of prevalence of the joint family system. At the same time, small families also exist—both due to single unit family types as well as because of migration-related causes where the parents may be residing in the village while the younger male working population may have migrated for work and the younger females may have migrated after marriage. In case of urban areas, family sizes are more consistently smaller because of predominance of single units families and preference for smaller families.

Years spent on Education (Education Level): As far as years spent on education are concerned, it was assumed that urban education levels would be higher than rural levels. In this case, the rural mean is 2.8, which is very low, while the urban mean is 14.96. This indicates that rural respondents are characterised predominantly by primary level education. Urban respondents, however, show more than fourteen years spent in education, on an average, which implies the graduate level. From different aspects like ability to pay for education, access to infrastructure related to education as well as level of information and awareness, the urban population is definitely better off than their rural counterparts.

Household Income: Income is another variable that shows differences in group means and variances. It is also likely to be the most important indicator for willingness to pay especially because, in general, most decisions are income-dependent in rural areas (more likely to be so in the case of a developing country). In this case, mean rural income is Rs 34,312 while mean urban income is Rs 1,93,537, which is at least 5.6 times more. This is largely due to the steady income from organised and service sector where the urban respondents are largely employed. Urban respondents in business sector are also more successful than their rural counterparts due to the availability of a wider and more diverse market and a population with higher paying capacities. Higher educational qualifications automatically result in better rate of employment and higher payments for the urban respondents. Also, the dependency ratios are lower in urban areas than in rural. Variances, however, are steady for both rural and urban incomes, both near about 6.5. Income levels are uniformly high in urban areas and uniformly low for rural areas.

Awareness of Levels of Pollution: Respondents in both rural and urban areas seem to be aware that the wetlands are not too polluted. The most important activities that are steadily beginning to pollute this ecosystem include some small manufacturing units dotting the fringes which do not follow any proper system for managing their wastes and pollutants. These units are still very small in number as well as in scale and therefore the level of pollution has not been too high, but given the chance to continue, an increased pollutant load would have a negative impact. These industries include leather, a number of small manufacturers of machine parts, brick kilns and hosiery factories.

Attitude towards Conservation: This variable sought to assess whether respondents have the awareness about conservation. It emerges that even though the respondents are not fully aware of the functions and services provided by the wetlands, most have the opinion that it must be protected and not allowed to degenerate or converted to other land uses. A clear consensus exists, as 80 per cent of the rural respondents have a positive opinion about conservation, while 95 per cent of the urban respondents support conservation. Some feel, however, that conversion of the wetlands to housing and infrastructure will provide them with employment opportunities.

Willingness to Use the Wetlands for Recreation: Sixty-two per cent of the rural respondents are not interested in using wetlands for their own recreation, but are of the opinion that urban population would use it and that it would encourage connectivity, communication and development of the area. thirty-eight per cent would like to use wetlands for recreation, but it would be impossible if entry is highly priced. The urban population, on the other hand is more keen on developing wetlands for recreation—91 per cent would like to enjoy the aesthetic beauty, boating and non-polluting activities.

CONCLUSION

This study was an attempt in establishing the preliminary relationship between the socio-economic parameters of a given population with its nature of preferences with regard to urban wetlands. Analysis of willingness to pay has emerged as an important survey-based valuation technique across the world and has been widely accepted as well as critically assessed for methodological flaws. In the study of assessing people's preferences for the Kolkata wetlands, it was essential to consider both rural and urban population as a large proportion of the wetlands lies in rural locations, which may undergo a transition towards urban land uses as they lie on the brink of the mega-city project of eastern Kolkata.

What emerges from the preliminary survey is that people are clear about the fact that wetlands are important to sustain, though all respondents are not aware of the total set of functions that these wetlands play in the life of the region and its people. Analysis of payment patterns show that people's preferences are clearly guided by their economic conditions or ability to pay. Also, specifically in case of rural population, where the linkages to the wetlands are stronger than urban areas, people are willing to part with a greater proportion of their income than urban respondents. This clearly shows the nature of user preferences. It is established that there are differences in people's attitude towards this important ecological entity. The differences may be explained by socio-economic conditions of the respondents and their understanding of the ecosystem. However, the fact remains that all stakeholders need to be sensitised on wetland issues. While this study finds apathy among people with greater ability to pay, it is also true that most of the urban respondents did not have adequate knowledge regarding the functions and services of the wetlands. It may be concluded that increased awareness may change this scenario. The dissemination of information on the importance of wetlands, their nature and characteristics should ideally be an essential goal. Environmental education for all direct and indirect stakeholders will help people appreciate this very important ecological entity. The holistic understanding of the ecosystem is expected to initiate the process of peoples' participation in wetland conservation and management.

NOTE

1. Corresponding author.

REFERENCES

Adger, W.N., P.M. Kelly and N.H. Tri. 2001. 'Costs and Benefits of Mangrove Conversion and Restoration' in R.K. Turner et al. (eds), *Economics of Coastal and Water Resources: Valuing Environmental Functions*, Netherlands: Kluwer Academic Publishers.

Bandyopadhyay, S., K. Narayanan and A. Ramanathan (forthcoming) 'Determinants of Willingness to Pay for Wetland Conservation: A Study of Kolkata Wetlands', *Studies in Regional Science*, 35(4).

Brouwer, R., I.H. Langford, I.J. Bateman and R.K. Turner. 2001. 'A Meta-Analysis of Wetland Contingent Valuation Studies' in R.K. Turner et al. (eds), *Economics of Coastal and Water Resources: Valuing Environmental Functions*, Netherlands: Kluwer Academic Publishers.

Chattopadhyay, H. 1990. *From Marsh to Township*, Calcutta: Department of History, Calcutta University and K.P. Bagchi & Co.

Chattopadhyay, K. 2003. 'Environmental Conservation and Valuation of East Calcutta Wetlands', in J. Parikh (ed), *Reconciling Environment and Economics: Executive Summaries of EERC Projects*, Mumbai: Indira Gandhi Institute of Development Research.

Chopra, K. 2002. 'Economic Valuation of Biodiversity: The Case of Keoladeo National Park', in G. Kadekodi (ed), *Environmental Economics in Practice: Selected Case Studies from India*, New Delhi: Oxford University Press.

Ghosh., D. and S. Sen. 1987. 'Ecological History of Calcutta's Wetland Conversion', *Environmental Conservation*, 14(3): 219–26. Autumn.

Government of Bengal. 1945. *Report on the Reclamation of Salt Lakes Area*, Department of Health and Local Self Government, Alipore, Bengal Government Press.

Hanley, N. and C.L. Spash. 1998. *Cost Benefit Analysis and the Environment*. Aldershot: Edward Elgar.

Markandya, A. and M.N. Murty. 2000. *A Cost-Benefit Analysis of the Ganga Action Plan – Cleaning Up the Ganges*. New Delhi: Oxford University Press.

Munshi, S.K. 1991. 'Calcutta: Land, Land Use and Land Market' in B. Dasgupta (ed), *Calcutta's Urban Future*, Government of West Bengal.

Murty, M.N., A.J. James and Smita Mishra. 1999. *The Economics of Water Pollution: The Indian Experience*. New Delhi: Oxford University Press.

Nath, K.J. 1995. 'Drainage, Sewage and Waste Disposal', in S. Chaudhuri (ed), *Calcutta: The Living City, Vol. II – The Present and Future*, Calcutta: Oxford University Press.

NOAA. 1993. 'Report of the NOAA Panel on Contingent Valuation', *US Federal Register*, 58(10): 4602–14.

Parikh, J.K. (ed). 2003. *Reconciling Environment and Economics: Executive Summaries of EERC Projects*, EERC, Mumbai: Indira Gandhi Institute of Development Research.

Prasad, L.P. Kumar and C.R. Babu. 2003. 'Valuation of Ecological Functions and Benefits: A Case Study of Wetland Ecosystems Along the Yamuna River Corridors of Delhi Region', in J. Parikh (ed), *Reconciling Environment and Economics: Executive Summaries of EERC Projects*, Mumbai: Indira Gandhi Institute of Development Research.

Rudra, K. 'Geomorphology of the South Bengal Delta' in S. Bandyopadhyay, et al. (eds), *Environment and Geomorphology* (in press).

Sikdar, P.K., S. Mondal, L.Saha, S.S. Saha and S. Banerjee. 2002. 'Environmental Impact Assessment of a Proposed Infotech Complex in East Calcutta Wetlands', *Environmentalist*, 22(3): 241–60.

Singh, S.P., B. Gopal and V. Kathuria. 2003. 'Integrating Management of Water Resources of Lake Nainital and its Watershed: An Environmental Economic Approach', in J. Parikh (ed), *Reconciling Environment and Economics: Executive Summaries of EERC Projects*, Mumbai: IGIDR.

Verma, M., N. Bakshi and R.K. Nair. 2003. 'Economic Valuation of Bhoj Wetlands for Sustainable Use', in J. Parikh (ed), *Reconciling Environment and Economics: Executive Summaries of EERC Projects*, Mumbai: IGIDR.

World Bank. 1995. *Restoring and Protecting the World's Lakes and Reservoirs*. Technical Paper No. 289.

10

Assessment and Valuation of Forest Revegetation[1]

Dibyendu Mondal,[2] *S. Singh* and *J.V. Dhameliya*

Abstract: While implementing a project of regenerating degraded lands (forests or otherwise), one often faces the question of justifying the benefits that would accrue from the initiative. In contrast when a natural forest is destroyed, one faces a very minimal compensation based on the timber available or just on the value of land with a small premium for the existing biomass or on compensatory afforestation of a barren landscape. This leaves us in a situation where answering only the value of the total flow of benefits from the forests does not give a true picture. There is a need to look for mech-anisms to value the forests in entirety.

This chapter is an attempt at valuation of forests using the Natural Resource Accounting System (NRAS) framework through a case study from Gujarat. The framework looks beyond the conventional system of accounting the direct investment vs flow of benefits to include existing stock and environmental benefits. This system attempts to calculate the total value or contribution to society at a given point of time. The framework helps us estimate the change in this value if a conservation action/destructive action is undertaken, how this change affects different stakeholders, that is, who the beneficiaries are and who the losers are, and, therefore, help in decision making. This can also help us in calculating the actual costs and benefits for taking up such an activity or even for the valuation/ compensation for the damage in case of a change in the land use.

INTRODUCTION

'How valuable is our forest?' is a question that can have several answers depending on how one perceives the resource. The dominant perception in the Forest Department (which manages most of the country's forests) is that it is a source of economic wealth, be it from timber or other forest produce. All that matters is yields and revenues, be it from timber or NTFP. Of late, a very light voice in the department seems to accept the need for conservation and biodiversity [Forest (Conservation) Act 1980 and National Forest Policy 1988]. In the local community's perceptions, forest means not only a source for clean and pure drinking water, meeting the day-to-day needs for food, fuel, fodder, shelter and culture, that is, a steady stream of returns, but also it is regulates the needs allowing no final harvest—ensuring that over-extraction does not take place and conserved for generations to come (Singh 2003). To an urbanite, forests mean aesthetics—places of recreation needed for fresh air, places for weekend solitude, reference areas for research and study, etc.

While implementing a project of regenerating the degraded lands (forests or otherwise), one often faces the question of justifying the benefits that would accrue from the initiative. We immediately tend to answer the question in terms of the total value of the current flow of benefits provided by that forest or in terms of the value of future flows of benefits like fuel wood, fodder, timber, leaf-litter, etc. or in terms of the value of conserving the forests for ecological benefits like fresh water, protection against soil erosion, a reservoir of floral or faunal biodiversity, refuges for rare plants, wildlife and fish species or in abstract terms as beauty, shade or habitats for important species, etc. (Reddy et al. 1997 and Perrings and Gadgil 2003). We more often than not tend to answer in terms of the direct flow of benefits from the resource or in terms of our relationship to the resource. Similar questions arise in situations when a natural forest is to be destroyed and the value of the standing resource is to be compensated. The major reason for excessive depletion and conversion of forest lands is the failure to account adequately for their non-market environmental values in decision making related to alternative use of the lands.

METHODOLOGIES FOR VALUATION

Faced with a similar question of justifying our work on revegetation of degraded lands in terms of cost-benefit analysis, we wanted to search for methodologies to capture the total value of the forests regenerated—to include and monitor the inflows and outflows, the total stock and also attempt to value the ecological goods and services it provides to humankind. The conventional accounting frameworks answered only a part of the question—the accounting for the inflows and outflows, more specifically, the transaction in terms of money (Reddy et al. 1997). This method is a gross underestimation of the value of the small plot of forest as it tends to consider natural resources as free unless extracted for production or sale.

An extension to the conventional accounting is valuing the forest in terms of the forecast of the potential benefits the forest could provide, especially timber (Reddy et al. 1997). The methods tried to calculate the expected stock at the end of a particular period and some of these tried to report on the growing stock each year. However, there are no specific references to any methodology in the literature that includes the existing stock of the inflows and outflows.

Of late, there has been an attempt to comprehensively value the forests for the services using various frameworks/methodologies (Hulkrantz 1992, National Research Council 1994, Barbier et al. 1997, Costanza et al. 1997, Saxena and Agrawal 1997, Adamowicz 2003, Pagiola 2004, The International Bank for Reconstruction and Development 2004). Usually the methodologies have attempted to consider the stock and inflows to the communities with an economic valuation of the ecological services it provides. Broadly two types of valuation methods—contingent valuation and economic valuation method are used depending on the issue. The contingent valuation method can be used to estimate both use and non-use values, and it is a widely used method for estimating non-use values. It is called 'contingent' valuation because people are asked to state their willingness to pay, contingent on a specific hypothetical scenario (Cooper et al. 2004) and description of the environmental service. The contingent valuation method involves directly asking people, in a survey (Whittington 1998), how much they would be willing to pay for specific environmental services or whether they are willing to accept compensation and willing to give up specific environmental services. Though the contingent valuation method is used to estimate economic values for all kinds of ecosystem and environmental services, it is also the most controversial of the non-market valuation methods.

On the other hand, the economic valuation is by far the most accepted method for valuation of natural resource services. The economic valuation is an attempt to assign quantitative values to the goods and services provided by environmental resources, whether or not market prices are available. The economic value of any good or service describes the value of the resource in providing such commodities, whether or not we actually make any payment. Thereby, economic valuation takes into account the current stock, the direct costs, the inflows to the communities and the environmental services provided by the resource.

Though we know intuitively that such resources may be important, in most cases we fail to describe the value of the resource nor do we ensure their judicious use. This is because the environmental resources we are dealing with are complex and multifunctional, and they provide innumerable goods and services to nature and human beings in particular. Further, we are talking of forest resources, which no longer remain only as local resources but are being treated more as national and global commons for the valuable services that they provide to humankind making them irreplaceable or which can only be replaced at a premium, compensating for the value provided to humankind and other organisms that depend on the same. Various methodologies have been suggested by different researchers for specific needs/objectives to value the forests, where some of them have attempted to reflect the values in the national account[3] while others have attempted to do so on a case-to-case basis to look into the cost and benefit of the activity.

There are several examples of forest resource accounts that include some non-market goods and services (Adamowicz et al. 2004; Hulkrantz 1992; Haener and Adamowicz 2000; Kriström and Skanberg 2001). Kriström and Skanberg (2001) have taken market and non-market accounts over time for forest resource accounting. In their study, they define the value of the capital stock of timber and non-timber goods (berries, etc.) while also defining the depreciation in the capital stock arising from environmental change to be included in the Green Net National Product (NNP). They have attempted to compute measures analogous to the appreciation/depreciation in timber accounts, valued recreational trips and the resultant study provides one of the most carefully constructed market and non-market accounting exercises. Anil Agrawal and Sunita Narain (1985), in the *Second Citizens' Report – The State of India's Environment 1984–85*, also argued that biomass is the basis for survival, the source of most income and protector of the environment and, therefore, must be added to the GNP and be rechristened as Gross Natural Product including the growing stock of the biomass.

IDENTIFICATION OF RELEVANT INDICATORS

Review of a number of existing literature revealed that different experts/researchers have opted for different set of parameters to suit the quantification needs (Anielski 1992; Costanza et al. 1992; Steiner et al. 2004). While all the methodologies have considered the valuation of the existing stock (either as potential timber or as the entire biomass that includes the twigs and leaves too), there are differences in the choice of parameters for the environmental services.

Costanza et al. (1992) in their study have included erosion control, soil formation, nutrient cycling, waste treatment, food production/NTFP, raw material/timber/fuel/fibre, genetic resources/biodiversity regulation, recreation and cultural parameters. Achim Steiner, Steven J. MacCormick and Jan Johnson (2004) in their methodology have attempted to include nutrient cycling, food production/NTFP, raw material/timber/fuel/fibre, genetic resources/biodiversity regulation, recreation, cultural parameters, carbon sequestration/air quality/climate, fresh water, human health, detoxification and natural hazard regulation in the calculation of the value of the forests.

Faced with the need for valuing the regenerated plots for the cost-benefit analysis of the project, we set out to evolve a framework, which can capture the cost and benefits from regeneration not just limited to cash inflow and outflow. Various contexts and the methodology used by different researchers have been reviewed to help us in building our framework for Natural Resource Accounting System (NRAS).[4] Extensive review of literature suggested that various parameters have been taken into consideration by different researchers. We attempted to make a global list of indicators that can be considered for valuation of the forests. The economic parameters that have been considered are fodder and other produce like NTFP, increase in biomass, employment, surface water, increased milk/livestock/agricultural production, reduction in migration, etc. The ecological parameters used are soil formation, nutrient cycle (soil fertility), checking of soil erosion, increase of biodiversity/variety, rate of natural regeneration, increased wildlife population, improvement in micro climate, climate regulation, atmospheric gas balance, pollination, habitation for birds and wildlife, preservation of endangered species, improvement in groundwater level, aesthetic value reducing pressure on existing forest, etc.

The framework for Natural Resource Accounting System (NRAS) is a framework that explains the interrelationships between the economy and environment. NRAS tries to set right the bias of conventional cost-benefit analysis by monitoring the environmental impact and helps develop indicators for valuation of ecological services (Ahmed et al. 1989). At the present stage of development of the framework, the following indicators have been selected for monitoring. Various parameters/indicators have been classified as direct costs, direct benefits, existing stock, indirect costs and indirect benefits and the methodology of accounting has been indicated. Human intervention/natural events are likely to change the status of the resources which can either facilitate the growth of the resource or disturb the resource. The NRAS framework (Table 10.1) helps in effectively monitoring these changes (ups and downs) in the status of the natural resource.

One would agree that there are a lot of other indicators under existing stock like ground cover (vegetation) and indirect benefits such as biodiversity, rate of regeneration, soil formation, habitation for birds and wildlife, water retention, aesthetics, etc. could also have been valued. We are in the process of identifying appropriate methodology for the quantification and valuation of such parameters.

The presentation of the valuation is done on an annual basis through the flow accounts (direct and indirect), stock account and balance-sheet. The flow account is prepared on an yearly basis indicating all costs and benefits and is almost similar to the profit and loss account of financial accounting. The stock accounting indicates the position of stock on a particular date for an individual item like biomass, environment (soil fertility and CO_2) and the balance sheet is prepared on the lines of financial accounting to indicate the present status of assets and liabilities on a particular date. This

Table 10.1 Accounting Framework and Various Parameters Covered

Income/Exp. Head	*Conventional Accounting*	*Economic Accounting*	*Environmental Accounting*
1. Direct Cost			
Land development	*	*	
Area regenerated	*	*	
Protection	*	*	
Institutional Development	*	*	
2. Indirect cost			
Crop foregone		*	
Others		*	
3. Existing Stock			
Standing tree biomass		*	
4. Direct Benefit			
Fodder	*	*	
Fuel wood	*	*	
NTFP		*	
5. Indirect Benefit			
Soil erosion control			*
Soil fertility increased		*	*
Carbon sequestration			*
Aesthetic Values			*

Note: * Parameters covered.

framework, therefore, instead of being a one-time valuation effort, is an yearly exercise to record the changes and the incremental growth in value of the resource.

OPERATIONALISING THE FRAMEWORK

This chapter is an attempt to introduce the framework for Natural Resource Accounting for discussion and outlines the process of economic valuation of the forest resource in the process of regeneration of a small patch of degraded grazing land.[5] Data has been collected for various parameters each year since the initiation of the study in 1997 and, therefore, helps in not only assessing the value of the forests at this point of time, but also shows the incremental value of the forests each year. The baseline information at the initiation of the regeneration process in 1991 also helped in showing the incremental improvement for certain parameters in 1997 and thereafter.

Accounting Inflows and Outflows

The process takes into account the direct costs, the current stock, the inflows to the communities and the environmental services provided by the resource. Regenerating an area/protecting an area involves certain direct costs like plantation and after-care and protection (in this context we include paying guards and watchmen to protect and maintain the area), and the institutional costs involving communities coming together for decision making, conflict resolution and maintenance of the resource. Figure 10.1 shows that the expenditure in the initial years (1991–96) was high due to

Figure 10.1 Year-wise Annual and Cumulative Expenditure Details

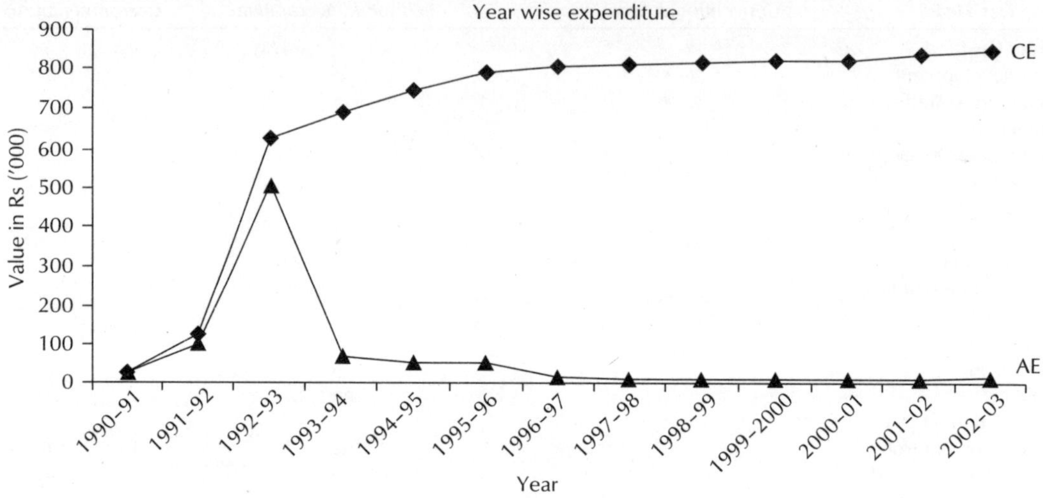

investments in regeneration and moisture conservation measures. Thereafter, only nominal recurring expenses were incurred. It may be noted that the initial expenditure in the process is high and tends to stagnate in most cases, where the investment is limited to protection and recurring expenditure by the institution.

This also includes the value of benefits forgone in the process of regeneration by the community in the initial period of protection. Such benefits are easily identifiable, as they often comprise marketable outputs (e.g., fodder, fuel wood, NTFP, etc.) and income sacrificed. The conventional accounting system fails to capture the benefits forgone in the total costs/investments in the process of regeneration.

Valuation at Market Price

As a result of the intervention, when the benefits begin to flow and are shared among the community, the outflow is monitored. The community, which has incurred costs in forgoing benefits during the initial stages, often has been observed to fix a very nominal price for the inflows/benefits shared or distributed within the village. In such a case, the conventional accounting tends to take into account only the price paid by the community as the actual value of the produce, which may not be true. The NRAS methodology tries to include the actual price of the commodity in the nearest market. For example, a bundle of fodder may cost Rs 0.50 in the village but the nearest market price of the same quantity could be Rs 2. The market price is taken into the accounting framework as the economic value (Figure 10.2) of the produce to give the actual representation of its worth. The figure shows the difference in value of the resource when calculated by conventional and economic methods. In case of conventional accounting, the value of the resource has been calculated at the nominal price fixed by the members while in case of economic accounting for the same resource, the prevailing market price has been taken into account.

Figure 10.2 Year-wise Value of the Resource under Different Accounting System

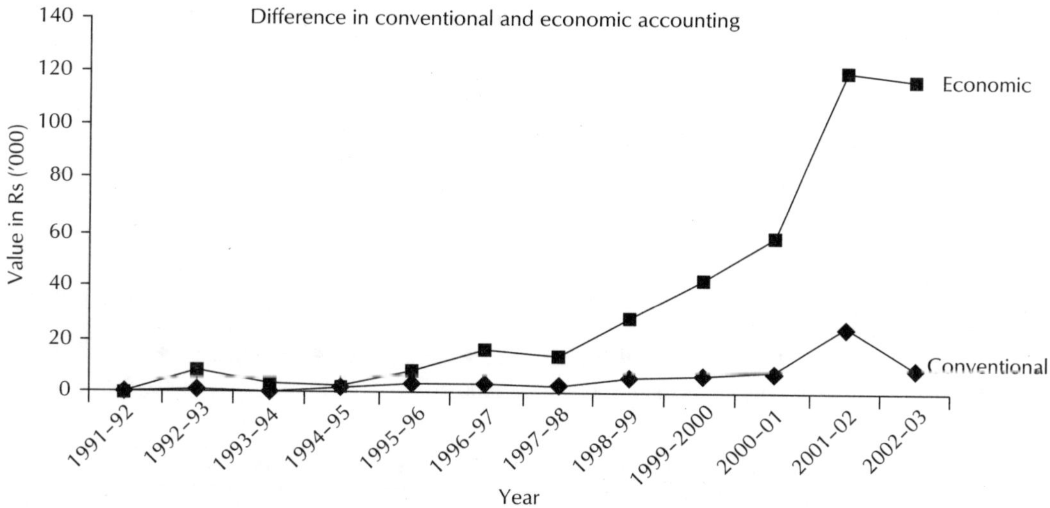

There are certain benefits enjoyed by the community for their use/consumption and therefore are not accounted under the conventional accounting methods as it does not involve any cash transaction; some examples could be open grazing of cattle in the plot during summer, collection of fuel wood, etc. But, the framework incorporates the economic value of the fodder/cattle feed consumed by cattle in the process of grazing. Similarly, the value of dry twigs as fuel wood and other NTFP (like gum, *Acacia nilotica* pods, fruits, etc.) have also been included as having an economic value.

Estimation and Valuation of Existing Stock

Valuation of the existing resource is an integral part of the accounting framework. The conventional accounting does not consider the value of the standing resource and tends to consider elements of nature—air, water, soil, plants, etc. to be free until they are converted into marketable products. Valuation of the forests involves the process of estimation of the existing biomass through standard methodologies taking 1 per cent of the total area using 20×20 sample plots. The amount of standing tree biomass (Figure 10.3) has been estimated by the weight and volume of timber and fuel wood.

The weight for single stemmed trees are estimated using the formula $W = a + (b \times D^2 \times H)$ (Chaturvedi and Khanna 1982) and for multiple stemmed trees using the formula $W = a + (b \times NS \times D^2 \times H)$ where W is weight in kilograms, D is diameter in decimetres measured at 50 cm above the ground, H is height in decimetres, NS is number of shoots and a and b are species specific constants.[6] The volume of the timber is calculated using the formula $V = \pi r^2 h$ where V is volume of the tree, r is the radius at breast height and h is height of the tree. The value from the sample plots is extrapolated for the total area to get the standing biomass in the entire plot. The estimation is repeated each year in the select plots to calculate the incremental growth (Figure 10.4) of biomass.

Figure 10.3 Year-wise Stock of Standing Biomass

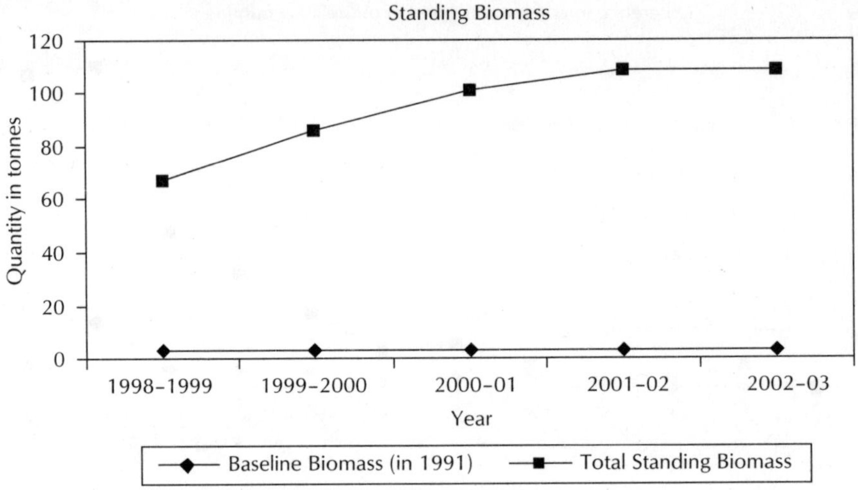

Figure 10.4 Year-wise Change in Incremental Growth of Standing Biomass

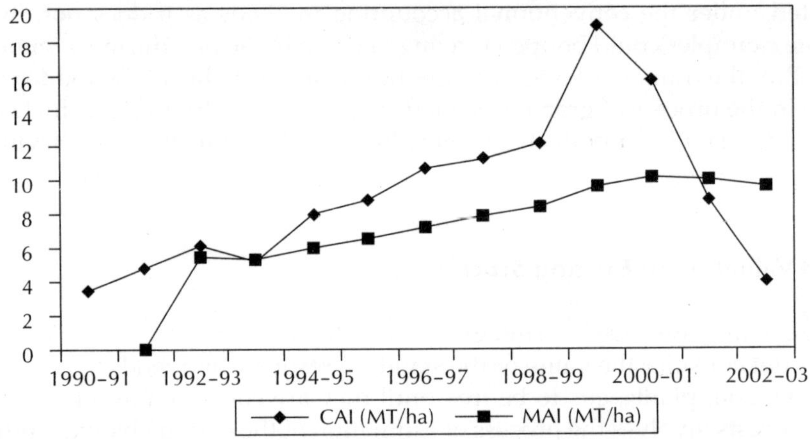

CAI = Current Annual Increment MAI = Mean Annual Increment

The valuation of the existing stock has been done through the market price method, where the resource is taken as equivalent to the market value of timber and fuel wood. All the trees above the circumference of forty-five centimetres have been considered as timber tree of which only 65 per cent have been valued at timber price and the rest 35 per cent are considered to be fuel wood. Further, all the trees of circumference less than forty-five centimetres have been valued at the market price of fuel wood.

As the wasteland regenerates, the biomass shows an increasing trend unless disturbed due to human interference or natural factors. A decline in the biomass is observed in 2001–02 and 2002–03 because of harvests at the community level. The Current Annual Increment (CAI) indicates the annual growth rate of the biomass, which is subject to variation due to a variety of natural factors such as droughts, spread of rain days, etc., or human factors such as harvests, excessive grazing, lopping, etc. The Mean Annual Increment (MAI) is a derived value taking into account the average CAI between two consecutive years and is more likely to show the average growth rate of biomass.

Valuing Environmental Services

The NRAS framework also includes valuing the environmental services that the plot of regenerated lands provide. As one observes, the regeneration helps in controlling soil erosion, improving soil fertility, soil formation, improving biodiversity, increasing moisture retention, sequestration of carbon besides also helping in regulation of microclimate. The forests also act as habitat for birds and wildlife, provide shade and have aesthetic values. Economic valuation of all these parameters is a difficult process and a variety of valuation techniques (Table 10.2) are used for the purpose.

Table 10.2 Existing Valuation Methods for Environmental Services

No.	*Valuation Methods*	*Brief Explanation of the Methodology*
1	Market Price Method	Estimates economic values for ecosystem products or services that are bought and sold in commercial markets.
2	Productivity Method	Estimates economic values for ecosystem products or services that contribute to the production of commercially marketed goods.
3	Hedonic Pricing Method	Estimates economic values for ecosystem or environmental services that directly affect market prices of some other good. Most commonly applied to variations in housing prices that reflect the value of local environmental attributes.
4	Travel Cost Method	Estimates economic values associated with ecosystems or sites that are used for recreation. Assumes that the value of a site is reflected in how much people are willing to pay to travel visit the site.
5	Damage Cost Avoided, Replacement Cost, and Substitute Cost Methods	Estimates economic values based on costs of avoided damages resulting from lost ecosystem services, costs of replacing ecosystem services, or costs of providing substitute services.
6	Contingent Valuation Method	Estimates economic values for virtually any ecosystem or environmental service. The most widely used method for estimating non-use, or 'passive use' values. Asks people to directly state their willingness to pay for specific environmental services, based on a hypothetical scenario.
7	Contingent Choice Method	Estimates economic values for virtually any ecosystem or environmental service. Based on asking people to make trade-offs among sets of ecosystem or environmental services or characteristics. Does not directly ask for willingness to pay—this is inferred from trade-offs that include cost as an attribute.
8	Benefit Transfer Method	Estimates economic values by transferring existing benefit estimates from studies already completed for another location or issue.

Valuation of change in soil fertility, carbon sequestration, soil erosion, improved biodiversity, shade and habitat for birds and wildlife as environmental services are being undertaken.[7] Valuation studies have been completed for two parameters—change in soil fertility and carbon sequestration. It is common knowledge that as the vegetation improves, there is an improvement in the quality of soil because of the decayed organic matter (humus), but it is difficult to value the improvement normally. The fertility of the soil (Table 10. 3) is measured based on the nitrogen, phosphorus and potassium (NPK) content, electrical conductivity and the pH value of the soil and any change in the soil fertility is calculated based on the replacement value method for NPK. It means for the changed amount of NPK, if one wants to replace the same quantity of nutrient by applying equivalent amount of fertiliser purchased from the market.[8] Sample from various parts of the plot[9] are taken each year and tested scientifically to estimate the changes.

Table 10.3 Year-wise Change in Soil Quality

Particulars	*1991*	*1998–99*	*1999–2000*	*2000–01*	*2001–02*	*2002–03*
Nitrogen (kg/ha)	419.56	1,388.8	1,452.80	1,361.92	1,913.79	1,501.44
Phosphorous (kg/ha)	41.00	61.84	87.02	82.52	52.50	48.33
Potash (kg/ha)	496.82	775.29	592.16	945.60	780.00	1,035.42
pH	7.75	7.65	7.67	7.36	7.31	7.78
EC (m. mhos/cm)	0.13	0.20	0.25	0.31	0.18	0.20
Total Replacement Value of NPK (Rs) for 25 hectares		**18,736.00**	**18,679.72**	**22,823.63**	**26,919.21**	**24,264.20**

The valuation of carbon sequestered uses the market cost method. There is a growing discussion on forests being carbon sinks through the absorption of carbon dioxide from the atmosphere by the process of photosynthesis. Article 3.3 of the Kyoto Protocol (UNFCCC) created the opportunity to establish new planted forests as carbon sinks and the carbon market is growing as Clean Development Mechanisms (CDM). The valuation of the carbon sequestered has been done based on the formula $C = 0.272\,X$ where C is the quantity of carbon sequestered and X is the wood biomass (green) above the ground. The coefficient 0.272 is derived from the relation given below:

Wood biomass (green) above the ground	X
Total (green) tree biomass	Y = 77% of X
Total (dry) tree biomass	Z = 46% of Y
Amount of carbon sequestration	C = 45% of Z

The total annual increment of carbon sequestered is given in the Table 10.4. The total carbon sequestered in the plot is 714 tonnes, which when valued at Rs 168/per metric tonnes (the average prices at the international level) is worth Rs 1.2 lakh.

Table 10.4 Year-wise Status of Carbon Sequestration

Particulars	*1991*	*1998–99*	*1999–2000*	*2000–01*	*2001–02*	*2002–03*
Quantity of carbon sequestration (tonnes/ha)	0.04	18.15	23.00	26.82	28.39	28.57
Amount of carbon sequestration (Rs '000)	0.17	76.23	96.90	112.64	119.23	119.90

Total Value of the Resource

The graphs (Figure 10.5) show the increase in the value of the forests with the quantification of each of the variables. This methodology therefore provides a means for measuring and comparing the various benefits of the forests and helps in improving the use and management of the resource.

The total value of the resource is the summation of the market value of the direct benefits, the value of the existing stock and the environmental services. As one would observe, the conventional

Figure 10.5 Year-wise Comparison of Value of the Resource under Different Accounting Procedures

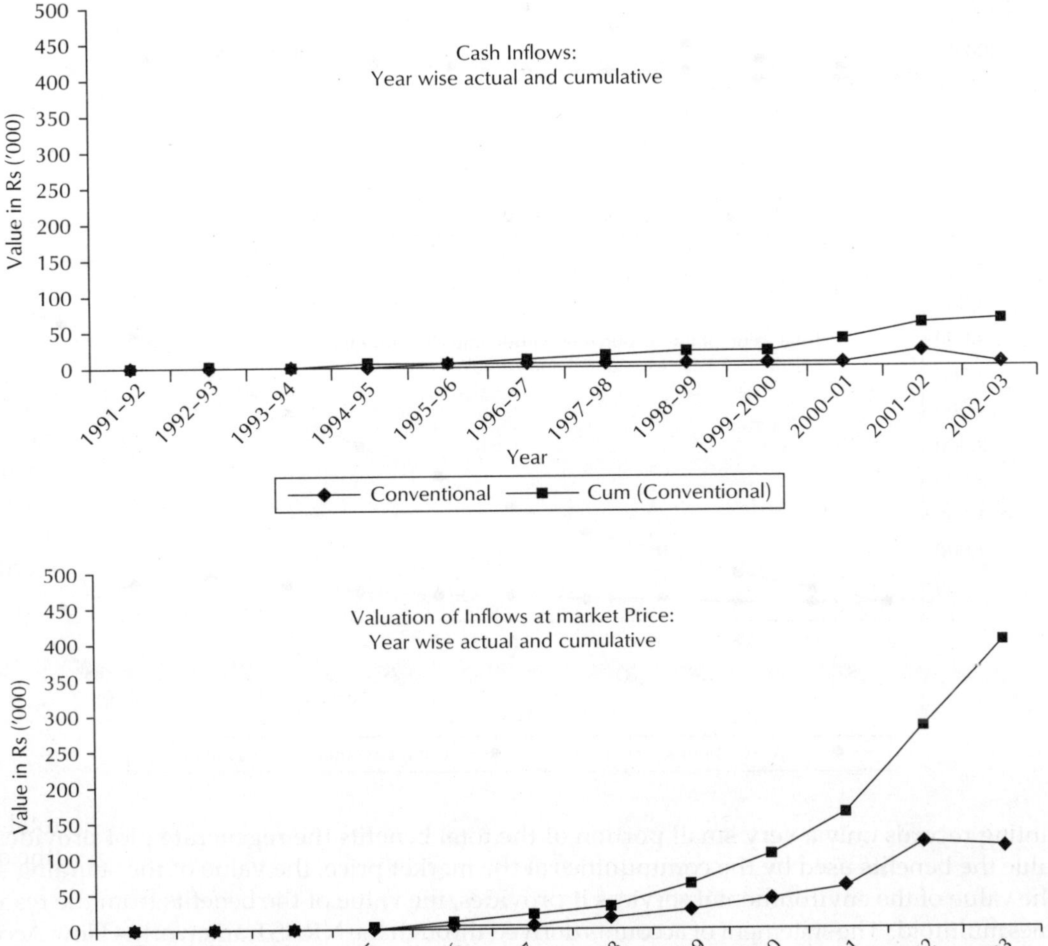

(*Figure 10.5 continued*)

(*Figure 10.5 continued*)

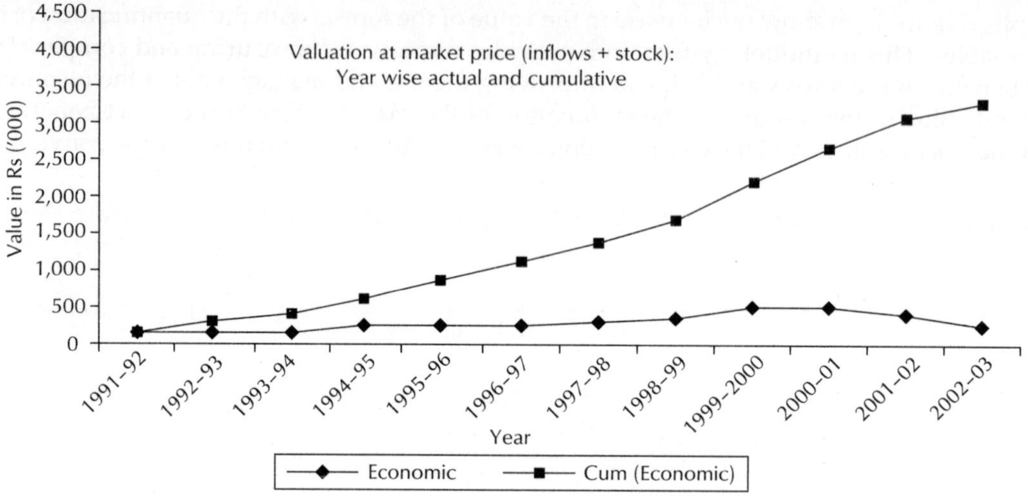

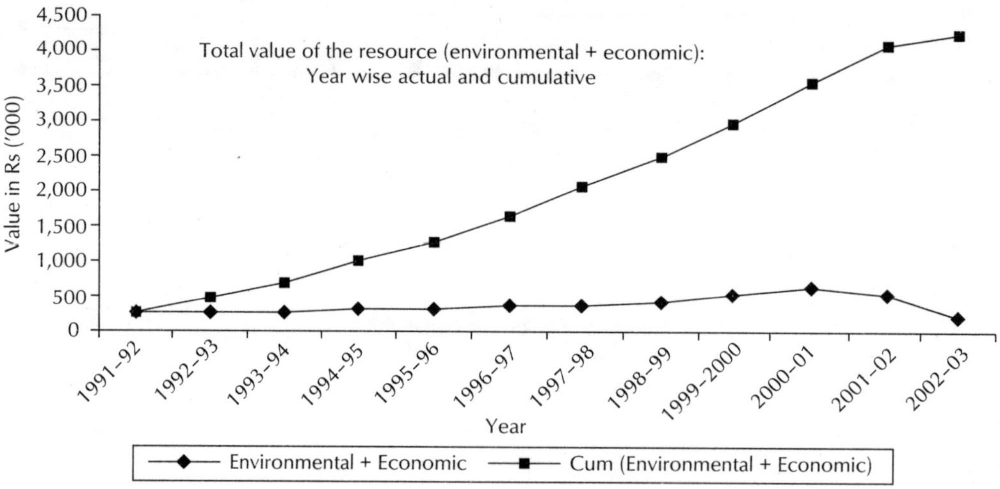

accounting records only a very small portion of the total benefits the regenerate plot provides. As we value the benefits used by the communities at the market price, the value of the standing stock and the value of the environmental services it provides, the value of the benefits from the resource becomes multifold. The statement of accounts derived through the NRAS framework—Flow Account and the Balance Sheet are attached in the annexures for reference. The total value of the resource as on 31 March, 2003 is about Rs 37.83 lakh.

The resource management institutions (TGCS in this case) often, therefore, face difficulty in cost-benefit analysis through the conventional accounting methods. The cumulative investment (Figure 10.6) in the twenty-five hectares of land is 8.51 lakhs for regeneration and protection of the

Figure 10.6 Year-wise Details of Investment and Benefits

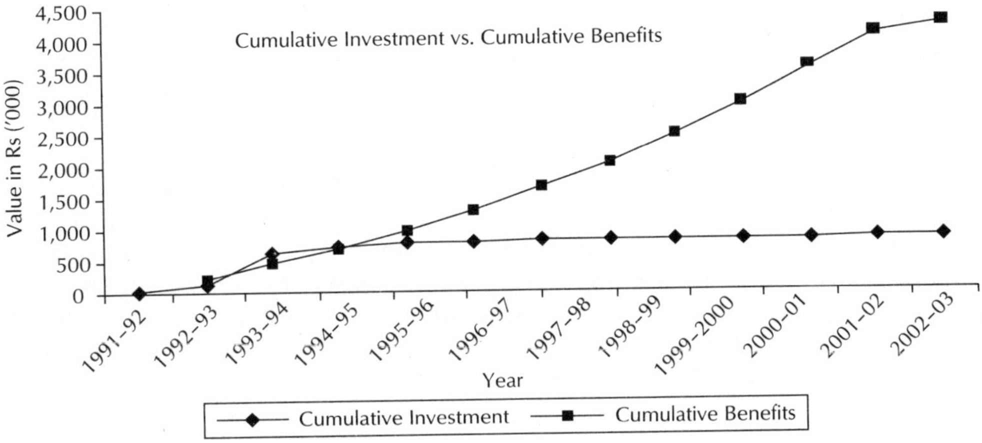

resource and the benefits in accordance with the conventional accounting is Rs 42.77 thousand, only indicating a massive loss making proposition. The accounting through the NRAS framework puts the value of the total benefit at Rs 37.83 lakh which is about four times the investment made in the plot. The valuation is still less as only a few parameters have been considered for valuation. The other parameters like soil formation, erosion control, biodiversity, habitat and aesthetic value are yet to be considered.

CONCLUSION

Natural Resource Accounting is evolving in India and many researchers and institutions are attempting to value the natural resources for the benefits they provide. The Framework for Natural Resource Accounting System (NRAS) is an attempt to further the discussion regarding the relevance of valuing the natural resources. By providing a means for measuring and comparing the various benefits of forests, NRAS can be a powerful tool to aid and improve decision making[10] and in the use and management of the forest resources.

As we observe in the case mentioned in the chapter, accounts maintained only in physical units do not enable policy makers to understand the impact of economic policies on natural resources thereby integrating resource considerations into economic decisions. The valuation of the resources using NRAS helps decision makers to take into consideration the actual value of the resource and not only the cash inflow and outflow. In this methodology, there is no conflict between accounting in physical and economic units because physical accounts are necessary prerequisites to economic accounts. NRAS therefore can be effective to help decision makers to decide on the conversion or conservation of the resource.

In addition, NRAS can be an effective tool to calculate the compensation for the resource when a particular piece of land is allocated or diverted for a particular purpose (say, industry or any development project). The present procedures in case of such diversion assume that forest resources are to provide the cost for compensatory afforestation which would be quite low in comparison to the actual value of the forest destroyed. This information could also be important for the community to effectively bargain for the loss.

It also makes a case for the inclusion of Natural Resource Accounting in the calculation of the Gross Domestic Product (GDP). Various countries have successfully integrated natural resource accounting to calculate national income. Though there is no consensus on how and what parameters and methodology could be used for accounting the environmental capital and the degradation and depletion of natural resources, however, we believe that putting in place such an accounting mechanism would help us value our resources better and take decisions according to the merit of the case.

However, a major difficulty facing valuation of a complex environmental system such as forests, is the insufficient information on and appropriate methodology for valuing different ecological benefits/parameters like biodiversity, habitat for wildlife, etc. Equally, it is difficult to provide a realistic value of the non-market environmental benefits or to get consistent results for the users of the resource using contingent valuation methods.

NOTES

1. The chapter is based on learning from the Tree Growers' Cooperative Project being implemented in the state.
2. Corresponding author—mondal@fes.org.in
3. Australia and US have integrated it into their National Accounts, many countries in Africa are attempting integrating the same in their National Accounts (e.g. Namibia, Zimbabwe, Mozambique).
4. This framework is a result of our continuous efforts at improving the methodology through the last ten years of practice. We have learnt a lot with our attempts at valuing the plot each year. The initial idea was worked upon by Indira Gandhi Institute of Development Research, Mumbai and Institute of Rural Management, Anand with the advice of Jack Ruitenbeek, Consultant, CIDA.
5. This attempt at valuation was taken up in Namnar Tree Growers' Cooperative Society (TGCS) supported by the Foundation for Ecological Security (FES) under its Tree Growers' Cooperative Project supported by the Canadian International Development Agency (CIDA). The work on the regeneration of the 25 hectares of wasteland was initiated in 1991. Namnar is a village in Lunawada Taluka, District Panchmahals, and is on the banks of river Mahi with 267 households. The plot where the regeneration and subsequent study was taken up is a grazing land leased to the TGCS by the Panchayat.
6. For example the value of a & b for Acacia nilotica (Stem + branch + twigs) is 0.0110 and 0.3928 respectively when oven dry and 0.0281 and 0.6872 respectively when measured green.
7. Change in soil fertility uses the replacement cost and substitute cost methods, the valuation of carbon sequestered uses the market cost method, soil erosion is calculated using damage cost avoided method and contingent valuation method (willingness to pay) is used for the valuation of improved biodiversity, shade and habitat for birds and wildlife as environmental services.
8. For the purpose of valuation of NPK the subsidised rate of fertiliser has been considered.
9. Collection of soil samples: A total of 5 plots are marked for collection of soil samples. From a soil sample plot, the soil is collected from 4–5 points, at each point a pit is dug up to 1 ft in the shape of V and the soils are collected from slices. From each pit around 0.5 kg of soil is collected and then we mix all five samples and retain 0.5 to 1 kg soil in a paper bag for testing. Each sample is tagged with date, time and collection point and sent for analysis.
10. To illustrate the use of the methodology, the District Registrar (Cooperatives) served liquidation notices to a few Tree Growers' Cooperatives taking into the fact that the cooperatives were at a loss or no/less monetary transactions, while

in reality, the institutions were doing well to protect the forest resource and meet their subsistence needs. On the presentation of the NRAS accounts to the Registrar, the department reverted the orders of liquidation and instructed the auditors to give special consideration to the Tree Growers' Cooperatives.

REFERENCES

Adamowicz, Wiktor. 2003. 'Economic Indicators of Sustainable Forest Management: Theory versus Practice', *Journal of Forest Economics*, 9(1): 27–40(14). 1 May 2003.

Adamowicz, W., P. Boxall, M. Haener, Y. Zhang, D. Dosman and J. Marois. 2004. 'An Assessment of the Impacts of Forest Management on Aboriginal Hunters: Evidence from Stated and Revealed Preference Data', *Forest Science*, 50(2): 139–52.

Anielski, Mark. 1992. 'Resource Accounting: Indicators of the Sustainability of Alberta's Forest Resources'. Paper Presented at the Second Meeting of the International Society of Ecological Economics, 3–6 August 1992, Stockholm.

Agrawal, Anil and Sunita Narain. 1985. *The state of India's Environment 1984–85 Second Citizens Report*. New Delhi: Centre for Science and Environment.

Ahmed, Yusuf, Salah El Serafy and Ernst Lutz. (eds) 1989. *Environmental Accounting for Sustainable Development*, Washington, D.C: The World Bank.

Barbier, Edward B., Mike Acreman and Duncan Knowler. 1997. *Economic Valuation of wetlands: A Guide for Policy Makers and Planners*. Department of Environmental Economics and Environmental Management, University of York, Ramsar Convention Bureau Institute of Hydrology, IUCN-The World Conservation Union.

Chaturvedi, A.N. and L.S. Khanna. 1982. *Forest Mensuration*, Dehra Dun, India: International Book Distributors.

Cooper, P., Gregory L. Poe, Ian J. Bateman. 2004. 'The Structure of Motivation for Contingent Values: A Case Study of Lake Water Quality Improvement', *Economical Economics*, 50: 69–82.

Costanza, R. and H. Daly. 1992. 'Natural Capital and Sustainable Development', *Conservation Biology*, 6(1): 37–46. March.

Costanza, R, Ralph d'Arge, Rudolf de Groot, Stephen Farber, Monica Grasso, Bruce Hannon, Karin Limburg, Shahid Naeem, Robert V. O'Neill, Jose Paruelo, Robert G. Raskin, Paul Sutton and Marjan van den Belt. 1997. 'The Value of the World's Ecosystem Services and Natural Capital', *Nature*, 387.

Forest (Conservation) Act. 1980, Ministry of Environment and Forest, Government of India, 25 October 1980.

Haener, M.K. and W.L. Adamowicz. 2000. 'Regional Forest Resource Accounting: A Northern Alberta case Study', *Canadian Journal of Forest Research*, 30: 264–73.

Hulkrantz, Lars. 1992. 'National Account of Timber and Forest Environmental Resources in Sweden', *Environmental and Resource Economics*, 2: 283–305.

Kristrom, B. and K. Skanberg. 2001. Monetary Forestry Accounting Including Environmental Goods and Services.

National Research Council. 1994. *Assigning Economic Value to Natural Resources* Washington, D.C: National Academy Press.

Pagiola, Stefano in collaboration with The Nature Conservancy and IUCN. 2004. *Assessing the Economic Value of Ecosystem Conservation*, Environmental Department Papers No 101. World Bank.

Perrings, C. and M. Gadgil. 2003. 'Conserving Biodiversity: Reconciling Local and Global Public Benefits', in Inge Kaul, Pedro Conceicao, Katell Le Goulven and Ronald U Mendoza (eds), *Providing Global Public Goods. Managing Globalization*, United Nations Development Programme. New York: Oxford University Press.

Reddy, Sudhakara B., Jyoti Parikh and Kirit S. Parikh. 1997. 'Framework for Economic, Social and Natural Resource Accounting for Land Regeneration' in J. Parikh and Sudhakara Reddy (eds), *Sustainable Regeneration of Degraded Lands*, pp. 107–36. New Delhi: Tata McGraw-Hill Publishing Company Limited.

Saxena, Rakesh and Rajesh Agrawal. 1997. 'An appropriate Socio-Eco-Environmental Accounting Framework Proposed for Tree Growers' Cooperatives With a Case Study', in J. Parikh and Sudhakara Reddy (eds), *Sustainable Regeneration of Degraded Lands*, pp. 137–66. New Delhi: Tata McGraw-Hill Publishing Company Limited.

Singh, Subrata. 2003. 'Conflicts & Disturbance—A reason to change: Lessons from Community based Natural Resource Management Institutions in Orissa, India', *International Journal on 'Forests, Trees and Livelihoods'*, 13(3): 233–46. United Kingdom.

Steiner, Achim, Steaven J. MacCormick and Jan Johnson. 2004. How Much is an Ecosystem Worth? Assessing the Value of Conservation (IUCN, The Nature Conservancy, The World Bank).

The International Bank for Reconstruction and Development/THE WORLD BANK 2004. How Much is an Ecosystem Worth? Assessing the Economic Value of Conservation.

ANNEXURE

Statement of Accounts

Annexure A Direct Flow Account of Namnar TGCS, 2002–03

			Economic (Rs)		Environmental
Items	*Physical*	*Financial (Rs)*	*Internal*	*External*	(Rs)
Benefit					
Fuelwood sales (tonnes)	18.85	9,425.00	18,850.00	0.00	0.00
Profit (from Tree biomass a/c)		0.00	3,15,945.54	0.00	0.00
Grant for Revegetation		7,510.00	7,510.00	0.00	0.00
Illegal Harvest (tonnes)	34.35	0.00	0.00	34,774.98	0.00
Minor forest produce (MFP)					
Babul pods + open grazing (tonnes)	540.00	0.00	0.00	54,000.00	0.00
Ber collection (kg)	95.00	0.00	0.00	475.00	0.00
Babul gum collection (kg)	120.00	0.00	0.00	4,800.00	0.00
Babul stick brush (no.)	54,750.00	0.00	0.00	2,737.50	0.00
Green foliage collection (tonnes)	9.00	0.00	0.00	1,350.00	0.00
Existing Stock					
Tree biomass a. Fuelwood (tonnes)	-391.43	0.00	-3,91,425.96	0.00	0.00
b. Timber (tonnes)	408.37	0.00	5,10,456.91	0.00	0.00
Total-(a)		**16,935.00**	**4,61,336.49**	**98,137.48**	**0.00**
Direct Net loss-(c) = (b–a) if b>a		0.00	0.00	0.00	0.00
Grand total-(a+c)	–	**16,935.00**	**4,61,336.49**	**98,137.48**	**0.00**
Cost					
Land revenue	–	2,315.00	2,315.00	0.00	0.00
Audit fee	–	0.00	0.00	0.00	0.00
Stationary	–	746.00	746.00	0.00	0.00

(*Annexure A continued*)

(Annexure A continued)

Items	*Physical*	*Financial (Rs)*	*Economic (Rs)*		Environmental (Rs)
			Internal	*External*	
Plantation & Aftercare	–	0.00	0.00	0.00	0.00
Salaries	–	4,350.00	4,350.00	0.00	0.00
Misc. exp.	–	1,325.00	1,325.00	0.00	0.00
Biogas exp.	–	4,500.00	4,500.00	0.00	0.00
Dead stock exp.	–	1,597.00	1,597.00	0.00	0.00
Economic loss due to formation of TGCS	–	0.00	9,425.00	98,137.48	0.00
Total-(b)		**14,833.00**	**24,258.00**	**98,137.48**	**0.00**
Direct Net benefit-(d) = (a–b) if a>b	–	2,102.00	4,37,078.49	0.00	0.00
Grand total-(b+d)		**16,935.00**	**4,61,336.49**	**98,137.48**	**0.00**

Annexure B Indirect Flow Account of Namnar TGCS, 2002–03

Items	*Physical*	*Financial (Rs)*	*Economic (Rs)*	*Environmental (Rs)*
Benefits				
Environmental gains				
Soil quality				
Nitrogen (Kg)	-10,403.75	0	0	-1,10,596.39
Phosphorous (Kg)	-105.13	0	0	-2,089.36
Potash (Kg)	6,444.16	0	0	45,699.85
Carbon sink (tonnes)	4.55	0	0	765.05
Soil loss (tonnes)	126.15	0	0	47,053.95
Employment -men		NE	NE	NE
-women		NE	NE	NE
Total-(a)	**0**	**0**	**0**	**-19,166.89**
Indirect Net loss-(c) = (b–a) if b>a		0.00	0.00	0.00
Grand total-(a+c)		**0**	**0**	**–19,166.89**
Costs				
Firewood loss (tonnes)				0
Open grazing lost (tonnes)	0	0	0	0
Total-(b)		**0.00**	**0.00**	**0.00**
Indirect Net benefit-(d) = (a–b) if a>b	0	0.00	0.00	-19,166.89
Grand total-(b+d)		**0**	**0**	**–19,166.89**

Note: NE = Not Estimated.

Annexure C Balance Sheet of Namnar TGCS as on 31 March 2003

Items	*Physical*	*Financial (Rs)*	*Economic (Rs)*	*Environmental (Rs)*
Assets				
Bank balance		21,388.19	21,388.19	0.00
Dead stock	–	4,704.00	4,704.00	0.00
Closing balance	–	183.60	183.60	0.00
Fuel wood-advance		14,295.00	14,295.00	0.00
Investment	–	2,200.00	2,200.00	0.00
Standing trees biomass (tonnes)	2,681.22	0.00	30,54,449.55	0.00
Environmental gains				
Soil quality (NPK-tonnes)	65.25	0	0	6,12,190.32
Carbon sink (tonnes)	720.80	0	0	1,21,093.64
Soil loss (tonnes)	–126.15	0	0	–47,053.95
Employment days	7,644	0.00	NE	NE
Direct net loss		0.00	0.00	0.00
Indirect net loss		0.00	0.00	0.00
Total Assets	**–**	**42,770.79**	**30,97,220.34**	**6,86,230.02**
Liabilities				
Share capital	–	4,660.00	4,660.00	0.00
Reserve fund	–	466.00	466.00	0.00
Depreciation	–	3,552.50	3,552.50	0.00
Direct net benefit		34,092.29	3,08,85,441.84	0.00
Indirect net benefit	–	0.00	0.00	6,86,230.02
Total Liabilities		**42,770.79**	**30,97,220.34**	**6,86,230.02**

Notes: NE=Not Estimated.

Total Value of the Regenerated Resource is 37.83 lakh.

11

Alternative Welfare Measures: Overview and Case Study of India

Brent Bleys

Abstract: This chapter outlines the different objections that have been put forward throughout the years to the use of the Gross Domestic Product (GDP) as a welfare measure. It is shown that some of the problems raised are even aggravated in developing countries. Three different approaches to alternative measurement of welfare are discussed. The first one involves extending the existing national accounts in order to arrive at a welfare measure that is theoretically more sound. The second approach evaluates the access to certain basic goods (e.g., education, health care, etc.) by the study of a set of social indicators. The third approach targets mental states more directly by means of survey data on reported subjective well-being. The Index of Sustainable Economic Welfare (ISEW), an indicator derived from the first type of approach, is discussed to a greater length. A case study presenting the available data on different types of alternative welfare measures is performed for India. This study reveals that economic growth is contributing substantially to human welfare.

INTRODUCTION

Two of the most widely used welfare measures are Gross Domestic Product (GDP) and GDP per capita. Although they were not originally designed to measure welfare, these indicators have become normative benchmarks for economic and even social performance. The figures of their growth rates are often the centre of political debate, as politicians tend to hide behind poor economic growth in times of depression or as they claim that the occasional strong economic growth is a direct result of their policy. Furthermore, GDP per capita is sometimes used to rank individual countries by state of development or welfare.

But looking at these economic indicators might just not be the same as measuring actual welfare since GDP measures only the value of marketed goods and services produced and consumed in an economy. The implicit assumption needed is that all economic growth adds to welfare, without making a distinction between the desirable and the undesirable, or between costs and gains (Cobb et al. 1995). Economic growth has become desirable by definition.

Yet, people today around the world experience more and more the negative consequences of economic growth—environmental degradation, stress, congestion, health-related problems, etc. These insights have led Goodwin (1997) to conclude that there exist different worldviews when it comes to the appraisal of economic growth, ranging from 'all economic growth contributes to well-being' to 'some economic growth may detract from well-being'. Within this context, Offer (2000) suggests that there might exist a curvilinear relationship between economic welfare and human welfare, in which the economic growth contributes more to human welfare in countries with a low standard of living.

The use of GDP as a welfare measure has been criticised since the early development of its underlying framework, the national accounts. Kuznets (1941), one of the founding fathers of the system of national accounts, expressed his concerns on this topic and highlighted four ambiguous terms in the definition of national income. These ambiguities point that, contrary to popular belief, GDP is not a value-free tool.

Criticism was also initiated by the social and the environmental movements—claims for the inclusion of natural capital in the national accounts and for adopting social indicators such as life expectancy and literacy rate as complements to the economic indicators, have grown through the years. A final category of criticism is of a more technical and methodological nature. It includes, for instance, the call for an equal treatment of similar expenditures by different agents within the national accounting framework.

England (1997) reviews the needs that have come forth from these different critiques on the GDP when thought of as a welfare measure. It is necessary:

- to specify the distinction between intermediate and final output;
- to distinguish between 'goods' and 'bads' in consumption expenditures;
- to account for asset depreciation in a comprehensive manner, including both manufactured and natural assets;
- to divide net output between consumption and capital accumulation;
- to take account of non-marketed goods and services (e.g. household services); and
- to take account of the welfare implications of various forms of social inequality.

Some of the problems specified earlier are even aggravated in developing countries. The omission of household services is an instance of this, since these services account for a larger part of the total economy in developing countries than they do in developed countries. Another problematic issue in developing countries is the process of data gathering. Ravallion (2003) finds that the per capita private consumption derived from the national accounts deviates significantly from the mean household income based on national sample surveys. In India, the ratio of the latter to the former is as low as 0.6 (Ravallion 2003).

Accepting the shortcomings of GDP and stating that this measure was never intended to be used as a welfare indicator is not the right way out, since this will not prevent GDP from being used in this way. Alternative measures exist already and they should be more supported by media, international organisations and policy makers. However, as long as these alternative measures lack wide acceptance as comprehensive indicators of welfare, GDP will be filling the role of welfare measure by default.

Offer (2000) distinguishes three different approaches to the 'alternative' measurement of welfare. The first one involves extending the existing national accounts in order to incorporate non-marketed goods and services and to allow for other adjustments (positive or negative) as well. A second approach starts from the idea that access to certain goods constitutes a precondition of welfare. This access is evaluated by the study of a set of social indicators. The last approach targets mental states directly, by means of survey data on self-reported subjective well-being and by research on the dynamics of hedonic experience.

Diener and Suh (1997) argue that it is necessary to use both social indicators and subjective well-being measures for the evaluation of a society, since these measures add substantially to the regnant economic indicators that are now favoured by policy makers. They claim that each approach to measuring welfare contains information that is not contained in the other measures.

The different approaches that were outlined by Offer (2000) will be analysed in the next section. One specific welfare measure, the Index of Sustainable Economic Welfare, will be highlighted in the third section, while the last section will give a review of the available data for India.

ALTERNATIVE WELFARE MEASURES

This section starts with a short summary of Veenhoven's findings (1996) on subjective welfare measures, after which social indicators, such as the Human Development Index and the Index of Social Progress, are briefly discussed. Finally the process of extending the existing economic accounts will be reviewed.

Subjective Welfare Measures

The basic premise of subjective welfare research is that in order to understand the welfare of an individual, it is important to directly measure the individual's cognitive and affective reactions to his or her whole life (Diener and Suh 1997). The important distinction with the other approaches is that subjective welfare measures are primarily concerned with the respondent's own internal judgement of welfare.

In separating subjective measures from the objective ones, Veenhoven (2002) makes a distinction between substance and assessment. Substance relates to the question of what is being measured, while assessment refers to the actual process of collecting the data. Measures are said to be objective when they are concerned with things which exist independent of subjective awareness and when the measurement is based on explicit criteria and performed by external observers. Subjective indicators measure subjective substances (like identity, happiness and trust) using subjective appraisal techniques such as self-reporting.

Veenhoven (2004) continues by dividing human well-being into four categories using a matrix that separates inner qualities from outer qualities vertically and life chances from life results horizontally. The four categories that appear in this matrix are: 'living in a good environment', 'being able to cope with life', 'being of worth for the world' and 'enjoying life'. For each of these categories, indicators (both objective and subjective) have been developed throughout the years to measure the progress made. An example of a subjective measure of the category 'enjoying life' is simply asking people how much they enjoy their life as a whole. Such questions can be framed in several ways, using different keywords and response formats.

Veenhoven (2004) rejects any attempt to aggregate objective measures of different categories into one single index (in order to get a picture of general well-being), since this process involves adding apples to oranges—there is no sense in combining 'chances' and 'outcomes'. He also claims that these attempts are incomplete as they are limited to only a few aspects of well-being.

According to Veenhoven (1996), the most comprehensive measure of well-being is how long and happily people live. This can be measured by combining data on length of life from civil registration with data on satisfaction with life as assessed in surveys. A simple measure, the Happy Life Expectancy (HLE) can be calculated by multiplying life expectancy with life satisfaction. Data on happy life years are available for sixty-seven countries in the 1990s (the number of countries covered continues to expand) and can be found on the World Happiness Database[1].

Veenhoven (2004) calls for the development of subjective welfare measures in developing countries, since there is a clear information deficit. Surveys on aspirations, needs and satisfactions of citizens could enrich the public policy debate and add to the quality of available information in these countries. Since these surveys are relatively cheap and of high value, the conduction should start as soon as possible (Veenhoven 2004).

Social Indicators

Land (1999) defines social indicators as statistical time series '... used to monitor the social system, helping to identify changes and to guide intervention to alter the course of social change'.

The social indicator movement emerged in the 1960s and was inspired by the idea that real welfare was not fully captured by economic indicators alone. After blooming in the 1970s, when several leading countries and international organisations published series of social indicators, the interest slowly waned during the 1980s. But now the movement has entered a new era with the development of summary social indicators. The purpose of these indicators is to summarise from different domains into a single index in order to get an idea of the progress of a country in terms of social conditions, both over time and compared to other countries (Sharpe 1999).

Two summary indices are discussed here: the Human Development Index and the Index of Social Progress.

Human Development Index

The Human Development Index (HDI), developed by the United Nations Development Programme (UNDP) is probably the best known composite index of social and economic well-being. The index was calculated for the first time in 1990.

The HDI keeps track of three dimensions that are considered important for human well-being (the indicator or indicators used are shown in brackets):

- a long and healthy life (life expectancy at birth);
- knowledge (adult literacy rate and gross enrolment ratio); and
- a decent standard of living (GDP per capita–purchasing power parities in US$).

Before the HDI itself is calculated, an index needs to be created for each of these dimensions. To calculate these indices, minimum and maximum values (goalposts) are chosen for each underlying indicator. Performance in each dimension is subsequently expressed as a value between 0 and 1. The HDI is then simply the average of the three dimension indices.

More recently, UNDP has developed some additional indices in order to reflect gender inequalities and poverty. The Human Poverty Index (HPI) measures deprivation in three basic dimensions of human development as captured in the HDI. Two alternative indices exist: HPI–1 which tracks poverty in developing countries, while HPI–2 is designed to measure deprivation in developed countries. In addition to the three basic dimensions, HPI–2 also captures social exclusion.

Gender inequalities are highlighted in the Gender-related Development Index (GDI) and the Gender Empowerment Measure (GEM). In the GDI methodology, female and male indices in each dimension are combined in a way that negatively affects the achievement between men and women. Focusing on women's opportunities rather than their capabilities, the GEM captures gender inequality in three important areas: political participation, economic participation and power over economic resources.

All the indices appear annually in the Human Development Report. The latest edition (UNDP 2004) can be found on the UNDP website.[2]

Index of Social Progress

Estes (1984) developed another multidimensional index for measuring social well-being: the Index of Social Progress (ISP). This index was originally designed to serve as a reliable tool for assessing shifts in the capacity of nations to provide for the basic needs of their populations and to facilitate the analysis of welfare-relevant data at regular intervals.

The updated methodology (Estes 1997) for the Weighted Index of Social Progress (WISP) aggregates forty-six social indicators into 10 sub-indices before arriving at the final index. The sub-categories of the ISP are: education, health status, women status, defence effort, economy, demography, geography, political participation, cultural diversity and welfare effort. The statistical weights for the exercise are derived through a two-stage varimax factor analysis in which each indicator and sub-index is analysed for its relative contribution toward explaining the variance associated with changes in social progress over time (Estes 1997). The latest effort (Estes 2003) provides WISP-scores in the year 2000 for 163 countries.

Estes (1997) claims that the WISP is a more comprehensive, valid and reliable instrument for assessing changes in social development over time than any of the other indices on national and international progress, such as the GDP and the HDI. Yet Osberg (2001) feels that the high complexity of the WISP calculation is limiting clear comparison with the other measures.

Extended Accounts

We can distinguish between two reasons for extending the conventional economic accounts. First, some changes are made in order to facilitate critiques of the methodology, such as the lack of taking capital depreciation into account (which has led to the calculation of the Net National Product) or the absence of natural capital stocks and flows (facilitated by the creation of complementary environmental satellite accounts such as the SEEA).

The second reason for extending the accounts is to provide a welfare measure that is theoretically more sound. Extended welfare accounts usually start out with the core of the System of National Accounts (SNA) and make adjustments on consumption and capital accounts; typically some commodities and services, which are not seen as final goods but as 'regrettable necessities', are eliminated. Finally, these accounts impute a value to sources of welfare from outside the market (such as household services).

The following subsection provides a historical overview of the most important contributions to the extension of economic accounts. The Measure of Economic Welfare, the Economic Aspects of Welfare, the Index of Sustainable Economic Welfare and the Genuine Progress Indicator will be discussed here.

Measure of Economic Welfare

One of the earliest efforts (Nordhaus and Tobin 1972) resulted in the Measure of Economic Welfare (MEW), a comprehensive measure of the annual real consumption of households. In the MEW index, consumption includes all goods and services, marketed or not, valued at market prices or at their equivalent in opportunity costs to consumers. Collective consumption is also included (as far as these expenditures are not considered as instrumental expenditures) and allowance is made for negative externalities such as environmental damage, and disamenities of congestion and urbanisation.

The corrections can be divided into three categories: (*i*) reclassification of GNP final expenditures; (*ii*) imputations for capital services, leisure and non-market work; and (*iii*) disamenities of urbanisation.

The study was undertaken to answer the following question: how good are measures of output for evaluating the growth of economic welfare? Nordhaus and Tobin (1972) conclude: 'Is growth obsolete? We think not. Although GNP and other national income aggregates are imperfect measures of welfare, the broad picture of secular progress which they convey remains after correction of their most obvious deficiencies.' Daly and Cobb (1989) disagree with these findings, claiming that the relatively close association between growth of per capita GNP and MEW disappears when the results are more carefully examined. They find that, when looking at specific time intervals or when adjusting some of the assumptions, the GNP is not a good proxy for welfare at all.

Economic Aspects of Welfare

A second attempt was made by Zolotas (1981) whose welfare measure, the Economic Aspects of Welfare (EAW), was constructed to depict the full range of actual changes in a society's quantifiable well-being, regardless of whether or not these changes were the outcome of market transactions. The EAW index also takes the private consumption expenditures as its starting point, while various other magnitudes are added or deducted according to whether they are positively or negatively related to economic welfare.

Negative adjustments are made for expenses on consumer durables, advertising, the depletion of natural resources, the rapid growth and the rising social cost of environmental pollution, the cost of commuting, private health and education outlays. Positive corrections include services from the stock of public capital, services from durable consumer goods, household services, leisure time and public sector services (relating mainly to expenditure on education and health).

Zolotas (1981) finds that his hypothesis, namely, that the economic aspects of social welfare are a diminishing function of economic growth in industrially mature societies, is confirmed by his empirical results: the EAW index rises at a lower rate than GNP. This is because the items deducted from private consumption grew faster than GNP during the same period. The results even overestimate the actual level of economic welfare, since the damage costs are higher than they are made to appear by the inadequacy of available data (Zolotas 1981).

The EAW differs from the MEW index by more sharply focusing on the current flow of goods and services and by largely ignoring capital accumulation and the issue of sustainability (Daly and Cobb 1989). The EAW measure also addresses the issue of environmental damages more directly than the index of Nordhaus and Tobin (1972), where there is only an imputation for urban disamenities.

The Index of Sustainable Economic Welfare and the Genuine Progress Indicator

Building upon the earlier efforts, Daly and Cobb (1989) constructed the Index of Sustainable Economic Welfare (ISEW). Like the previous indices, the ISEW starts with the personal consumption expenditures and adjusts this figure for such factors as income distribution, net capital growth, resource depletion, environmental damage and the value of unpaid household labour. The index was later revised by Cobb and Cobb (1994) and some parts of the methodology were updated.

In 1995, Redefining Progress[3] elaborated further on the ISEW framework to arrive at a new index for measuring economic welfare: the Genuine Progress Indicator (GPI). This measure adds a number of new categories to the ISEW: the value of volunteer work, costs of crime and family breakdown, loss of leisure time, cost of underemployment and cost of ozone depletion. The GPI has already been compiled in the United States (Anielski and Rowe 1999; Venetoulis and Cobb 2004) and in Australia (Hamilton and Denniss 2000).

The ISEW will be discussed to a greater length in the following section.

INDEX OF SUSTAINABLE ECONOMIC WELFARE

Daly and Cobb (1989) claim that 'as GNP does not come close enough to measuring economic welfare, its continued use as if it were a significant indicator of economic well-being is an egregious instance of the fallacy of misplaced concreteness.' In order to offer an alternative, they have created the Index of Sustainable Economic Welfare (ISEW) based on earlier research in the field of extending the economic accounts.

England (1997) considers the ISEW as the only effort so far to integrate all critiques on the GDP when used as a welfare measure (see the introduction of this chapter) into one accounting scheme. The index touches the welfare effects of both macro-economic activity and social inequality, and takes into account the effects of economic growth on the environment.

In what follows, I will discuss the theoretical underpinnings of the ISEW, look more closely into its methodology, review the results of international studies and investigate the strengths and weaknesses of the index.

Theory

A combination of four elements can be thought of as essential in the theoretical framework underlying the ISEW:

- Hicksian income;
- extension of the capital concept;
- inclusion of non-market flows; and
- defensive expenditures.

The central criterion for defining the concept of income has been well stated by Hicks (1939): 'we ought to define a man's income as the maximum value which he can consume during a week, and still expect to be as well off at the end of the week as he was at the beginning.' The same basic idea of income holds at the national level and for annual time periods (Daly and Cobb 1989). It is worth to note that the central defining characteristic of income is sustainability, as the total capital stock should be kept intact.

Another important aspect of measuring sustainable economic welfare is the extension of the capital concept. When reviewing the different interpretations of capital, Fisher (1965) found that they generally had three elements in common:

- capital has a productive capacity;
- capital generates dividends for the future; and
- capital only includes factors that themselves have been produced in the economic system.

These factors are today still functioning as a guideline for the classification of capital. In this sense, only man-made goods, such as machinery and buildings, will classify as capital. But when the third condition is relaxed, a broader definition of capital will emerge (Stymne 2000). This would allow the capital concept to be extended in order to include natural capital and human capital.

A third issue is the inclusion of non-market flows such as household labour and environmental degradation (Jackson et al. 1997). This allows for externalities to be incorporated in the ISEW framework.

Finally, the concept of 'defensive expenditures' (expenditures that are made to offset a decrease in welfare), which was introduced by Leipert (1989), can be used to distinguish 'good' from 'bad' when consumption expenditures are to be evaluated. Common examples of defensive expenditures include: locks and security systems, hospitals bills from car accidents, personal water filters, etc.

Methodology

Daly and Cobb (1989) first described the ISEW in their influential book *For the Common Good: Redirecting the Economy Toward Community, the Environment, and a Sustainable Future*. The methodology was described in great detail and the index was compiled for the United States (1950–86). Five years later, Cobb and Cobb (1994) revised the methodology, adding new items to it and incorporating new valuation methods. Their work became the foundation for the international research in measuring economic welfare.

In the following years, the ISEW was calculated in many countries: Germany (Diefenbacher 1994), the Netherlands (Rosenberg et al. 1995), Sweden (Jackson and Stymne 1996), the United Kingdom (Jackson et al. 1997), Austria (Stockhammer et al. 1997), Italy (Guenno and Tiezzi 1998), Chile (Castaneda 1999) and Poland (Gil and Sleszynski 2003). In each country, minor adaptations were made to the original methodology in order to overcome problems with data availability or to pay attention to country-specific issues.

The basic methodology used to calculate the ISEW can be summarised as follows (Jackson et al. 1997):

ISEW = personal consumption
-losses from income inequality
+domestic labour
+non-defensive public expenditures
-defensive private expenditures
+capital adjustments
-costs of environmental degradation
-depreciation of natural capital.

Income inequality is factored in on the assumption that an additional and equal amount of money adds more to the welfare of a poor family than it does to a rich family. Inequalities should, therefore, not be considered separate from the magnitude of economic welfare (Daly and Cobb 1989). In the original ISEW, the Gini coefficient is used to adjust personal expenditures to income inequality. Subsequent studies performed in other countries favour the use of the Atkinson index (e.g. Jackson and Stymne 1996) as it explicitly states the preference for an equal distribution of income.

The idea of including the production of services by members of a household is intuitively compelling but the calculation of the imputation is not very straightforward (Daly and Cobb 1989). Problems arise from the definition of household labour, its measurement and its valuation. The total number of hours spent on housework is derived through surveys on time use, while valuation is based on a shadow price (average wage rate of domestic workers).

Public expenditures are considered to be mainly defensive: only half of the expenditures on higher education and on health are considered to enhance welfare (Cobb and Cobb 1994). Although this methodology is used in most of the international ISEW studies, some exceptions exist, for instance, in the calculation of the GPI for Australia (Hamilton and Denniss 2000).

Defensive private expenses are to be subtracted from the consumption base, since they have already been included. The ISEW (Cobb and Cobb 1994) makes adjustments for expenditures on health and education, costs of commuting, personal expenditures on pollution control and costs of car accidents.

Capital adjustments are needed in order to deal properly with consumer durables and to keep track of net capital growth. Durable consumer goods should be regarded as a capital stock—the annual expenditures on these goods are not important but the annual services delivered by this stock are. In order to calculate the net capital growth, Daly and Cobb (1989) advocate the introduction of a 'growth requirement' that is defined as the growth of capital necessary to compensate for depreciation and population growth. Annual net capital growth equals the annual capital growth minus the growth requirement. Besides this calculation, Daly and Cobb also include a category (net investment position) that takes into account whether the source of capital can be sustained on the assumption that sustainability requires long-term national self-reliance.

The ISEW keeps track of environmental damage in two ways. First, there are items that reflect the direct effects of water, air, and noise pollutions, based on estimates of emission trends and damage costs. A second category consists of estimates of long-term environmental damage from climate change and ozone depletion. These damages are assumed to be cumulative and directly related to energy consumption and consumption of CFCs respectively.

Natural capital and the availability of natural resources are also incorporated in the ISEW since current depletion impoverishes future generations. Daly and Cobb (1989) reject the idea of discounting the effects of resource depletion on the future and instead propose the view that 'any reduction in economic welfare in the future below the level currently enjoyed should be counted as if the cost occurred in the present'. In the ISEW, an estimate of the amount that would need to be set aside in a perpetual income stream in order to compensate future generations for the loss of services from non-renewable energy resources, is deducted. In addition, the value of the loss of biological resources such as wetlands and farmlands is factored in.

Human capital and the value of leisure are explicitly excluded from the ISEW framework. Daly and Cobb (1989) recognise the importance of human capital in sustainable economic welfare, but the validity of measuring inputs such as expenditures on medical care or on schooling to derive meaningful estimates of the stock of human capital is questioned. The imputation for leisure is omitted 'because of the dubious calculations involved in it and [because] it would outweigh all other components in a measure of welfare' (Daly and Cobb 1989).

In practice, the ISEW is the result of a lengthy series of adjustments to the personal consumption expenditures. For a more detailed review of the methodology, Cobb and Cobb (1994) or Jackson et al. (1997) can be consulted.

Results

A common finding among the series of international studies devoted to measuring economic welfare is the growing divergence between GDP per capita and ISEW per capita during the last two decades. In many countries, this divergence can be explained by an increasing income inequality, rising costs of resource depletion and escalating long-term environmental costs. During the 1980s and 1990s, economic welfare levels-off or starts declining in most countries. Figures 11.1 to 11. 4 present the findings of the ISEW or GPI studies in the United States, the United Kingdom, Sweden and Australia.

Max-Neef (1995) finds in these results a confirmation of his 'Threshold Hypothesis': 'for every society there seems to be a period in which economic growth (as conventionally measured) brings about an improvement in the quality of life, but only up to a point—the threshold point—beyond which, if there is more economic growth, quality of life may begin to deteriorate'.

All the evidence points to the fact that GDP or GDP per capita should not be used to measure welfare: economic growth does not always guarantee a rise in welfare.

Figure 11.1 Divergence between Per Capita GDP and ISEW (United States)

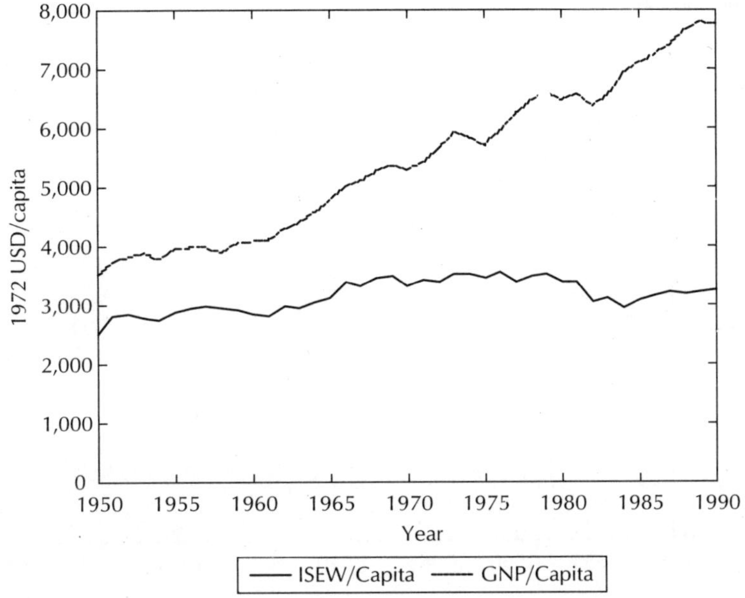

Figure 11.2 Divergence between Per Capita GDP and ISEW (United Kingdom)

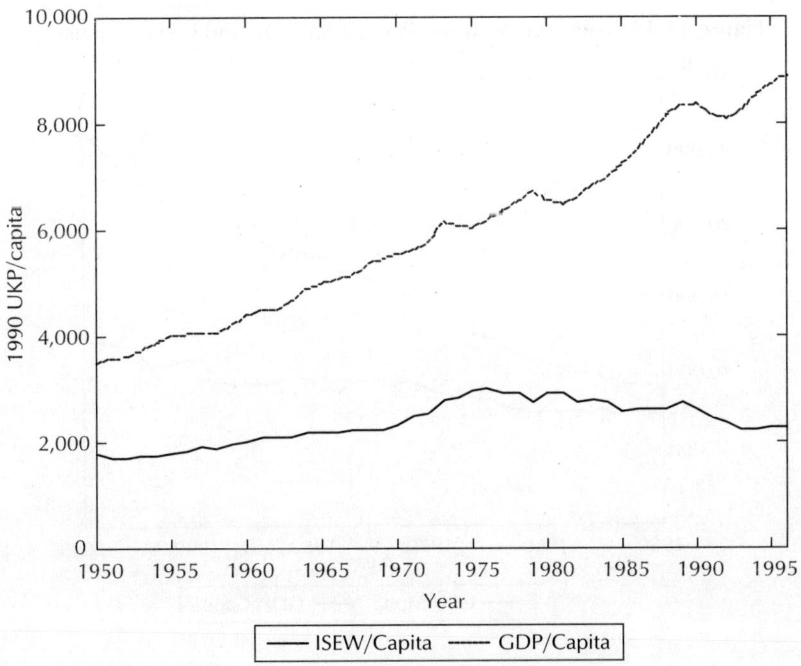

Figure 11.3 Divergence between Per Capita GDP and ISEW (Sweden)

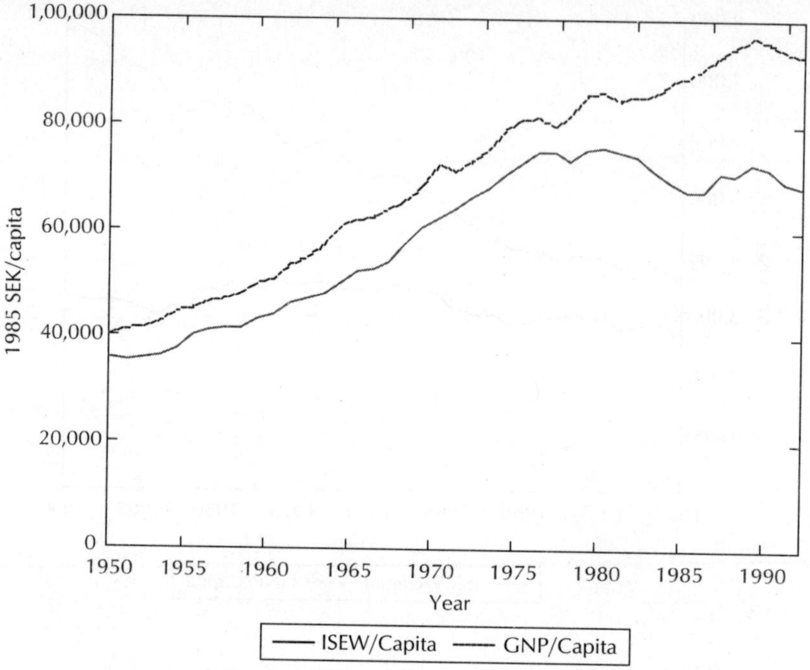

Figure 11.4 Divergence between Per Capita GDP and GPI (Australia)

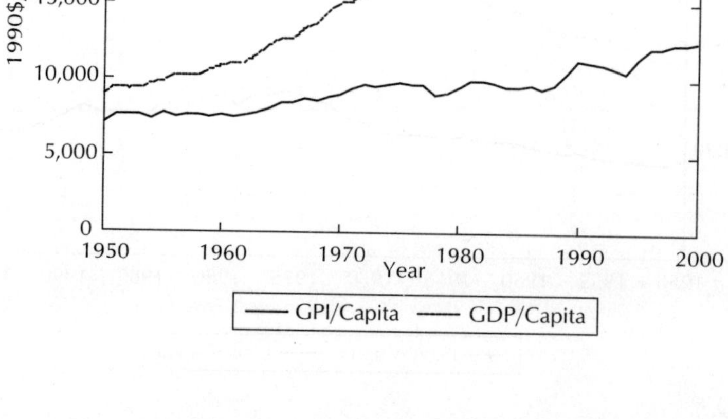

Discussion

Daly and Cobb (1989) consider the ISEW to have a high value for policy-making. They claim that policies governed by the index can truly stimulate economic welfare since the ISEW highlights policy areas that should receive most attention. These could be: reducing income inequality, investing more to sustain the economy into the future, taking measures to control environmental pollution, etc. But Daly and Cobb also acknowledge the caveats and limitations of their index. Their most important concern is that the base of the ISEW relies on consumption. Although consumption is certainly a more appropriate measure of welfare than production, Daly and Cobb consider it still to be questionable given the diminishing welfare returns of increases in consumption. They are also unhappy about the fact that the ISEW does not take any account of the relative level of wealth or consumption.

Neumayer (1999) argues that the ISEW lacks a sound theoretical foundation and that the index is arbitrary in the components it includes or excludes as contributors to welfare. Lawn (2003) offers a solution to the first critique by pointing out that the income concept of Fisher can be used as a theoretical foundation. Neumayer's second objection is refuted by Hamilton and Denniss (2000) who claim that the selection of adjustments within the ISEW framework is a direct result of the process that identifies the deficiencies of the GDP as a measure of welfare.

According to Neumayer (1999), the authors of the ISEW commit the mistake of methodological inconsistency in two respects:

- the ISEW cannot function both as an indicator of current welfare and an indicator of sustainability at the same time: what affects current well-being need not affect sustainability and vice versa; and
- the index is not an indicator of strong sustainability, but one of weak sustainability, since the ISEW framework allows for perfect substitution among different types of capital.

Neumayer (2004) proposes the use of the Human Development Index in combination with the Genuine Savings indicator to solve the first inconsistency.

Neumayer (1999, 2000) and Crafts (2002) criticise the valuation methods of various items in the ISEW framework, claiming that adjustments to their methodologies would remove the general finding of the studies on the index (decline in sustainable economic welfare during the past decennia). The two components that have attracted the greatest critique are non-renewable resource depletion and long-term environmental damage.

This sensitivity of the ISEW to the underlying assumptions within its framework and its valuation methods has led Neumayer (2000) to conclude that the threshold hypothesis, as defined by Max-Neef, fails to materialise and that the growing gap between ISEW, on the one hand, and GDP, on the other, 'might be an artifact of highly contestable methodological assumptions'. Neumayer (1999) concludes that '[there remain] doubts about the policy relevance of an ISEW measure that necessarily rests on arbitrary assumptions and can be shown to be invalid as a reliable indicator for welfare and sustainability'. Although this might be too strong a conclusion, it should be noted that many other authors (e.g. Rosenberg et al. 1995) appreciate the ISEW more as a first step to a better measure for welfare than as an ideal indicator of national welfare.

Compiling an Index of Sustainable Economic Welfare is still a valuable exercise, given the importance of its underlying rationale (economic growth and welfare are different concepts) and the potential of the index as a communication tool. The ISEW offers an ideal way to make people understand the different effects that economic growth has on human welfare. The index is also appreciated because it provides an empirical translation of the critiques on the GDP as a welfare measure.

As there is no widespread consensus on the methodology of the ISEW, international acceptance is still some way off. Efforts at a national level are helpful in screening different welfare issues and their valuation methods. Compiling ISEWs allows for the start of a learning process, which can eventually lead to a methodology that is internationally agreed upon.

However, policy decisions should not be guided by one single indicator: neither the GDP nor the ISEW offers a complete picture of a society. For an exhaustive report on welfare, accompanying indicator systems would be necessary (Stockhammer et al. 1997). These might include the other alternative welfare indicators mentioned earlier in this chapter (social indicators and subjective welfare indicators).

In the next section we will analyse India's progress during the last forty years based on the indicators discussed in this chapter.

CASE STUDY FOR INDIA

When we look at the two most widely used welfare measures, the GDP and the GDP per capita, India has scored quite well during the last four decades (Figure 11.5). The level of GDP went up by more than 400 per cent between 1960 and 2000, while GDP per capita more than doubled during the

Figure 11.5 Economic Growth and Growth in Per Capita GDP in India (1960–2002)

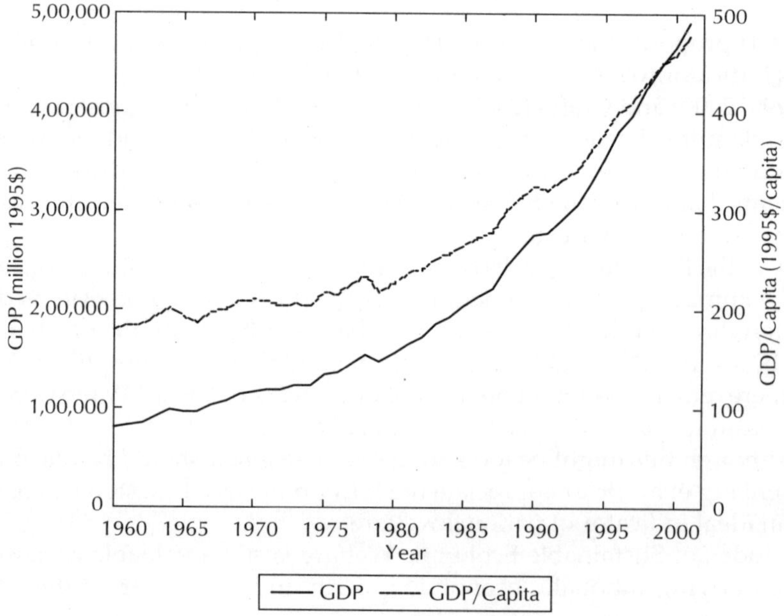

same period[4]. When we look at the annual percentage growth rate, we notice that, since the 1980s, India comes out better than many developed countries. These standard economic indicators show a very positive evolution, but how is this evolution reflected in the other types of indicators?

Examining some of the most prominent social indicators, we find a similar trend. Life expectancy at birth rose from 38.7 in 1952 to 63.9 in 2002. The adult illiteracy rate fell from 66.9 per cent in 1970 to 43.5 per cent in 1998, while gross enrolment ratios in secondary education increased by 5 per cent between 1990 and 1996.[5] These trends are also reflected in the Human Development Index (HDI), which rose from 0.411 in 1975 to 0.595 in 2002.[6] The Weighted Index of Social Progress (WISP) shows a positive evolution as well. The index almost doubled during the period under consideration: it went up from 19 in 1970 to 34 in 2000 (Estes 2003). The trends of both the HDI and the WISP are depicted in Figure 11.6.

Figure 11.6 Evolution of the Human Development Index and the Weighted Index of Social Progress for India

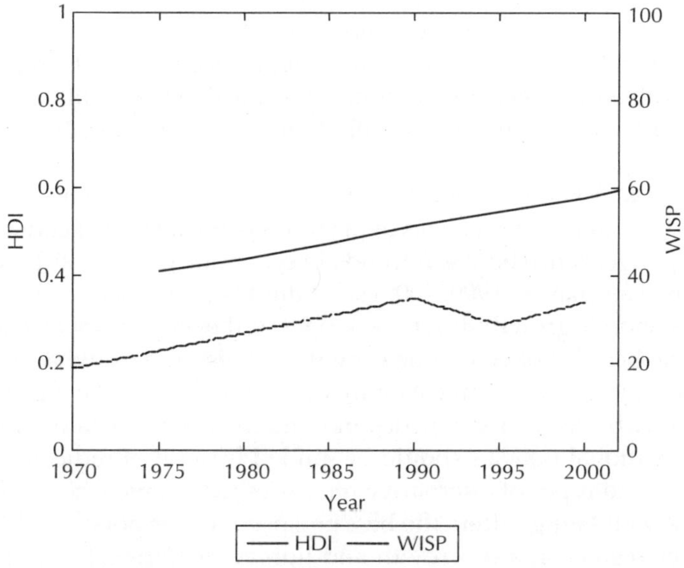

These findings clearly indicate that the economic growth had a positive effect on these social indicators. It is safe to say that the increase in economic welfare contributes to overall human welfare in India. But there is still a long road ahead: India scores relatively low in comparison to other countries. In 2002 India ranked 127th in out of 177 countries on the HDI, while it ranked 111th out of 163 countries on the WISP two years earlier (Estes 2003).

Looking at the available subjective welfare measures offers a second way to analyse the economic progress made. With an average Life Satisfaction of 5.7 points out of 10 and a Happy Life Years score of 35.7, India scores average compared to the other countries.[7] Unfortunately, there have not been many historical happiness assessments within India and the ones that have been conducted show opposite trends. This makes well-founded conclusions about the interaction between economic growth and subjective welfare impossible.

The case of alternative economic welfare measures is even worse. Neither the Index of Sustainable Economic Welfare nor the Genuine Progress Indicator has been compiled for India. Given the amount of time and data needed to construct an Index of Sustainable Economic Welfare or Genuine Progress Indicator, it is not difficult to understand why these efforts are lagging behind. Data availability problems exist even in developed countries: the lack of historical time series of specific indicators needed for the exercise is a recurrent problem in these countries.

CONCLUSION

A number of alternative approaches to measuring welfare have been developed during the past decennia. Each of these approaches has generated specific indicators to evaluate the welfare implications of the economic progress that was made. Unfortunately only a few of these alternative welfare measures have gained widespread acceptance (e.g. the Human Development Index). It is important that all alternative approaches are supported, since they each contain information that is not captured by the other ones (Diener and Suh 1997). Furthermore, different indicators focus on different aspects of human well-being and, so, all of them are needed in order to obtain a complete picture of a society's progress.

In developed countries, it is not clear whether economic growth is still contributing to human welfare and, if so, to what extent. The analysis of the Index of Sustainable Economic Welfare suggests that there is a growing divergence between trends in economic growth and economic welfare. Yet the objections raised by Neumayer (1999, 2000) shed doubts upon these results. There is, however, still a consensus that economic growth should no longer be the only policy objective, as people today not only experience the benefits of economic growth, but also the drawbacks (for example, stress and environmental pollution). Alternative welfare measures, such as the ISEW, can be used to help people realise these drawbacks, as they incorporate more aspects of human well-being than just the economic sphere. National policies should be guided by a set of indicators, of which the GDP is only one. A mixture of all types of alternative measures (economic indices, social indicators and measures of subjective well-being) offers the best perspectives for policy-making.

The relationship between economic growth and human welfare is less complex in developing countries. According to Offer (2000), policies aimed at GDP growth can add significantly to human welfare in these countries, since the accompanying rise in standards of living allows for improvements in the social conditions of their populations (education, health services, etc.).

As India has a relatively low GDP per capita, the process of economic growth comes with high increases in human welfare. This can be found in the case study performed in the previous section, as economic growth is translated into rises in the available alternative measures such as the Human Development Index and the Weighted Index of Social Progress. The benefits of the economic growth in India clearly outweigh the drawbacks. But India still has a long road ahead as many improvements are needed in the social and economic conditions of the country. Policy makers should, however, not just focus on the quantity of economic growth as the quality of the growth is also important. The question as to whether the growth can be sustained in the future is essential in this regard.

NOTES

1. http://www2.eur.nl/fsw/research/happiness/
2. http://hdr.undp.org/reports/
3. http://www.rprogress.org/
4. World Development Indicators 2004 CD-ROM (World Bank Group)
5. UNEP Geo Data Portal—http://geodata.grid.unep.ch/
6. UNDP Human Development Report Data—http://hdr.undp.org/statistics/data/
7. World Happiness Database—http://www.eur.nl/fsw/research/happiness/

REFERENCES

Anielski, Mark and Jonathan Rowe. 1999. *The Genuine Progress Indicator – 1998. Update: Data and Methodology*. San Francisco: Redefining Progress.

Castaneda., Beatriz. 1999. 'An Index of Sustainable Economic Welfare (ISEW) for Chile', *Ecological Economics*, 28: 231–44.

Cobb, Clifford, Ted Halstead and Jonathan Rowe. 1995. 'If the GDP Is Up, Why Is America Down?', *Atlantic Monthly*, October: 59–78.

Cobb, Clifford and John Cobb. 1994. *The Green National Product*. Lanham: University Press of America.

Crafts, Nicolas. 2002. 'UK Real National Income, 1950–98: Some Grounds for Optimism', *National Institute Economic Review*, July: 87–95.

Daly, Herman and John Jr. Cobb. 1989. *For the Common Good. Redirecting the Economy Toward Community, the Environment and a Sustainable Future*. Boston: Beacon Press.

Diefenbacher, Hans. 1994. 'The Index of Sustainable Economic Welfare: A Case Study of the Federal Republic of Germany', in COBB, Clifford and John COBB, Jr. (eds), *The Green National Product: A Proposed Index of Sustainable Economic Welfare*, pp. 215–45. Lanham: University Press of America.

Diener, Ed and Eunkook Suh. 1997. 'Measuring Quality of Life: Economic, Social and Subjective Indicators', *Social Indicators Research*, 40: 189–216.

England, Richard. 1997. 'Alternatives to Gross National Product: a Critical Survey', in Frank Ackerman, et al. (eds), *Human Well-Being and Economic Goals*, pp. 373–405. Washington DC: Island Press.

Estes, Richard. 1984. *The Social Progress of Nations*. New York: Praeger Publishers.

———. 1997. 'Social Development Trends in Europe, 1970–94: Development Prospects for the New Europe', in *Social Indicators Research*, 42: 1–19.

———. 2003. *At the Crossroads: Development Challenges of the New Century*. Dordrecht: Kluwer Academic Publishers.

Fisher, Irving. 1965. *The Nature of Capital and Income*. New York: Kelley.

Gil, Sebastian and Jerzy Sleszynski. 2003. 'An Index of Sustainable Economic Welfare for Poland', *Sustainable Development*, 11: 47–55.

Goodwin, Neva. 1997. 'Interdisciplinary Perspectives on Well-Being—Overview Essay', in Frank Ackerman et al. (eds), *Human Well-Being and Economic Goals*, pp. 1–14. Washington DC: Island Press.

Guenno, Giorgio and Silvia Tiezzi. 1998. *The Index of Sustainable Economic Welfare (ISEW) for Italy*. Milan: Fondazione ENI Enrico Mattei.

Hamilton, Clive and Richard Denniss. 2000. *Tracking Well-Being in Australia: The Genuine Progress Indicator 2000*. Canberra: The Australian Institute.

Hicks, John R. 1939. *Value and Capital*. Oxford: Clarendon Press.

Jackson, Tim and Susanna Stymne. 1996. *Sustainable Economic Welfare in Sweden: A Pilot Index 1950–1992*. Stockholm: Stockholm Environmental Institute.

Jackson, Tim, Nic Marks, Jon Ralls and Susanna Stymne. 1997. *Sustainable Ecomonic Welfare in the UK, 1950–1996*. London: New Economics Foundation.

Kuznets, Simon. 1941. *National Income and Its Composition, 1919–1938: Volume 1*. New York: National Bureau of Economic Research.

Land, Kenneth. 1999. 'Social Indicators', in Edgar Borgatta and Rhonda Montgomery (eds), *Encyclopedia of Sociology*, pp. 1844–50. New York: Macmillan.

Lawn, Philip. 2003. 'A Theoretical Foundation to Support the Index of Sustainable Economic Welfare (ISEW), Genuine Progress Indicator (GPI), and Other Related Indices', *Ecological Economics*, 44: 105–18.

Leipert, Christian. 1989. 'Social Costs of the Economic Process and National Accounts: The Example of Defensive Expenditures', *Journal of Interdisciplinary Economics*, 3(2): 27–46.

Max-Neef, Manfred. 1995. 'Economic Growth and Quality of Life: A Threshold Hypothesis', *Ecological Economics*, 15: 115–18.

Neumayer, Eric. 1999. 'The ISEW: Not an Index of Sustainable Economic Welfare', *Social Indicators Research*, 48: 77–101.

———. 2000. 'On the Methodology of ISEW, GPI and Related Measures: Some Constructive Suggestions and Some Doubt on the 'Threshold' Hypothesis', *Ecological Economics*, 34: 347–61.

———. 2004. *Sustainability and Well-Being Indicators*. WIDER Research Paper No. 2004/23. Helsinki: United Nations University—World Institute for Development Economics Research.

Nordhaus, William and James Tobin. 1972. 'Is Growth Obsolete?' *Economic Growth*, National Bureau for Economic Research. New York: Columbia University Press, pp. 1–80.

Offer, Avner. 2000. *'Economic Welfare Measures and Human Well-Being'*. Discussion Papers in Economic and Social History n°34. Oxford: University of Oxford.

Osberg, Lars. 2001. 'Needs and Wants: What is Social Progress and How Should it Be Measured?', in Keith Banting, Andrew Sharpe and France Sthilaire (eds), *The Review of Economic Performance and Social Progress*, pp. 23–42. Montreal: Institute for Research on Public Policy and the Centre for the Study of Living Standards.

Ravallion Martin. 2003. 'Measuring Aggregate Welfare in Developing Countries: How Well do National Accounts and Surveys Agree?', *The Review of Economics and Statistics*, 85(3): 645–52.

Rosenberg, David, Tammo Oegema and Marcel Bovy. 1995. *ISEW for the Netherlands: Preliminary Results and Some Proposals for Further Research*. Amsterdam: Instituut voor Milieu- en Systeemanalyse Amsterdam (IMSA).

Sharpe, Andrew. 1999. *A Survey of Indicators of Economic and Social Well-Being*. Ottawa: Canadian Policy Research Networks.

Stockhammer, Engelbert, Harald Hochreiter, Bernhard Obermayr and Klaus Steiner. 1997. 'The Index of Sustainable Economic Welfare (ISEW) as an Alternative to GDP in Measuring Economic Welfare. The results of the Austrian (revised) ISEW calculation 1955–1992'. *Ecological Economics*, 21: 19–34.

Stymne, Susanna. 2000. *Measuring Sustainable Economic Welfare*. Centre for Environmental Strategy. Guildford: University of Surrey.

UNDP. 2004. *Human Development Report 2004. Cultural Liberty in Today's Diverse World*. New York: United Nations Development Programme (UNDP).

Veenhoven, Ruut. 1996. 'Happy Life Expectancy: A Comprehensive Measure of Quality-Of-Life in Nations', *Social Indicators Research*, 39: 1–58.

———. 2002. 'Why Social Policy Needs Subjective Indicators', *Social Indicators Research*, 58: 33–45.

———. 2004. *Subjective Measures of Well-Being*. WIDER Discussion Paper No. 2004/07. Helsinki: United Nations University—World Institute for Development Economics Research.

Venetoulis, Jason and Cliff Cobb. 2004. *The Genuine Progress Indicator 1950–2002: 2004 Update*. San Francisco: Redefining Progress.

Zolotas, Xenophon. 1981. *Economic Growth and Declining Social Welfare*. New York: New York University Press.

12

Storm Protection Value of Mangroves in Coastal Orissa

Saudamini Das

Abstract: One of the most prominent ecological services of the mangrove forests has been their protection to the inland residents from the ravages of cyclones. It is widely recognised and evidences have been found to prove that areas having mangrove cover have received less damage from cyclones than unprotected areas. However, there is hardly any attempt to systematically and scientifically quantify this critical service of the mangrove ecosystem. This may be a reason of mangrove protection or plantation receiving little patronage from the government compared to other coastal plantation schemes, particularly casuarinas, in highly cyclone prone areas. The chapter describes a methodology that can be used to capture the storm protection value of mangrove and casuarina forests. The methodology is based on scientific and meteorological parameters and theories. The storm protection value being conditional for the probability of cyclone hitting the coastline, this methodology would be using the cyclone probability of the coastline to calculate the value of mangroves and casuarinas trees. The results are expected to give a valuable guidance for coastal zone management and maintenance of cyclone shelterbelt plantations.

INTRODUCTION

Mangroves are the salt tolerant tropical and sub-tropical forests that grow in the inter-tidal areas and estuary mouths of the coastal zone. These forests are habitat to a wide variety of marine and terrestrial flora and fauna and also provide a range of provisioning, supporting, regulating and cultural services to mankind (MA 2003). Mangrove loss or conversion of mangrove areas into acquacultural developmental or anthropogenic uses is common worldwide with the most prominent global drivers being aquaculture, agricultural practices and population growth of low and middle income countries (Barbier and Cox 2004).

Mangrove forests act as a safety net to coastal poor (Ruintenbeck, 1994; Barbier and Sathirathai 2001) by providing both the sources of livelihood as well as protection to their lives and properties in the face of natural calamities like cyclone. Mangrove loss increases the vulnerability of these people and more so, if the coastal areas are dominated by poor living in mud houses. The glaring example is the colossal loss of human life, livestock and property in the backward state of Orissa in India due to the cyclone of 1999 that originated in the Indian Ocean and had its landfall near the Paradeep port off Orissa coast. If the mangrove forests would have been there, bestowed by nature in the first place, the loss, probably, would have been much less.

Storm protection function of mangroves is a well recognised protective service both in tropical and sub-tropical areas and there is reported evidence of areas protected by mangroves incurring less damage due to cyclone than the unprotected areas (International Federation of Red Cross and Red Crescent News, 19 June 2002; Badola 2002; Tynkknen 2000; Fosberg 1971). Mangrove trees provide windbreak and the strong criss-crossed roots of mangrove trees break the wave energy and reduce the height as well as the speed as the storm surges. Research has shown that for every mile of vegetative wetland (not necessarily mangrove), storm surge height can be reduced by one foot (Rockel M. 1991).

However, there has neither been any attempt to evaluate this function of mangroves systematically nor to link up this value to the probability of cyclone hitting the coastal areas or to the storm surge vulnerability of the coast. The present chapter describes a methodology that can be used to evaluate this function and calculate the probabilistic or expected value of mangroves.

CYCLONES AND COASTAL ORISSA

The climatically sensitive state of Orissa is a frequent witness to the ravages of cyclones and super cyclones and the accompanying calamities resulting from them. In Indian subcontinent, the annual frequency of cyclones crossing the east coast of the country is very high compared to the west coast and along the east coast, it is again highest for the coast of Orissa as evident from Tables 12.1 and 12.2.

Compared to 445 cyclonic disturbances in the east coast of the country, the west coast experienced only ninety-five disturbances during 1891–1990 and the state of Orissa faced maximum severe cyclones compared to other coastal states of India.

Again, of the twelve most devastating cyclonic storms that made landfall in east coast of India in the last 263 years, the state of Orissa was witness to five of them whereas the state of Andhra Pradesh to four and West Bengal to three.

The cyclone tracks published by the Meteorological department of the Government of India show a particular coastal stretch of Orissa (Coast of Jagatsinghpur, Kendrapara, Bhadrakh and Baleshwar districts) to be more vulnerable to cyclones as maximum cyclones originating in the months of June to September have their landfall in these areas.

Along with the high frequency of cyclones, these areas also have a high vulnerability index due to high population density and high storm surge vulnerability. Various indices calculated by him also corroborate the fact that districts along the eastern coast are more vulnerable than those on the western coast. Out of the forty-nine coastal districts of India, Baleshwar and Cuttack (these districts were further subdivided into Baleshwar, Bhadrakh, Kendrapara and Jagatsinghpur by Government of Orissa in 1993) of Orissa and North and South 24 Parganas of West Bengal were found to be the most vulnerable districts with the only difference being that the districts in Orissa have

Table 12.1 Monthly Distribution of Cyclonic Storms for the Period 1891–1990 for the East and West Coast of India

Month	*BOB*	*AS*	*Month*	*BOB*	*AS*
January	6	1	August	27	6
February	1	0	September	40	6
March	4	1	October	82	17
April	21	6	November	99	15
May	51	19	December	42	4
June	39	18	***Total***	**454**	**95**
July	42	2			

Source: Damage Potential of Tropical Cyclones, Indian Meteorological Department, Govt. of India, 2002.
Note: BOB: Bay of Bengal; AS: Arabian Sea.

Table 12.2 Total Number of Cyclonic Storms (CS) and Severe Cyclonic Storms (SCS) Crossing Different Coasts of India (1891–2000)

State	*No. of CS and SCS*
West Bengal	69
Orissa	98
Andhra Pradesh	79
Tamil Nadu	62
Karnataka	2
Maharashhtra and Goa	18
Gujarat	28
Kerala	3
Total	**359**

Source: G.S. Mandal, WMO 2001, "Tropical Cyclones and their forecasting and warning system in North Indian Ocean: WMO TD No. 430, Tep No. 28, WMO, Geneva.
Note: The frequencies don't include the occurrence of depressions or deep depressions or cyclones of less than 62 KMPH wind speed.

a slightly higher ability to bounce back after a disaster than those in 24 Parganas. In Orissa, though historically Baleshwar was the most vulnerable district, with global climate change, Cuttack (presently Kendrapara and Jagatsinghpur) is being predicted to be the most vulnerable one (Shukla et al. 2003).

This coastal region was bestowed with thick mangrove forests by nature[1] which were destroyed due to developmental factors like establishment of Paradeep port, roads, fishing ports etc., anthropogenic activities like human settlement, agriculture, betel vine farms etc., and also due to establishment of aquaculture farms both by private and public sector units. At present, this region has approximately 216 sq km of mangrove forests (both dense and degraded) in different patches (Anon 2004). Because of their strategic location in a highly cyclone prone area and the storm protection value of mangroves being conditional to the frequency of cyclones crossing the coast line, an evaluation of the storm protection function of the mangroves of Orissa coast could give the most approximate value of this protective service by them.

THE CASUARINA FORESTS OF ORISSA

The entire coastline of Orissa has basically two types of forests; mangroves in the low—lying swampy areas, and casuarina trees on the slightly higher sandy beaches. The casuarina forests that dot the coastline of Orissa were planted under the special scheme of coastal shelter belt plantation to protect the coastal areas from the ravages of cyclonic storms. Like the super cyclone of 1999, the state of Orissa was ravaged by a severe cyclonic storm in October 1971 that had brought heavy loss of human and cattle life and damage to property. The Ministry of Irrigation and Power, Government of India had appointed a cyclone distress mitigation committee headed by P. Koteswaram, DG of Observatories, Indian Meteorological Department. The Committee made fifty-nine recommendations of which the 32nd measure was to build coastal bunds along the tidal inundated vulnerable areas and to plant forests to a depth of about one kilometer from the coastline to have them act as a wind breaker, hence preventing soil erosion. The afforestation programme was implemented with great sincerity under the coastal shelterbelt plantation scheme that started in 1974 and was stopped in 1984 after covering 7,880.59 hectares of sandy coast. Under this scheme, casuarina, cashew and coconut plantations were raised along the sea coast and on the estuaries of major rivers of the then Cuttack and Baleshwar districts. Though there is hardly any research on the contribution of these forests to micro or macro livelihood support base, the initial aims of these plantations were to protect coastal areas from cyclone and sand casting, supply firewood and improve social environment. However, the windbreak or cyclone buffer function of these forests will be evaluated by the present methodology and compared with that of mangroves. Mangrove trees and casuarina trees are complementary to each other since areas where mangroves grow are not suitable for casuarinas and vice versa. Hence, the locations having casuarinas will have no mangroves in their coastline.

REVIEW OF LITERATURE

Most of the literature on mangrove forests have focused on three different issues:

1. Identifying the drivers of mangrove loss [Parks et al. 1994; Dewalt et al. 1996; Primavera 2000; Barbier and Cox 2004].
2. Comparing the economic value of mangrove preservation to mangrove conversion [Christensen 1982; Dugan 1990; Lal 1990; Ruitenbeck 1994; Naylor and Drew 1998; Gilbert and Janssen 1998; Janssen and Padilla 1999; Nickerson 1999].
3. Modelling and valuing the mangrove fishery linkage (Sathirathai 1998; Barbier and Strand 1998; Ronnback 1999; Barbier and Sathirathai 2004].

Evaluation of other critical functions of mangroves like storm protection, shore line stabilisation, flood control, etc. have received scanty attention so far. The storm protection value of mangroves is mostly being equated to the cost of erecting a protective structure on the coastline (Chan et al. 1993). However, this measure is hypothetical and is not based on an any survey or statistical data. Moreover, erecting a wall on the coastline will segregate the inter-dependent coastal ecosystem that may have severe ecological impact in future. Badola (2002) equated the storm protection value of mangroves to the difference in average damage suffered per household due to cyclones in a mangrove protected village compared to that of an unprotected village. This study is on a very small scale and since the villages of Badola's area of study are not so homogeneous, the result has little use for policy making.

Sathirathai (1998), though mentioned the protective function of mangroves to be both windbreaking and shore line stabilisation, and has actually equated the protection value of mangroves to the shore line stabilisation function, leaving the windbreaking function unvalued. Peter Espuet (2000) wrote about the heavy payment by insurance companies and more expenditure by Public Works Department in mangrove cleared areas compared to mangrove protected areas after cyclones. The difference in insurance claims (or insurance premium) and expenditure of public works department (or tax rates) could be a way to capture the storm protection value of mangroves. But this is possible only if all properties are insured and all areas are homogeneously developed and hence cannot be applied to under-developed countries.

On the question of appropriate methodology to measure this value, Spanink and Beukering (1997) argued the inappropriateness of either WTP or WTA measures as suggested by James (1991) to measure the expected damages or increased risk due to the loss of mangrove protection to people. They suggested the hedonic price method and defensive expenditure to be the more appropriate approaches to value this protective services, though their applicability to a particular area is conditional upon the presence of a developed house market or to the degree of substitutability between the defensive measures and the protection provided by the mangroves. In poor underdeveloped countries, the absence of the above mentioned requirements limits the applicability of hedonic price method or defensive expenditure approaches.

The present methodology is based on the actual damages suffered due to cyclones in mangrove protected and mangrove cleared areas and hence doesn't suffer from any subjective bias or doesn't require the existence of any market or the substitutability or complementarity between any economic measures.

THE METHODOLOGY

As mentioned earlier, storm protection value of mangroves is linked to the probability of storms hitting the coastal areas and any methodology trying to capture this value should take this into account. Accordingly, the present methodology consists of two parts. Part one elaborates the modelling and the estimation of the damages to life and property due to cyclones. Part two estimates the expected value of the mangrove and casuarina forests using the results of part one and the probability of cyclones of different intensities hitting the coastline.

(*i*) *Modelling and Estimation of Damage Function:* Damages due to cyclone are caused by gale winds, storm surges and flooding due to torrential rains, but the present research will consider damages only due to wind and storm surges. Damages due to flood are excluded because of time, data constraints and the requirement of a different model.

Total Damage: Damages to property (only houses), loss of human lives and livestock due to high wind and storm surge.

Location I = Village Panchayats. Damages at any location i, due to cyclones depend on the wind velocity at the location and the property/life at risk.

$D_i = f(V_i, P_i)$ where D_i is total wind and surge damages at location i, V_i is velocity of wind at location i, and P_i is property at risk (can be represented by population or some measure of income) of location i.

Wind velocity at a location is dependent on the distance of the location inland, d_{oi}; the distance of the location from the path of the storm, d_{1i}; the intensity of the storm at landfall, g; and the nature of intervening terrain between landfall and the location or within the bounded area of the location and landfall, l_i.

$$\therefore V_i = F(d_{oi}, d_{1i}, g, l_i) \tag{2}$$

F(.) is a non-linear function as wind velocity decreases with distance from the path of the storm and distance inland. Hence an exponential function is used for F(.).

$$V_i = g \exp.(\beta_1 d_{oi} + \beta_2 d_{1I} + \beta_3 l_i) \tag{3}$$

Wind damage is directly proportional to square of wind velocity2 and damages due to storm surge are proportional to the height of storm surge that again depends on the square of wind velocity.[2] Total damage is again directly proportional to property/life at risk or P.

Hence $$D_i = (\alpha + \eta\varepsilon)V^{\gamma}_i.\ \phi P_i = \lambda V_i^{\gamma} P_i \tag{4}$$

where α represents the proportionality between wind damage and wind velocity; η represents that between storm surge damage and height of surge; ε that between height of surge and wind velocity and ϕ between total damage and property/life at risk.

Again $\alpha, \eta, \varepsilon > 0$, $o < \phi < 1$, $\gamma \approx 2$ and $\lambda = \alpha\phi$.

In the area of study, the intervening area between the land fall and location has either mangrove or casuarina forests. Hence l_i is defined as the km of mangrove (M) or casuarina (C) forests on the inland distance of the location I or $l_i = d_{oi} M + d_{oi} C$. Casuarina forests grow on the sandy beaches which are at a higher topography than the sea level and the mangroves grow on the swampy low lying areas of the sea coast.

Hence if $M > 0 \Rightarrow C = 0$
or $C > 0 \Rightarrow M = 0$

Now equation (3) can be written as

$$V_i = g \exp.(\beta_1 d_{oi} + \beta_2 d_{1i} + \beta_3 d_{oi} M + \beta_4 d_{oi} C) \tag{5}$$

where M is defined as width (km) of mangrove forest, W_m, multiplied by health index of the forest, h_m

or $M = h_m W_m$

where $0 < h_m < 1$ and W_m is width of the forest patch in kilometers traversed by the storm.[3]

C is simply the width of casuarina forest in kilometers on the inland distance traversed by the storm.[4]

Equation (4) can be written as:

$$D_i = \lambda V_i^{\gamma} P_i = \lambda\, [g \exp (\beta_1 d_{oi} + \beta_2 d_{1i} + \beta_3\, d_{oi}\, M + \beta_4\, d_{oi} C]^{\gamma} P_i$$

or $$(D/P)_i = \lambda\, g^{\gamma} \exp. (\beta_1\, \gamma d_{oi} + \beta_2 \gamma d_{1i} + \beta_3\, \gamma d_{oi} M + \beta_4 \gamma d_{oi} C).$$

or $$(D/P)_i = \lambda g^{\gamma} \exp. (\beta_1'\, d_{oi} + \beta_2'\, d_{1i} + \beta_3'\, d_{oi} M + \beta_4'\, d_{oi} C) \quad (6)$$

where $\beta_1^1 = \gamma\beta_1,\ \beta_2^{\ 1} = \gamma\beta_2 \ldots$ etc.

As per meteorological calculation, the damages on the left of convergence zone being 1.5 times higher than the damages on the right of the convergence zone, equation (6) will be divided by 1.5 for all areas left of cyclone path[5].

$$(D/P)_i = \frac{1}{\sigma_i}\left[\lambda g^{\gamma} \exp.(\beta_1' d_{oi} + \beta_2' d_{1i} + \beta_3' d_{oi} M + \beta_4' d_{oi} C)\right] \quad (7)$$

where $\sigma_i = 1.5$ for areas on the left of cyclone path and $\sigma_i = 1$ for areas lying on the right of cyclone path.

Equation (7) expresses the proportional damage at any location or in a particular village panchayat to be dependent on the wind intensity of the storm at landfall and other physical factors like distance in land, distance from the path of the storm, the mangrove or casuarina forests on the coastline and whether it is at the left or right of convergence zone etc.

This equation will be estimated using the cross-sectional data of different village panchayats that were affected by the super cyclone that crossed the Orissa coast in October–November 1999 by making a logarithmic transformation of the variables.

(*ii*) Estimation of the cyclone probability of the area under study and the expected value of the mangrove and the casuarina forest:

The particular coastline of Orissa in the area under study is quite prone to cyclones and the probability of a cyclone having its landfall in this area is very high.

Let $\Pi_i\, (d_{1i}, g)$ be the probability of a cyclone with landfall wind intensity g, passing within d_1 km of location i, hitting the coastline.

Hence the expected storm damage per unit of property at risk at i will be

$$E(D/P)_i = \frac{\lambda}{\sigma_i} \exp.(\beta_1'\, d_{oi} + \beta_3'\, d_{oi} M + \beta_4'\, d_{oi} C)$$
$$\int\!\!\int g^{\gamma} \exp.(\beta_2' d_{1i})\, \Pi_i\, (d_{1i}, g)\, dg\, dd_{1i} \quad (7)$$

If the mangroves are over-exploited or are cleared, the expected damage or the risk to properties at any location due to cyclones will increase.

Hence the marginal expected per unit damage effect at any location i, due to mangrove loss or over-exploitation is

$$-\left[\frac{\partial E(D/P)_i}{\partial h_m} + \frac{\partial E(D/P)_i}{\partial W_m}\right] (\text{as } M = h_m W_m)$$

$$= \frac{\lambda}{\sigma_i} \exp.(\beta_1' d_{oi} + \beta_3' d_{oi} M + \beta_4' d_{oi} C)$$

$$\left[\int\int g^{\gamma} \exp.(\beta_2' d_{1i}) \Pi_i(d_{1i}, g) dg, dd_{1i}\right] \left[\beta_3' d_{oi}(W_m + h_m)\right]$$

$$= \frac{\lambda}{\sigma_i} \exp.(\beta_1' d_{oi} + \beta_3' d_{oi} M + \beta_4' d_{oi} C) \;\; \beta_3' d_{oi}(W_m + h_m)$$

$$\int\int g^{\gamma} \exp.(\beta_2' d_{1i}) \;\; \Pi_i(d_{1i}, g) dg, dd_{1i} \tag{8}$$

Equation (8) shows the marginal damage effect of mangrove loss due to both over-exploitation of forest and clearing of every kilometer of the forest.

The marginal damage effect of casuarina loss will be:

$$\frac{\partial E(D/P)_i}{\partial C} = (\beta_4' d_{oi}) \frac{\lambda}{\sigma_i} \exp.(\beta_1' d_{oi} + \beta_3' d_{oi} M + \beta_4' d_{oi} C)$$

$$\int\int g^{\gamma} \exp.(\beta_1' d_{1i}) \Pi_i(d_{1i}, g) \, dg, dd_{1i} \tag{9}$$

Then equation (8) and (9) will be valued by using the estimated parameters of equation (7) for every location that is, village panchayats and the cyclone probability function $\Pi_I(d_{1i}, g)$.This function will be represented by discrete probability distribution of storms of different intensities hitting the area of study in the last 100 years. These in formations are available with the Indian Meteorological office, Pune.

For storm protection value of mangrove, the marginal damage of every village panchayat having mangrove protection will be multiplied by the property/life at risk of the respective panchayats and will be added together to get the marginal value of a km of mangroves per year.

Similarly the value of a km of casuarina forest will be calculated by adding the marginal damages of panchayats having casuarina protection.

Cyclone Probability of Orissa Coast

The Indian meteorological department categorises the cyclonic disturbances into seven different categories as given in Table 12.3.

The damage potential of the 1st two categories of disturbances is very negligible, whereas it increases at an ascending rate with the latter categories. The meteorological department of

Table 12.3 Categorisation of Cyclonic Disturbances

Types of Disturbances	*Meteorological Name*	*Associated Wind Speed*
Low Pressure Area	L	< 17 knots
Depression	DD	17–27 knots
Deep Depression	DD	28–33 knots
Cyclonic Storm	CS	34–47 knots
Severe Cyclonic Storm	SCS	48–63 knots
Very severe cyclonic storm	VSCS	64–119 knots
Super cyclonic storms	SUCS	≥ 120 knots

Note: 1 knot = 1.86 km per hour (kmph).

Government of India has published the path of three types of cyclonic disturbances over the Indian ocean and Indian subcontinent for the period 1891–1990.

The cyclone probability for the coast of Orissa has been calculated from these cyclone track records. Though DD and CS are very frequent on Orissa coast, the occurrences of catastrophic disturbances like VSCS and SUCS are rare. In between 1737 and 2000 (263 years), the state of Orissa witnessed only five catastrophic cyclones.

The mangrove forests are found in Jagatsinghpur, Kendrapara and Bhadrakh district whereas casuarina forests are found in all the coastal districts but the cyclone frequency of these districts are not uniform. Hence the cyclone probability to be used in the valuation of these forests is calculated separately. For mangroves, it is the annual probability of cyclone per km of coastline of Bhadrakh, Jagatsinghpur and Kendrapara districts, whereas for casuarina, it is per km of coastline of the above three and Puri district. The cyclone track records show a uniform cyclone vulnerability of the locations of Bhadrakh, Jagatsinghpur and Kendrapara districts whereas the cyclone frequency is very low for Puri. Table 12.4 shows the frequencies of different cyclonic disturbances for the coast of Bhadrakh, Jagatsinghpur and Kendrapara districts.

The combined coastline of the above three districts being 185 km and that of Puri being 155 km, the annual probability per km of coastline is shown in Table 12.5.

The annual frequencies per kilometer presented in column 5 and 6 and the associated mid-point wind speed of different cyclone categories will be used to estimate the expected damage to properties due to mangrove or casuarina loss.

Estimation of Expected Damage

The marginal expected damage to properties at any village panchayat i, due to mangrove loss or over-exploitation of mangrove has been shown to be equal to:

$$-\left[\frac{\partial E(D/P)_i}{\partial h_m} + \frac{\partial E(D/P)_i}{\partial W_m}\right]$$

$$= \frac{\lambda}{\sigma_i} \exp.\,(\beta_1' d_{oi} + \beta_3' d_{oi} M + \beta_4' d_{oi} C)$$

Table 12.4 Monthly Frequency Distribution of Cyclonic Disturbances in the Coast of Jagatsinghpur, Kendrapara and Bhadrakh Districts of Orissa State for the Period 1891–1990

Cyclone Type	*January to April*	*May*	*June*	*Jul.*	*Aug.*	*Sept.*	*Oct.*	*Nov.*	*Dec.*	*Total*
DD	0	0	10	10	15	7	1	1	0	**44**
CS	0	1	7	11	14	7	1	1	0	**42**
SCS	0	2	1	3	2	0	5	1	0	**14**
VSCS	0	1	0	0	0	0	0	0	0	**1**

Source: Track Records of depressions and storms over Indian ocean, Indian Meteorological Department, Government of India, New Delhi.

Table 12.5 Annual Probability of Cyclone per km of Selected Coastline of Orissa Coast

	Wind Criteria in kmph	*Mid Print Wind Speed*	*Total Frequency in 100 Years*	*Annual Frequency per km*	
				a	b
DD	52–61	56.54	44	0.0023784	0.0018333
CS	62–88	75	42	0.0022703	0.00175
SCS	89–117	103	14	0.0007568	0.0005833
VSCS	119–221	170	1	0.0000541	0.0000417

Source: Table 12.4.
Notes: (a) For coastline of Bhadrakh, Kendrapara and Jagatsinghpur (185 km).
(b) For coastline of Bhadrakh, Jagatsinghpur, Kendrapara and Puri (240 km).

$$\left[\int\int g^{\gamma}\exp.(\beta_2' d_{1i})\Pi_i(d_{1i},g)dg,dd_{1i}\right]\;\left[\beta_3' d_{oi}(W_m+h_m)\right]$$

$$=\frac{\lambda}{\sigma_i}\exp.(\beta_1' d_{oi}+\beta_3' d_{oi}M+\beta_4' d_{oi}C)\;\;\beta_3' d_{oi}(W_m+h_m)$$

$$\int\int g^{\gamma}\exp.(\beta_2' d_{1i})\;\;\Pi_i(d_{1i},g)dg,dd_{1i} \tag{8}$$

The calculation of this value needs the estimated coefficients of damage function, the β_s; the distance inland; the existence of either mangrove or casuarina on the coastline; the cyclone probability function and the range of integration for the cyclone types and damage diameter area.

The cyclone probability function is represented by the discrete probability distribution described in Table 12.5. Though the damage potential of DD is very little due to non-availability of detailed data on landfall wind speed of different cyclones, excluding them would have made the data points of discrete probability distribution very small. As the area of study included four districts and the super cyclone of October 1999 had affected a vast area of more than 200 km radius, the range of integration for d_1 is taken to be 0–100 km. As the damage function has not been estimated for the area of study, some approximate values of β coefficients are used to get some approximate result. Suppose[6]:

β_1 = -0.1150
β_2 = -0.0486
β_3 = 0.0017
β_4 = 0.0006
λ = 5.687

If there is a village panchayat on the right side of the path of the storm 5 km away from the sea coast and having 2 km of dense mangrove forest on the inland distance, the per capita marginal value of mangrove for that location will be as follows:

$$MV\,(d_{oi}, M) = MV\,(5, 2)$$

$$=\frac{5.687}{\sigma_i=1}\exp.(-0.1150\times 5+0.0017\times 5\times 1+0.0006\times 0)$$

$$\int_0^{100} \exp.(-0.0486d_1)dd_{1i.}\, 2(\Sigma g^{\gamma}\Pi)\ \left[0.0017(5+1)\right]$$

$$= 5.687\ \ \exp.(-0.5750+0.0085)\int_0^{100} \exp.(-0.0486d_1)dd_{1i}\ 2\Sigma g^{\gamma}\Pi(0.0102)$$

As the location is having mangroves in the coastline, C = 0 and the discrete storm probability is multiplied by 2 as any location will be affected if a cyclone passes through either its left or right side.

The mangrove being dense mangrove, $W_m = 5$ and $h_m = 1$.

Using the discrete probability distribution of Table 12.4 and $\gamma = 2$, $2\Sigma g^{\lambda}\Pi = 59.910532$.

Putting the respective values, the MV (5, 2) = Rs 40.2664.

The loss of one km of mangrove increases the expected damage to properties by Rs 40.2664 per capita. If there are 5,000 people living in that panchayat, the value of a km of mangrove to that panchayat comes out to be Rs 2,01,332. Thus, the value of a km of mangrove to every panchayat getting the protection can be calculated and summed up. Similarly, the value of casuarina forests will be calculated.

CONCLUSION

Mangrove forests are the critical media through which the land and the sea interact with each other. They are located in those low lying swampy areas of the coast line where the sea poses a potential threat to human life during a cyclone. These forests protect people and properties from the fury of cyclone and hence are the most important gift from nature to those vulnerable people.

Northern coastal Orissa, particularly the coastal areas above the 20ºN latitude is highly cyclone prone and witnesses one cyclone per year on an average. These coastal areas, with an average altitude of less than 5 metres, were bestowed with thick mangrove forests, that were degraded for various reasons. The state government has planted casuarina trees as cyclone buffers over the entire coastline of the state whereas the mangrove regeneration is yet to receive the due attention. It may be due to the lack of quantification of this ecological service of the mangroves.

Unlike other indirect services of mangroves, an evaluation of storm protection function requires meteorological inputs and any correct assessment of this service needs a sound methodology based on proper meteorological and scientific facts and theories. The methodology described in the present chapter can give correct assessment of the protection provided by the mangrove forests and the casuarina forests to the people of coastal Orissa during the super cyclone of October 1999. These results will have important policy uses and can help different governments at different levels in maintaining a safe coastal zone.

ACKNOWLEDGEMENTS

The author is deeply grateful to Ajay Kumar, Reader, Department of Mathematics of the University of Delhi and Charan Singh, Meteorologist, Indian Meteorological Department, Government of India, New Delhi, for their valuable guidance.

NOTES

1. District Gazetteers of Cuttuck districts, 1939 and 1996.
2. Data provided by the meteorological department of the Government of India proves this. The data and the result are available with the researcher.
3. Typically as defined earlier, W_m should have been the sq km of mangrove forests in the rectangular area between landfall, the location, distance inland and distance from path of the storm. But mangroves grow only in coastal zone and provide protection to areas lying inland.
4. Casuarina forests are being grown and maintained by the government and the forests are in uniform health in the entire coastline, whereas the mangroves don't have uniform health in the study area. In some areas they are in very good ecological health and in others, in highly degraded form.
5. Based on discussion with meteorologists.
6. These values are based on S. Farber's work.

REFERENCES

Anon. 2004. *Status of Mangrove in Orissa.* Wildlife Organisation, Forest Department, Government of Orissa.

Badola, Ruchi. 2002. 'Valuation of the Bhitarkanika Mangrove Ecosystem for Ecological Security and Sustainable Resource Use'. EERC Working Paper Series, WB-1.

Barbier, Edward B. 2000. 'Valuing the Environment as Input: Review of Applications to Mangrove-Fishery Linkages', Ecological Economics, Vol. 35(1): 47–61.

Barbier, Edward B. and Ivar Strand. 1998. 'Valuing Mangrove-Fishery Linkages: A Case Study of Campeche, Mexico', *Environmental and Resource Economics,* 12(2): 151–66.

Barbier, E.B. and S. Sathirathai. 2001. 'Valuing Mangrove Conservation, Southern Thailand', *Contemporary Economic Policy,* 19(2): 109–22.

Barbier, E.B. and Mark Cox. 2004. 'Introduction: Global Mangrove Loss and Economic Development', in E.B. Barbier and S. Sathirathai (ed.), *Shrimp Farming and Mangrove loss in Thailand,* UK: Edward Elgar.

Barbier, E.B. and S. Sathirathai. 2004. *Shrimp Farming and Mangrove Loss in Thailand.* United Kingdom: Edward Elgar.

Chan, H.T. et al. 1993. 'The Socio-economic, Ecological and Environmental Values of Mangrove Ecosystems in Malaysia and Their Present State of Conservation', in B.F. Clough (ed.), *The Economic and Environmental Values of Mangrove Forests and their Present State of Conservation in South-East Asia/Pacific Region, Vol.1,* pp. 41–81. International Society for Mangrove Ecosystems, International Tropical Timber Organisation and Japan International Association for Mangroves, Okinawa.

Christensen, B.O. 1982. 'Management and Utilisation of Mangroves in Asia and the Pacific', FAO Environment Paper, 3, Rome.

Dewalt, Ballie R., Philippe Vergne and Mark Hardin. 1996. 'Shrimp Acquacuture Development and the Environment: People, Mangroves and Fisheries on the Gulf of Fonseca, Honduras', *World Development.* 24(7): 1193–208.

Dixon, John A., et al. 1994. *Economic Analysis of Environmental Impacts* (2nd edn). London : Earthscan Publication in Association with the Asian Development Bank and the World Bank.

Dugan, P.J. (ed.) 1990. 'Wetland Conservation: A Review of Current Issues and Required Action', Gland, Switzerland: World Conservation Union.

Farber, Stephen. 1987. 'The Value of Coastal Wetlands for Protection of Property Against Hurricane Wind Damage', *Journal of Environmental Economics and Management*, 14: 143–51.

Fosberg, F.R. 1971. 'Mangroves Verses Tidal Waves', *Biological Conservation*, 4: 38–39.

Gilbert, Alison and Ron Janssen. 1998. 'Use of Environmental Functions to Communicate the Values of a Mangrove Ecosystem under Different Management Regimes', *Ecological Economics*, 25(3): 323–46.

IMD. 2000. 'Tracks of Cyclones in Bay of Bengal and Arabian sea 1890–1990'. Indian Meteorological Department, Govt. of India.

———. 2002. 'Damage Potential of Tropical Cyclones'. Indian Meteorological Department, Government of India.

International Red Cross. 2002. *International Federation of Red Cross and Red Crescent News*, 19 June.

James, R.F. 1991. The Valuation of Wetlands; Approaches, Methods and Issues, PHP/AWB, Sumatra Wetland Project No. 29 Asian.

Janssen, Ron and Jose, E. Padilla. 1999. 'Preservation or Conversation of Valuation and Evaluation of a Mangrove Forest in the Philippines', *Environmental and Resource Economics*, 14(3): 297–331.

Lal, P.N. 1990. 'Conservation or Conversion of Mangroves in Fiji'. Environment and Policy Institute, Occasional Paper, II (Honululu, East-West Centre).

Mohanty, Nimai Charan. 1982. 'Mangroves of Orissa' (in Oriya Language), Cutuck, Orissa: Project Swarajya.

MA (Millennium Ecosystem Assessment), Ecosystems and Human Well-being: A Framework for Assessment (2003), Washington, DC: Island Press.

Naylor, Rosamond, Mark Drew. 1998. 'Valuing Mangrove Resources in Kosrae, Micronesia', *Environmental and Development Economics*, 3(4): 471–90.

Nickerson, Donna J. 1999. 'Trade-offs of Mangrove Area Development in the Philippines', *Ecological Economics*, 28(2): 279–98.

Parks, Peter J. and Manuel Bonifaz. 1994. 'Non-sustainable Use of Renewable Resource: Mangrove Deforestation and Mericulture in Ecuador', *Marine Resource Economics*, 9(1): 1–18.

Peter, Espuet. 2000. 'Environment and Development in Coastal Region and in Small Islands', *Jamaican Gleaner*, 2 February.

Primavera, J.H. 2000. 'Development and Conservation of Philippine Mangroves: Institutional Issues', *Ecological Economics*, 35(1): 91–106.

Rockel M. 1991. 'Economic Valuation of Wetlands'. Discussion Paper 065, Louisiana, US: Institute of Environmental Communication.

Ronnback, Patrik. 1999. 'The Ecological Basis for Economic Value of Seafood Production Supported by Mangrove Ecosystem', *Ecological Economics*, 29(2): 235–52.

Ruitenbeek, H.J. 1994. 'Modelling Economy-Ecology Linkages in Mangroves: Economic Evidence for Promoting Conservation in Bintuin Bay, Indonesia', *Ecological Economics*, 10: 233–47.

Sathirathai, Suthawan. 1998. Economic Valuation of Mangroves and the Role of Local Communities in the Conservation of Natural Resources: A case study of Surat Thani, South of Thailand. EEPSEA Research Report. http:703.116-43-43-77/publications.research1/ACF9E.html.

Senapati N. and Kumar, D.C. 1977. Orissa District Gazzettier, Puri, Gazettier Unit, Department of Revenue, Government of Orissa.

Spanink, F. and Pieter van Beukering. 1997. 'Economic Valuation of Mangroves Ecosystems; Potential and Limitations'. CREED Working paper No. 14.

Shukla, P.R. et al. 2003. Climate Change and India: Vulnerability Assessment and Adaptation. New Delhi: University Press.

Tynkknen, O. 2000. 'Orissa Cyclone: A Natural Phenomenon or a Sign of Things to Come', *Friends of Earth*, Finland.

Section 4

Community and Natural Resources Management

13

Community Participation in Joint Forest Planning and Management in Karnataka

P. Sudha[1], *V. Ramprasad, P.R. Bhat, R. Jagannatha Rao* and *N.H. Ravindranath*

INTRODUCTION

People have always participated in forest management. The tradition of regarding forests near human habitation as a common resource and nurturing them has been noted since historical times. Lands close to the settlements were enough to satisfy the needs of the people, therefore forests remote from habitation generally were never over-exploited and every village and hamlet, where subsistence agriculture was practiced, ensured optimum utilisation of forest resources that did not exceed the carrying capacity of the forest. The sanctity attached to forests is representative of the importance the people earlier placed on the forests. One such example is the Sacred Grove where harvest and disturbances of any kind were or in some parts are still prohibited. These forest patches represent primary forests and are of high conservation value (Malhotra et al. 2001).

The capitalist British empire stabilised and spread its influence on the colonies by exploitation of virgin forests for teak, spices and rubber based on the justification that the natives were quite unable to manage their own forests without British help. The process of colonisation attached not only a commercial value to timber but also unscrupulously aided the exploitation of forest resources. The alienation of people from their forests had begun.

The forest people now felled forests – the place where their gods resided, to guarantee continuous supply of timber to Her Majesty's expanding empire. The large tracts of village lands, forests and common lands that were left untouched around habitations degraded due to the lack of institutional arrangements for regulation and management in the absence of the time-tested practices. To further aggravate the degradation, people increasingly lost touch with ancient practices that could have ensured the sustainable usage of forests. The native knowledge on forest management had been found wanting!

The Forest Department under the newly-formed government of independent India severely restricted the people's rights to collect timber and NTFPs from the 'State owned forests'. The numerous (forgotten) people's resistances to regain their forest rights gained voice with the Chipko Movement. A few cases of success in people's management of forests together with the Forest Department and the realisation of the fact that they (Forest Department) could not manage forests in seclusion and arrest encroachments culminated in the decision that sought to experiment with the Joint Forest Management.

Evolution of Joint Forest Planning and Management in Karnataka

The First Forest Act of 1865, based on a resolution on Forest Policy by German agriculturist Voelcker maintained that it would be necessary to restrict the access of rural communities to Protected and (especially) Reserved Forests in order to better guard against soil erosion and against a possible threat to agricultural production (Jewitt 2001).

The Forest Policy, 1952, was independent India's first forest policy which stated among other things, that a third of the land area should be brought under forest cover. The policy aimed at massive afforestation activities primarily to meet the national needs for the defence and industrial sectors and to generate revenue.

The National Commission on Agriulture, 1976 emphasised the production of industrial wood from forests. The NCA recommended that Forest Corporations be created to attract institutional finance. Forestry activities on poor sites was decried and recommended the promotion of social forestry plantations on non-forest lands to reduce the pressure on natural forests and to meet the local needs for forest produce. A World Bank-aided project was taken up to promote Social Forestry activities in Karnataka. Though the project successfully created plantation assets with high technical inputs, concrete strategies to promote people's institutions for planning, implementation and management of social forests were lacking. The assets created under social forestry could not be sustained due to the inadequacy of mechanisms for sharing not only the responsibilities but also the benefits.

The Forest Conservation Act of 1980 was passed in order to reduce the indiscriminate diversion of forest land for non-forestry purposes and which helped regulate and control land use changes. Mandatory rules for compensatory plantations in case of forest diversion and requirement of Central government approval for such activities were the highlights of this act.

The Ministry of Environment and Forests, Government of India based on Forest Policy of 1988 issued a circular in 1990 outlining the concept of Joint Forest Management and the need for its speedy implementation throughout the country. A significant departure from the previous policies by envisaging people's participation in management of forests was observed. The Policy also sought to create a massive people's movement with involvement of women for achieving the objectives of maintaining environmental stability and increasing tree cover and to minimise the pressure on existing forests.

On the basis of this circular and the 1988 National Forest Policy, the Government of Karnataka launched a programme called Joint Forest Planning and Management (JFPM) by issuing a Government Order on 12 April 1993 to enable the formation of Village Forest Committees (VFCs) in partnership with the Karnataka Forest Department. The provisions of the Government Order were vigorously pursued while implementing the Department of International Development of United Kingdom (DFID) assisted Western Ghats Forestry Project (WGFP). Currently about 3,799 VFCs are managing 3.44 lakh ha of JFPM area largely initiated under two major externally funded

Table 13.1 Extent of JFPM Area in the Eastern Plains Region

No.	*Forest Division*	*No. of VFCs*	*No. of VFC Members*	*Extent of Forest Area Coming under VFCs' Jurisdiction*	*Extent of Older Plantations before Formation of VFC*	*Extent of Plantations Raised under JBIC*
1	Bangalore	816	79,270	56,612	13,869	18,142
2	Belgaum	419	21,775	90,172	6,110	7,823
3	Bellary	502	56,427	53,113	11,766	20,583
4	Chamarajanagar	75	9,869	6,955	449	2,140
5	Chickmaglur	34	1,651	1,594	894	338
6	Dharwad	268	40,357	42,157	5,277	8,141
7	Gulburga	282	27,111	13,871	5,915	8,058
8	Mysore	265	27,898	9,093	6,421	2,629
9	Hassan	438	33,578	38,523	36,241	15,045
10	Shimoga	62	5,604	11,899	2,287	1,334
	Total	**3,161**	**3,03,540**	**3,23,989**	**89,229**	**84,233**

projects—Western Ghats Forestry and Environment Project (WGFEP) funded by Department for International Development (DFID) for the period of 1992–93 to 1999–2000 and the Japan Bank for International Cooperation (JBIC) assisted Forestry and Environment Project being implemented for Eastern Plains of Karnataka (FEPEP) for 1996 to 2005 (Ravindranth and Sudha 2004). The first JFPM order was issued in 1993, which was amended in 1996. A new order was issued in 2002, which strengthened the programme in the state. Thus the JFPM policies of the state are continuously being evolved to provide better incentives to the community to participate in the programme.

Incentives for Community Participation

Policies in JFPM are being strengthened to promote community participation in development and management of forestlands. The key question that can be raised is if these policies are providing enough incentives that promote community participation.

In this chapter, an attempt is being made to understand the JFPM programme and policies that provide social, economic and environmental incentives to the village communities to participate and help sustain the programme.

The social incentive to the village community is to provide a platform for them to communicate and interact (among themselves and with the Forest department) for the betterment of the village and their individual well-being. Financial incentives at regular intervals along with hopes of long-term benefits, sustain the interest of the communities in JFPM. Intangible impact of JFPM on water table, soil conditions, air quality, forest cover etc. produces visible results of community efforts thus promoting further the cause of environmental protection. Does the JFPM provide for these incentives? If so, does the JFPM policy have provisions for such broad-based incentives? The following sections describe the policies that have provided the incentives to the community to sustain their interest and the community's response to the impact of JFPM.

METHODOLOGY

A study was conducted to understand the impact of JFPM on the village community as perceived by them. Group discussions with all the stakeholders were conducted in 226 and 269 villages of the Western Ghats and Eastern Plains region respectively. The study was based on field studies conducted in four forest divisions in the Western Ghats and six forest divisions of the Eastern Plains.

RESULTS

Policies Promoting Community Participation

To promote community participation, JFPM seeks to create a semi-autonomous democratic committee known as the Village Forest Committee (VFC). According to the 2002 JFPM guidelines, the VFCs can be constituted for a part of a village, a village or group of villages. In areas that are predominantly inhabited by tribals, JFPM can be practiced in all the forests irrespective of crown density, compared to non-tribal areas where it is restricted to areas below 25 per cent crown cover. All the adults in the village interested in conservation, development and management of forests are eligible to become members of the VFC and every member is entitled to get an equal share of produce/proceeds. The VFC members elect a Management committee constituting ten elected members of whom two each are SC/ST members, landless labourers and artisans and four are general body members. Of these ten members, half are women in each of the categories.

The village forest committee thus promotes community participation in three ways:

(*a*) By its democratic system, every individual member of a village has a right towards forest management and the rules that govern the functioning of VFC help provide common platform to address individual issues. Thus, equality in membership irrespective of caste, creed and sex is established. The management committee (elected body) constituting fourteen members has to meet once every three months to manage the affairs of the VFC and the general body (of all members) of the VFC meets annually to discuss and decide issues of JFPM.

(*b*) The VFCs thus formed requires that it is legally recognised and does not remain a non-statutory body. To provide legal backup, the VFC's can be registered as Associations under the Karnataka Forest Act.

(*c*) Women members are to be represented equally as the men in the management committee. This provides for enough teeth to address gender and equity issues.

The VFC is also involved in preparation of micro plan and assists the Forest Department in implementation of the plan, and takes the responsibility for protecting and managing the plantation assets after three years of formation. The main role of the Forest Department is to enhance the skill and capacity of VFC members to enable them to take up planting works at the formation stage itself.

Rural Development through Entry Point Activity

One of the main activities of the JFPM programme especially in the Eastern Plains is the Entry Point Activities (EPA) conducted by the Karnataka Forest Department. Involvement of local communities in forestry programmes is a desirable objective, though difficult to achieve because forest activities have long gestation periods and are, therefore, accorded low priority by the local community. Forestry programmes are usually considered to be 'government' programmes; where implementation and protection is seen as the responsibility of the Forest Department. Most of the plantation programmes entail closure of the area, which increases hardship to the local people. Therefore, cooperation of all stakeholders is necessary for successful protection, management and benefit-sharing of plantations.

The EPA helps overcome the community barrier and breaks the ice between the Forest Department and the community. The main objective of entry point activities is to elicit the willing participation of the communities in JFPM and win the trust and confidence of the people. The main objectives of EPA are to mobilise all stakeholders of the community and compensate the community for the loss due to closure of forest areas, especially those areas where traditional rights to forest produce is in vogue. The community collectively identify EPA that may include creation of community assets to be maintained by them. The EPA are carried out through the Village Development Fund that provides a formal financial basis for initiating and maintaining developmental activities as may be decided by the community as a whole.

Keeping in view the crucial motivational role of entry point activities in the afforestation projects, Rs 35,000 per VFC was allocated under the Eastern Plains project. Under the project, EPA has been implemented in 2029 VFCs (Table 13.2) accounting for 65 per cent of the VFCs formed in the project area. Maximum VFCs with EPA was in Bangalore, Gulburga and Chickmagalur forest circle. Belgaum and Shimoga has least number of VFCs with EPA activities. The EPA has promoted income generation activities such as pisciculture, training for tailoring to women self-help groups, NTFP processing etc. The JFPM programme has provided the impetus for the community to interact and discuss the issues of priority and concerns to the village.

Community Organisation

The impact of JFPM is also evident in the enhancement of community organisation including women empowerment and leadership development. The communities perceive an improvement especially in leadership development, better community organisation and communication due to JFPM. Empowerment of women was perceived especially in the Eastern Plains due to the self help groups formed under the Stree Shakthi programme. These have also had a positive impact on their involvement in VFC activities.

Protection practices adopted in the JFPM areas have had a direct impact on productivity of grass in the Eastern Plains, which is an important forest product. Most of the VFCs reported an increase in grass productivity in the JFPM area, though there was a decline in the Western Ghats area due to closure of tree canopy and the dominance of Acacia plantations. Illegal extraction of timber and fuelwood has considerably reduced due to protection measures adopted by the community in

Table 13.2 Entry Point Activities Taken Up under the Eastern Plains Afforestation Project

Forest Circle	*No. of VFCs*	*No. of VFCs with Entry Point Activities*	*Percentage of VFCs with Entry Point Activities*
Bangalore	735	654	89
Belgaum	441	107	24
Bellary	505	383	76
Chickmagalur	37	31	84
Dharwad	265	162	61
Gulburga	283	240	85
Mysore	265	169	64
Shimoga	58	16	28
Total	**3,100**	**2,029**	**65**

many of the VFCs in the Western Ghats region (Ravindranath and Sudha 2004). It has also been successful in evicting encroachers on forestland to a large extent, especially in the Western Ghats region.

Relationship with the Forest Department

Over the years the village communities have always viewed the Forest Department as policing the forests to deny them their collection of subsistence needs. The Forest Department had also viewed all sections of the village communities as destroyers of the forests. This new faith in each other and protection of the forests hand-in-hand is the beginning of a social dimension in forest management. Implementation of further forest management programmes will be easier due to this now well-established relationship.

Involvement of community in forestry activities and awareness about forest protection has increased due to JFPM. Especially in the Western Ghats region, the most perceived impact of JFPM as quoted by them is increased involvement by them in forestry activities. There has been a decrease in illegal cutting of trees and encroachment of forestland due to protection effects of the community. Thus, social impact has been the major incentive in the Western Ghats region for the community.

JFPM has fostered a better relationship between the community and the Forest Department compared to pre-JFPM times. In the Western Ghats region, 91 per cent of the VFCs and in the Eastern Plains, 94 per cent of the VFCs sampled, reported an improved relationship (Figure 13.1).

Overall, the social impact of JPFM has been noteworthy in the Western Ghats region, probably due to the longer duration of the project. The major impact has been with regard to better involvement of the community in forestry activities thus reducing the incidence of illicit cutting, smuggling and encroachment. Thus, overall, 44 per cent of the VFCs in Western Ghats region perceived social impact due to JFPM (Table 13.3). In the Eastern Plains, the social impact has not been as much pronounced where only 15 per cent of the VFC's perceive a social impact due to JFPM.

Financial Incentives

The major financial incentives from JFPM are through benefit-sharing between the Forest Department and the VFC from the sale of timber and through forest produces such as fodder, fuelwood and NTFPs such as fruits, seeds, flowers etc. from the JFPM area.

Figure 13.1 Major Social Impact of JFPM as perceived by the Community in Karnataka

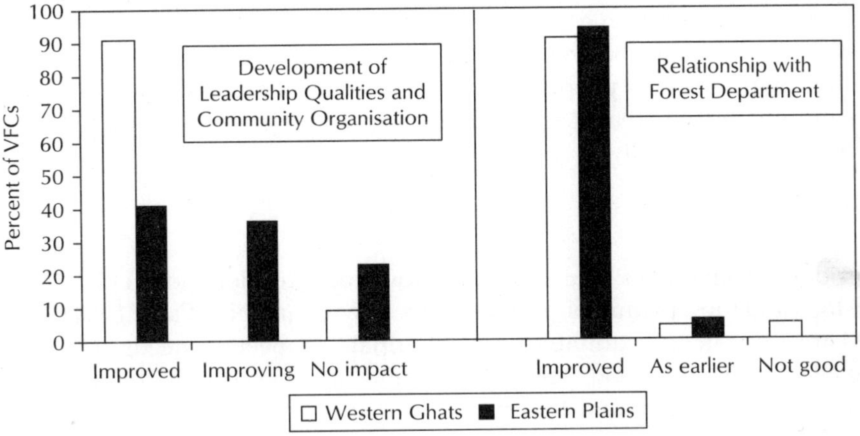

Table 13.3 Community Response to Impact of JFPM

Impact of JFPM	Per cent of VFCs	
	Western Ghats	Eastern Plains
Social Impact	44	15
Involvement of village communities in forestry	11	–
Illegal cutting and smuggling decrease in	11	–
Decrease of encroachment	7	–
Increase in unity and cooperation among community	6	11
Rural development	6	4
Better relationship with the Forest Department	3	–
Economic impact	13	50
Increased availability of fuelwood and leaves	12	–
Increase in grass production	–	24
Economic benefits[1]	1	26
Environmental impact	6	16
Greening of area	4	9
Improvement for posterity	1	4
Water table increase, improvement in soil conditions	1	3
No impact*	37	19
Total	**100**	**100**

Notes: [1] includes income and employment generation.
* No discernable impact or the community could not perceive any impact.

Benefit-Sharing from JFPM Area

To enhance the income of VFCs, the Government of Karnataka Order 2002 specifies a share of the benefits from the JFPM area to the VFC. In the plantations raised after the formation of VFC, 90 per cent of the share from NTFPs will be provided to the VFC. From timber plantations, 75 per cent share from the harvest will be provided to the VFC. To provide further financial benefits to the community, the Forest Department has included plantations that were established before formation of the VFC as JFPM area and the VFC members are eligible for a share of the benefits accrued from this area. The proceeds from natural trees in degraded forests and older plantations raised before formation of VFC will be shared equally between the VFC and the Government. To enhance community participation and support, it is proposed to even bring valuable species like sandal and rose, which grow naturally in JFM areas, for benefit-sharing and the proceeds are to be shared between Government and the VFC equally.

Flow of Products

The flow of products from the JFPM area includes fuelwood, fodder and NTFPs. The VFC members are entitled to lops and tops, prunings, grasses, leaves, twigs and NTFPs for *bona fide* use, free of cost from the JFPM areas on fair and equitable basis, essentially to meet the needs of the local population. Prior to disposal of fruits, firewood, poles, timber and final harvest, the requirements of the local villagers is treated as a priority and the surplus is disposed of by the VFC through open public auction.

According to the study, increase in availability of fuelwood was observed in large number of VFCs especially in the Western Ghats region and in the Eastern Plains with older plantations. Increase in grass productivity in the JFPM area has been a major incentive in the Eastern Plains, while in the Western Ghats there has been a decrease in grass productivity due to closure of canopy. The increase in the Eastern Plains has been due to closure of the plantation area from grazing and protection from fire incidence. In fact, in the Eastern Plains, the major impact of JFPM has been an increase in grass productivity (24 per cent VFCs). (Figure 13.2)

In 72 per cent of the VFCs, the Forest Department utilised local labour for plantation activities and in 16 per cent of the VFCs (Figure 13.2), other employment generation activities were implemented. Under the FEPEP, EPA has promoted income generation activities such as pisciculture, tailoring, NTFP processing, that have provided economic benefits to the community. The major impact of

Figure 13.2 Impact on Flow of Forest Products from Plantations (Pls) and Natural Forests (NF) under JFPM, as Perceived by the Local Community in Karnataka

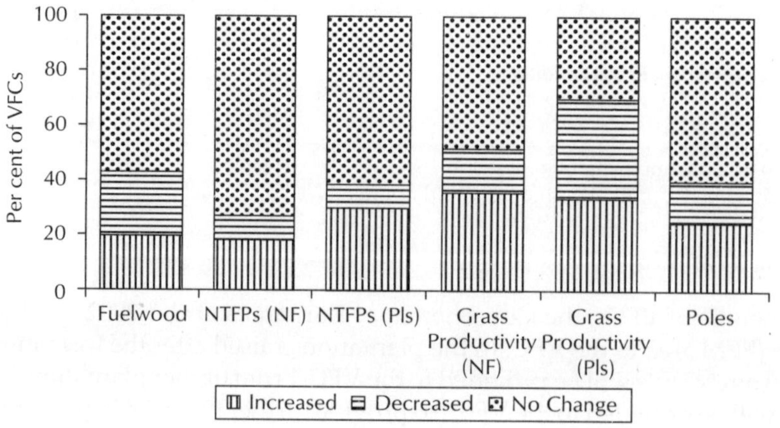

JFPM in both project areas has been the economic benefit (Figure 13.3) through the wage labour component whereas in Eastern plains the benefits by EPA was also perceived.

Figure 13.3 Major Economic Impact of JFPM as perceived by the Community in Karnataka

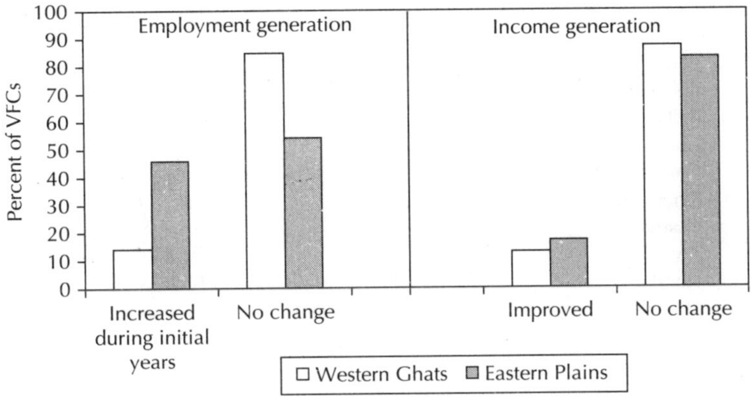

Financial Sustainability

To ensure sustained flow of financial benefits to the community, the 2002 JFPM Order specifies that out of the benefits accruing to the VFC, at least 50 per cent will go to the Village Forest Development Fund (VFDF). The money has to be used for expenditure on all forestry-related activities defined in the microplan and the MoU. Expenditure is mainly towards promotion of tree growth in the JFPM areas, nursery raising, afforestation, forest protection works, expenditure incidental to harvest forest produce and expenditure towards promotion of JFPM in particular and forestry in general.

The balance 50 per cent may either be equitably shared by the members of the VFC as dividends or be deposited in a Village Development Fund (VDF) for financing developmental works in the village as approved by the VFC. The VDF can be used by the VFC for developmental activities other than forestry. While spending the VDF, the VFC shall decide the development needs of the village and execute the works under the supervision of the concerned Gram Panchayats as these institutions have the experience and functionaries for undertaking village development works.

Financial Incentives to the Community

In the Western Ghats and the Eastern Plains, Karnataka, plantations raised various programmes like social forestry; also degraded forests have been included as JFPM area from which the VFC receives 50 per cent of the income out of timber sale proceeds. In 122 VFCs in Karnataka, benefit-sharing has taken place.

In the Western Ghats region, twenty-seven VFCs have shared benefits and the total revenue generated from JFPM area is Rs 23,64,880. The total revenue received by the VFCs is Rs 11,88,843 (50 per cent from timber sale and 90 per cent from NTFPs) of which 98.8 per cent is from timber and 1.2 per cent from NTFPs. The average revenue per VFC is about Rs 44,031 (Table 13.4).

In the Eastern Plains, harvest of timber and revenue from NTFPs occurred in ninety-five VFCs amounting to a total revenue of Rs 88,35,331. The average financial benefit that has accrued to the VFCs is

Table 13.4 Income Generation from JFPM Area in the Western Ghats and Eastern Plains Region

		Income Generated from Plantations after Formation of VFCs		*Extent of Older Plantations in the Jurisdiction of VFCs Ready for Harvesting (ha)*
No.	*Forest Division*	*Small Timber*	*Minor Forest Produce*	
Eastern Plains region				
1	Bangalore	6,36,700	1,46,168	642
2	Belgaum	–	–	151
3	Bellary		1,84,473	552
4	Chamarajanagar	–	–	–
5	Chickmagalur	–	–	155
6	Dharwad	–	–	35
7	Gulburga	–	1,35,868	661
8	Mysore	85,000	7,050	75
9	Hassan	75,53,092	86,980	3,445
10	Shimoga	–	–	315
Total		**82,74,792**	**5,60,539**	**6,031**
Western Ghats region				
1	Karwar	8,95,910		
2	Honavar	8,25,696		
3	Sirsi	6,27,266		
4	Haliyal		28,814	
Total		**23,48,872**	**28,814**	

Note: Income generation is the total revenue generated from JFPM area, of which 50 per cent of the share will be VFC's if generated from plantations established before VFC formation and 75 per cent if from plantations formed after formation of VFC.

Rs 48,41,881 (50 per cent from timber sale and 90 per cent of NTFPs) of which timber sale accounted for 89 per cent and NTFPs, 11 per cent. The average income to the VFCs amount to be Rs 48,862 per VFC.

Thus JFPM has provided village communities with greater legitimate access to benefits accruing from the forest and has, importantly, augmented their livelihood, besides enabling them to create village funds for undertaking developmental activities. In the Western Ghats project area, most of the plantations established under the WGFP during 1993–97 are due for harvest, which amount to 22,000 ha with projected benefits of 2,80,000 per VFC excluding the lops and tops. In addition a potential 15,000 ha under social forestry can be included under JFPM (Sudha et al. 2005). In the Eastern Plains, about 6,031 ha of older plantation under JFPM are ready for harvest. Thus the VFCs can expect substantial returns in the coming years.

Environmental Incentives

Vegetation Improvement

A study conducted (Sudha et al. 2004) in Karnataka to understand the ecological impact of JFPM based on community perspective showed that nearly 75–80 per cent of the VFCs reported an increase in canopy cover and tree density, 50 per cent VFCs reported increase in species richness and about 57 per cent VFCs perceived no change in the availability of fuelwood.

With regard to fodder availability, in the Western Ghats 60 per cent of the VFCs perceived a decrease in fodder availability due to closure of tree canopy inhibiting grass growth. In the Eastern plains, the converse is true where 42 per cent of the VFCs reported increase largely due to closure of area to grazing. (Figure 13.4). Six per cent of the VFCs in Western Ghats and 16 per cent in the Eastern Plains reported environmental benefits from the JFPM programme such as increase in water table (Table 13.3) after establishment of plantations under the JFPM programme.

Figure 13.4 Change in Density, Species Richness and Canopy Cover in the JFPM Area, as perceived by the Community

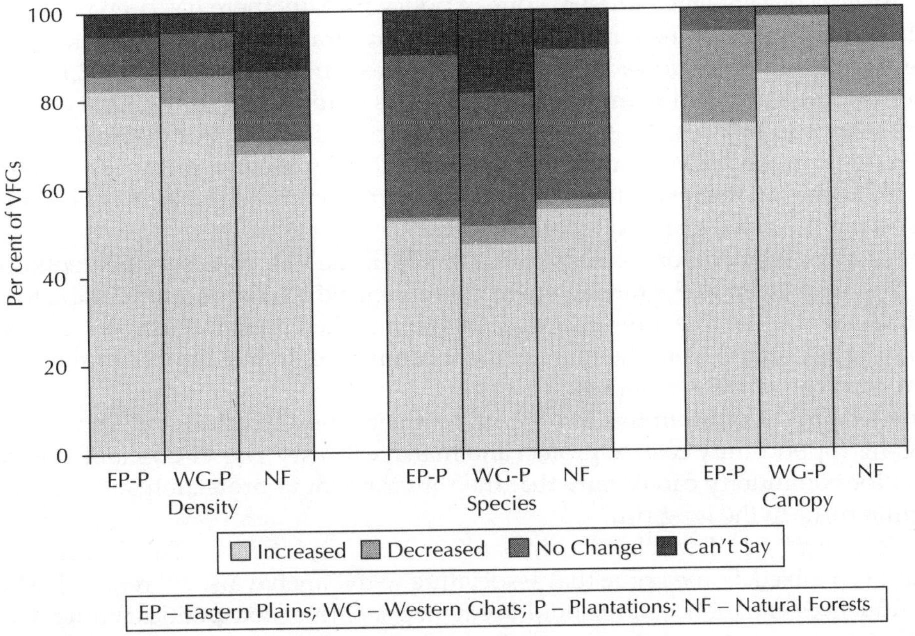

DISCUSSION

Institutional Sustainability

The 2002 JFPM order of Karnataka to a large extent, addresses most issues that usually ensure sustainability of a participatory programme. The order addresses important aspects such as absolute participation of the village community, gender and equity concerns and financial sustainability. It is important to understand the extent to which the policies have influenced the performance of the programme. Preliminary studies in Karnataka showed mixed performance of JFPM at the VFC level based on functioning, performance and impact of JFPM in the Western Ghats and Eastern Plains (Sudha, et al. 2004). Based on the analysis of responses of the VFC president and the management committee members, at the state level, in 29 per cent of VFCs the impact of JFPM was visualised as good, moderate in 38 per cent of VFCs and no discernable impact was felt in 33 per cent of VFCs (Figure 13.3). In the Western Ghats, 32 per cent of VFCs reported the impact as *good*, a majority (46 per cent) as *moderate* and 11 per cent could not conclusively perceive any impact.

In the Eastern Plains, 26 per cent of VFCs reported the impact as *good*, 30 per cent as *moderate* and in majority (44 per cent) of VFCs, the impact was *not discernable*, as the programme is still new. Compared to any other development programmes linked to forestry activities, JFPM has made a significant impact. Though JFPM policy provides the incentive, the programme is still on the horizon. It is essential to understand the various issues governing JFPM, to implement the programme sustainably and expand it further by involving forest fringe communities in an effective way.

Enhanced participation of communities: In spite of policy backup, there has been poor participation of marginalised group and women in the VFCs—a serious drawback in JFPM evolution and development needs policy interventions. According to studies, on an average only 12 per cent of men and 6 per cent women population are members of VFCs (Sudha et al. 2004). One way of ensuring their participation is to link resource use to membership of the VFC. For example, if they need to collect fuelwood from the JFPM area, they are to compulsorily become members of the VFC. Also, on the lines of Panchayati Raj institutions, the management committee posts may be made available to women and marginalised group on roster system.

Another major impediment for the viability of the VFC is the VFC members' perception of a strong presence of the department in the management committee, which needs a rethinking to modify the existing arrangements. The treasurer post may be given to the Forest Department staff and a joint-secretary may be appointed from the management committee. In the absence of the Secretary, the joint secretary can convene the meetings.

Legal status for VFCs: Communities have to invest their time and effort and also sacrifice several short-term gains (opportunity cost) to protect and manage forests. The VFCs should get legal recognition so that the community can be sure that their investment in protection and regeneration will provide them returns in the long run.

Linkage to Panchayati Raj: The relationship between VFCs and the local Panchayats from some of the states has been mixed. Some argue that associating with Panchayats will render JFM vulnerable to political influences and, hence, JFM should remain a separate, non-political entity. Others opine that promoting synergy between Panchayati Raj and JFM will lead to pooling of resources, which can be efficiently used to promote overall village development, including forest resources. Though there are no ready solutions, grassroots support, including that of Panchayati Raj institutions at the village level, is required to sustain JFPM, and linkages with other rural development programmes such as animal husbandry, small-scale enterprise development etc. need to be established. Also, in Karnataka, the membership structure of the Grama Sabha and the VFC are the same and hence is easier for them to link up. An added advantage when compared to Gram Panchayat is that while formally the Gram Panchayat Chairman has to convene the meeting, who could be from a neighbouring village, for a VFC, the President is from within the village and it is easier to call for meetings.

Capacity building: According to a study, in Karnataka, the awareness of village community and more so of the management committee about JFPM is lacking (Rao et al. 2004). They need to be educated about the programme and made aware of their roles and responsibilities and the need to actively participate in VFC activities. NGOs must motivate the communities to cooperate wholeheartedly for the success and sustenance of their local JFPM venture.

The capacity building activities like training, skill development, marketing support, value addition activities and such other inputs are to be continuously provided. With closure of externally aided projects along with non-availability of financial support from the Forest Department, capacity building programmes are not conducted periodically. This needs to be addressed by linking them to the Forest Development Agencies. Also the government's share of the profit may be used for this purpose. The funding from the state and district sector plans also should be made available.

Participatory approach: It is gratifying to note that there has been a perceptible attitudinal change among the community and the front line staff in accepting the participatory forest management as the most effective way of protecting the forests. The front line staff is to be further trained and prepared for the change in the management pattern, procedures and systems. Reward and recognition are necessary to encourage the front line staff of Forest Department to further the movement. The political and bureaucratic support to the JFPM process needs to be secured for enhancing participatory role in forest management.

Role of NGOs: NGOs help build the rapport between the community and the Forest Department and enable implementation of JFPM. They have played a significant role in creating awareness, building institutions and in promoting community participation. VFCs have performed better where NGOs have played an active role (Ravindranath, et al. 2000). They need to be encouraged, as they play an important role in developing community organisations such as SHGs and help to negotiate between conflicting groups in the village. It is also seen that wherever local NGOs are involved in the JFPM programme, there is formation of a large number of SHGs, mainly for women and the poor, which leads to their better articulation (Ramalingegowda and Shivanagowda 2001). Local NGOs working on forestry issues should be preferably involved effectively, from the beginning of the JFPM project. At the same time, school teachers, youth leaders and knowledgeable elderly village members should also be given the task of working with the Forest Department and the community, so as to reduce the dependence on NGOs and make JFPM more participatory.

Promotion of VFC federations: The Forest Department and NGOs should facilitate formation of federations of VFCs at block, forest range, division/circle level. These federations should have institutional recognition, legal standing, and decision-making powers. Federations could strengthen the VFCs and resolve to promote JFPM, resolve conflicts, enhance the community's negotiating powers with the Forest Department and other developmental agencies, help NTFP processing and marketing, enable equitable sharing of resources among and within VFCs, promote capacity-building etc. The Government should also link up the primary producers through these federations to the end-users through the wholesale merchants.

Financial Sustainability

Periodical flow of benefits: Forest protection and regeneration often leads to flow of income in the long run, such as from timber sales, but to sustain community participation, it is very important to ensure periodic flow of benefits, particularly in the short-term as well. Most of the financial benefit is from timber sales which has occurred only once in most of the VFCs. A sustained annual flow of benefits from the forest area is what the communities would seek. This may require promotion of grass production and annual and perennial NTFP species.

NTFP based enterprise: Timber requires a long gestation period and does not provide income in the short-term. NTFP based forestry, on the contrary, would provide a regular income. The focus of JFPM should, therefore, be on sustainable production and management of NTFPs, which will be the main source of earnings from the forests for the beneficiaries. The important factors that state Forest departments need to consider are: inclusion of NTFP species during afforestation, design of harvest and processing methods for the NTFPs, and developing marketing linkages for the NTFPs with consumer industries.

Adding value to the NTFPs by developing simple technologies to process them can ensure sustained economic flows to the beneficiaries. Product development and quality control mechanisms need to be put in place to ensure competitive marketing of produce. Creating federations of VFCs at a taluk, forest division or district level would be ideal for pooling resources and creating linkages with the consumer industries. Value addition to NTFPs and sale of the products by establishing linkages with consumer markets will enable sustained flow of products and monetary benefits. The government should not have monopoly over marketing the NTFPs, nor must it allow traders and mills to monopolise. Trade in NTFPs should be decentralised gradually so as to encourage healthy competition (Saxena 1997).

Link to rural development: The success of the JFPM programme is predominantly dependent on ensuring sustained economic returns to the beneficiaries. Initial financial support through EPA should be a prerequisite for starting JFPM in a village and is to be linked with rural development activities in the village through Panchayats, the FDA and other line departments. One such concept is water costing especially in drought prone areas. Check dams and gully checks can be created by the VFC and water stored. The farmers can utilise the water for their crops at a nominal price per acre, thereby generating income for the VFCs. Another entry point activity that could be initiated is *revolving fund concept* for the VFC that can be utilised as a micro fund for SHGs at a nominal rate of interest to create income generation activities. This will also lead to linkage of SHGs to the VFCs and enhance participation of women in VFC activities. The interest earned on the revolving fund could be used as salary to a watchman for watch and ward functioning of the JFPM area. Strategies also have to be developed by the government to promote poverty alleviation programmes for the landless and the artisans. For example mulberry cultivation can be promoted especially for the landless, which could generate income for the VFC; besides the landless can also take up sericulture.

Once the forest has stabilised, enough monetary resources have to be generated to sustain the JFPM programme. The state guidelines support the policy of ploughing back into the programme a part of the benefits accruing from the forests. Provisions must be incorporated to enable funds to be generated from among the community to match the seed money provided by the government so as to strengthen the concept of joint ownership and increase the sense of responsibility among the community members and sustain their interest in JFPM. Development of forest resources under JFPM should also be linked to development of agroforestry, water conservation, agriculture and livestock development, in order to enhance the synergies between the various facets of rural community living.

Environmental Sustainability

Vegetation management: The JFPM programme has led to an increase in vegetation cover in all the VFCs mainly due to afforestation of exotic species such as eucalyptus and acacia in Western Ghats.

Emphasis on management and rejuvenation of natural forests should be promoted, which to an extent has been done in the Eastern Plains. Mere protection may lead to regeneration of local species in some regions. In case of protection and enrichment planting in areas with root stocks, the species that will regenerate will be those that have endured fire and browsing; for example, in teak dominated areas, the species regenerate along with a few associated species. Rehabilitation of these areas with associated native species has to be done through enrichment planting. Many of the silvicultural practices could adversely affect biodiversity conservation, for example, selective propagation of commercial species. Promotion of natural regeneration is a cost-effective, participatory and biodiversity-friendly approach. Thus, in areas with potential for regeneration, plantation forestry should not be practiced. Thus capacity-building programmes should also include training the community in management of rootstocks in areas where regeneration could be promoted.

The afforestation model should be adopted where natural regeneration potential is low, with no rootstock and poor soil status. In areas under afforestation, the main thrust should be to meet the requirements of the community. Planting multipurpose species as practiced in the Eastern Plains region where thirteen different afforestation models have been developed and implemented, promote not only biodiversity but also meet local needs.

Water and soil conservation: JFPM programme will not be sustainable without any intangible benefits to the community. Soil and water conservation practices if adopted, can enhance these benefits through increased available water, tree biomass productivity and crop productivity. These are labour intensive (gully checks), but enhance regeneration and growth rates of grass and tree species. Community participation and voluntary labour could reduce these costs to a large extent.

Silvicultural practices: Silviculture and management systems, which have hitherto concentrated on timber production, need to change focus to production of fuelwood, fodder and NTFPs to meet the demands of the community under the JFPM system. Forest regeneration under JFPM requires species choice to meet the diverse local needs, silvicultural practices to enable grass production, harvest of fuelwood, grass and NTFPs at periodic intervals. The species choice and silvicultural practices should also be compatible with local soil, rainfall and topography along with specific community needs.

The community has to be involved at every stage of decision-making, implementation and management of the JFPM area. However, it is important to consider who in the community are involved in decisions on species choice. If economically sound section of the community is consulted, their choice of species may be different from that of women, the landless and the artisans. The community also requires orientation of site conditions to match the suitability of species for afforestation.

Interventions such as improving the quality of planting stock and developing plantations using hybrids, cloning and biotechnological methods have to be implemented that can improve production of fuelwood, pole and timber species such as Eucalyptus, Acacia auriculiformis, Acacia nilotica and Teak. These can be planted on community lands and farmlands. High productivity will reduce the rotation period of the plantation and provide short-term benefits to the VFC, at periodic intervals. This will sustain the interest of the community in the programme.

Sustaining JFPM

The JFPM policies have been very progressive. But it is important to ensure that the Forest department officials have the will and the commitment to implement JFPM so that the implementation is at par with the progressive policies.

Sustainable management of forests is only possible by providing the forest-dependent communities with means of securing their livelihood. Clearly defined, secure tenurial rights are a fundamental requirement for encouraging sustainable use of forests. In the absence of tenurial security, people choose short-term gains, leading to the degradation of natural resources. It has also been recognised that to ensure continued participation, VFCs need legal standing, assured usufruct rights, attractive share in income and control on decision-making. To achieve this, there is need for not only policies but also awareness and capacity-building among communities as well as the Forest Department staff, to translate the policies into action. In the long-term, there is a need to integrate forest management with overall sustained rural development. Thus an integrated village landscape approach is the need of the hour.

ACKNOWLEDGEMENTS

We acknowledge Ford Foundation, New Delhi for the financial assistance, Ministry of Environment and Forest, New Delhi for their support to CES and Karnataka Forest Department for their help in conducting the study. We thank the staff of CES at Bangalore and at Sirsi for their help at various stages of the study.

NOTE

1. Corresponding author.

REFERENCES

Government of Karnataka. 2002. *Proceedings of the Government of Karnataka on JFPM*. Forest, Environment and Ecology Development, Bangalore: Government of Karnataka.

Jagannatha Rao R., P. Sudha, P.R. Bhat, G.T. Hegde and N.H. Ravindranath. 2004. 'Awareness and Capacity Building Programmes under Joint Forest Management: Case Study of Karnataka', in N.H. Ravindranath and P. Sudha (eds), *Joint Forest Management in India. Spread, Performance and Impacts*. Peace: Universities Press.

Jewitt. S. 2001. *Environment, Knowledge and Gender: Local Development in India's Jharkhand*. Aldershot: Ashgate.

KFD. 1999. Western Ghats Forestry Project. Project Experience and Achievements 1992–2000, Bangalore: Karnataka Forest Department.

Malhotra, K.C. 1995. 'Biodiversity conservation and community development in Southwest Bengal', in H. Wood, M. McDaniel and K. Warner (eds), *Community Development and Conservation of Forest Biodiversity through Community* Forestry, Bangkok: RECOFTC.

Malhotra, K.C., Y. Gokhale, S. Chaterjee and S. Srivastava. 2001. *Cultural and Ecological Dimensons of Sacred Groves in India*. Indian National Science Academy, New Delhi and Indira Gandhi Rashtriya Manav Sangrahalaya, Bhopal.

Ramalingegowda, S.K. and B. Shivanagowda. 2001. 'The Status of JFPM Programme in Karnataka and Role of Forest Department in Sustainable Forest Management by Communities and Stakeholders', *Myforest*, 37(4): 603–10.

Rao J.R., K.S. Murali and N.H. Ravindranath. 2002. 'Enabling, Empowering: Joint Forest Planning and Management in Karnataka: Current Status and Future Potential', *Wasteland News*, 17(3): 14–27. February–April.

Ravindranath, N.H. and P. Sudha. 2004. *Joint Forest Management in India. Spread, Performance and Impacts*. Universities Press.

Ravindranath, N.H., K.S., Murali, I.K. Murthy, P. Sudha, S. Palit and K.C. Malhotra 2000. 'Summary and Conclusions' in *Joint Forest Management and Community Forestry in India*, Oxford and IBH Publishing Co. Pvt. Ltd.

Saxena. N.C.A. 1997. A *Saga of Participatory Forest Management in India*, Indonesia: CIFOR.

Sudha, P., J.R. Rao, B.C. Nagaraja, P.R. Bhat and N.H Ravindranath. 2004. 'Gender and Equity Concerns in Forest Management: Genesis and Experience from Karnataka', *Economic and Political Weekly*.

14

Sustainability and Food Security Issues among Farmers in Tamil Nadu

H.N. Chanakya[1], *S.T. Somashekara Reddy* and *P. Balachandra*

Abstract: Resource poor farmers (RPF) living below poverty line constitute about a third of the rural population in many Indian states. In this chapter, we seek to study the status of food security and vulnerability among RPF members of self help groups (SHG) who have adapted and used appropriate technology and eco-friendly inputs in agriculture. The expected outcome is an increased availability of food grains to the RPF family. This higher availability manifests as higher quantities of food intake if farmers have not incurred any credit to be repaid with harvested grain. We seek to quantify food consumption and food security manifestations within this process of change among some of the economically poorest farmers in nine districts of Tamil Nadu (TN). The food consumption pattern was measured after three years of practicing eco-friendly technologies. The average daily per capita food consumption of the study sample during peak drought was only 4 per cent lower than that of the TN state average for a non-drought year, suggesting it to be a sign of improvement in food availability and food (calorie) security. This suggests that promoting SHG mechanisms can effectively address the problem of food security among RPFs. The chapter also discusses the impact of demographic features such as the male:female ratio, able:aged ratio, adult:children ratio, young males/females: adult males/females ratio on the daily per capita food consumption. It was observed that increases in daily per capita food intake were always accompanied by an increase in the share of own grains and use of millets in the overall food basket. The major component of the food basket however is dominated by rice even among these rain-fed farmers who raise non-rice crops. This dependence on an externally raised crop is considered non-sustainable. Thus, the important policy implication is that there is a need to significantly increase the non-rice component of the food basket to ensure long-term sustainability among these rain-fed farmers. Providing short-term credit to SHGs, group based monitoring and promoting use of low external eco-friendly inputs are some of the other policy initiatives that could enable the RPFs to achieve a high level of daily per capita food consumption and concomitant food security components.

INTRODUCTION

Background

RPF in rural India lead a subsistence life with inadequate access to secure food resources and it manifests largely during recurring droughts. Their land holdings are meagre, and during drought the food production is minimal. Further, the employment opportunities as daily wagers are non-existent during these harsh times. Even the national surveys conducted frequently in rural areas, which estimate poverty and nutrition status, have substantiated this. Such surveys attempt to identify that fraction of the population and corresponding income levels whose daily diet comprises of less than 2,400 kCals/capita/day. This calorie intake cut-off for an average rural mildly working adult (60 kg body weight) is a convenient measure and is often used to indirectly indicate nutritional sufficiency. People falling below this level are considered to be suffering from both poverty and malnutrition (Gopalan et al. 2000). They may constitute a third of the total population in some states (Kozel and Parker 2003). In South India, TN is reported to have the largest number of such calorie challenged people as well as nutritionally challenged in the form of chronic anaemia among women (Radhakrishna 2002a) and children. The average calorie intake in the TN is reported to have fallen (Meenakshi and Vishwanathan 2003) from 1861 (in 1983) to 1,826 kCal/capita/day (1999–2000). Similar figures are available for head count ratio (Meenakshi and Vishwanathan 2003 74.6 in 1983 and 78.7 in 1999–2000). NSS data (NSSO 2001a) shows that among various sections in rural areas of TN, monthly per capita expenditure (MPCE) for thirty days is lowest among the agricultural labour (Rs 270) and followed by self-employed in agriculture (Rs 310), only marginally better than Madhya Pradesh (Meenakshi and Vishwanathan 2003). This being the case, rural TN offers an interesting case for an examination of calorie intake and its distribution especially in a condition that is conducive to invoke food security threats, namely recurring droughts.

Origin of Food Security Threats

Food security threats are manifest to a greater extent and severity among RPF in the rain-fed belt. Frequently occurring meteorological and agricultural droughts (insufficient and missed rains during crop growth) have gradually pushed a large segment of these RPFs from transitional food insecurity (in the past) to chronic food insecure situations at present (LEISA 1999; Chanakya and Reddy 2002). Among these RPF, the phenomena of male migration, decreasing nutrition, strong daily wage dependencies, high levels of credit etc. have already emerged (Watts 1987) taking this segment of population very high up on the scale of vulnerability (food insecurity; Watts 1987). Various forms of coping strategies are visible among this group and are discussed later on in the chapter.

Typical Food Substitution Strategies

The reports of the National Sample Survey of India (2001b) appear to validate the findings of Report 472 (NSSO 2001a), that rice is entirely purchased among a large segment of farmers in spite of the fact that TN is known for high rice production. Rice is usually purchased in large quantities from the (PDS). In contrast, it reports highest dependency on homegrown food inputs among vegetables and fruits. Potential substitutions of purchased inputs for homegrown foods are interesting to study for their impact on food consumption patterns and as a coping strategy. There is also a need to know whether homegrown cereals and millets provide greater food and nutrition in times of scarcity (droughts), how they are distributed among family members, how families with more children or women respond to scarce food etc.

The Purpose of this Study

We seek to determine improvements in food security and vulnerability among SHG members who have adapted and used appropriate and eco-friendly technology inputs in agriculture for periods between 1–3 years in about eighty villages of Tamil Nadu. The expected outcome is a sustainable increased availability of food grains to the RPF family. We seek to quantify food consumption and food security manifestations within this process of change. These RPF form a relatively homogenous group within a narrow range of resource access. For studying such a group from South India, there were few precedents with regards to the methodology, rigour needed, understanding the factors influencing measurements etc. (Maxwell and Frankenberger 1992) and therefore we could derive little benefit from past studies on such a group. The RPFs have used SHG concepts primarily to escape falling into a credit trap as well as to further and widen their options for food security. From a total size of 100 such SHGs, we have sampled thirteen SHGs falling within chronically drought affected areas.

Food Security Attributes Measurement and Categorisation

Three attributes of measurable food security were examined in the study:

1. Calorie nutrition sufficiency.
2. Variety of food basket—rice, millets, pulses, vegetables and fruits (milk and meat are ignored being insignificant in the study population).
3. Access and entitlement indicators—where these came from and how much control the farmer has on these sources (own, purchased, public sources and PDS).

The data collection was made during peak stress periods for food security, namely, a year after the previous harvest. The analyses therefore included component of timeliness of food access within the concept of food security. However, issues of risk and insurance were not measured in this study.

Calorie sufficiency was estimated directly from the Daily Per Capita Food Consumption (DPCFC) and the Daily Per Capita Energy Intake, DPCEI

$$\text{DPCEI} = [(\text{cereals} + \text{millets} + \text{pulses}) \times 4 \text{ kCals/g}]$$

The presence of a certain minimum pulse intake is considered to cover concepts of sufficiency of protein in food—its sufficiency is not considered in detail. From the per capita daily food intake, the daily calorie intake was grouped as below. The daily calorie intake was expected to occur over four distinct ranges as follows:

Food Intake (g/d)	*Calorie Range (kCal/d)*	*Category*
a. <350 g/d	<1,400/d	severe deficiency
b. 351–450	1,400–1,800/d	sub maintenance (survival)
c. 451–600	1,800–2,400/d	adequate
d. >600	>2,401/d	Unchallenged/comfortable

The study sample comprises adults in the range of 45 to 55 kg for men and 35 to 45 kg for women and thus the average adult weight is considered to be 45 kg. The average daily calorie need (sufficiency) for this group is estimated to be 1,800 kCal/capita/d (= 45*2,400/60). A 25 per cent deviation from these corrected to the nearest 50 g fraction formed the basis of the classes indicated above and tables presented later on.

METHODOLOGY: DATA COLLECTION AND ANALYSES

Characteristics and Choice of the Study Sample

In eighty-two villages across ten districts of TN, RPF initiated SHGs have networked to form the Low External Input and Sustainable Agriculture (LEISA) Network. An NGO is associated with each of the SHG. Members of these SHGs using principles of appropriate technology, organic farming and participatory technology development (Chanakya 1999) attempt to attain family level food security and overcome credit trap leading to undesirable consequences including suicides (Jha 2003 and Deol 2003) etc. The SHGs routinely measure and monitor the agricultural and household inputs. It was thus easy to obtain reliable data on daily food consumption. Women (most affected [Radhakrishna 2002b] by food security threats) constitute nearly 35 per cent of the SHGs. Usually the husband and wife together discuss at the SHG where significant discussions and efforts are carried out on the modes of reducing family expenditure. All members usually represent the poorest of the village. This group was reasonably well aware about food security (LEISA 1999) and therefore formed a good study sample.

Internal Criteria for the Sample Group

In rural India, there are serious difficulties in accurate identification of the poor and the vulnerable (Harris 1987) based on income levels and occupation. Dependence on Common Property Resources (CPR) brings in seasonal variations (NSSO 2000) in vulnerability. This study sample of SHGs was reasonably homogenous and selected on a set of criteria set by RPFs themselves as indicated below:

- land holding less than a hectare;
- the family needs to be residing in a hut or be a recipient of a house from government;
- needs to be a ration card holder which is exclusive for the poor;
- family members find alternative income as wage labourers;
- the income of the family from other sources needs to be less than Rs 2,400 per annum;
- member should not own any irrigation facilities; and
- farmers with tank irrigation facilities are included only when such holdings are less than 0.05 ha.

Each SHG had a maximum size limited to twenty members. These individual SHG members provided data inputs in open SHG meetings. Such uncontested information was regarded as information that met verification and approval of the peers or fellow members of the SHG. There were thus no compulsions or chances to give any false data. More importantly, the SHGs were spread across nine districts in Tamil Nadu and the entire data was collected in a span of five days.

Data Collection

Data on the household consumption of food articles were obtained by the recall method. The respondents, namely all the members of the RPFs' SHGs were asked to list all the previous days' family household spending in a SHG meeting. This was administered by a staff member of the associated Non-Governmental Organisation (NGO) during the first week of September 2002 without any prior intimation to the SHG. The data collected by this was analysed and was found to have a lot of skew and extremes. This was reported to each of the SHGs and the requirement for accurate estimates was discussed. Subsequently, during the third week of October 2002, these SHGs were requested to monitor their food consumption on a specific date during this week. The data reported by each of the families was approved as normal daily expenses when verified by the rest of the SHGs and recorded. This data was then collated and analysed. Food grains in the typical daily diet, namely rice, millets, pulses, fruits and vegetables used on that particular day were monitored for each of the farm along with its source (from own stocks/farm land, purchased, from common property resources or from Public Distribution Systems (PDS). These were recorded source-wise at the SHG meetings and used for final analyses. Thus, data was obtained from 260 families belonging to thirteen SHGs. Even after so much care during data collection, data from twenty-three families was found to be unreliable and not considered for further analysis.

The timing of data recording was set to match with the time of least grain stock among the RPF, namely in the middle of a cropping season (South West [SW] monsoon region) or early cropping period (for farmers in the North East [NE] monsoon belt). It was envisaged that RPF families recording a high level of per capita daily food consumption indicated sufficient food stock to overcome drought risk and consequent low vulnerability. A low vulnerability in turn indicated a successful intervention in terms of a sustainable mode of agricultural production aimed at food security. Similarly, a higher dependence on a food basket provided from own sources indicated internalisation of the sustainability principles.

Data Analysis

The analysis of the sample characteristics have been done in two ways:

(*i*) The sample population of all the villages combined is divided into various groups formed, based on different levels of average per capita food consumption per day. For arriving at per capita food consumption, we have used only the consumption of rice, millets and pulses (excluding vegetables and fruits). This measures the ability of a given group of people to meet their daily calorie intake. With this we could form six groups of sample population falling under different per capita food per day ranges (Table 14.1).

(*ii*) In the second instance, we grouped the sample villages into different per capita food per day range (including rice, millets and pulses) for further analysis. However, in this case we could form only five classes of per capita food ranging from 300 to 550 g/d (e.g. Table 14.8).

The relationship between DPCFC and various factors has been obtained by using standard curve fitting method. The following relationships have been analysed: DPCFC and family size, and ratios like adult-child, male-female, able-aged, own-bought and food-vegetable.

Table 14.1 Demographic Information for Various Ranges of Per Capita Food Consumption (No.)

Per Capita Food Range (g/d)	No. of Families	Children (1–14 Years)		Adult (14–60 Years)		Aged (60+ Years)		Total
		Male	Female	Male	Female	Male	Female	
250–350	33	32	32	48	43	13	16	184
350–400	55	40	41	78	86	10	15	270
400–450	53	35	39	88	78	8	3	251
450–500	34	31	25	48	43	13	17	177
500–550	33	17	14	51	49	2	5	138
550–650	29	14	14	44	38	5	7	122
Total (No.)	**237**	**169**	**165**	**357**	**337**	**51**	**63**	**1,142**

RESULTS AND DISCUSSION

Sample Population and Food Consumption Pattern

Demographic Characteristics and Food Consumption Pattern

It may be observed (Table 14. 1) that as the food consumption (range) increases there is a perceptible decrease in both (*a*) number of families and (*b*) population belonging to the higher food consumption ranges. Although this study sample represents the poorest strata of the farming society in Tamil Nadu, the average per capita energy intake is 1,750 kCals/capita/d. When compared to the TN average of 1,826 kCals/capita/d for a non-drought year, it is only 4 per cent lower. This observation is encouraging in terms of a rise in the calorie intake level among the study sample—reaching from possibly near starvation level to a level close to the state average. From data in Table 14.1 it may be observed that over 60 per cent of the families fall above the class 400–450 g/capita/d (1,600–1,800 kCals/capita/d). This indicates that not all have benefited equally by the interventions. About 40 per cent are still food security/calorie sufficiency-wise threatened.

A greater fraction of the study sample is in the age category of 14–60 (considered to be able bodied adults for the purpose of potential for daily wages). The family composition is reasonably well distributed between male and female population among all age groups and per capita food consumption levels. The spread is not significantly skewed in favour of any one sex or age group. There was no observable domination of male population in the chosen sample with a female: male ratio of 979:1,000.

Family Composition across Food Consumption Ranges

Another interesting feature is the decrease in the fraction of young and old population with increasing per capita food consumption range (Table 14.2). Even the family size decreases with increasing food consumption range (with only one exception: the 450–500 g/capita/d range). The fraction of able bodied adults within a family, however, remains constant with only minor fluctuations. Various interpretations may be given for such an observation.

Table 14.2 Average Family Composition among Various Ranges of Per Capita Food Consumption (No.)

Per Capita Food Range (g/d)	*Children (1–14 Years)*		*Adult (14–60 Years)*		*Aged (60+ Years)*		
	Male	*Female*	*Male*	*Female*	*Male*	*Female*	*Total*
250–350	0.97	0.97	1.48	1.33	0.39	0.48	5.62
350–400	0.75	0.78	1.51	1.67	0.18	0.27	5.16
400–450	0.68	0.74	1.66	1.45	0.15	0.06	4.74
450–500	0.91	0.74	1.41	1.26	0.38	0.50	5.20
500–550	0.52	0.42	1.55	1.48	0.06	0.15	4.18
550–650	0.48	0.48	1.52	1.31	0.17	0.24	4.20
Average	**0.71**	**0.70**	**1.51**	**1.42**	**0.22**	**0.27**	**4.82**

- The obvious interpretation could be the relatively less food needs of children compared to adults. The families dominated by children exhibit lower average per capita food consumption.
- Another interpretation is that, the size of the food basket being the same across the sample, a decrease in family size along with a predominance of adults in the family leads to better earning capacities and fewer young and aged (dependent) mouths to feed. A higher proportion of potentially working/earning members concomitantly raises per capita food availability and hence higher consumption in these sample families.
- Finally the families with higher per capita food consumption may belong to a marginally higher income or food producing strata making available more food through other means.

With the data available, we are unable to quantify to what extent each of these three factors have influenced the per capita daily food intake to rise among predominantly able bodied in the study sample.

Most of the RPF families in the study group raise crops on their land and also work as agricultural labour to supplement their income. Being predominantly rain-fed farmers, they mostly raise millets and a few other crops. The productivity of their rain-fed land is limited with the existing technology and low inputs. This combination is expected to provide only for a part of their total food requirement. Members of this group also work as wage labourers and hence a part of their daily food is also expected to come from purchases made within the village and from PDS. However, the opportunities for employment are finite and limited to the agricultural season. This being the case not all the members of the family will find employment. The daily food budget/stock of the family becomes finite. As a consequence, we expect that the larger the family size (brought about by an increase in the non-earning/able bodied members), the per capita food consumption will fall significantly—more intensely under drought conditions. In case the drought lasts longer or for larger family sizes, there is a need to augment the family's internal mechanism to increase its food availability or become prepared for migration.

The main food consumption per family (rice, millet and pulses, Tables 14.3 to 14.5) indicates that the families belonging to higher per capita food range have access to greater quantities of food even though their family sizes are small. This suggests that members of this group are better off. Rice dominates the food basket in all the per capita food ranges (Table 14.3). Millets form the second most important food input for these families. In the per capita food ranges of 450–500 and 550–650 g/d, the per capita consumption of millets and pulses are significantly high compared to

Table 14.3 Total Food Consumption per day for the Sample Families (in kg)

Per Capita Food Range (g/d)	*Rice*	*Millets*	*Pulses*	*Main Food per Family*	*Vegetables*	*Fruits*	*Total Food*
250–350	37.50	13.00	4.95	1.68	11.00	5.25	71.70
350–400	74.00	19.00	7.90	1.83	23.55	4.50	128.95
400–450	82.75	15.50	7.85	2.00	24.95	1.50	132.55
450–500	44.25	28.50	9.20	2.41	8.25	1.75	91.95
500–550	61.50	7.50	3.65	2.20	12.35	1.25	86.25
550–650	44.50	22.50	7.00	2.55	10.25	1.75	86.00
Total	**344.50**	**106.00**	**40.55**	**2.07**	**90.35**	**16.00**	**597.40**

Table 14.4 Daily Per Capita Total Food Consumption among Sample Families* (in g)

Per Capita Food Intake Range (g/d)	*Rice*	*Millets*	*Pulses*	*Vegetables*	*Fruits*	*Total*
250–350	203.80	70.65	26.90	59.78	28.53	389.66
350–400	274.07	70.37	29.26	87.22	16.67	477.59
400–450	329.68	61.75	31.27	99.40	5.98	528.08
450–500	250.00	161.02	51.98	46.61	9.89	519.05
500–550	445.65	54.35	26.45	89.49	9.06	625.00
550–650	364.75	184.43	57.38	84.02	14.34	704.92
Total	**301.66**	**92.82**	**35.51**	**79.12**	**14.01**	**523.12**

Note: * Please note total shown in the last column includes fruits and vegetables and hence exceeds the food range in the first column. Food consumption range uses only total grains for computing calorie sufficiency.

Table 14.5 Share of Food Items in Daily Per Capita Food Consumption for the Sample Families (%)

Per Capita Food Range (g/d)	*Rice*	*Millets*	*Pulses*	*Vegetables*	*Fruits*
250–350	52.30	18.13	6.90	15.34	7.32
350–400	57.39	14.73	6.13	18.26	3.49
400–450	62.43	11.69	5.92	18.82	1.13
450–500	48.12	31.00	10.01	8.97	1.90
500–550	71.30	8.70	4.23	14.32	1.45
550–650	51.74	26.16	8.14	11.92	2.03
Total	**57.67**	**17.74**	**6.79**	**15.12**	**2.68**

other groups (Tables 14.4 and 14.5). The families belonging to these two categories have tried to maintain a proper mix of food basket. Vegetable and fruits are important supplements to the main food. It is surprising to observe the slight domination of bought (either PDS or market) component of food items in the sample (Table 14.6 and 14.7). This indicates that whatever the crops raised in their own land, is sufficient to meet only partial food needs.

Sample Villages and Food Consumption Pattern

To study the food consumption patterns in the sample villages, we summarised the data based on different per capita food ranges for the chosen villages and the district to which they belong.

Table 14.6 Composition of Daily Food Consumption among Sample Families (in kg)

Per Capita Food Range (g/d)	*Rice*		*Millets*		*Pulses*		*Vegetables*		*Fruits*	
	Own	*Bought*	*Own*	*Bought*	*Own*	*Bought*	*Own*	*Bought*	*Own*	*Bought*
250–350	9.50	28.00	6.50	6.50	1.25	3.70	0.75	10.25	0.25	5.00
350–400	28.25	45.75	14.50	4.50	1.85	6.05	2.25	20.80	0.50	4.00
400–450	45.50	37.25	10.25	5.25	1.00	6.85	8.75	16.20	0.00	1.65
450–500	21.50	22.75	18.25	10.25	1.00	8.20	1.25	5.00	0.50	1.25
500–550	19.50	42.00	5.00	2.50	0.60	3.05	2.25	9.60	0.00	1.60
550–650	25.50	19.00	12.00	10.50	2.70	4.30	2.75	7.00	1.00	1.20
Total	**149.75**	**194.75**	**66.50**	**39.50**	**8.40**	**32.15**	**18.00**	**68.85**	**2.25**	**14.70**

Table 14.7 Composition of Daily Per Capita Food Consumption among Sample Families (in g)

Per Capita Food Range (g/d)	*Rice*		*Millets*		*Pulses*		*Vegetables*		*Fruits*	
	Own	*Bought*	*Own*	*Bought*	*Own*	*Bought*	*Own*	*Bought*	*Own*	*Bought*
250–350	51.63	152.17	35.33	35.33	6.79	20.11	4.08	55.71	1.36	27.17
350–400	104.63	169.44	53.70	16.67	6.85	22.41	8.33	77.04	1.85	14.81
400–450	181.27	148.41	40.84	20.92	3.98	27.29	34.86	64.54	0.00	6.57
450–500	121.47	128.53	103.1	57.91	5.65	46.33	7.06	28.25	2.82	7.06
500–550	141.30	304.35	36.23	18.12	4.35	22.10	16.30	69.57	0.00	11.59
550–650	209.02	155.74	98.36	86.07	22.13	35.25	22.54	57.38	8.20	9.84
Total	**134.89**	**176.44**	**61.26**	**39.17**	**8.29**	**28.91**	**15.53**	**58.75**	**2.37**	**12.84**

From Table 14.8, it may be observed that the Pahukkal village in Kancheepuram district, which belongs to 300–350 gm per capita food range (severe deficiency category), exhibits relative dominance of older people in the sample families (about 16 per cent). Another interesting feature is the clear domination of female children and absence of older people in the sample of Mettupatti village in Pudukottai district. Both the villages belong to low DPCFC range. This indicates that some relationships between these observed phenomena and per capita food consumption is not quite clear at this stage. However, the later sections present discussions on these issues.

Further, sample families of Pahukkal village completely depend on bought food and share of vegetables and fruits in the food basket are high compared to other sample villages (Table 14.9).

However, sample families of Mettupatti village on the other hand depend mostly on food products grown on their own land. Even the sample families of villages belonging to per capita food range of 450–500 g/d show more dependency on homegrown foods (Table 14.9). The total food consumption and the per capita food consumption for sample villages are presented in Tables 14.10 and 14.11. Rice is a dominant food item in all the villages. However, the villages in the per capita food ranges of 400–450 and 450–500 g/d consume significant quantity of millets and pulses in relation to rice (Table 14.12). Relative share of vegetables and fruits in the per capita food consumption is very high in Pahukkal village.

Overall the analysis shows that there is a very strong dependence on rice that has become the staple food for most of the villages studied, irrespective of whether this crop is suitable for or grown in the village by the farmers etc. Even in areas which are typically suited for millets, rice has replaced

Table 14.8 Demographic Details in Sample Villages

Per Capita Food Range (g/d)	Villages	Districts	Families	Sample Population	Children (1–14 Years)		Adult (14–60 Years)		Aged (60+ Years)	
					Male	*Female*	*Male*	*Female*	*Male*	*Female*
300–350	Pahukkal	Kancheepuram	20	109	24	21	24	23	10	7
350–400	Mettupatti	Pudukottai	14	57	1	7	24	25	0	0
400–450	T. Narayanapuram, Thailakoundanur-1, Panangkattupakkam, Varadharajapuram, Konampatti-1,	Madurai, Salem, Kancheepuram,								
450–500	Kuttikinathur	Trichy, Erode	113	533	76	65	178	167	20	27
	Konampatti-2, Kelaparai, Thailakoundanur-2, Vedarpuliankulam	Madurai, Salem, Dharmapuri	72	354	59	62	99	90	20	24
500–550	Ozhindhiyapet	Villupuram	18	89	9	10	32	32	1	5
Total			**237**	**1,142**	**169**	**165**	**357**	**337**	**51**	**63**

Table 14.9 Food Consumption Pattern in Sample Villages (g/d)

		Rice		*Millets*		*Pulses*		*Vegetables*		*Fruits*	
Villages	*Districts*	*Own*	*Bought*	*Own*	*Bought*	*Own*	*Bought*	*Own*	*Bought*	*Own*	*Bought*
Pahukkal	Kancheepuram	5,000	28,000	0	1,500	0	2,550	0	16,000	0	7,000
Mettupatti	Pudukottai	16,750	1,000	3,000	0	0	1,700	0	5,750	0	0
T. Narayanapuram, Thailakoundanur-1, Panangkattupakkam, Varadharajapuram, Konampatti-1, Kuttikinathur	Madurai, Salem, Kancheepuram, Trichy, Erode	60,500	98,750	28,250	17,000	4,850	12,000	7,000	26,850	250	1,500
Konampatti-2, Kelaparai, Thailakoundanur-2, Vedarpuliankulam	Madurai, Salem, Dharmapuri	57,000	38,500	32,750	16,000	3,550	14,150	10,250	11,750	2,000	5,250
Ozhindhiyapet	Villupuram	10,500	28,500	2,500	5,000	0	1,750	750	8,500	0	950

Table 14.10 Total Food Consumption in Sample Villages (in kg)

Per Capita Food Range (g/d)	*Villages*	*Districts*	*Rice*	*Millets*	*Pulse*	*Vegetables*	*Fruits*	*Total*
300–350	Pahukkal	Kancheepuram	33	1.5	2.55	16	7	60.05
350–400	Mettupatti	Pudukottai	17.75	3	1.7	5.75	0	28.2
400–450	T. Narayanapuram, Thailakoundanura1, Panangkattupakkam, Varadharajapuram, Konampatti-1, Kuttikinathur	Madurai, Salem, Kancheepuram, Trichy, Erode	159.25	45.25	16.85	33.85	1.75	256.95
450–500	Konampatti-2, Kelaparai, Thailakoundanur-2, Vedarpuliankulam	Madurai, Salem, Dharmapuri,	95.5	48.75	17.7	22	7.25	191.2
500–550	Ozhindhiyapet	Villupuram	39	7.5	1.75	9.25	0.95	58.45
Total			**344.5**	**106**	**40.55**	**86.85**	**16.95**	**594.85**

Table 14.11 Daily Per Capita Food Consumption for the Sample Villages (g/d)

Per Capita Food Range (g/d)	*Villages*	*Districts*	*Rice*	*Millets*	*Pulse*	*Vegetables*	*Fruits*	*Total*
300–350	Pahukkal	Kancheepuram	302.75	13.76	23.39	146.79	64.22	550.92
350–400	Mettupatti	Pudukottai	311.40	52.63	29.82	100.88	0.00	494.74
400–450	T. Narayanapuram, Thailakoundanur-1, Panangkattupakkam, Varadharajapuram, Konampatti-1, Kuttikinathur	Madurai, Salem, Kancheepuram, Trichy, Erode	298.78	84.90	31.61	63.51	3.28	482.08
450–500	Konampatti-2, Kelaparai, Thailakoundanur-2, Vedarpuliankulam	Madurai, Salem, Dharmapuri,	269.77	137.71	50.00	62.15	20.48	540.11
500–550	Ozhindhiyapet	Villupuram	438.20	84.27	19.66	103.93	10.67	656.74
Total			**301.66**	**92.82**	**35.51**	**76.05**	**14.84**	**520.88**

millets in the daily food basket. Correspondingly, it is expected that there is little drive to raise crops suitable to the area (rainfall, agro-climatic zone and soil). Vegetables and fruit distribution among the villages is highly skewed—the pattern is not easily determined. Fluctuations in the share of pulses in the food basket are small across the samples and the districts.

Analyses of Factors Influencing Daily Per Capita Food Consumption Levels

The relationships between various factors and DPCFC have been obtained using standard curve fitting method. For this purpose, first, the average per capita food consumption per day was estimated

Table 14.12 Share of Food Items in Per Capita Food Consumption per day for the Sample Villages

Per Capita Food Range (g/d)	*Villages*	*Districts*	*Rice*	*Millets*	*Pulses*	*Vegetables*	*Fruits*
300–350	Pahukkal	Kancheepuram	54.95	2.50	4.25	26.64	11.66
350–400	Mettupatti	Pudukottai	62.94	10.64	6.03	20.39	0.00
400–450	T. Narayanapuram, Thailakoundanur-1, Panangkattupakkam, Varadharajapuram, Konampatti-1, Kuttikinathur	Madurai, Salem, Kancheepuram, Trichy, Erode	61.98	17.61	6.56	13.17	0.68
450–500	Konampatti-2, Kelaparai, Thailakoundanur-2, Vedarpuliankulam	Madurai, Salem, Dharmapuri	49.95	25.50	9.26	11.51	3.79
500–550	Ozhindhiyapet	Villupuram	66.72	12.83	2.99	15.83	1.63
Total			**57.91**	**17.82**	**6.82**	**14.60**	**2.85**

for the families belonging to each of the per capita food ranges as shown in Table 14.1. Thus, six DPCFC estimates were obtained for the six groups of per capita food ranges. Similarly, average values of all the factors were estimated for the same six per capita food ranges. Then, these factor values were plotted against the average values of DPCFC to obtain the relationships.

Family Size

The relationship between family size and per capita food consumption was determined by plotting these two factors for the sample population (Figure 14.1). A strong negative correlation (R^2 value of 0.79) is visible which suggests that with family sizes greater than 4.75, these sample families will begin facing calorie malnutrition during droughts. The daily calorie intake will fall below the targeted 1,800 kCals above this family size and has the potential to cause various health and nutrition related problems if this situation is prolonged. It may be recalled that the data for the above relation has been obtained during peak 'drought affected' situation. These relationships need to be worked out for non-drought and chronic drought situations in order to determine how and to what extent food security is compromised under drought. These current calorie sufficiency values of per capita daily food intake under a prolonged drought spell suggest significant levels of resilience among the RPF-SHG families.

Adult-Child Ratio

The adult child ratio is expected to influence the DPCFC in two ways, namely (*a*) children consume lesser food than adults and (*b*) the food basket is divided among a larger number of dependent persons when children are more in number (low adult:child ratio). A strong positive relationship seen in Figure 14.2 is a combination of these two influences. For this group, families with adult: child

Figure 14.1 Per Capita Daily Food Consumption vs Family Size

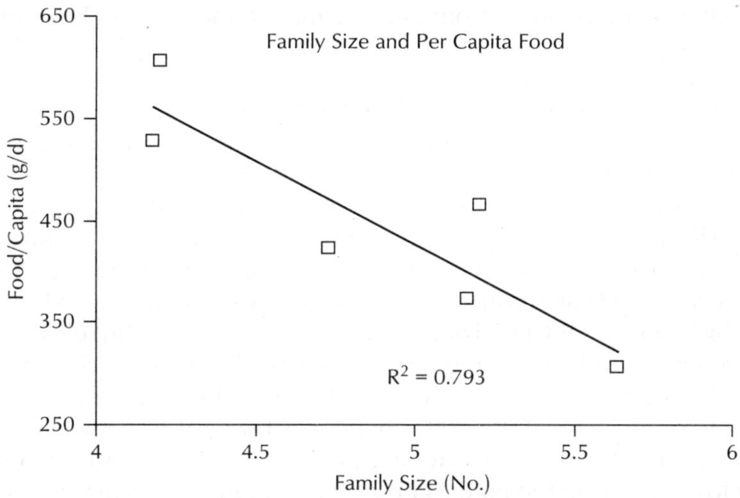

Figure 14. 2 Influence of Adult to Child Ratio on Per Capita Food Consumption

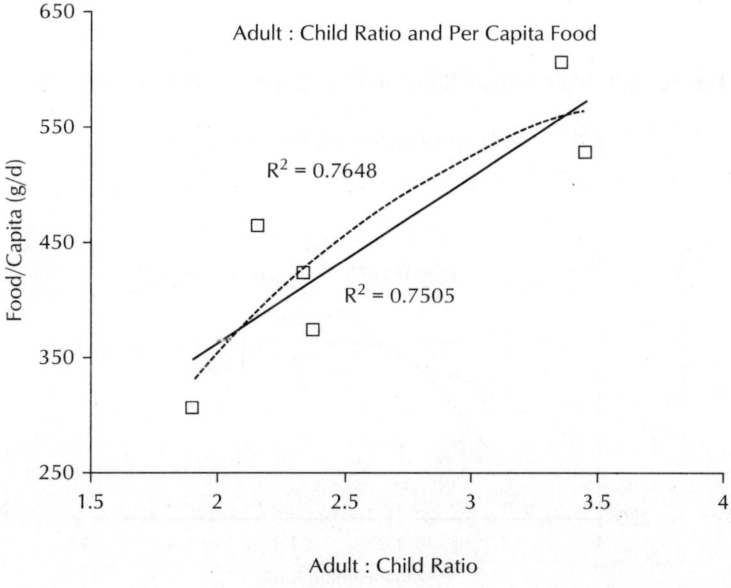

ratios of above 2.5 only exhibit DPCFC of above sub maintenance calorie nutrition. It also suggests that families with greater than one child are vulnerable under drought conditions. This results in a condition of food insecurity bringing DPCFC to sub-maintenance levels.

Male-Female Ratio

The DPCFC has two types of influences from varying male-female ratios. First, males are expected to have higher per capita food consumption over females (in a general population and to a lesser extent for working population). Second, it is a convention to believe that males have a greater opportunity for labour/wages and employment potential. Also the bread winner concept creates greater opportunities for men to work and 'provide for' the families. In many parts of Tamil Nadu, female foeticide is often reported. Here a higher female:male ratio is considered a liability in terms of providing for food, dowry etc. In most parts of India, adult male wages are generally higher than female wages. All these tend to suggest that families with a greater number of males stand a better chance to survive droughts. This is expected to be manifest as a higher DPCFC even during a drought year and even when local resources are on the verge of exhaustion. Male-female (MF) ratio is plotted in Figure 14.3 against DPCFC. There was no strong relationship observed except that the trend line suggests a positive relationship between the two. The DPCFC rises only gradually with increasing MF ratio. For every one unit rise in MF ratio, the DPCFC rises by 300 g. With available field data, it is difficult to suggest that this 300 g increase is brought about solely by a higher level of per capita food intake by adult males. A greater depth of data and analysis is required to understand this phenomenon. However, from the pattern observed in Figure 14.3, just as is normally believed, families with lower MF ratio (more females) are likely to be more vulnerable during droughts. Various factors that influence this phenomenon need to be studied in detail. This also means that when droughts affect such areas and family available food stock is limited, men will have a higher need for migration.

Figure 14.3 Male-Female Ratio and Per Capita Food Consumption

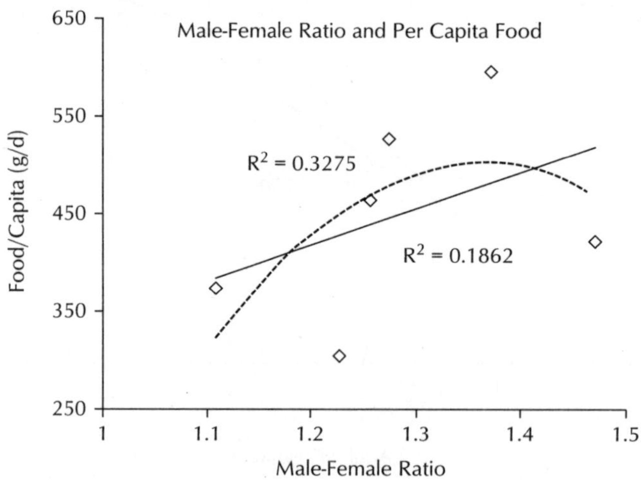

Able-Aged Ratio

The presence of a large fraction of aged population is expected to signify reasonable levels of health within the study sample. They constitute about 10 per cent of the sample population (Tables 14.1 and 14.2).

Significant aged population is seen in two DPCFC classes—the lowest and the middle range (Tables 14.1 and 14.2). It is believed that under normal conditions, a large aged person component in the family will only be marginally productive and contribute to a lesser extent to the family income/food basket. On the other hand, during droughts when there is very little labour/daily wage opportunities within villages, the overall food basket of the family would have shrunk (due to low wages), due to exhausting of family food stocks, inadequate coping strategies etc. In the presence of a large fraction of aged persons, the limited size of the food basket will have to be shared by a greater number of people. It is expected to introduce stressed conditions earlier at the onset of drought and manifest them to an even more severe extent as the drought progresses. Coping strategies will inevitably lead to migration. Figure 14.4 shows the able:aged ratio of various families plotted against DPCFC. This data was collected during a peak drought period and therefore a strong positive relationship has been obtained between the able:aged ratio and the DPCFC—higher the able:aged ratio, greater is the DPCFC or food and nutrition security during a drought condition. We expect that the slope of this relationship will reduce or level off during adequate food supply/security periods or when the able:aged ratios are quite high. A reasonably strong positive relationship is seen. The slope suggested that an able:aged ratio greater than 1.75 seems to be needed to keep the food security above sub maintenance levels when aged persons are in the family. Once again the contribution of a higher DPCFC of the able persons influencing this trend needs to be determined by a detailed study.

Figure 14.4 Aged Persons Ratio vs Per Capita Food Consumption

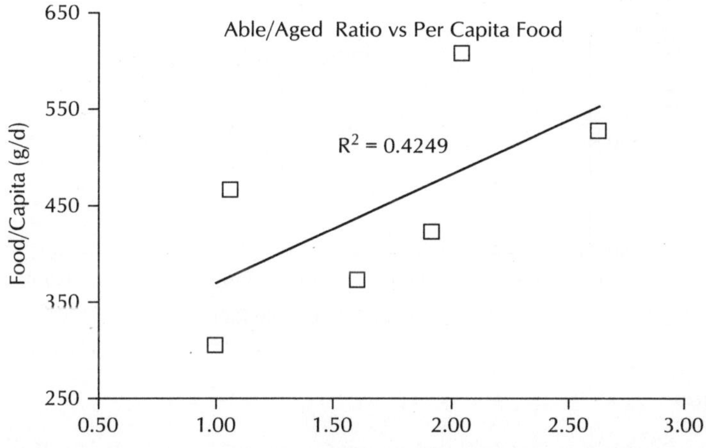

Gender Bias and Girl Child Skew

From various field discussions as well as everyday observations, we identified that families with more females, especially girl children are careful and thrift, and we also showed that MF ratio significantly influences the DPCFC. In order to quantify the gender bias in terms of food security we compare the DPCFC under various ranges of young girl children in proportion to adult females as well as young male population vs adult male population (able bodied males). These results are presented in Figures 14.5 and 14.6. There is generally a negative relation for both the parameters. Increases in young male:adult male (YM:AM) and young female:adult female (YF:AF) ratios generally reduce the family's DPCFC. The slope YM:AM is more gentle compared to YF:AF slope. This shows that the drop in DPCFC is more rapid with the increase in female children compared to the domination of male children in the family. YF:AF levels above 0.63 tends to bring families to sub-maintenance levels while YM:AM ratio of 0.75 tends to bring the family into sub-maintenance levels of food insecurity (<350 g/capita/d). There is thus a slight bias towards the male child in the family of RPFs. This is perhaps the first time that such a parameter is taken for this study sample. In the absence of an appropriate baseline, it is not possible to judge whether this difference is significant or caused by biological factors. It is necessary to measure on this parameter more frequently to arrive at a reasonable conclusion.

Figure 14.5 Young Female Children and their Influence on Food Consumption

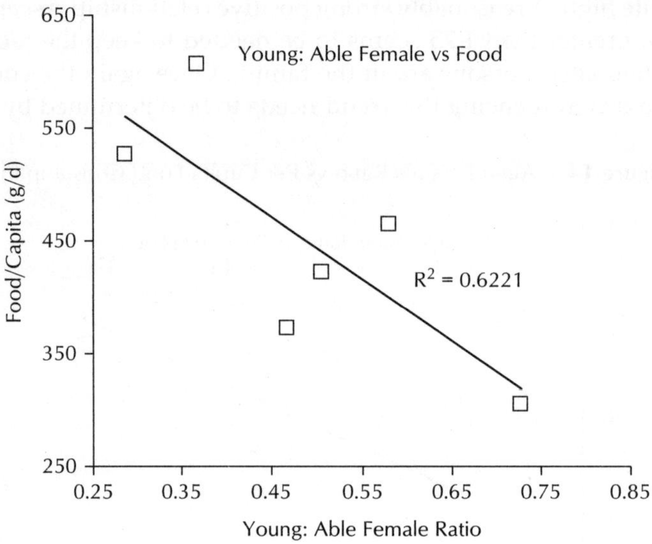

Own-Bought Grain Ratio

This group comprises mainly typical resource poor farmers (75 per cent small and marginal farmers and 25 per cent landless farmers). We expect that the RPF family will go through yearly (transitional) cycles of sufficiency (for a few months immediately after crop harvest) and insufficiency (or sub maintenance calorie nutrition) during periods when:

(*a*) family stocks of grains are depleted;
(*b*) a large fraction of the daily grains is bought from wages; and
(*c*) family stocks are low and coping mechanism of voluntary food reduction to sub-maintenance levels sets in etc.

Figure 14.6 Young Male Children and their Influence on Food Consumption

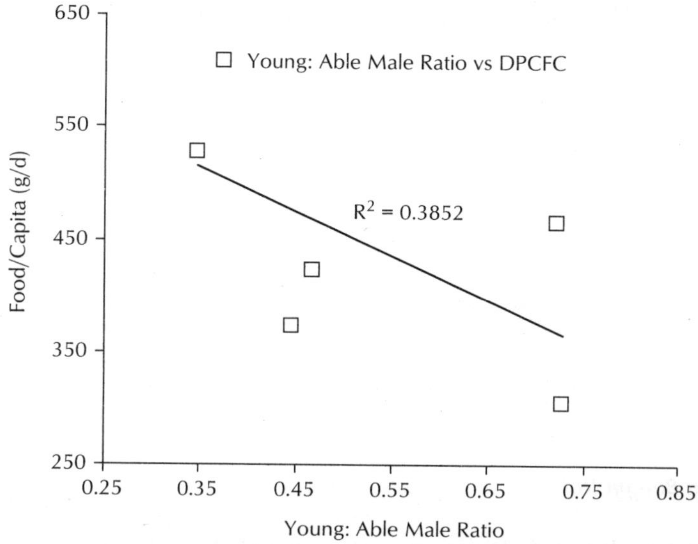

The presence of a significant quantity of grains from within family sources/stocks in the daily diet during periods temporally away from harvest time or during drought periods is an indication that such families are vulnerable to a lesser extent and enjoy a higher level of food security. The extent of other factors is expected to be location specific. The quantities of food grains (cereals, millets, pulses, vegetables, fruits etc) raised from own resources, purchased from PDS and collected from CPRs were quantified (Tables 14.6 and 14.7).

As a part of the interventions, these RPFs run SHGs attempted to raise a greater quantity of millets and cereals for home consumption on their land and reduce the fraction of purchased food grains. The purchased food grains are rarely bought for cash and are most often taken on credit. A greater fraction of grain arising from home grown sources indicates a certain degree of food security and self-sufficiency—as opposed to purchase of grain from market/PDS systems for cash or credit. In times of drought and in the absence of local employment, daily wage source etc., a significant component of homegrown food items in the food basket is indicative of a lesser degree of vulnerability to drought first and credit next and finally a higher level of food security. The ratio of grain obtained from own sources over that bought for cash/credit from local sources or PDS in the daily food basket then becomes a good indicator of a lower vulnerability arising from better insurance and lower credit.

A higher own/bought ratio coupled with calorie sufficiency even during periods of drought will suggest greater extent of temporally spread food security arising from greater self-sufficiency and other forms of insurance and risk aversion.

The ratio of own food grains to bought food in the daily food basket (rice, millets and pulses) is plotted against DPCFC in Figure 14.7. The result proves the hypothesis that own food provides higher food security. All increases in DPCFC are accompanied by an increase in the fraction of food from own sources or vice versa.

Figure 14.7 Ratio of Food from Own Sources/Bought vs Per Capita Food Consumption

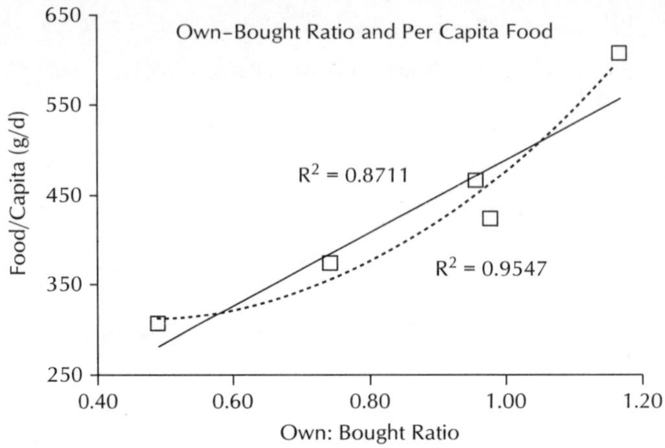

Impact of Higher Own Bought Ratio

These farmer groups are all rain-fed farmers and recurring droughts have driven them into a credit trap. Credit is first taken to purchase agro-inputs for food crops. A part of the harvested grain is initially traded to cancel the credit. However, when droughts occur, there is lower harvested grain, lower stocks left after cancelling the credit and during the oncoming season credit is taken to meet family food grain needs also. This spirals with every drought year putting farmers deeper into debt. Escaping such credit trap is one of the main objectives of this farmers' network. Increasing own sources of food in the daily food basket overcomes a major pull into the credit trap. Food security provided through internal mechanisms and from local resources is thus much sought after. Greater the farmers' control on resources of production, greater will be his share of the product (food) and food security implied therein. The farmers in the network have achieved this in two ways, first, substituting local materials for purchased agro inputs and second by increasing the share of own food grains in the daily food basket.

There was no baseline numbers measured for this parameter when the farmers of the LEISA network initiated the planned interventions in terms of:

(*a*) use of appropriate technology inputs;
(*b*) eco-friendly and organic agriculture techniques; and
(*c*) credit aversion strategies.

We therefore attempt to compare this current data (drought year 2002 for the network farmers) with TN state average for year 2000 (no drought). The following observations may be made:

(*a*) Extrapolating the results presented in Figure 14.7 suggests that these farmer can support only a low level of DPCFC (c. 150 g/cap/d) when own sources of food grain in the daily food basket is nil.
(*b*) All increases in DPCFC have been possible by an increased share of own food grains in the food basket.

(*c*) The average per capita food consumption in the network is just 4 per cent lower than that of the state average of 1,826 kCals/capita/d for a non-drought year.

(*d*) Over 60 per cent of the network of RPF farmer families have achieved calorie sufficiency levels at/above sub maintenance levels measured during peak drought.

(*e*) There is not enough data spread to determine limits to substitutions between own and purchased sources of food items. With increasing per capita food used, the share of own sources gradually increases. In other words, greater share of own food sources increases the food intake and consequent food security. Therefore for this kind of target group, higher levels of food security can come only by mechanisms that enable farmers to internally raise the outputs of grain and food basket. The obvious policy implication is to enhance this capability in farmers such that a minimum threshold of food security is assured within the system.

(*f*) Figure 14.8 and Tables 14.6 and 14.7 show the various components of the food basket—rice, millets and pulses together provide the basic nutrition and later on food security. Rice is the single largest food component accounting for over 60 per cent of the total food grains in the daily diet. This study group consists of largely resource poor rain-fed small and marginal farmers. Depending upon rice to meet a large part of their calorie nutrition and food security is thus not a sustainable option. This level of rice use is now dependent on imports into this ecosystem from another nearby location. Such imports into the ecosystem are not conducive to long term food security and ecosystem sustainability. Various measures need to be taken up internally within the SHGs where locally suited grains and millets are raised at these rain-fed locations and food habits gradually switched to more homegrown cereals and millets.

Figure 14.8 The Proportion of Various Components of the Food Basket (Rice, Millet and Pulses) Change with Increasing Per Capita Daily Food Consumption

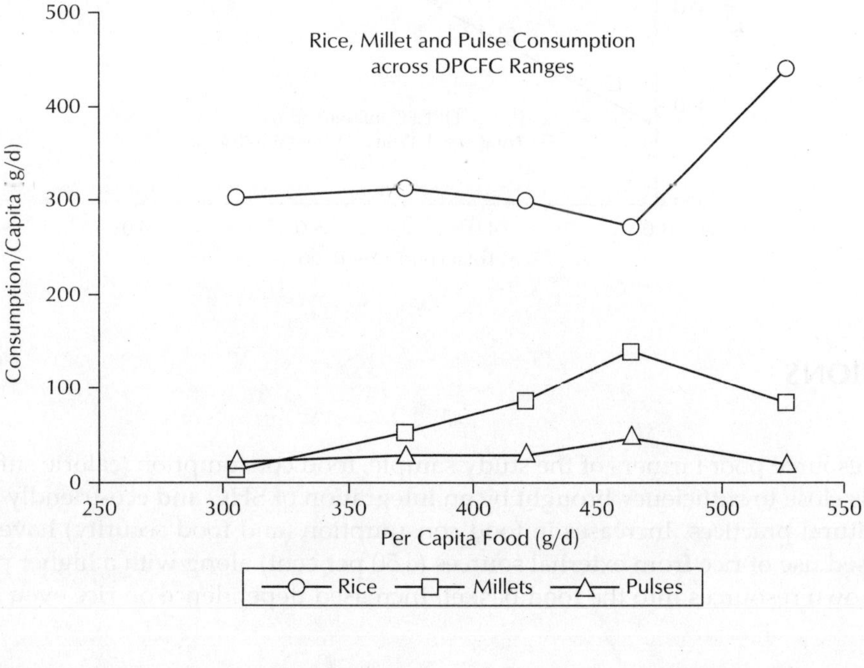

(*g*) From Figure 14.8 and Tables 14.4 & 14.5 it is seen that increases in the DFCPC is accompanied with a corresponding increase in the millet component. Table 14.3 shows that with increasing DFCPC, millets obtained from own sources increase significantly (>60 per cent). Nutrition/food security is thus enhanced significantly by an increase in millets raised on own land resources. Policy and enabling measures to increase food availability/security will therefore have to come from enabling an increase in millet production on own land.

Food-Vegetable Ratio

The daily intake of vegetables and fruits and its relation to the DPCFC is determined in Figure 14.9. A very strong relation between food:vegetable ratio vs the DPCFC may be noted ($r^2 = 0.98$). This tells us that with the increase in the DPCFC, there is a corresponding increase in the consumption of vegetables and fruits. In other words, the relative share of vegetables and fruits remain the same in the food basket irrespective of increase in the consumption of rice, millet and cereals. It appears that the sample group of farmers does not view vegetables and fruits as compensatory items in lieu of main food items. They are used as add-on items to increase the quality of food intake.

Figure 14.9 Total Grain/Total Fruit and Vegetable Ratio vs Per Capita Daily Food Consumption

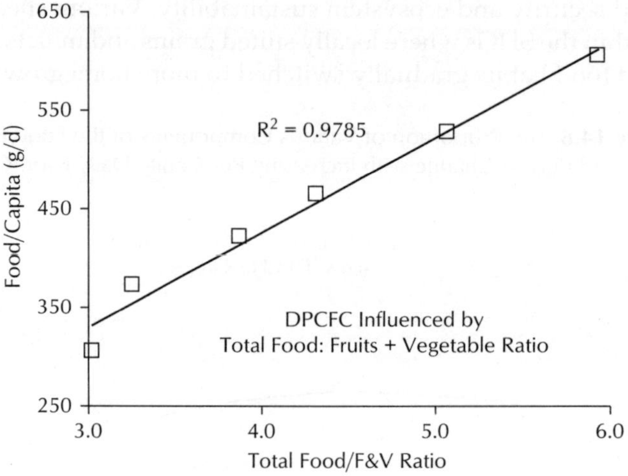

CONCLUSIONS

Among the resource poor farmers of the study sample, food consumption (calorie sufficiency) has risen to levels close to sufficiency brought by an integration of SHG and eco-friendly low external input agricultural practices. Increases in food consumption (and food security) have come about by an increased use of rice from external sources (c.50 per cent) along with a higher proportion of millets from own resources into the food basket. Increased dependence on rice even among these rain-fed farmers and zones (>50 per cent of food basket) is not sustainable in the long run and needs to be reversed. Policy implications of these observations are that food security enhancement efforts need to emphasise the use of locally raised cereals and millets from own resources in order to enhance the sustainability of such food security increases. Vegetables and fruit consumption measured as a ratio to the total cereal, millet and pulses show a good indication of food security (calorie sufficiency under drought conditions) and may be used as a quick measure of food security for a limited purpose. Providing short-term credit to self help groups, group based monitoring and promoting use of eco-friendly and low external inputs are some of the other policy initiatives that could enable the RPFs to achieve a high level of daily per capita food consumption and concomitant food security components.

ACKNOWLEDGEMENTS

This study was possible due to the active support of the LEISA staff as well as farmers who patiently provided all the required answers during the participatory surveys. Their help is gratefully acknowledged.

NOTE

1. Corresponding author.

REFERENCES

Chanakya, H.N. 1999. 'PTD Emerging in South India—an overview', *LEISA-India Newsletter*. 1: 8–12.

Chanakya, H.N. and S.T Somashekara Reddy. 2002. 'A Better Livelihood with Every Drop. Tamil Nadu Peasants Vision of Water and Livelihood Nexus', in Danielle Morley (ed.), *Perspectives on freshwater*, UNED EarthSummit 2002. Forum on Freshwater Report. http://www.earthsummit2002.org/freshwater/FreshwaterReport.pdf.

Deol, Rajesh. 2003. 'Farmers on Warpath', *Deccan Herald*, Sunday, 23 March.

Gopalan, C., B.V. Ramashastry and S.C. Balasubramanian. (eds) 2000. *Nutritive Value of Indian Foods*. Revised Edition. Hyderabad: NIN.

Harris, B. 1987. 'Poverty in India—Micro level Evidence'. Paper presented at Workshop on Poverty in India, October 1987. Oxford: Queen Elizabeth House.

Jha, Prem Shankar. 2003. 'Agriculture in Distress'. *The Hindu*, 25 February 2003.

Kozel, V. and B. Parker. 2003. *A Vrofile and Diagnostic of Poverty in Uttar Pradesh, Economic and Political Weekly*. 25 Jan: 385–404.

LEISA Network Plan Proposal. 1999. 'Low External Input and Sustainable Agriculture Network'. Trichy, Tamil Nadu.

Maxwell, S. and T.R. Frankenberger. 1992. Household Food Security: Concepts, Indicators, Measurements—A Technical Review. UNICEF-IFAD.

Meenakshi, J.V. and B. Vishwanathan. 2003. *Calorie Deprivation in Rural India 1983 to 1999–2000*, Economic and Political Weekly. 25 Jan: 369–75.

National Sample Survey Organisation. 2000. *Common Property Resources in India*. Jan–June 1998. NSS 54th Round.

———. 2001a. *Differences in Level of Consumption among Socio-economic Groups*, September 2001. NSSO-Report 472.

National Sample Survey Organization. 2001b. *Sources of Household Consumption in India 1999–2000*, September 2001. NSSO Report 474.
Radhakrishna, R. 2002a. 'Food and Nutrition Security in Tamil Nadu – 2002', in K. Parikh and R. Radhakrishna (eds), India Development Report 2002. Oxford.
———. 2002b. 'Food and Nutrition Security in Karnataka', in K. Parikh and R. Radhakrishna (eds), *India Development Report 2002*. Oxford. Chapter 7, pp. 153–76.
Watts. 1987. c.f. Maxwell S. and Frankenberger T.R. (eds), 'Household Food Security: Concepts, Indicators, Measurements – A Technical Review, UNICEF-IFAD. p. 92.

15

Collective Action in an Embedded Choice Model: A Study of Fishing Cooperatives in Calcutta

Zakir Husain[1]

Abstract: Literature has traditionally used the rational choice model to explain collective action. This approach explains the successes and failures of collective action in terms of net benefits of collective action. Typically, however, the rationality of agents is bound with actions depending upon perceived and not actual opportunity costs. Perceptions of costs and benefits, however, are formed by the historic, socio-cultural and economic context. It is necessary to identify the relevant contextual factors and analyse how they interact with the decision making process of individuals to determine the successes and failures of collective action.

This chapter shows how the historic and economic context may shape the objective function and constraints of agents to determine their collective decisions. Based on the experiences of two fishermen's cooperative in Calcutta, an optimisation model, incorporating the perceptions of agents, is constructed to analyse labour supply by its members. Choice in this model is therefore *embedded* in the specific context of the actors in which actors are situated. The context may mould behaviour to yield equilibrium choices and comparative static results that are not consistent with the rational choice model.

The model explains why members supply just enough labour supply so as to ensure the survival of the cooperative. It is also shown that changes such as increasing benefits to collective action may paradoxically reduce labour supply to the cooperative.

INTRODUCTION

From an economic perspective, individual activities can be divided into two categories: those directed towards generating private benefits and those providing collective benefits. Although the latter category includes only a small proportion of human activities, they are also important both from an individual and social perspective. Instances of such activities include contribution of labour, capital and other resources towards the provisioning of public goods, managing of natural resources through a common property regime and participation in a cooperative or social security system.

Most attempts to analyse these activities are based on the rational choice model. The main criticism levelled against this approach has been that it is based on an 'undersocialised' approach that minimises the importance of socio-political forces and exogenous economic changes (Fields 1979). This results in an incomplete analysis of collective action, especially its dynamics (Steins et al. 2000). To remedy such a shortcoming of rational choice model, it is necessary to incorporate contextual factors so that the rational choice model is embedded within a proper socio-political, historical and economic context (McCay 2002; Husain and Bhattacharya 2004b). While the works of Edwards and Steins (1999b), McCay (2002) and Husain and Bhattacharya (2004b) have attempted to develop a framework examining how contextual factors interact with the physical characteristics of the resource, market parameters and technological processes, there has not been any attempt to incorporate contextual factors into a formal utilitarian framework. This task is attempted in this chapter.

The chapter is organised as follows. The rational choice literature on collective action is critically reviewed, followed by a description of the characteristics of the alternative embedded (or situated) choice model. This chapter focuses on how the interplay of historical processes and the economic context determines the objective function and constraints faced by economic agents and hence the success of collective action. After presenting a brief history of the cooperatives, a simple theoretical model has been used to examine the role of contextual factors in explaining labour supply to the cooperative. Case studies of two cooperatives in Calcutta, India, that is, Captain Bheri Fishermen's Cooperative and Bon Hooghly Fishermen's Cooperative, are used to empirically illustrate the contextual model of rational choice in this chapter.

COLLECTIVE ACTION, RATIONAL CHOICE AND CONTEXTUAL ANALYSIS

The rational choice model is based on the consideration of the relative costs-benefits of economic and social actions. Initial contributors to the literature on collective action argued that the costs of collective action was borne by the person undertaking the action, while benefits of such actions dissipated within the group (Olson 1962). Hence, private costs would exceed private benefits, encouraging individuals to free ride. Collective action would occur only if some agents benefitted so much from the collective action that they were prepared to contribute even if others free-rode on them. Olson's thesis stimulated research in various collective action arenas. In the literature on common pool resources, for instance, Hardin (1968) has used this argument to show that natural resources jointly used by a group would inevitably be overexploited and degraded. Popkin (1977) used Olson's frame of argument to refute Scott's (1976) writings on the importance of social security systems in Southeast Asia. Their argument was later presented in terms of the Prisoner's Dilemma Game within the game theoretic literature.

Olson's argument, however, fails to explain the numerous instances of successful collective action observed in empirical studies (e.g. Wade 1988; Berkes 1989; Chopra et al. 1990; Singh, 1994). Subsequent writers therefore attempted to explain why collective action was a viable action to individuals. Again, basing their theoretical framework on the rational choice model, these writers argued that repetitive interactions in a closed system with interactions between agents in different arenas would tend to remove the incentives to free ride and hence the constraints to cooperation

(Kimball 1988; Coate and Ravallion 1989; Ostrom 1990; Fafchamps 1992). This argument was supported by the developments in iterative game theory, showing that the 'prisoners' with common interests could learn to cooperate (Axelrod 1986; Balland and Platteau 1996; Bowles and Gintis 1998; Taylor 1977).

Some Limitations of Rational Choice Model

Despite the conflicting conclusions, the two approaches to collective action presented above share a common ground. That is, they are both based on the rational choice model. The basic premise of this model is that economic decisions are based on an analysiss of the 'actual' costs and benefits of decisions. But this premise has serious problems limiting its validity in general contexts. Simon (1957) has argued, for instance, that individuals are rational but only in a limited sense. This is because their decisions are based, on not actual costs-benefits, but *perceived* costs-benefits. These perceptions determine the bounds to rationality (the bounded rationality approach). Thus the rational choice approach abstracts from the context in which the agent is situated while deciding on his course of action.

In view of the role of perception in bounded rationality, Klooster (2000) and McCay (2002) argue that it is necessary to consider the social nature of perception – the process through which individuals determine the nature and seriousness of the collective problem is strongly influenced by economic and socio-cultural factors – and then evaluate the available options. McCay and Jentoft (1998) point out that the rational choice model is 'contextually thin' because it minimises the considerations of the historical, political, economic and socio-cultural processes that determine perceptions. This failure leads to an inability to fully analyse the successes and failures of collective action. Simultaneously, dynamic changes in the history, political, economic and socio-cultural context that shape collective action are also ignored. As a result, the roles of evolutionary processes in the collective arena are ignored or inadequately explained (Steins et al. 2000).

The limitations of the rational choice model has led to the development of the embedded or situated choice approach based on a more realistic proposition that individual perceptions of costs and benefits are embedded in the history, politics, and the socio-cultural and economic background of the agents. Therefore, to understand the actions of agents, it is necessary to understand how these perceptions determine the bounds to rationality by influencing the perceived objective function and constraints. The development of 'contextual analysis' by researchers on common pool resources (Edwards and Steins 1999b; McCay 2002) marks an important conceptual breakthrough in collective action literature.

Contextual Factors

Contextual factors are defined as the 'dynamic forces constituted in the user groups' social, cultural, economic, political, technological and institutional environment' (Edwards and Steins 1999a). These forces determine the benefits of collective action as well as its opportunity and transaction costs (Edwards and Steins 1999a; Husain and Bhattacharya 2004b). For analytical convenience and in terms of the intimacy of effects, the contextual forces can be divided into two categories – exogenous and endogenous.

The endogenous variables have an immediate impact on the choice set of the community. They may affect terms of access to the resource itself, or some suitable alternative, or the demand for the resource by consumers. Examples of such endogenous contextual factors include characteristics of the local ecological system, cultural characteristics of the community, extent of development of transport facilities linking the community to nearby markets, etc. The exogenous variables are beyond the control of the resource users. They affect the CPR system by influencing its ability to produce the benefits and by influencing the demand for the benefits from the resource. Instances for these exogenous contextual factors include changes in world market prices, government policy relating to resource use and developments in labour market outside the community.[2]

Embedded Choice Model

The need to incorporate contextual factors calls for a shift in the conceptual foundations away from the rational choice model to the 'situated' or 'embedded' rational choice model (McCay 2002). The new model argues that the situation or context in which the choice has to be made determines the rationality of a choice. For instance, social psychologists (O'Riordan 1976; Ajzen and Fishbein 1977; Hill 1981) have shown that the response of resource users to an environmental problem will be conditioned by the interaction among information flows, experience, awareness, concern for others and norms. Simultaneously socio-political forces both mould cognition and define the set of behavioural responses. O'Riordan (1976) suggests that a plausible behavioural model should conceptualise a transactional arrangement in which the individual negotiates with his environment in a symbiotic manner, each influencing the other. Responding to a collective action problem then becomes a game in which individual tests both the environmental response and the reaction of the socio-political system to which he belongs. This allows him to identify the limits of his abilities and the social and institutional constraints on his actions (Figure 15.1).

Figure 15.1 Interaction between Contextual Factors and Behaviour

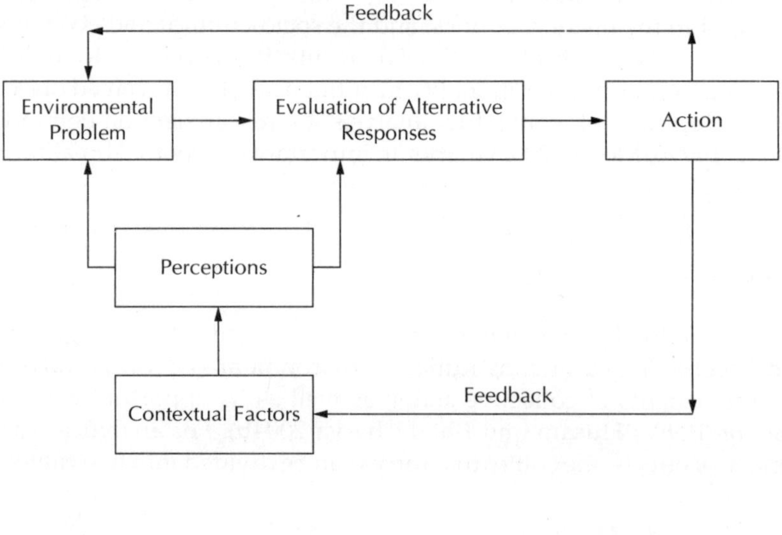

This chapter incorporates contextual factors into an optimisation model of household behaviour to examine the decision to supply labour to the cooperative or to free ride. It is shown that the structure of the problem—the nature of the objective function and the constraints faced by the agent is determined by the historical and economic context in which the community is situated. Although the member makes his choice rationally, it is bounded rationality as his decision-making process is shaped by the context in which he is situated or embedded.

A BRIEF LOOK AT THE COOPERATIVES

This section presents a brief history of the two cooperatives. This is based on focus group interviews and general group discussions with cooperative members. In addition, individual-based survey of fifty per cent of the members using a questionnaire was also undertaken. The financial performance and structure of the cooperatives is analysed using the audited reports.

Captain Bheri Fishermen's Cooperative

The Captain Bheri Fishermen's Cooperative Society is among India's oldest surviving cooperatives. It is located adjacent to the Eastern Metropolitan Bye Pass to the south of Chingrihata. Initially, the cooperative was oriented towards meeting the subsistence requirements of its members. In the 1970s, some contextual developments changed the nature of the undertaking.

After 1971, the influx of refugees from East Pakistan into this area led to deforestation of the area to satisfy the demand for fuelwood. The clearing of the forestland opened up the surrounding areas for urban development and led to the emergence of wholesale markets in Chingrihata. It also led to better fish growth, as the shade was reduced (leading to greater fish mobility) and deposit of leafy debris in the waterbodies decreased.

Around this time, the creation of a drainage system led to the release of sewage water into these waterbodies. This sewage water contained nutrients that could be consumed by the fish. A new production process—sewage-based fishery emerged. The polluted water was a rich source of nutrients, which was a better substitute for the cheap natural fish food or the more expensive artificial variety. Simultaneously, fish culture reduced the pollution level of the water to acceptable levels, so that it could be subsequently used for irrigation. The process of production, therefore, converted the public 'bad' (the sewage water) to a public good (the irrigation water).

These developments changed the form of cooperating. The working became more cohesive and formal; commercial motives became more dominant. This led to accelerated growth of the cooperative—in terms of revenue, profit and productivity. Over time, the cooperative flourished. The cooperative won four National Productivity Awards between 1989 and 1992.[3] This encouraged the cooperative to diversify their activities to providing recreational services. Paddle boats were introduced. The banks were promoted as picnic sites. Licences were sold to food vendors. All these increased revenues to the cooperative.

This was followed by a dramatic decline in fortunes. Between 1994 to 95, the revenue from pisciculture fell sharply; losses replaced profits; debts mounted; and labour supply from cooperative members tapered off. Failure to pay wages led to mass absenteeism; this further affected functioning of the cooperative. Simultaneously, vandalism and anti-social activities forced the Fisheries Directorate to close down boating and allied activities.

From 1999 to 2000, however, the cooperative is attempting a turnaround. Although outbreak of an epidemic amongst the fish affected revenues in 1999–2000, in the following years a small profit was posted.

Bon Hooghly Fishermen's Cooperative

The cooperative was founded by migrants from East Pakistan (currently Bangladesh). The traditional occupation of the migrant individuals varied—but a dominant section (sixteen individuals) were fishermen by caste. Forced by political disturbances and riots, the fishermen migrated to India. Initially, they used to catch fish individually or in small groups.

Around 1972, Jamuna Bhowmick, wife of one of these fishermen, Gobinda Bhowmick, heard a radio programme extolling the benefits from a fishermen's cooperative established at Mudiali, Calcutta. Deeply impressed, she informed her husband and Biren Sarkar, tutor of their children, about the programme. Gobinda Bhaumik and Biren Sarkar talked the issue over with other refugees and decided to form a cooperative based on three water bodies in their locality.[4] One of the lakes was situated in Noahpara, where these refugees traditionally caught fish; the other two were adjacent lakes at Bon Hooghly, near the Indian Statistical Institute. All three had been degraded because of eutrophication and did not generate any significant benefit for the neighbourhood.

Initially the catch was low as the founding members lacked skill. Therefore, the members contacted some traditional fishermen who had settled in the suburbs in Calcutta and invited them to join them. A Cooperative was formally set up in 1974 under the name of Bon Hooghly Fishermen Cooperative Society. The total number of members increased to seventy-eight, of whom six members have since died. Membership is by inheritance; it can also be transferred to persons nominated by the member. No new individual from outside the community can otherwise be inducted into the society.

Initially, the activity of the cooperative was confined to fishing. Later on, in the early 1990s, boating was introduced to supplement revenue. Gross revenue from fish sales have increased steadily, especially from the late 1980s. Though there was a decline in 1997–98 and 1998–99, revenue again picked up in 1999–2000. Boating has also been providing steadily increasing returns. The Cooperative has operated at a profit except for three years. From the early 1990s, net profit has been high. However, in 1998–99 and 1999–2000, the level of profit has shown a steep decline.

Contextual Factors and Labour Supply in the Cooperatives

Members supply labour to the cooperative, the collective choice arena, to appropriate the flow of benefits generated by the common pool resource (the waterbodies). In an analysis of the rationality of such labour supply along the lines of the rational choice, theorists would focus on costs-benefits determined by market and technological parameters and the physical characteristics of the waterbodies using the framework suggested by Oakerson (1986, 1990). However, it is also necessary to study the bounds on rationality—how the socio-cultural characteristics of the community, their historical formation and their economic context determine opportunity costs of collective action and perceptions of such costs, and how they change over time. This will explain the success or failure of collective action.

Now what are the contextual factors that influence the supply of labour to undertake collective action? The objective of collective action and the socio-economic conditions of the community are important in this regard. The communities studied in this chapter are characterised by poverty and vulnerability. Income levels are very low in both cases—in case of Captain Bheri, for instance, per capita income is only Rs 509. In the case of the Bon Hooghly cooperative, members had been dispossessed of their homestead land and other physical assets by the political disturbances and riots. In such communities, survival motives have been found to be dominant concerns in explaining the direction of collective action (Husain and Bhattacharya 2004b). Contextual factors define, *inter alia*, the range of alternatives available to individuals to augment their income from outside the collective action arena. Commercial growth and residential expansion in the Baranagar and Chingrihata has led to the development of a labour market in these two areas. In Chingrihata, for instance, there are electric bulb and glove manufacturing units. In both areas, there is an informal sector where individuals can work as porters, transporters, tailors and petty traders. At the same time, the simultaneous influx of migrants and population growth has led to excess supply of labour in this market. As a result, wages have been depressed and uncertainties created. Nevertheless, the labour market provides an alternative means of livelihood. Hence it determines the opportunity costs of collective action and ultimately determines whether such action will be undertaken.[5]

LABOUR ALLOCATION IN A SITUATED CHOICE MODEL

This chapter examines the relation between contextual factors in the form of employment alternatives in a nearby labour market and collective action. The starting proposition of this chapter is that individuals seek to ensure a secure income flow with which they can consume a minimum level of goods and services (including leisure). The problem facing them is to determine how much labour to allocate to the collective action arena and how much to the labour market. The individual's choice has been situated in the context of his poverty and vulnerability, and the conditions prevailing in the labour market.

Now the problem of determining individual supply of labour to the cooperative has two perspectives:

1. *Collective perspective*: The individual's supply of labour is to a collective choice arena. The returns to the individual will, therefore, depend upon the supply of labour from *all* individuals.
2. *Individualistic perspective*: This considers the effect of individual supply of labour on utility of the family. The supply of labour by other individuals is assumed by the individuals to be given (Cournot–Nash assumption).

Evidence from the survey indicates that community consists of homogenous individuals. This implies that individual choice functions, constraints and expectations will tend to converge. Thus, the analysis can be undertaken in terms of a typical individual. The starting proposition is that the collective dimension of the problem will determine the rewards associated with the set of feasible labour supply decisions by the individual. Using a diagrammatic approach, this chapter examines how an individual chooses between these alternatives based on their associated pay-offs.

The Collective Dimension

Since the labour supplied is directed towards attaining a collective outcome, it is expected that the pay-off associated with a particular level of labour supply by the individual is not unique but depends upon the supply of labour by other individuals. This implies that the situation can be represented using a game theoretic matrix. If there are M possible levels of labour supply by the household to the cooperative (denoted by l_{ci}, when l_c denotes labour supply to the cooperative with i = 1, 2, …M, indicating level of labour supply), the relation between each choice and the pay-off can be represented as below.

Individual x assumes that the other individuals supply the same amount of labour. Now, when l_{ci} = M, supply of labour to the cooperative is highest; when l_{ci} = 1, supply of labour to the cooperative is lowest. Suppose, individual x supplies 3 units of labour. Then he will prefer that other individuals supply *at least* 3 units of labour or more. If other individuals supply 3 units of labour, then the burden of work is distributed evenly. If they supply more labour (say 4), then x is able to free ride to some extent on others. But if other individuals supply less labour (say 2), then the greater burden of work falls on x; it is now the other individuals who are free riding on him. Generalising, it can be said that if individual x supplies i units of labour then he will want the others to supply j units

Figure 15.2 Collective Supply of Labour and Pay-offs

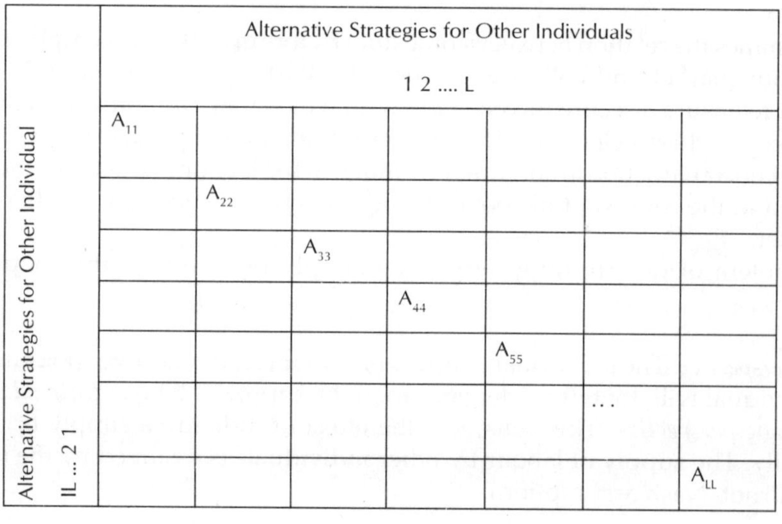

of labour, when $j \geq i$. Alternately, it can be argued that, given supply of labour by other individuals = j, individual x will chose i such that $i \leq j$.

Since this is true for all individuals, agents are confronted with the classical Prisoner's Dilemma situation. With simultaneous play and absence of communication, a sub-optimal equilibrium will result with all players choosing 1. However, there are certain reasons why this process will not occur in reality.

Ostrom (1990) has shown that the presence of institutions satisfying some characteristics (referred to as 'design principles' in CPR theory) affect individual incentives in such a manner that a socially optimal solution will emerge. Based on other studies of the evolution of cooperation in a PDG situation (Acheson 1998; Balland and Platteau 1996; Bowles and Gintis 1998), it is possible to identify the forces that will ensure cooperation:

1. The production process requires supply of labour not in isolation but in close proximity to one another. As a result, there is no possibility of 'hidden action' by any single individual—since all labour has to be supplied publicly (in front of other members), the other members will notice any shirking (Acheson 1998; Bowles and Gintis 1998).
2. There is constant interaction over time between the individuals in the collective arena. The labour allocation decision has to be repeated every day; this allows scope to teach deviators to cooperate. In game theoretic terms, the strategy need not be invariant—it is possible to learn the optimal strategy through a trial and error process (Acheson 1998).
3. Interaction between the players occurs not only within the collective arena, but in other spheres also. This enables the creation of reputations—anticipations of what the other players will do in given circumstances. It also provides further avenues for teaching through sanctioning and rewarding in other spheres. For instance, a habitual defector may find himself cut off from the social network relations existing between individuals. This denies him help during a crisis. Individuals who normally cooperate, on the other hand, have good reputations; easy availability of help during any crisis acts as a reinforcement to further cooperation (Acheson 1998; Bowles and Gintis 1998).
4. The lack of alternatives outside the cooperative creates a closed system on which the individuals become totally dependent. By removing the possibility of leaving the cooperative, the possibility for the adoption of hit-and-run strategies is eliminated. An effect similar to parochialism is created (Bowles and Gintis 1998).

In such a situation, the veil of anonymity enabling a player to get away with defection is absent; all actions are public. Further, both sanctioning and the credibility of the threat are serious enough to keep potential defectors in line. However, this does not rule out defection in the resultant equilibrium. Both theoretical (Taylor 1977; Balland & Platteau 1996) and empirical (Steins 1999; Husain and Bhattacharya 2002) studies have shown that a certain amount of defection can be 'tolerated' by co-operating players in equilibrium. For instance, it has been observed that those members who are in distress are permitted to defect without reprisals in each sub-game (Husain and Bhattacharya 2004a) so that the composition of the defecting group changes in each sub-game.

Thus, in the equilibrium of each sub-game, the players are partitioned into two groups—a group of players who defect and another group who cooperate. This implies that all members of the second group chose to contribute the same level of labour (say i), while members of the first group chose j—such that $i < j$. The equilibrium is repeated in the successive sub-games until the number of players defecting crosses a certain threshold level. In that case, cooperation unravels to a stable equilibrium characterised by universal defection. In terms of the matrix, the community moves up the diagonal towards a_{11}.

Framework of Embedded Choice

Having decided to cooperate (that is, supply the same amount of labour to the cooperative), individuals have to decide on this level. It has been seen that members can supply $l_1, l_2, \ldots l_M$ labour units to the collective action arena; the associated pay-offs from the cooperative to these alternatives are a_{11}, $a_{22}, \ldots a_{MM}$. The members will now rank these alternatives and choose the alternative yielding highest benefits. However, the choice is not made in a vacuum but in a specific historic and economic situation. This context shapes constraints and determines the parameters of the optimisation problem.

It had been pointed out earlier that members also supply labour to the nearby informal sector. So their choice consists of allocating work hours between three alternative uses—enjoying leisure (l), working in the cooperative (l_c) and working in the labour market in his neighbourhood (l_s). For convenience it is assumed that the member has already decided how much leisure he will enjoy. This reduces the problem to allocating total labour hours in a day, less leisure hours, between the labour market and the cooperative: $24 - l = L \text{ (say)} = l_c + l_s$.

The returns to the alternative uses of labour are as follows. In the labour market, the worker gets an income of w but this has to be discounted by the uncertainty of getting work. Suppose the probability of getting work is q. His expected income from the secondary job is qwl_s. In addition he also gets a wage from the cooperative. This is w_c, so that his returns from supplying labour to the cooperative should be $w_c l_c$. In reality he is paid a fixed daily amount, irrespective of the work done. Let this amount be W. Thus, total income to any member is: $W + qwl_s$.[6] This income is used to purchase subsistence goods (X) at a price p.

This, however, is not a complete description of the expenditure constraint. Consider the labour-consumption space. Consider the position where the entire labour time (L)[7] is supplied to the cooperative. This means that the individuals are supplying L = m units of labour to the cooperative. The individual income will be W, represented by point A in the diagram. Now, households can increase their income, and hence expenditure, by reducing labour supply to the cooperative and diverting it to the labour market. This is denoted by a movement up the budget line AB (Figure 15.3).

Now, whether the cooperative is able to pay the wages W to the members depends upon the revenue obtained by the cooperative. If the supply of labour falls below a critical level, say L_c^*, then the cooperative will be unable to generate sufficient revenue to pay its members. For simplification, it is assumed that the cooperative stops payments to its members. So any reduction in the labour supplied to the cooperative below L_c^* will be accompanied by a drop in the income level by W. This is shown in terms of a downward shift in the budget line of the individuals.[8]

Mathematically the budget line can be represented as follows:

$$
\begin{aligned}
p.X &= W && \text{for } l_c = L \\
&= W + q\,w\,(L - l_c), && \text{for } L > l_c \geq L_c^* \\
&= q\,w\,(L - l_c), && \text{for } l_c < L_c^*
\end{aligned}
\qquad (1)
$$

Figure 15.3 Budget Line of the Individual

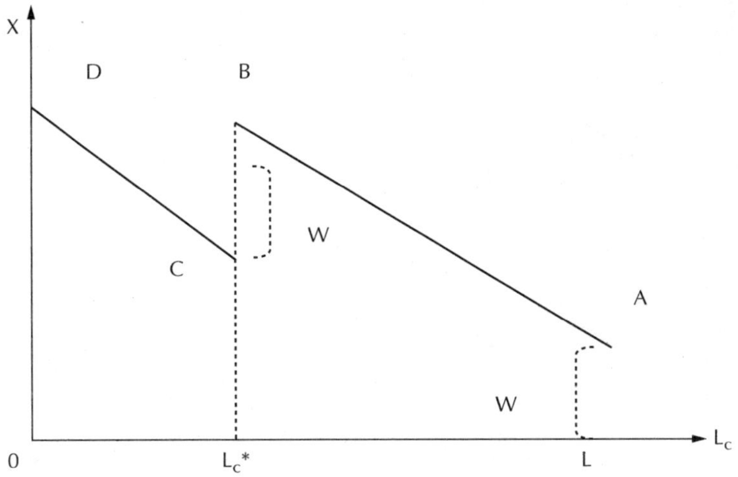

The objective of the individual is to maximise utility. It is assumed,

$$U = U(l_c, X) \text{ with } U_l < 0, U_X > 0, U_{ll} > 0, U_{XX} < 0 \quad (2)$$

The indifference curves will be positively sloped. Further, as agents move upwards, keeping l_c constant, utility levels will increase. Convexity of the indifference curves to the horizontal axis is also assumed. Given the objective function and constraints, the optimal labour supply by a representative individual is determined based on a graphical exposition.[9]

In Figure 15.4(a), a corner solution occurs, with the individuals diverting their entire labour to their secondary occupations. In terms of the matrix (Figure 15.2), individuals move up the diagonal to defect simultaneously and choose not to supply any labour to the cooperative. Collective action is not a viable proposition to the individuals and the cooperative fails. Note that *all* individuals have decided to reduce labour supply together. This behaviour is different from the individual attempts to free ride as discussed earlier. In the earlier case, it was a decision taken by a member *individually*; in the present case, however, it is the *group that takes the decision* to reduce labour supply to the collective arena. To distinguish these two types of decisions, they are referred to as *individual defection* and *group defection* respectively.

Alternately, a situation represented by Figure 15.4(b) can occur. Here the highest indifference curve is tangent to the budget line at the kink. This again leads to a corner solution—this time at the kink, with the individuals supplying L_c^* labour to the cooperative and $(L - L_c^*)$ labour to the secondary market.

Context and Equilibrium

The survey indicates that the cooperatives are in the equilibrium as described by Figure 14.4(b). But what are the reasons for this choice? To answer this question, it is necessary to consider how

Figure15.4(a) Equilibrium of the Individual

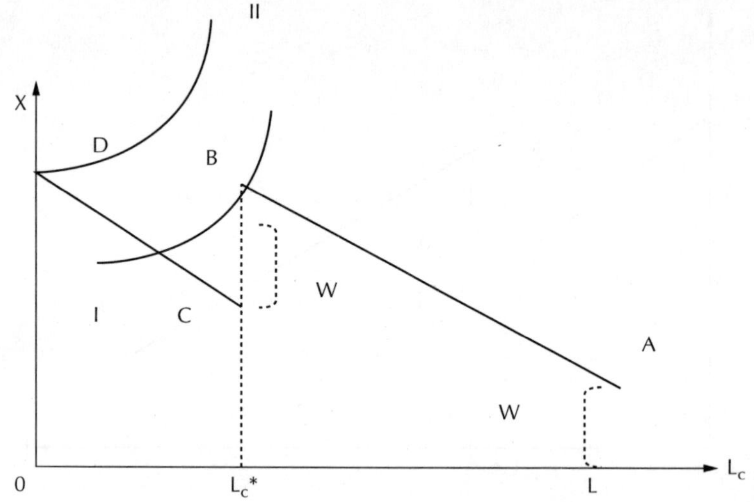

Figure 15.4(b) Equilibrium of the Individual

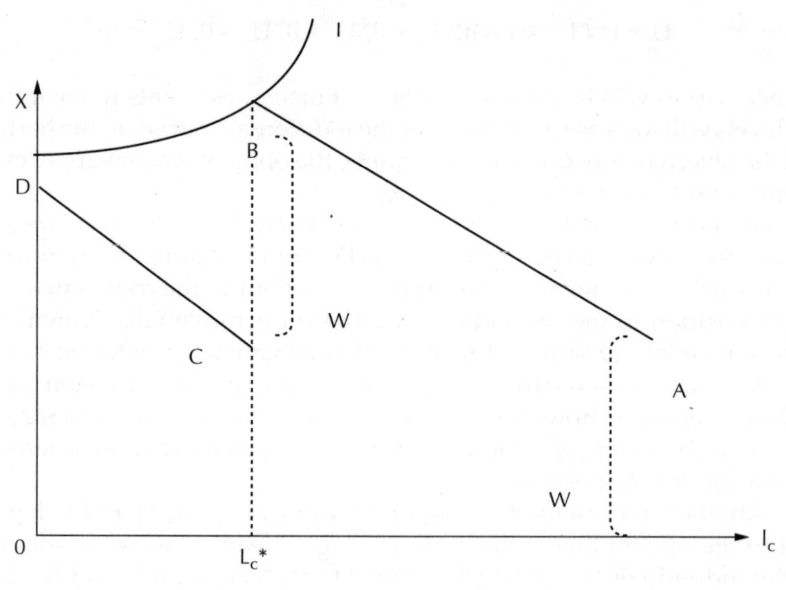

the historic and economic context shape the nature of the collective identity and preferences of the members.

Although the cooperative is a formal organisation consisting of a limited number of individuals, forces stronger than the legal structure imparts the collective identity of the members. Specifically, it is the presence of a collective consciousness that gives the community its collective identity. Community exists among individuals who share 'common *interests,* common *identification… growing out of shared characteristic*' (Ascher 1995: 83). The belief in shared identities and common experiences reduce individuality of community members—they become more willing to cooperate over formal decisions to manage and conserve resources.

The members of Captain Bheri are long time residents in the Chingrihata area. Their shared past has created common interests and shared characteristics. Further, the common problems faced by them in their daily struggle for existence means that they have a common future. This has engendered what Ostrom (1990) calls mutually recurrent expectations.

Similarly, the members of the Bon Hooghly cooperative had gone through a similar experience and were facing a common problem. They had all been dislocated and re-settled in a culturally new locality. All of them were facing the desperate problem of earning a secure income flow. Although some of them had obtained work, the income earned by them was low and had to be supplemented by other means.

Louderdale et al. (1984) have shown in a game theoretic framework that the perception of a common threat may increase group solidarity. The creation of a group identity encourages co-operation because members of a social group tend to regard other members favourably and believe them to be trustworthy, honest and cooperative. Other members of the same group are expected, therefore, to reciprocate cooperative behaviour. In addition, inclusion within a social group reduces social distances between members so that they make less distinction between their own and other's welfare.

Along with the historical context, economic opportunities outside the domain of the resource were another important variable affecting behaviour of individuals. The labour market was extremely 'tight'—in the sense that unemployment was high, and employment and income insecure. The uncertainty and risk of loss of livelihood created a situation where transaction costs in the form of search costs and insecurity of job reduces the efficacy of the labour market as an allocative mechanism. Individuals were no longer guided by market principles (in the form of seeking sectors with highest returns) in allocating labour; instead members incorporate security considerations into their objective functions. Thus the economic context in which the choice of the members was embedded strengthened the historical base for collective action by determining the target of the cooperative.

Specifically, both forces acted to depress the Marginal Rate of Substitution of labour supply (to the cooperative) for consumption. This can be defined as the ratio of Marginal Disutility (MDU) of labour supply to the Marginal Utility (MU) of consumption (that is MDU_{lc}/MU_{x}). Since MU of consumption will be high for households close to the subsistence level, while the sense of community and uncertainties in the external labour market implies a low absolute value of MDU of labour supply to the cooperative, MRS will be low. This implies that the indifference map will consist of relatively flat indifference curves. Consequently, equilibrium will occur at the kink rather than at the corner. While respondents were unwilling to state the extent of their dependence on the labour market, this study found that about 50 per cent of their income was obtained by working in part-time jobs.

COMPARATIVE STATICS IN EMBEDDED CHOICE MODEL

Now, over time, there may be parametric changes in the appropriation methods, in wage levels, etc. These changes can both increase or decrease net benefits of collective action. Once again, however, the effect of such changes will depend upon the context in which the agents are embedded.

The complexities and surprises in undertaking comparative static analysis is illustrated by examining the impact of a change common to both the cooperatives—the introduction of paddle boating. The following subsection shows how the interaction of parametric changes with contextual variables may cause unexpected changes in the collective arena.

Effect of Introducing Boating Facilities

In the 1990s, both the cooperatives diversified their activities and introduced boating as a sideline. While the purchase of boats required a large capital outlay, it represents a sunk cost; operating costs of boating, on the other hand, are negligible. Day guards usually double up as ticket collectors; other operating costs include painting and minor repairs, which do not mean any large amount. On the other hand, boating generates additional revenue for the cooperative. It also generates positive externalities in the form of better fish growth due to circulation of dissolved oxygen in the water and movement of fish owing to the waves created during paddling. Introduction of boating facilities, thereby, increases the net benefits from the cooperative. All these can be expected to strengthen the base for cooperation. However, if the effect of contextual factors on perceptions is incorporated into the comparative static exercise, the expected result is quite different.

The introduction of paddling facilities in the Bon Hooghly Cooperative increased profit. But if revenue from paddling is deducted from the total profit (to examine the cooperative simply as a fishing unit) the positive profit is reduced to losses (Table 15.1). The change was even more dramatic in the case of Captain Bheri Cooperative. Although there was no decline in the revenue from fish sales, there was a sharp decrease in the level of profits (Table 15.2).

Paddle Boating and Performance of Cooperatives

To understand the reasons underlying this unanticipated change it is necessary to refer to the graphical treatment undertaken earlier.

It had been seen that, *ceteris paribus*, the introduction of paddle boating increases revenue of the cooperative while operating costs remain constant. Consequently, the financial position of the cooperative improves. This can benefit the members in two ways—through a distribution of the increased surplus among members or through a wage hike. Both will cause an upward shift in the budget line. Now the West Bengal Cooperative Societies Rules, 1987, places restrictions on the distribution of profits amongst members. Section 114(1) of the Rules states: 'In every cooperative society with shares, dividend may be declared up to a maximum of 12 per cent per annum…'. Even this payment has to be approved by the Registrar of Cooperative Societies, West Bengal.[10] On the other hand, wages are sticky and tend to be hiked after about a decade. This is because the cooperative management is generally more risk averse than members are, and tend to view increases in cooperative revenue as temporary gains. Such increases are interpreted as a permanent increase only if revenues are maintained at the new level for several years at a stretch.

Thus, the increase in profits as a result of introducing paddle boating did not lead to an upward shift in the budget line. Instead, the improved financial position of the organisations led to an extension of the budget line from AB to AB′ so that the kink in the budget occurs at a lower level

Table 15.1 Trends in Net Profit and Pisiculture (in Rs)—Bon Hooghly Cooperative

Year	*Net Profit*	*Revenue from Fish Sales*	*Revenue from Boating*	*Wages to Hired Workers*	*Profit to Pisciculture*
1993–94	58,011	15,07,050	20,348	41,775	37,663
1994–95	64,730	21,17,682	30,247	40,500	34,483
1995–96	25,310	18,31,720	98,325	38,775	-73,015
1996–97	54,841	19,89,203	89,050	7,52,585	-34,209
1997–98	46,298	19,51,512	63,720	7,38,878	-17,422
1998–99	4,421	16,67,995	72,155	6,82,532	-67,734
1999–2000	2,196	22,88,275	82,910	8,98,414	-80,714

Table 15.2 Trends in Revenue from Fish Sales and Profit (in Rs)—Captain Bheri

Year	*Revenue from Fish Sales*	*Cooperative Profit*	*Casual Workers*		*Overtime to Members*
			Wages	*No. of Mandays*	
1991–92	9,85,777	-73,806	36,136	602	–
1992–93	14,28,972	55,991	48,263	804	–
1993–94	17,07,799	-2,04,341	57,535	959	19,240
1994–95	14,05,333	-23,305	73,920	1,232	27,215
1995–96	n.a.	n.a.	n.a.	n.a.	n.a.
1996–97	22,98,711	-25,838	72,920	1,215	9,500
1997–98	13,50,314	-1,83,032	83,740	1,396	12,175
1998–99	12,82,424	-1,33,555	51,517	859	1,100
1999–2000	14,80,410	-1,00,960	61,075	1,018	125
2000–01	17,53,626	-1,72,054	1,01,767	1,696	–

of l_c^* (the critical labour supply corresponding to the kink falls from l_{c1}^* to l_{c2}^*). This is shown in Figure 15.5.

It is obvious that the new equilibrium will occur at a lower level of labour supply (B′)—corresponding to the kink in the modified budget line. Thus, labour supply remains at the threshold just ensuring that the organisation survives; but now, with the introduction of an additional source of revenue, this survival can be ensured with a lower level of labour supply. So the introduction of paddle boating paradoxically weakened the base for cooperation. Consequently, *all* members simultaneously reduce their labour supply to the cooperative, diverting the labour thus released to the informal sector and thereby increasing total income. This is an instance of what was referred to as collective defection as manifested by households moving upwards along the diagonal (Figure 15.2).

Such a change affects the nature of the cooperative. Fishing may become less important, as the cooperative becomes increasingly dependent on boating and allied activities. But the viability of the cooperative as a vehicle to attain some collective target (in this case, ensuring a secure income flow) remains as it continues to provide a means to generate income. A serious problem will occur if the members over-estimate the revenue expected from boating and reduce labour supply below the threshold labour supply level (l_c^*). This will affect the viability of the cooperative and reduce its ability to meet wage liabilities. Initially, reserves will be used to meet wage bills; but after a time, delays in wage payments will occur. If the cooperative pays only a part of the wages due to each worker, this will be equivalent to a wage cut. This is represented by a downward parallel shift in

Figure 15.5 Effect of Introducing Boating Facilities

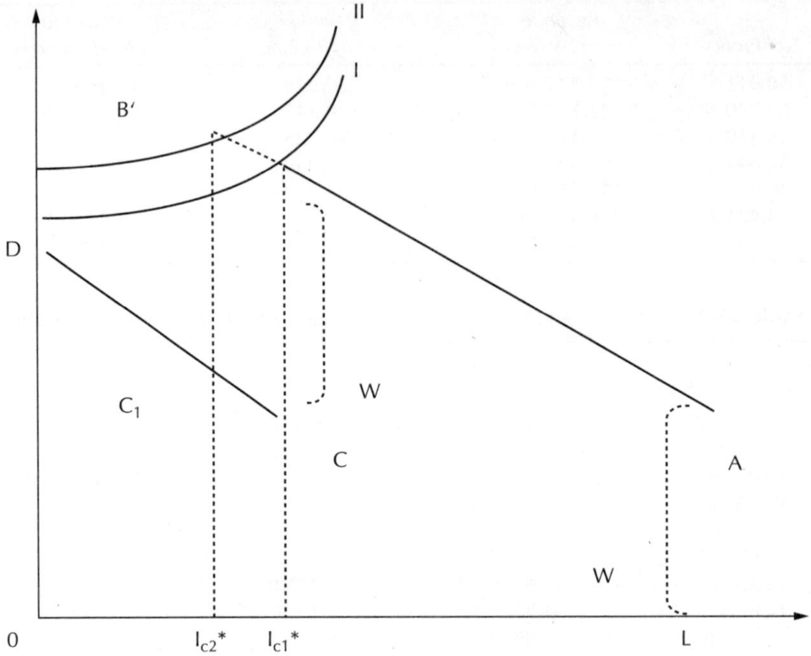

the budget line from AB′ to A′B″; the CD segment, however, remains unaffected. If there is a sufficiently large amount of wages withheld by the cooperative—so that the new budget line (A′B″) falls to below III (Figure 15.6)—the optimal level of labour supply to the cooperative will fall from l_{c2}^* to 0. The survival of the cooperative is now threatened.

Interpreting Financial Performance using Formal Framework

Group discussions with members and the management of the two cooperatives revealed that a sharp reduction in labour supply to the cooperative occurred after the introduction of boating facilities. The management tried to compensate this by hiring daily workers and paying overtime allowance to the members. As a result revenue from fish sales was maintained but the increased wage bill led to a decline in the profits from fishing.

In the case of the Bon Hooghly Cooperative, there was a reduction in the labour supply to the cooperative from l_{c1}^* to l_{c2}^* (as in Figure 15.5). Since labour supply did not fall below the critical level, the cooperative continued to function—although the basic activity of the cooperative changed from pisciculture to boating. Simultaneously, returns from boating compensated for the decline in profit from pisciculture (preventing the budget line shifting downwards as in Figure 15.6). This created a situation where the net profit of the cooperative remained positive even though pisciculture became a losing line of activity (Table 15.1).

In the case of Captain Bheri, however, the situation depicted in Figure 15.6 occurred. Possibly owing to an over-estimation of the financial viability of the cooperative, the decline in labour supply

Figure 15.6 Effect of Wage Arrears

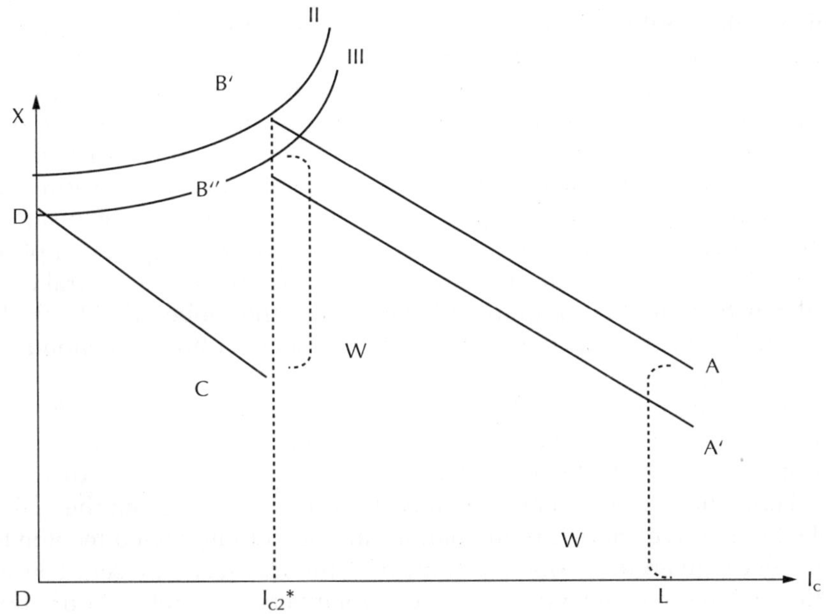

was even more marked than in the Bon Hooghly cooperative (beyond l_{c2}^*). There was an even greater shift to hired labour, increasing costs significantly. As a result, in Captain Bheri, the fall in profits from pisciculture was greater than the increased revenue from boating.[11] This affected the ability of the cooperative to pay wages (arrears in salary exceeded actual wage disbursals in 1991–92 and 1992–93) and led to a downward shift in the budget line (to A′B″, as in Figure 15.6). This caused a further fall in labour supply. Some workers even stopped coming to the cooperative for months on end, that is $l_c = 0$.

In the mid-1990s, the Directorate of Fisheries intervened to ban paddle boating. This reduced revenues and further affected the ability of the cooperative to meet wage liabilities. While the withdrawal of paddle boating returned the kink to l_{c1}^*, by now the increasing wage arrears had shifted the perceived budget line down well below III so that labour supply to the cooperative was virtually stopped (Figure 15.6). It was only later on, in 2000–01, that the declining fortunes of the cooperative could be arrested with a sanctioning of absenteeism through the strict implementation of 'no work no pay' principle. This is leading to a revival in the fortunes of the cooperative.[12]

CONCLUSION

To sum up, the equilibrium labour supply by the cooperative members is delicately poised at the threshold level, which is defined as the minimum labour supply required to keep the cooperative going. In the long run, the pressures of survival force the members to constantly search for avenues to increase their income. This may affect the evolution of the cooperative society in various ways.

For instance, it may change the form of the society (Bon Hooghly) or it may even threaten the survival of the society (Captain Bheri).

However, this result should not be generalised—it is specific to this model being generated by the particular combination of contextual factors influencing the perceptions of agents. For instance, to the extent that immediate needs are not so pressing and members incorporate projections about the future in the decision-making process, members may realise that supplying labour above l_c^* to the cooperative may reduce their short-term income (and consumption) but increase their long-run income, consumption and utility. Such cooperatives may represent a viable form of organisation.

With regard to policy making, this chapter warns that introducing organisational and other changes on the basis of costs-benefits from collective action overlooking the complex nature of the interaction of the collective arena with the historic, economic and socio-cultural context in which actors are located may have adverse or unintended effects on the cooperative. Such changes should be introduced carefully ensuring a consistency with the socio-cultural, economic and historical setting of the target group.

One organisational change, however, that policy makers can certainly think about implementing in the context of the present study is with regard to the incentive structure of cooperatives in West Bengal. At present the West Bengal Cooperative Society does not allow for distribution of profit amongst its members. The reason is that retention of profit will strengthen the viability of the co-operative, enable it to survive short-run fluctuations and lead to increased income to members in the long run. However, given the economic context of the members, present concerns dominate future possibilities; this has created a disincentive to members to supply labour above the critical level thereby affecting the target of attaining long-run viability. Relaxing the limits on dividend pay-outs will allow the members to share in increased profits of the cooperative and provide an incentive to increase labour supply.

NOTES

1. The survey of Captain Bheri was funded under a UGC Minor Research Project. The author is grateful to the Directorate of Fisheries (Govt. of West Bengal) for permitting the survey and to Madhumita Mukherjee (Deputy Director, Fisheries Directorate) for her active interest and support. The survey of Bon Hooghly Cooperative was part of an IGIDR project funded by the World Bank and Ministry of Environment & Forests undertaken in collaboration with N. Bhattacharya Kalyani University.

   This chapter was presented at the International Conference on 'Environment and Development' organised by School of International Studies, Jawaharlal Nehru University, New Delhi on 7–8 April, 2005 and at the Fourth Biennial Conference of the Indian Society for Ecological Economics on 'Ecology and Human Well-Being' at Indira Gandhi Institute of Development Research, Mumbai on 3–4 June, 2005. The views expressed by the author do not claim to represent the opinion of the Government of India.
2. It should be noted that the distinction between local and remote contextual factors is made for the purpose of analytical convenience only. In reality we have what Edwards & Steins (1999b) call a 'contextual factors continuum'. This refers to the series of relationships linking remote with local contextual factors and the entire set of contextual factors with the individual choice set. The conceptualisation of a continuum implies that remote and local contextual factors may mutually reinforce changes in the resource regime.
3. First award in 1989 and 1990, and second award in 1991 and 1992.
4. This is a case of learning in the context of imperfect information.

5. It is not possible to generalise the nature of causality between poverty, vulnerability, contextual factors and collective action. For instance in Bankura, collective action to deepen a pond supplying drinking water was inhibited by opportunities in the local labour market. But, in Purulia, it was found that the absence of such opportunities forced villagers to intensify their rate of exploitation of local forest resources (Bhattacharya and Husain 2002). The precise nature of these alternatives and the extent to which they are supplementary or complementary to collective action in achieving the primary objective of the community members are , therefore, very important.
6. There is no fixed relation between cooperative income (W) and wages from informal sector (w_s)—in some cases W was higher; in other cooperatives, W is lower.
7. L, therefore, corresponds to M, defined earlier as the maximum possible labour supply to the cooperative.
8. In reality there will be a series of L^*_c. As l_c is reduced below each L^*_c, the cooperative enforces a wage cut, lowering the budget line parallel to the earlier budget line.
9. Since the constraints are non-linear, classical optimisation techniques will not apply. Kuhn-Tucker conditions will have to be used. The results of the optimisation and specially the comparative static exercises will be both cumbersome and will fail to provide results that can be interpreted conveniently. So we have relied on a graphical exposition and avoided a mathematical approach.
10. The rationale is to prevent the short-sighted policy of reckless distribution of surplus and allow the organisation to become financially strong.
11. The difficulties of monitoring hired labour and managerial inefficiencies that crept in at about this time also contributed to the decline in profits.
12. In 2001–2002, however, a fish epidemic broke out leading to heavy losses.

REFERENCES

Acheson, J. 1998. 'Lobster Traps Limits: A Solution to a Communal Action Plan', *Human Organization*, 57(1): 43–52.

Ascher, W. 1995. *Communities and Sustainable Forestry in Developing Countries*. San Francisco: ICS Press.

Ajzen, I. and M. Fishbein. 1977. 'Attitude-Behaviour Relations: A Theoretical Analysis and Review of Empirical Research', *Psychological Bulletin*, 84(5): 888–918.

Axelrod, R. 1986. 'An evolutionary approach to norms', *American Political Science Review*, 80(4): 1095–111.

Balland, J.M. and J.P Platteau. 1996. *Halting Degradation of Natural Resources: Is there a Role for Rural Communities?* Oxford: FAO and Clarendon Press.

Berkes, F. (ed.), 1989. *Common Property Resources: Ecology and Community-Based Sustainable Development*. Belhaven Press: London.

Bhattacharya, R.N. and Z. Husain. 2002. 'Evolution of CPR Institutions: An Analysis of the Environment-Economy Interaction'. Project Report. Mumbai: Indira Gandhi Institute of Development Research.

Bowles, S. and H. Gintis. 1998. 'How Communities Govern: The Structural Basis of Stochastic Norms', in A. Ben-Ner and L. Putterman (eds), *Economic Values and Organisations*. Cambridge: Cambridge University Press.

Chopra, K., G.K. Kadekodi and M.N. Murty 1990. *Participatory Development: People and Common Property Resources*. New Delhi: Sage Publications.

Coate, S. and M. Ravallion 1989. 'Reciprocity without Commitment: Characteristics and Performance of Informal Risk Sharing Arrangements'. Discussion Paper 96. Warwick Economics Summer Research Workshop.

Edwards, V.M. and N.A. Steins. 1999a. 'Special Issue Introduction: The Importance of Context in Common Pool Research', *Journal of Environmental Policy and Management*, 16: 195–204.

—— 1999b. 'A Framework for Analyzing Contextual Factors in Common Pool Research', *Journal of Environmental Policy and Management*, 16: 205–21.

Fafchamps, M. 1992. 'Solidarity Networks in Preindustrial Societies: Rational Peasants with a Moral Economy', *Economic Development and Cultural Change*, 41: 147–75.

Fields, A.J. 1979. 'On the Explanation of Rules Using Rational Choice Models', *Journal of Economic Issues*, XIII (1): 49–72.

Hardin, G. 1968. 'The Tragedy of the Commons', *Science*, 162: 1243–248. 12 December.

Hill, R.J. 1981. 'Attitudes and Behaviour', in M. Rosenberg and R.H. Turner (eds), *Social Psychology: Sociological Perspectives*, pp. 347–76. New York: Basic Books Inc.

Husain, Z. and R.N. Bhattacharya. 2002. 'Rationality, Norms and Co-operation—The Evolution of CPR Institutions', in R. Kasnobis (ed.), *Issues in Sustainable Development*, pp. 47–67. New Delhi: University Grants Commission.

—— 2004a. 'Attitudes and Institutions: Contrasting Experiences with JFM in India', *Environment and Development Economics*, 9(4): 563–77.

—— 2004b. 'Contextual Factors and Common Pool Resources: Evolution of a Fishermen's Cooperative', *Ecological Economics*, 50: 201–17.

Kimball, M.S. 1988. 'Farmer's Cooperatives as Behaviour towards Risk', *American Economic Review*, 78(1): 224–32.

Klooster, D. 2000. 'Institutional Choice, Community and Struggle: A Case Study of Forest Co-Management in Mexico', *World Development*, 28(1): 1–20.

Louderdale, P., P.P. Smith-Cunnier, J. Parker, and J. Inveriarty, 1984. 'External Threat and the Definition of Deviation', *Journal of Personal and Social Psychology*, 46(5): 1058–68.

McCay, B.J. and S. Jentoft. 1998. 'Market or Community Failure? Critical Perspectives on Common Property Management', *Human Organization*, 57(1): 21–29.

McCay, B.J. 2002. 'Emergence of Institutions for the Commons: Contexts, Situations and Events', in National Research Council, *The Drama of the Commons*, pp. 361–402. Washington D.C: National Academic Press.

Oakerson, R.J. 1986. 'A Model for the Analysis of Common Property Problems'. Working Paper No. R86–13. Workshop in Political Theory and Policy Analysis, Indiana University.

—— 1990. 'Analyzing the Commons: A Framework'. Working Paper No. W90–9. Workshop in Political Theory and Policy Analysis, Indiana University.

O'Riordan, T. 1976. 'Attitudes, Behaviour and Environmental Policy Issues', in I. Altman and Joachim F. Wohwill (eds), *Human Behaviour and Environment: Advances in Theory and Research*, Vol. 1. New York: Plenum Press.

Olson, M. 1965. *The Logic of Collective Action: Public Goods and the Theory of Groups*. Cambridge: Harvard University Press.

Ostrom, E. 1990. *Governing the Commons*. Cambridge: Cambridge University Press.

—— 1999. 'Institutional Rational Choice: An Assessment of the Institutional Analysis and Development Framework', in P.A. Sabatier (ed.), *Theories of the Policy Process*, pp. 5–71. CO: Westview Press.

Pinkerton, E. and M. Weistein 1995. *Fisheries that Work: Sustainability Through Community-Based Management*, Vancouver, BC: The David Suzuki Foundation.

Popkin, S. 1979. *The Rational Peasant*. Berkeley: University of California Press.

Scott, J. 1976. *The Moral Economy of the Peasant*. New Haven: Yale University.

Simon, H. 1957. *Administrative Behaviour*, 2/e, New York: Macmillan.

Singh, K. 1994. *Managing Common Pool Resources: Principles and Case Studies*. New Delhi: Oxford University Press.

Steins, N.A. 1999. *All Hands on Deck: An Interactive Perspective on Complex Common-pool Resource Management Based on Case Studies in the Coastal Waters of the Isle of Wight ,UK., Connemara , Ireland and the Dutch Wadden Sea*. Den Haag, Netherlands: Cipp-Data Koninklijke Bibliotheek.

Steins, N.A., V.M. Edwards and N. Rolling. 2000. 'Redesigned Principles for CPR Theories: An Interactive Perspective to CPR Theory'. Paper presented at 8th Conference of International Association for the Study of Common Property Resources, Indiana University.

Taylor, M. 1977. *Anarchy and Cooperation*. London: John Wiley and Sons.

Wade, R. 1988. *Village Republics: Economic Conditions for Collective Action in South India*. Cambridge: Cambridge University Press.

16

Environment and Community Action: A Case Study of Shifting Cultivation[1]

Subhash Chandra Srivastava

Abstract: Shifting cultivation is the primary source of livelihood of the tribal communities in Northeast India. While the system has been treated as a system of food security, it is also held responsible for the environmental degradation in the region. The present study finds that contrary to common perception of stagnation about the system, it has undergone significant changes in cropping pattern, mostly in favour of market economy. This has come out largely as a response to ever growing population, declining productivity and reducing land availability. The case of Tangkhul Nagas, the predominant community in Ukhrul district of Manipur, reveals that the tribal communities have their own understanding about the ecological problems arising out of the practice of shifting cultivation and have their own priorities for sustainable development without affecting their traditional fabric of life and culture. The chapter concludes with the communities' preference of programmes and policies for sustainable development including the planning for land use and re-forestation. The local populace also has a very clear understanding of the problems they are likely to face in the process and the need for the help in the form of training and finances.

INTRODUCTION

Presumed to have originated in the Neolithic period (Maithani BP and S C Srivastava 1997), shifting cultivation (also known as *jhum* cultivation) is still the most prevalent form of agriculture practiced by nearly two hundred tribal groups in the hilly areas of Northeast India (Singh 1994). Geographically only about 25 per cent of the total area of Northeast is under cultivation (IFAD 1994) and shifting cultivation alone accounts for 25 per cent of the total geographical area (Northeast Council Secretariat 1995). Due to geographical and security reasons, the tribes always preferred to live in the hill tops in small groups surrounded by vast tracts of forest. Consequently *jhum* had been the most rational and viable method of farming and deforestation, the most immediate way to access land for cultivation. With adequately long rotation cycle of 15–20 years, allowing for natural recuperation of fertility of the soil, the practice has remained sufficiently scientific. Consequently, *jhum*ing supplemented by hunting and gathering of wild nuts, fruits, etc., was the main source of subsistence. Forest, in other words had been the key to the livelihood of the tribal communities in the entire Northeastern region of India. With the growth of population and reduction in area available for *jhum*ing, the rotation cycle has gone down to around 5 years. As a result, the fertility of soil and production from *jhum* is constantly declining. The system is perceived to have been stagnant in terms of technology and operations over the years. On the other hand, we have evidences to show that despite the declining productivity and reducing *jhum* cycle, the overall productivity (per hectare) in shifting cultivation is quite high and, in a number of instances, even higher compared to plain and settled cultivation (Srivastava 1999). Today, the environmentalists interpret the *jhum* system as hazardous for its impact in terms of progressive degradation of soil, environment and production base. At the same time, it leads to deforestation and degraded forests. The perils of deforestation are being felt mainly in the form of high velocity wind and shortage of drinking water. Despite these negative influences and the attempts made by the governments of Northeastern states as well as by Government of India and its different agencies, the system continues unabated for the inherent merits of food security, topographical viability and equity implications.

In view of the above, the present chapter addresses itself to following issues:

- the extent of dependency of the tribal communities on different types of agriculture systems and the trends in area under cultivation;
- perceived changes in shifting cultivation system from the point of technology and environment; and
- understanding the community responses to these changes and exploring the proposed community action/preferences for a sustainable future.

METHODOLOGY

The study is primarily based on structured group interviews with different sections of the society in the selected watersheds and draws primary data from the community responses on the dynamics of changes in the society over the last ten to fifteen years. From this view point, the present study is basically qualitative in nature and based on the community's perceptions of changes in the land availability and environment as captured in terms of deforestation or degradation and community preferences for future actions.

Selection of Watersheds

In the absence of remote sensing data, selection of watersheds was done in consultation with the field agencies of agriculture, horticulture, forest and soil conservation departments of Manipur Government which have the experience of implementing watershed management projects of Central/state government. Following criteria was adopted for the selection of watersheds:

1. Size and area: size should range between 100–200 households and the area of the micro-watershed should range between 400–1,000 hectares.

2. Degree of land degradation: 40 to 60 per cent of the land under watershed should be under open forest.
3. Intensity of *jhum* cultivation: 40–60 per cent of the total households should be engaged in *jhum* cultivation.
4. Short *jhum* Cycle: 4–6 years.
5. Pressure on land: inductively 40 to 60 per cent of the cultivators should have one to two hectares of land including *jhum*, terraced and settled cultivation.

STUDY AREA

The present study covered five micro-watersheds covering ten villages with a population of 1,090 households. Four of these watersheds belong to the community of Tangkhul Naga and one belongs to Thadou Kuki tribe (Annexure 1). Significant differences were found in the socio-institutional arrangements of the two communities; but with regard to production system, the differences are almost negligible.

Of these 1,090 families, 954 families were Naga families and 136 Kukis. 976 families were found to be engaged in agriculture and belong to the category of cultivators (including *Jhoomia*). Thus, 89.21 per cent of the total families covered are dependent on agriculture directly and actively engaged in cultivation for their livelihood.

DEPENDENCE ON DIFFERENT CULTIVATION SYSTEM

From the point of view of dependency of families on cultivation system, three types are taken into account:

- purely *jhum* dependent;
- *jhum*-cum-settled; and
- purely settled cultivation system (this type is studied mainly to understand the process of transformation of purely *jhum* dependent to purely settled cultivation type).

Table 16.1 shows the absolute and relative distribution of the households under study in terms of their dependency on different types of cultivation systems.

JHOOM CULTIVATION

Fifty-three per cent of the total families are solely dependent on *jhum* cultivation (Table 16.1). The proportion of dependent families is minimum in Kanlui Lok (29.41 per cent) followed by Wuira Lok

Table 16.1 Household Dependency on Agricultural Systems

No.	Name of Watershed	Shifting-cultivation	Shifting Cum Settled Cultivation	Settled Cultivation	Total Number of Households
1.	Wuira Lok	85(34.00)	143(57.20)	22(8.80)	250(100)
2.	Kanlui Lok	55(29.41)	87(46.43)	45(24.06)	187(100)
3.	Ramlikang	165(58.92)	50(17.86)	65(23.21)	280(100)
4.	Yokong Lok	113(84.96)	20(15.04)	–	133(100)
5.	Heitup Lok	99(78.57)	22(17.46)	5(3.97)	126(100)
6.	**Total**	**517(52.97)**	**322(32.99)**	**137(14.04)**	**976(100)**

Source: NIRD Field Surveys.
Note: Figures in parentheses show percentages to respective total.

(34.00 per cent). Both the watersheds are located on Ukhrul-Jasami-Nagaland road and well connected with the state and private transport system. Contact with the outside world and demonstration effect of valley cultivation were two major factors behind the growth of terracing in the two watersheds.

Cropping Pattern in *Jhum* Cultivation

Despite the common perception of obsolete and near stagnant system, the *jhum* system has undergone considerable changes over the last 10–15 years in terms of the production structure and crop-mix. Two of the oldest *jhum* products—millets and cotton are now rarely grown. Incidentally, only two watersheds where these commodities are still grown happen to be where *jhum* dependency is maximum—Yoking Lok and Heitup Lok. The full range of *jhum* products can be described as below:

Foodgrains: Paddy, pulses, maize, millets, rice-bean, cowpea, French bean, green pea and gram. Most of these products except for paddy, maize, millets and gram are new. Thus, it appears that, within the food crops, pulses have been affected positively and a number of new varieties have come up.

Vegetables: Over the last 10–15 years, *jhum* cultivation has undergone a 'Vegetable-Revolution'. From a mere pumpkin, the vegetables now include cabbage, cauliflower, mustard leaf, brinjal, tomato, ginger, chilly, *arvi* (in Heitup Lok only) and potato. It is a 'vegetable revolution' in the sense that in the face of growing shortage of foodgrains, vegetables have become their major strength for consumption as well as for sale. All these crops are sold in the market and income earned is used for covering up for the paddy deficit.

Fruits and Plantation Crops: Papaya, banana, citrus fruits, orange, grapes, guava and tree bean. All these products have also entered the market.

Cash Crops: Cotton, sugarcane and tobacco are main cash crops, which by definition are marketed. Coffee is also grown in most of the watersheds but on kitchen gardening basis and its use is confined to domestic consumption only.

Trends in Area Under Cultivation

Following major points emerge from the interviews with the different sections of population in the watershed area:

1. The 'net' area under *jhum* cultivation has increased continuously over the last 10–15 years and community forest has been the main source of new land. Thus, with the increase in area under *jhum*, the area under forests has declined.
2. Loss in the productivity of land and increase in population have been two major reasons for increase in area under *jhum*.
3. Though the net area under *jhum* has increased, the size of the plot and land holding per family has declined. Currently the average size of plot varies from 0.50 hectare to 0.98 hectares. This average area varies from watershed to watershed and also large variability is found in the smallest as well as the biggest sizes of the plot. The variability is found to be maximum in Heitup Lok (Kuki community) where the largest size of *jhum* plot is six times (1.5 ha) that of the smallest size (0.25 ha).

Impact of Reduction in the Size of Holding

The reduction in the size of holding also means reduced availability of land for agriculture over 10–15 years of time. This has led to:

(*a*) *Emergence of food crisis*: One of the important consequences of reducing size of holdings and declining productivity has been the emergence of food-crises. None of the watersheds was found to be self-sufficient with regard to paddy—the main food crop. Their paddy production is sufficient for them for about 4–5 months of the year. For the rest of the period, they have two sources to cover up for the deficit: (*i*) government supplies under PDS through FPS, and (*ii*) purchase from the market. During the deficit period almost every family buys 2–3 kg of paddy per day from the market as an emergency resort to sort out food crisis.

(*b*) *Reduction in jhum cycle*: Another important impact of reducing plot size and increase in population has been the reduction in the *jhum* cycle to 4–5 years now from the earlier 10–15 years average, which in turn means increase in cropping intensity.

(*c*) *Increase in cropping intensity*: The above changes have led to a change in cropping pattern. The same plot of land is now cultivated for three years before being left as fallow instead of just one year. Moreover the proportion of area under paddy to total area under cultivation seems to be declining. This is clear from the three-yearly cultivation system of a plot. In this three-yearly cultivation system, paddy occupies only 33 per cent of the total area under cultivation. Usually during first year, only paddy is grown, which also means a shift from the mixed to mono-cropping system. In some cases, during the first year pulses are grown, followed by paddy and maize in the second year and vegetables mixed with maize in the third year. Although, apparently, the 'net' land available per family has been reduced, but since the same plot of land is cultivated for three years, this increase in 'gross' availability of land must have off-set the 'net' reduction. The impact of this on the availability of land has to be studied separately and more carefully.

(*d*) *Diversification and Commercialisation*: The change in cropping pattern due to reduced availability of land and more intensive cultivation has ultimately led to a highly diversified production structure within shifting cultivation which has commercial implications. Almost all the vegetables and plantation crops have entered the market though their degree varies from place to place.

- The most important impact of reduced availability of land is the emergence of more permanent use of land in the sense that (*a*) same plot is used for three years instead of one year; and (*b*) longer duration crops are now also grown, especially horticultural crops.
- Over the years under study, the ownership structure and land relations have undergone a lot of changes, especially in the three watersheds of Kanlui Lok, Wuira Lok and Ramlikong, in the favour of a few, affecting the egalitarian structure of society.
- Private ownership (security of tenancy) has increased with terracing and other more permanent uses of land. Since terracing involves investment in land and can be done only by those who are capable of doing so, the benefits of such privatisation can not accrue to the common man. Moreover, due to higher productivity on terraced plots, the disparities within the society has grown in favour of a few.
- Over the years, with the growth in population, fragmentation of holdings has taken place leading to reduction in the plot-size. This reduction in the size of holdings has further implications:

  Terracing became easier and privatisation increased.

  Some of the holdings became uneconomical with selling of (private) land taking place. Thus, due to fragmentation, the number of private landholders has gone up (amount of total land remaining the same) and at the same time selling of land has also increased. Since the number of sellers is larger than number of buyers, the number of families who once enjoyed private ownership of land by inheritance have now become landless (in terms of private holding). This process has led to concentration of the land resources.

  Increased intensity of cultivation, reduction in *jhum* cycle, changes in cropping pattern, commercialisation of *jhum*, increase in wage labour and growth of alternative economic vocations have taken place.

Forest and Shifting Cultivation

Forest plays a major role in the domestic economy of every household. Fuelwood and fodder are two main products of forest used by every household. Besides, timber cutting and selling (Kanlui Lok, Wuira Lok, Yokong Lok and Ramlikong) are also prevalent. Heitup Lok is the only watershed where timber is not being cut and sold (due to general deficiency). Honey, herbal medicines, bamboo, citrus and other fruits including *amla* are the major forest products collected and used and occasionally also sold. In a few instances, fuelwood is also sold (Heitup Lok).

Both male and female are engaged in collecting forest products in their free time and, on an average, they have to travel a distance of 2–4 km for this.

Trends in Deforestation

Table 16.2 shows the relationship between the *jhum* cycle and forest degradation. The area under degraded forest is above 70 per cent in all the watersheds except for in Yokong Lok and Kanlui Lok. While in Yokong Lok, the lower area under degraded forest is because of the longer *jhum* cycle, in Kanlui lok it is mainly because of a larger village area. The study reveals deforestation to be a direct phenomenon related with *jhum*ing, that is, the smaller the *jhum* cycle, the larger is the degradation and vice versa, as can be observed from Table 16.2.

Table 16.2 *Jhum* Cycle and Degradation

Watershed	*Watershed Area under Degraded Forest (%)*	*Length of Jhum Cycle (Years)*
Wuira Lok	72.20	4–5
Kanlui Lok	49.03	3–4
Raml	75.28	4–5
Yokong Lok	09.35	10–12
Heitup Lok	80.07	3–4

Effects of Deforestation: Perception and Response

Communities feel that deforestation has led to an ecological imbalance. Problems like high velocity wind, water scarcity and reduction in the supplies from the forest (basically fuelwood) have emerged due to reduction in forest cover. This realisation has led some of the villages to restore the forests. This has been found in three watersheds, Kanlui Lok, Ramlikong and Wuira Lok. The communities are adopting the following measures:

- Ban on cutting of timber: This has been done through resolutions passed by village authorities. The timber business at one time played a big role in raising the standard of living of people in Kanlui Lok. Now, with the depletion of forests, their standard of living is again sliding back which they are trying to sustain by unauthorised means like collecting 'transport tax', etc., forcibly.
- Privatisation of land for forestry: This is most illustrative in Ramlikong, where after realising the emergence of water scarcity, the village authority (local level people's institution which has the right to decide the land use and its allocation among the villagers) allotted some land to every household for growing trees. Thus, each household is mandated to grow four to five trees around its house. This arrangement has resulted into restoring the water balance within six to seven years of time and resolving the water crisis (Phadang village).
- Social plantation, especially by women societies (Marem village in Wuira Lok and Kharasom cc village in Kanlui Lok).
- Protecting the forests through vigilance by Youth Clubs (Kanlui Lok).

Despite the realisation of the effects of deforestation, the communities feel that the cutting of forest can not be stopped unless alternative income generating activities are given.

ALTERNATIVES TO SHIFTING CULTIVATION

In the face of reducing size of landholding, declining productivity and degradation, following alternatives to *jhum* have been suggested by the communities in different watersheds:

1. Development of wet-terraced cultivation.
2. Soil conservation and horticultural plantation.
3. Livestock development.
4. Development of small and cottage industries based on raw materials available locally like bamboo, cotton etc.
5. Food and fruit processing industries.
6. All the foregoing points should be considered and preceded by (*a*) awareness building among the communities; and (*b*) training for skill development among locally available manpower.

COMMUNITY ACTION PLAN

PRA exercises with the villagers reveal that land use planning has to be the starting point for any programme combining *jhum* control and development. The following are the main results of the PRA exercises regarding the land use planning:

1. Mixed land use, keeping trees in the higher ridges, horticulture crops with half moon terraces in the middle portion and field crops in the lower terraces/contour bunds are ideal. However, this approach has its own limitations under the given land tenure conditions.
2. The terrace risers, which constitute 35–40 per cent of land area can be effectively utilised for raising perennial fodder grasses and legumes. These fodder crops along with other fodder crops raised with horticultural crops and trees as well as the crop residues can provide substantial feed for livestock in addition to helping conservation of soil.
3. Crop planning in the watershed should adapt to proper sequencing of crops as water/moisture availability is more in lower portions than in the upper slopes. Hence, rice cultivation should be avoided in the upper slopes.
4. Cultivation of perennial plantation crops such as rubber, tea and coffee, black pepper, cashew, fruits, etc., should be done depending on agro-climatic and market conditions. In such an approach, food security issue needs to be tackled through appropriate measures.
5. Farming systems approach based on land capability classification would help in optimum utilisation of the soil and water resources for sustainable yields and benefits to the farmers.
6. Frequent demonstrations of new technologies of land use and crop production in both irrigated and unirrigated conditions within the watershed.

7. Regeneration of degraded *jhum* lands through planting of appropriate tree shrubs and plant species. Plantation of horticulture crops and broom grass can also prove helpful.
8. Improvement approach is needed where longer *jhum* cycle is prevalent.
9. Massive education and awareness programmes are required especially with a focus on village heads, clan leaders and other local elected or traditional leaders with a view to influencing local opinion about *jhum*.

The success of these measures will however depend on:

- direct involvement of the farmers in decision-making and execution;
- secured land tenure;
- use of appropriate mix of technologies including soil conservation, water harvesting, topo-sequencing etc; and
- organisational innovations for securing inter-department coordination.

CONCLUDING OBSERVATIONS

Shifting cultivation has been and continues to be the major farming system of the tribal communities in the five selected micro-watersheds of the Ukhrul district of Manipur. Although during the last 15–20 years, there has been a decline in the number of families dependent solely on the *jhum* system, this decline has only been in a relative sense. In absolute sense, the total number of families dependent on *jhum* cultivation today are more than what used to be fifteen years ago, mainly due to population growth.

The population growth during the last two decades has resulted into lower availability of land, reducing plot size per household and reducing *jhum* cycle. This has also led to productivity decline, soil erosion and deforestation as most of the new *jhum* lands are developed by cutting the forest. The tribal communities have shown remarkable adjustment with these negative trends associated with the *jhum* system. As a result, over the years the system of shifting cultivation has undergone significant changes, especially as far as cropping pattern is concerned. At present, there seems to be a trend of using old *jhum* plots for growing cash crops, catering to the markets. This has come out largely as a response to ever growing population, declining productivity and reducing land availability.

Although significant differences are found in the productivity per plot between *jhum* and terraced cultivation, productivity being invariably higher in the terraced cultivation, *jhum* has continued over the years mainly because of three of its main advantages—food security, equity in land distribution and use of land otherwise not suitable to any other form of cultivation.

The major factors behind their sustenance are bigger size of the plots, private ownership and the farmers' capacity to invest. Yet, due to a larger range of crops, also meaning greater food security and some income (as vegetables are now being grown for the markets also), the *jhum* system has sustained over the years.

The process of change from shifting to terraced cultivation has not been very smooth and has affected the land relations having significant bearing on the accessibility of land to the poor and causing managerial problems for the local village level institutions.

Over the years, there have been two changes in this traditional system—with the growth of terracing, land under 'private' ownership (security of tenancy) has increased; and with the emergence of village authority, the chief's discretionary powers have reduced.

In the face of the emerging system, development of terraces with assured irrigation facilities are the top priorities of the communities. This has to be followed by a proper land use planning. The fulfillment of both these priorities is likely to involve some gestation period. In the interim, therefore, supportive livestock schemes, small scale food and fruit processing units, cottage industries based on local raw materials and government food security programmes, especially in the form of wage employment programmes, should be undertaken on a short to medium-term basis.

NOTE

1. The views expressed in the chapter do not necessarily reflect the views of the organisation with which the author is associated.

REFERENCES

International Fund for Agriculture Development (IFAD). 1994. 'Northeastern Community Resource Management Report for Upland Areas'.

Maithani, B.P. 1996. 'Poverty Alleviation through Growth: Policy Options for Northeast', (mimeograph), Guwahati: National Institute of Rural development, Northeastern Regional Centre.

Maithani, B.P. and S.C. Srivastava. 1997. 'Socio-Economic Production System of Tribal Communities—Manipur'. Report submitted to International Fund for Agricultural Development, National Institute of Rural development, Guwahati: Northeastern regional Centre.

Northeast Council Secretariat, 1995. *Basic Statistics*, Shilong.

Singh, S.K. 1994. *People of India, An Introduction*. Volume I. National series. Anthropological Survey of India. Delhi: Oxford University Press.

Srivastava, S. C. 1999. 'Levels and Structure of Development: An Inter-district Study of Northeast India', Hyderabad: National Institute of Rural Development.

Annexure 1 List of Watersheds and Villages Covered

Name of Watershed	*Villages Covered*
Heitup Lok	Jalembung, Shangkai, Lamai Chimpau
Kanlui Lok	Kharasom CC, Kharasom Lacho and New Kharasom
Ramlikong	Phadang
Wuira Lok	Kalhang, Kuirei, Marem
Yokong Lok	Mawai

Section 5

Policy Reforms and Sustainable Development

17

Globalisation and Sustainable Development: Economic and Environmental Conflicts

Prabha Panth

Abstract: This chapter discusses the various areas of conflict between economic and environmental globalisation. Economic globalisation spreads a uniform pattern of development worldwide. At the same time environmental problems are also getting globalised. To tackle this global environmental destruction, global environmental laws and regulations are established that have to be implemented across all countries. This is environmental globalisation. However, the tenets, practices and policies of economic globalisation clash with those of environmental globalisation. In this conflict, economic globalisation prevails and the same environmentally destructive type of development is continuing all over the world. After examining a number of cases, the chapter concludes that there is a mismatch between economic and environmental globalisation and unless economic globalisation is coordinated with environmental globalisation and they work towards a single goal, it will become very difficult to achieve sustainable development worldwide.

INTRODUCTION

The achievement of sustainable development is a global endeavour that requires a massive reorganisation of the world's economic policies, institutions and technology, and can be achieved only if there is coordination and cooperation of all sectors and all countries. It requires a universal effort and cannot be achieved if just one country follows a sustainable path; for the gains to the world environment due to one country's sustainable development will be cancelled out by the environmentally destructive practices of others. On 'Spaceship Earth' all have to sink or swim together.

The move towards sustainable development should involve a gradual change to eco-friendly techniques and commodities, and the phasing out of environmentally destructive ones. In this regard, globalisation policies should put pressure on all countries to slowly abandon their unsustainable economic development and encourage the growth of sustainable practices. Until sustainable practices are an integral part of the developmental plans of world economies, mere environmental tinkering at the edges of economic policy cannot lead to sustainable development. Nor can sustainable development be achieved merely by a single ministry within a country, trying to clean up the environmental mess created by other sectors. Development will continue to be unsustainable, unless the policies of all sectors such as planning, trade, industry, agriculture, etc. integrate sustainable practices with their objectives. As a starting point, environmental conservation, pollution control, etc. could be taken up. But the long-term move towards sustainable development can be realised only when there is a concentrated effort to change the mode of production and consumption to eco-friendly development.

But the globalisation policies are seriously deficient in this outlook. Both at the national and international levels, environmental policies are subservient to economic policy and environmental regulations are often bent or ignored to accommodate economic development. This dichotomy between economic and environmental policies is all the more apparent in the case of globalisation policies. Industrial development and the lifestyles of the 'advanced' countries have become universal with the advent of economic globalisation policies. But the present form of development of the 'advanced' countries is unsustainable and is the chief reason for the environmental degradation of the Earth (Meadows et al. 1972). By universalising this form of production and consumption, environmental degradation is also being globalised, with almost all countries of the world following a common pattern of economic development that is uniformly unsustainable and environmentally destructive.

To counter global environmental destruction, international agencies such as the United Nations Environment Programme (UNEP) are struggling to introduce measures to ensure that all countries of the world follow some general environmental norms and policies to achieve sustainable development. Environmental laws being universal, this can be called as environmental globalisation.

Thus economies of the world are subject to two forms of globalisation:

1. Economic globalisation
2. Environmental globalisation.

While economic globalisation proposed by the World Trade Organisation (WTO) is patently unsustainable, environmental globalisation attempts to direct nations towards sustainable development. But these two forms of globalisation conflict with each other—for many WTO policies affect the environment adversely, while a number of UNEP directives infringe on WTO regulations.[1] So, following the directives of one leads to violating those of the other.

This is the dilemma faced by the developing countries—should they follow the dictates of environmental globalisation or those of economic globalisation? And how should they modify their own national laws to satisfy these contradictory prescriptions? Since they are dependent on aid, technology and other forms of economic help from the international agencies and industrial countries, this variance in the two global polices makes it difficult for them to chose one form of globalisation without breaking the rules of the other. In both cases they face the threats of sanctions. Further, developed countries use both types of globalisation to exploit them, so that the less developed countries are getting the worst of both bargains.

In India also, this conflict between economic and environmental policies has become sharper since the advent of the globalisation and liberalisation era and is mirrored in the conflicts between her economic and environmental regimes.

This chapter examines different types of inconsistencies between the policies of environmental globalisation and economic globalisation in the context of sustainable development, with particular reference to developing countries like India. It points out how economic and environmental globalisation are at odds with each other and suggests that unless the two policies work in tandem, both at the national and international levels, it will be impossible to achieve sustainable development globally.

CONFLICTS BETWEEN ECONOMIC AND ENVIRONMENTAL GLOBALISATION

The need for sustainable development in the world[2] was recognised around the 1980s, at about the same time as the advent of economic globalisation. Various international environmental conferences stressed the need for environmental protection and the need to move towards an alternative form of development based on eco-friendly methods of production and consumption. However, economic globalisation policies are not geared to sustainable development; instead they advocate the universal spread of the same form of unsustainable development, with a few environmental caveats.[3]

Surprisingly, although the United Nations (UN) is the fountainhead for both types of globalisation, it is unable to coordinate and unify its global economic and environmental policies towards the ultimate goal of sustainable development. This is fundamentally due to the fact that the objectives and policies of WTO are in opposition to those of UNEP—for, while the former wants the present mode of industrial development to be spread worldwide, the latter is calling for sustainable development. But more often than not, economic globalisation is usually given greater priority, while environmental rules are put on the back burner, so as not to discourage industrial development. The result—unsustainable development worldwide.

This opposing nature of economic and environmental globalisation policies can be traced to differences in their theoretical base, their priorities and their method of implementation. Since the ideologies of economic and environmental globalisation differ, it is not surprising that their policies and impact also clash with one another. The various reasons for the conflicts between the two forms of globalisation will now be discussed.

Theoretical Conflicts between Economic and Environmental Globalisation

Free markets vs Government controls: Theoretically, economic globalisation is based on 'free markets'. Founded on Neo-classical economic theory, it assumes that economic efficiency can be achieved through free markets at the national level and free trade at the international level. Therefore, it calls for the reduction of all types of governmental controls, both internal and international, to allow the working of a free market.

But environmental degradation is caused by 'market failure'. Environmental economists point out that the free market does not include 'environmental externalities'. Firms and other economic agents do not include environmental and pollution impact costs in their economic decisions so that environmental degradation continues unheeded and uncorrected. Hence, environmental economists argue that it is necessary to introduce severe governmental regulations to ensure that the ambient environment is not affected negatively by the working of free markets. This applies both to local and international markets. Therefore, freeing the market through economic globalisation will only increase the cases of market failure and free trade will freely permit worldwide environmental degradation.

Herein lies the conflict—while economic globalisation wants free markets, environmental globalisation pleads for stricter government intervention and control over the market. So the very theoretical base of economic and environmental globalisation is conflicting—the former wanting more liberalisation of the economy and the latter wanting more government controls. An economy subject to both forms of globalisation will be puzzled about which one of these it has to follow—free markets or government regulation.

Kuznet's Curve: It has been pointed out that some types of pollutants follow a Kuznet's inverse U-shaped relationship with income (Hettige et al. 1997). As the Gross National Product (GNP) of a country increases, its pollution also rises and reaches a maximum. But after a certain level of income, environmental rules become more stringent, environmental regulation increases and pollution emissions fall. Thus according to this theory, pollution in less developed countries is of a transient and self-correcting nature, and as the critical level of income is crossed, pollution (and other environmental damages) will automatically decrease. We can interpret this as follows: since economic globalisation will increase the income of less developed countries, once this crucial income level is crossed, it will automatically cause their levels of pollution to fall. Hence the theory implies that environmental controls are not needed in less developed countries as economic development will itself take care of the environmental problems.

The Kuznet's curve theory assumes that there is no harm if developing countries continue to pollute up to the critical income level, after which the pollution levels will automatically start falling. But there are reasons why this may not hold true:

1. Since the present form of development is unsustainable, it is doubtful if most of the third world countries can even reach this level of income, before the Earth runs out of environmental resources.
2. It ignores the fact that environmental damages are cumulative and irreversible. So by the time (if at all) the less developed countries reach this 'critical' income level, the world environment might be damaged irreparably.
3. Whatever the level of income, governments will not voluntarily impose environmental norms nor will the industrial sector permit them to do so. Therefore, there is nothing automatic about either the imposition of government regulation or of pollution decreasing after a certain level of income.
4. The less developed countries are now forced into an unsustainable development path, which they may not be able to correct even if they cross the 'critical' income level.
5. Finally, there is no universal proof of the Kuznet's curve. The theory is based on only a few pollutants—such as sulphur dioxide and urban SPM. But municipal wastes or carbon dioxide emissions are found to be rising monotonically with PCY (World Development Report 1992).

Therefore it is necessary to have global environmental laws, which stipulate that all countries, whether with high or low incomes, should take protective measures from the present itself.[4] If they do, then a lot of environmental damage can be avoided. The pressure to introduce sustainable development practices necessarily comes from outside—for, although governments can ignore or suppress local people, it will be more difficult to withstand international pressures. We can conclude that the move towards environmental globalisation is due to the fact that countries cannot afford to wait to cross the critical income limit (if any) before undertaking environmental protection, for by then it may be too late, for them as well as for the global environment. They have to take action now, to avoid the environmental pitfalls of industrial development, and to initiate sustainable development as early as possible.

Pollution Havens vs Haloes: The environmental impact of Foreign Direct Investment (FDI) in less developed countries is yet another theoretical area of controversy. According to the supporters of economic globalisation, FDI introduces the latest environmental friendly technologies and commodities and trade regimes that impose stricter environmental regulations on less developed countries. These measures will improve environmental conditions in developing countries. Thus supporters declare that FDI wears a 'pollution halo' and economic globalisation leads to more eco-friendly industrial development (Wheeler 2000).

Opponents of economic globalisation, on the other hand, point out that FDI from developed countries tend to be in 'dirty' industries. Since pollution regulation is stricter in industrialised countries and pollution abatement costs are higher, polluting industries in developed countries tend to relocate in the more environmentally-lenient Third World. So it is argued that economic globalisation has converted less developed countries into 'pollution havens' for the dirty industries of the West, leading to increase in their environmental deterioration.[5] Since the rules of WTO are very stringent, less developed countries cannot use environmental norms to prevent the dirty industries from being set up. Therefore, they argue that economic globalisation has led to the worsening of the environment in less developed countries, while improving the environment of the developed countries.[6] (Low and Yeats 1992; Mani and Wheeler 1997).

The pollution haven vs haloes debate has been raging for some time now, but the bottom line is that the present form of industrial globalisation is itself not directed towards sustainable development. Environmental impact being global, the world's population suffers regardless of who pollutes or where the dirty industries are located. If pollution havens are being established, then the world suffers through the destruction of the environment of less developed countries. Similarly, if pollution haloes exist, then the developed countries retain the polluting industries, which again destroy the global environment. At the global level, it does not matter where the polluting industry (such as a nuclear reactor) is set up—ultimately if one part of the world suffers, then the entire world suffers—and global development will be unsustainable.

Policy Conflicts between Economic and Environmental Globalisation

The policies flowing out of the divergent theories of economic and environmental globalisation are naturally conflicting. It is well recognised that the policies of WTO frequently violate the UNEP and Multilateral Environmental Agreements (MEAs). Not only are the two forms of globalisation working against each other but they are also biased favourably towards the advanced countries and work against the interests of the developing countries. The industrial countries are quick to use the environmental and the economic globalisation regulations against the developing countries to gain economic advantages from them. A few such cases are discussed below:

Discrimination: Although WTO does not prevent national governments from taxing environmentally harmful goods, these are restricted only to domestic products and processes, and 'characteristics' of imported products. WTO does not permit the application of national environmental standards to production processes of the exporting countries.

However, environmental regulation does require discrimination between eco-friendly and unfriendly methods of products and production (Sampson 2000). Only if the importing country can use environmentally safe goods and techniques, can it hope to introduce sustainable development. But this violates the WTO rationale of avoiding all types of 'discrimination', so as to prevent countries from giving any type of special trading advantages to others, or discriminate against a particular product because of the manner in which it was produced. Less developed countries are now caught in a cleft stick: if they ignore international environmental regulations, then environmental sanctions will fall on their exports affecting their foreign exchange earnings. If, however, they want to comply with international environmental standards, it will increase their costs of production and make their products uncompetitive in the world market.

Meanwhile, industrial countries are quietly sending their own environmentally hazardous products and industries to developing countries, which are too weak to prevent this form of exploitation.

Trade Sanctions: Industrial countries have played the environmental card to ban many products of developing countries. For example—garments from India that used azo-dyes were banned in Europe, as also prawns and fish imports in USA on the grounds that 'turtle excluding' devices were not used. Less developed countries are badly affected by such bans as, for example, the ban of ivory trade that has affected many African nations. Here environmental globalisation works against economic globalisation.

But advanced countries are getting the benefits of both economic and environmental globalisation. They are powerful enough to continue selling environmentally unsafe and banned products to developing countries as well as establishing banned industries therein. For instance, many banned pesticides are being exported to less developed countries and most of the chemical industries have also been shifted from developed countries.[7] The difference in the trade related environmental controls by developed as against developing countries is most glaring in the case of Genetically Modified Organisations (GMOs). Powerful Multi National Companies (MNCs) like Monsanto not only resist all efforts by the importing countries against GMOs but also threaten them with trade sanctions if they refuse to import and use them.[8]

MEAs and WTO: The WTO rules on trade run counter to the UNEP Multilateral Environmental Agreements. For example, although USA was still not a party to the Cartagena Biosafety Protocol, (not having ratified the UN Convention on Biological Diversity), it still threatened to ban the imports from Thailand in 2001 if Thailand introduced laws requiring labelling of GE food products. It also forced the Thai government to raise the contamination threshold of GMOs from 3 per cent to 5 per cent, leaving serious loopholes in the labelling law. The following year South Korea came under similar pressure. The US government tried to force the South Korean government to raise the threshold from 3 per cent to 5 per cent in its labelling laws. Sri Lanka and Bolivia are other less developed countries that have faced the wrath of USA for banning imports of GE foods (Greenpeace 2003). There are many other examples of such coercion by advanced countries against less developed countries (Jha 1999; Perez 2002).

Trade Related Intellectual Property Rights (TRIPS): Another feature of the WTO that is environmentally unsustainable is the TRIPS agreement that defines how products can be protected from piracy. TRIPS aims to prevent imitation of products and protects those who have invented, discovered or introduced them. This discrimination is obvious when it is noticed that action is taken against less developed countries that break the TRIPS laws, especially of the powerful TNCs, while the same rule is not applied when firms from developed countries plunder the products of less developed countries. Many less developed countries object to this bias in TRIPS and its partisan spirit towards the rich nations (Oxfam: Policy Paper 2001).

TRIPS laws were broken by firms from developed countries, who tried to patent products and processes known for generations to indigenous people of some developing countries. TNCs of the western countries particularly USA, applied for patents for *Neem*, turmeric and *Basmati* from India and Pakistan and jasmine rice from Thailand. Only after expensive litigation against the powerful US firms, could they be freed of the patents. Although the Cartegena Protocol on Biodiversity involves protection of the species and processes of indigenous communities, and their takeover by foreign firms, the developed countries state this to be against the WTO's policy of free and unrestricted trade. This illustrates a case where an environmental law—the conservation of bio-diversity—is opposed by the free trade laws of WTO.[9]

There are many such cases, where the WTO policies work contrary to the global environmental policies. The result is that neither the economic nor the environmental globalisation policies have been able to ensure environmental protection, in the short-run, or direct the world economies towards sustainable development, in the long-run.

Conflict between World Financial Bodies and Environmental Globalisation

World bodies like the World Bank (WB), the International Monetary Fund (IMF), etc. are the key lenders to less developed countries and responsible for ushering economic globalisation. But their economic policies violate environmental globalisation and they have been criticised for sponsoring environmentally damaging projects, at the behest of powerful transnational companies and countries, that force them to perpetuate their own environmentally damaging type of development. At the same time, increasing pressures from world communities and affected countries to conserve the environment is forcing them to temper their loans with environmental norms. In the conflict, these financial bodies are still leaning towards unsustainable development, while paying lip service to environmental protection. Therefore, the lending policies of the global financial institutions often contravene UNEP's environmental norms.

The World Bank: A report by WB's Inspection Panel concluded that its staff violated many of its own policies, including those on resettlement of indigenous people, environmental assessment and project supervision. The environmental consequences of its programmes have been severe. Much of the Bank's $22 billion annual lending to projects and programmes are in environmentally sensitive areas such as energy, agriculture and transport. The Bank also failed to achieve its stated goals of poverty reduction and 'sustainable development.' Furthermore, the WB's 'environmental' lending often serves as little more than a camouflage for other environmentally destructive projects. The Bank's support to the highly environmentally destructive Narmada Sagar Sarovar Project in India is well known. It withdrew only after worldwide protests were made against its environmental and social impact.

According to the Bank's official report in 2003, villagers in the state of Jharkhand, Eastern India suffered harm to their livelihoods as a result of a World Bank-backed coal-mining project. The Environmental Management Plan for the Parej East mine had suggested that only about half of the 253-hectare mine area could be reclaimed for agricultural land after mining, while the rest would be filled up with water. But the villagers were not warned that the water would be poisoned by contact with coal and other sediments (Down to Earth 2004).

After facing immense flak for its neglect of environmental matters when granting loans, especially in less developed countries, the WB has since the past decade, implemented a few environmental regulations in its agenda. For instance, an Environmental Assessment Report is now to be submitted for projects financed by it. However, the Bank managed to get out of environmental embarrassments by stating that since its loans were given to financial intermediaries, it was up to them to ensure environmental compliance in the projects they undertook. Now, the Bank is also offering a few environmental training programmes, etc. But the WB's projects themselves being unsustainable, its environmental regulations are far too little to negate the larger environmental destruction of its projects.

The IMF: The IMF's policy aims to correct short-term balance of payment problems without regard to the long-term impact of such measures. The IMF, it has been found, lacks the expertise to deal with social and environmental issues of the widely differing economic situations in each country. The IMF's continual disregard for people and the environment in borrowing countries has undermined the foundations of sustainable development in many countries.

The Global Environment Facility (GEF): An international financial mechanism to fund projects intended to benefit the global environment, it is largely managed by the WB. As such, GEF projects suffer from the same problems as the Bank's regular lending operations. The GEF follows a top-down, ineffective approach of dealing with problems related to climate change, biodiversity, international waters and ozone depletion. The WB requires that many GEF projects be attached to its large regular loans, which often are at odds with protecting the global environment. Also, the USA being the chief donor, dictates policies with strings attached that will help its own economy and trade (Greenpeace website). The contributions from developed countries to the GEF have been falling over the years—USA's contribution fell from $7.23 million in 1999 to $6.5 million in 2001. UK, Germany, Finland, Sweden, Norway and Denmark have also reduced their contributions. The fund is used to finance environmental projects in less developed countries. This decrease in contributions shows that the top industrial nations are losing interest in promoting sustainable development in developing countries (UNEP Annual Report 2001).

The few cases cited above show that the move towards sustainable development is dogged by inconsistencies in the economic and environmental globalisation policies. While economic globalisation is being implemented with gusto, environmental globalisation is still unable to control environmental degradation on a large scale, nor is it able to direct global development towards a sustainable path.

Further, the advanced countries use both economic and environmental globalisation rules as it suits them to exploit the developing countries.

GLOBALISATION AND SUSTAINABLE DEVELOPMENT IN INDIA

In the case of India also, the same conflicts dog the economic and environmental globalisation of the country. Economic globalisation policies are oriented towards unsustainable development and dominate and even violate environmental globalisation policies. The economic policy of India, after economic globalisation, has not been remodelled towards sustainable development, but only includes some environmental mopping up of the damages created by industrial development. Environmental policy in India is too weak to counter the unsustainable development introduced by the WTO regime. Some of the areas of conflict are now discussed.

History of Globalisation in India

Although environmental globalisation was initiated much before economic globalisation in India (1972 and 1990 respectively), her economic policies do not reflect a move towards sustainable development. There is very little coordination between her economic and environmental regulations and economic policies are still being given greater priority over environmental ones.

History of Economic Globalisation

India initiated economic globalisation in the 1990s due to her massive international debt, foreign exchange crises and a stagnant economy. The usual regulations of free trade, liberalisation and privatisation of the economy were imposed on India as well. Therefore, economic globalisation recommendations were included in the national economic policies. As such, the strict control over the economy, that was a feature of the planning era in India, was replaced by a more liberal economy.

History of Environmental Globalisation: Environmental protection was enshrined in the Indian Constitution[10] much before the advent of economic globalisation. Also India participated in the first UN Conference on the Human Environment, Stockholm, 1972, and so environmental globalisation in India preceded economic globalisation by almost two decades. Following the UNEP's environmental governance and its recommendations, the Government of India set up a Department of Environment and initiated legislation to control environmental pollution. Various Environmental Acts were put into force in the 1970 and 1980, much before the economic globalisation and liberalisation policy was started.

Since environmental globalisation came first, we would expect that environmental protection and sustainable development would be strong ingredients of India's economic globalisation policies as well. But that is not so for here again environmental controls are often flouted; priority is given to economic globalisation, encouraging an unsustainable type of economic development.

Conflicts in Economic and Environmental Organisation and Laws

Just as the UN is unable to coordinate the working of its economic and environmental branches, the Government of India is also unable to coordinate the economic and environmental globalisation of the country and there is a conflict between the two. While the departments of Trade, Industry, Agriculture, Planning and Finance have jumped headlong into industrial development, with just a few environmental rules, the Environmental Ministry, under the duress of the UNEP, the International Environmental Agreements as well as local pressures, is vainly trying to implement stricter environmental controls through government regulation. But the economic sectors being more powerful have succeeded in adopting economic globalisation with its environmentally harmful pattern of development, disregarding environmental norms (Panth 1997).

Environmental Laws in India: Although environmental protection directives were enshrined in the Constitution, environmental regulation in India was taken up seriously only after the Stockholm Conference, with the passing of many Environmental Acts and the creation of the Ministry of Environment and Forest (MoEF) and the Pollution Control Boards (PCBs).

Unfortunately, environmental laws are 'soft laws,' mere guidelines, so that taking punitive action against defaulters is difficult. For example, though industrial pollution is continuing unabatedly, the onus is on the victims to sue the polluters. Since the victims are usually the poor and helpless, with little knowledge of the law, and the polluters are powerful industrialists, not all the culprits are caught or punished for their environmental misbehaviour. Although recently the PCBs have been given the authority to close down polluting units, lack of staff, inefficiency, corruption and threats, make it difficult for them to cope with the powerful forces behind the polluting firms.

The same lack of authority makes other environmental laws such as the Forest Act, the Wildlife Act, etc., ineffective. Economic incentives override environmental laws, with government, foreign and private sector's industrial and development activities given more priority than environmental protection. But deregulation of commercial activity under economic globalisation should not mean dilution of environmental standards. The enforcement of environmental standards must be regarded as a fundamental part of the agenda of economic reforms. Unfortunately, this has not been the case (Sudarshan 1996).

State governments, especially, are flouting environmental norms in a bid to attract foreign investment under economic globalisation. For example, in 1995, the Rajasthan Govt. issued a directive to denotify forest lands, to allow mining by local and foreign mining companies, thus defying the Forest Conservation Act. The famous Narayana Sarovar Sanctuary of Gujarat was forcibly denotified at the instance of a Japanese investor who wanted to set up a highly polluting cement factory there (Kothari 1998).

A WB study (1993) put the value of environmental damages in India at about US $9.7 billion per year, or 4.5 per cent of GDP of 1992. However, the VIIth Plan's allocation to Science, Technology and Environment was a low 1.4 per cent of total plan outlay, which increased to 2.1 per cent in the VIIIth Plan and 3 per cent in the IXth Plan. Since three sectors share this amount, the percentage share to Environment is even lower. The low budgetary allocation to the environmental sector, compared to the higher environmental damage cost, shows the low priority given by the Central Government to environmental conservation. This is in spite of the fact that data analysis shows that environmental degradation is increasing faster than GDP. According to the estimates made by the Centre for Science and Environment (CSE), while GDP increased 2.5 times between 1975 and 1995, industrial pollution quadrupled and vehicular pollution increased 8 times (Down To Earth 1999).

Conflicts in Policy in India

Economic vs Environmental Globalisation Policy: The New Economic Policy (NEP) introduced in India at the WB's directive, does not give much prominence to environmental protection except for a few guidelines. These are not substantial enough to ensure environmental protection. Nor is the NEP directed towards sustainable development; on the contrary it has opened the doors to the influx of unsustainable and environmentally dangerous techniques and goods from the industrial countries. A few of the environmental stipulations in the NEP are given below:

Environmental Impact Assessment (EIA): Though compulsory for both Indian and foreign firms with investment up to $12.5 million, it is often disregarded. Though 29 industries of Schedule I require environmental clearance from the Central Government, there is pressure on the MoEF to reduce the number of industries and to raise the limit to $37.5 million in power and other infrastructure industries. These are some of the most environmentally destructive activities. For other projects, there is a pressure to raise the limit to $25 million and clearance is being asked for from State PCBs rather than the central PCB. But state governments, anxious to develop and attract investment, try to dilute the EIA still further (MoEF Annual Report 2002). Therefore, environmental clearance of projects has now become easier. While this facilitates economic globalisation, it creates greater environmental damages and encourages an unsustainable form of development by watering down environmental clearance rules.

Industrial Location: The NEP permits the setting up of polluting industries without environmental clearance (except for those subject to compulsory licensing) within 25 km of the periphery of cities with population less than 1 million, and in other prior designated industrial areas. This regulation has extended to the areas of pollution near small cities like Ratlam in Madhya Pradesh Small-scale industries are exempted from locational restrictions, except those subject to local land use and zoning laws. The result is that these areas have developed into pollution hotspots—such as Patencheru near Hyderabad and the Ankaleshwar belt or the so-called 'Golden Corridor' in Gujarat, which are highly polluted, with excessive concentrations of toxic chemicals, in air and water.

Fiscal Incentives: Five-year tax holiday and other incentives are offered to industrial units set up in backward regions of India, especially the Northeast regions. But now the levels of pollution and natural resources degradation in the backward regions of Mizoram and Assam have increased tremendously.

Patents: The Patent rules under TRIPS also discriminate against less developed countries. India's native plant resources such as *Neem*, *Basmati*, wheat and turmeric were on the verge of being patented by US firms. But at the same time, as said earlier, India cannot prevent GMOs and GM technology, which are known to be environmentally dangerous, from being imported.

Conflicts in Practice

If sustainable development has to be achieved, then the New Economic Policy should have encouraged environmentally friendly products and processes through trade, privatisation, etc. Instead, it is permitting trade and production of more polluting and damaging products, many of which have been banned in the developed countries. We shall now look at some examples of how economic practices are working in contravention to both national and global environmental protection and safety regulations.

Free Trade: Free trade is now rife in many banned products—with many threatened plant and animal species, such as wild orchids removed from the list of banned exports. A general category of plants, plant portions and derivatives obtained from the wild, were also removed from the negative lists of exports (Kothari 1998). This is a breach of the Plant Protection Varieties rules of the MoEF and encourages foreign control over our indigenous species.

Toxic wastes imports are allowed under the Free Trade Regime, which violates the Basel Protocol on Trade in Hazardous Wastes. According to Greenpeace, hundreds of tonnes of plastics and metals such as lead, copper and other wastes are coming into India from countries like Australia, Canada, UK and USA, apparently for recycling. An Indian company, Futura Industries of Tamil Nadu, has imported 10,000 metric tonnes of plastic wastes since 1992, out of which only 30–40 per cent could be recycled. In 1992–93, imports of lead acid battery wastes from Australia increased nearly three-folds from 126,000 kg to 346,000 kg (Kothari 1998).

FDI: Out of the thirty-five high priority areas in which FDI is allowed, at least ten are in environmentally harmful industries such as chemicals, fertilisers, pesticides, pharmaceuticals, cement, and paper and pulp industries. Only two are in environmentally friendly areas—alternate energy systems and bio-insecticides. Bulk drug units were relocated into India, after USA banned their local production in 1985. Now bulk drug factories are the main causes of pollution in the hotspots of Patencheru and Jeedimetla in Hyderabad, due to their uncontrolled and untreated effluents.

The share of dirty industries in total FDI was 51 per cent in 1991–2000. Of these 27.4 per cent was in Energy, 4.5 per cent Chemicals, 7.5 per cent Transport, 5.5 per cent Metallurgy and 3.5 per cent in Food Processing—all classified as Red or most polluting industries, while Hotels and Tourism having 1.7 per cent and Textiles 1.2 per cent, came under Orange industries (CPCB Annual Report 2003). Also the largest inflow to approvals of foreign investment was in the dirty industries chiefly chemicals with 37.7 per cent. (SIA [FDI Data Cell] 2003).

The above data suggests that India may be turning into a 'pollution-haven' for dirty industries. Other writers have also found empirical evidence that does point to a 'pollution-haven' effect in India (Bhattacharya 2002; Mani and Wheeler 1997). Bhattacharya (2003) has shown that there has been a rise of dirty products in exports compared to imports, since the start of the globalisation process in India, implying a 'haven' effect.

De-licensing: Many goods produced by the public sector were de-licensed to allow private and foreign firms to produce them. These include polluting industries such as mining, power production, chemicals, etc. Though the law suggests that foreign firms should obtain licences, the NEP does not prevent them from establishing hazardous industries in India. For example, Hindustan Lever Limited, a branch of Unilever—a US-Dutch company, started a mercury thermometer factory in the pure environs of Kodaikanal after its main factory was closed down in USA. The thermometers were made for export. The resulting level of mercury poisoning was almost 250 times the permitted limits in this region and adversely affected the environment (Greenpeace website).

Choice of Goods: The NEP does not prevent the production of commodities banned by International Organisations such as the World Health Organisation (WHO) and UN. For example, foreign companies like Bayer of Germany and DOW of USA are producing pesticides banned by the POPs UNEP 1995 treaty[11] (*India Environment Issues* 2001). But POPs and trade in hazardous substances are banned under India's environmental policy.

While the State Pollution Control Boards (SPCBs) are struggling to find ways to control vehicular pollution in metro regions, Euro II vehicles, that are now passé in the West due to the introduction of Euro IV norms, are being imported and produced here. Nor does the NEP insist that only green vehicles be produced and imported, but is permitting diesel and petrol vehicles, thus missing a chance to shift to a more sustainable development path. This has resulted in the same type of urban pollution that characterises Western development.

Although foreign firms should obtain environmental clearance by showing their Environmental Impact Statements, these are usually scuttled through and they are allowed to start production. A case in point is that of Coca Cola, which gave a wrong EIA about its water requirements at Kasargod, Kerala. But its total water usage has depleted the groundwater resulting in water scarcity and drought conditions in the district (Down to Earth 2004).

Energy Projects and GHG emissions: At a time when there is increasing pressure from countries like USA on India and China to control their CO_2 emissions,[12] thermal power production by foreign firms through captive coal mines has been given priority in India. Out of the total FDI approved in 1991–2000, 27.4 per cent were in commercial energy, of which 15.4 per cent were in power and 12 per cent in oil refineries (SIA [FDI Data Cell] 2003). Power production through fossil fuels, contravenes the Kyoto Protocol on GHG that calls for the phasing out of fossil fuels and a shift to renewables. Renewable energy is still extremely expensive and if the same amount of subsidies given to fossil fuels are given to it, then there is a chance to shift to a more sustainable type of development. But the economic globalisation policy does not encourage green products such as—renewable over non-renewable energy, or organic over chemicals, or biodegradable over non-biodegradable plastics.

Agriculture: Economic globalisation is converting agricultural lands for export-oriented crops, which are again meant for urban and foreign markets. The food security of the country is threatened with the patent regime, which tends to usurp the common knowledge base of the people, with the appropriation of the seeds, patents on bio-diversity, people's knowledge and the monopolisation of the agribusiness in the hands of a few conglomerates from the developed nations.

Agricultural production for export has led to monoculture, for example in Punjab, *Basmati* for exports has replaced hundreds of other rice varieties. Pepsicola's tomato sauce requires only a particular variety of tomato, which has replaced all other local varieties. Such conditions lead to loss of biodiversity and loss of generic species, in the name of economic globalisation.

Land and Coastal Regulations: There is a move to change the land acquisition act with corresponding changes in the laws regarding displacement and rehabilitation. The purpose is to make land more easily available for industrial companies under globalisation. A bill has been passed by both the houses of Parliament to change the coastal regulation zone to facilitate environmentally hazardous aquaculture farming. Aquaculture, while earning foreign exchange, affects the coastal ecosystems, due to monoculture and the use of pesticides. A National Environmental Engineering Research Institute (NEERI) study showed that aquaculture led to a loss of Rs 142 crores due to fall in rice production, pollution of salt pans, wage losses to farmers and of local fisheries in Tamil Nadu in 1996 (Kothari 1998). It has also resulted in heavy pollution and degradation of pristine shorelines and lakes such as Pulicat and Chilkur lakes on the East Coast.

Foreign Industries Environmental Records

According to the 'pollution haloes' thesis, foreign firms bring with them stricter environmental regulations and follow the strict environmental rules of their home countries. But the record of foreign firms belies this theory.

The classic case is that of the Union Carbide factory in Bhopal, which not only did not adhere to the environmental norms of India, but also grossly violated international safety measures for hazardous substances. Coupled with that, the compensation paid to the victims of MIT gas was meagre—only $470 million compared to the $3.5 billion that was asked by the Indian government. Recently the DOW Company purchased Union Carbide but refused to shoulder its liabilities. On the contrary, the company was ready to sue the victims for staging a peaceful protest in Mumbai (Greenpeace website).

In the factory premises in Bhopal, twenty-one dangerous chemical wastes continue to seep into the ground water. The chemical stockpile of 2000 mt remains in the compound, affecting the lives of about five residential areas surrounding the factory (Greenpeace 1999). DOW Chemicals that has bought out Union Carbide, refused to either pay compensation or clean up the wastes. It was only after continued struggle that the US courts ordered DOW to cleanse the area of toxic wastes in July 2004, twenty years after the tragedy. Even then the Madhya Pradesh (MP) government, dragged its feet till almost the last day, until there was huge public outcry against the delay (Greenpeace website). This case shows the environmental callousness of large MNCs and the ineffectiveness of local Indian government rules to control them. In spite of this gigantic industrial tragedy, the economic policy in India still does not give weightage to industrial and environmental safety measures.

The Soft Drinks Case: The giant MNCs—Coke and Pepsi, finding loopholes in the Indian environmental laws on safe levels of pesticides in food, did not remove dangerous pesticide residues from their bottled drinks in their Indian factories. This is contrary to the claims of the pollution haloes thesis that states that the environmental performance of international companies is carried over to the host countries. In that case, there should have been no pesticides in the colas and bottled water of TNCs. But the American and European standards for pesticides in food items (which are set at zero levels) were not followed. Samples tested by the Centre for Science and Environment, New Delhi, found that the levels of pesticides in many of the samples of bottled water from TNCs exceeded EEC standards—nearly twenty-three times for Aquafina of Pepsi, to more nearly 109 times for Kinley of Coca-Cola (Down to Earth 2003). Lindane, a deadly insecticide, was thirty-five times more than EEC norms in Coca-Cola samples, while DDT was about sixteen times higher in Pepsi-Cola. High levels of other pesticides were also found in the samples. But samples of the two drinks from USA showed no trace of the pesticides[13] (Down to Earth 2003).

Other foreign investors breaking environmental regulations in India include DOW chemicals, Atochem, Kumaia, Bayer, Mitsubishi, etc. These companies have set up their agro-chemical plants in India to produce banned or restricted products such as pesticides, without proper environmental safeguards. (Jha 1999) Due to their powerful international links, pressures from donor countries like the USA, European countries, as well as the WB and IMF, such firms are let off the hook. The Indian government is too weak to deal with such violators and it is up to the NGOs, local agitations and the Green Bench to bring them to book. All this shows that economic globalisation policy in India is not only breaking environmental rules, but also not introducing or initiating sustainable development in the country.

Conflict within States

Many state governments are diluting environmental norms to attract FDI. They prefer economic development over environmental protection and are blatantly contravening environmental rules to attract foreign and private investment. Jha (1999) calls this the 'race to the bottom effect' in India. A number of cases can be citied:

Haryana, constituted a high-power committee to take spot decisions on foreign investments, NRI projects and 100 per cent export-oriented projects and to obtain environmental clearance from the PCB in 15 days. This contravenes the environmental impact assessment rules that require three months of deliberations before environmental clearance can be given to any project.

Punjab has also set up a committee to provide land 'off the shelf' and for clearance of FDI projects within 24 hours of submission of the proposal. Kerala has set up a green channel to expedite environmental clearances, while Rajastan has reduced the number of industries to be inspected from fifteen to three, for both Indian and foreign companies (Kothari 1998).

In Orissa, the Bhittar Kanika Sanctuary, the largest nesting place in the world for the endangered Olive Ridley turtles, was truncated to accommodate foreign trawling jetties and yards (Kothari 1998).

The Andhra Pradesh Government has taken over tribal lands and given mining leases on them to about fifteen mining companies, both Indian and foreign, in the ecologically fragile Eastern Ghats region.

Du Pont-Thapar

This joint venture project was envisaged to build a chemical factory in Goa to produce Nylon 6.6. This is a highly hazardous industry, which would have affected the rich ecology of the Goan beaches. But the government gave land at zero cost as well as other infrastructure. Only after the locals protested, did the firm shift to Tamil Nadu, where economic incentives were given to it. Thus state governments' urge for industrial growth has led them to bend many environmental considerations and MoEF rules.

CONCLUSION

Theoretical and practical conflicts between economic and environmental globalisation has led to the domination of the former over the latter. Although sustainable development has to be speedily implemented, there is little indication that the economic globalisation policies, worldwide as well as in India, are directed towards its achievement. Globalisation has not led to a uniform consensus on ways of achieving sustainable development. Instead, economic globalisation is forcefully implementing unsustainable development methods worldwide. Environmental globalisation and environmental measures introduced in economic globalisation are only end-of-the-pipeline cleanups. There is no move to have a quantum shift towards green, sustainable and environmentally friendly methods of production globally and to phase out the environmentally destructive mode of development. Instead, environmental policy is used (if at all) only to mop up the damages caused by the present form of development rather than as a means to change the mode of development towards sustainable methods.

With the awareness of the environmental crisis and with the knowledge that the entire world would be adversely affected if the world environment collapses, globalisation could have been the key element in the introduction of sustainable practices everywhere. But short-term and vested interests of powerful TNCs and industrial countries have made certain that there is no digression to a sustainable path, which would affect their own profits. Environmental globalisation, trying vainly to chip away the edges of environmental destruction worldwide, is unable to direct the world economies towards sustainable development.

Since sustainable technology and commodities are not available in all sectors and need to be evolved, both economic and environmental policies should unite to encourage the growth of, and research in alternative methods of production and consumption that are environmentally safe. But both economic and environmental globalisation do not show this perspective. Unless both economic and environmental globalisation policies recognise the fact that a global effort is needed to undertake the move towards sustainable development, short-term economic interests will always dominate and it will be impossible to achieve sustainable development.

India, as other less developed countries, is in a quandary about whether to follow economic or environmental globalisation. If it wants to follow the former, then it will end up with massive environmental destruction; and if it wants to follow the latter, it needs monetary, technological and other types of help to implement environmental protection. Also since alternative techniques are still not available in the majority of economic activities and industries, the environmentally unsafe path still has to be followed. This earns the condemnation of both types of globalisation—the economic globalisation group faulting India for not following WTO and WB rules, and the environmental globalisation band for continuing on an environmentally destructive path and contributing to global environmental degradation.

Difficult choices have to be made. If economic globalisation is used as a platform for introducing environmental globalisation, then it is possible to move towards an environmentally sustainable path. If the same amount of subsidies, incentives, coercion, research as well as free trade, that is directed at present to unsustainable development, is instead directed to green industries and products, then there is hope that the environmentally destructive path of the present globalisation regime will be averted and sustainable development will become a reality.

NOTES

1. For example, while 27.4 per cent of total FDI approved in India between 1991–2002 was in the energy sector, including power and oil refineries, the negotiations on climate change call for the phasing out of fossil fuels to reduce GHGs.
2. Sustainable development was first mentioned in the WCS of 1983, and later popularised at the WCED by Brundtland in his book *Our Common Future*, 1987.
3. Although the WTO has a section on trade and environment and holds many meetings, such as the Symposium on Trade and Environment, to discuss the sustainable nature of trade policies, it is not able to achieve any accord with environmental regulations nor is it able to construct a trade and liberalisation policy that can lead to sustainable development (Voon 2000; Simpson 2000).
4. For example, USA and other developed countries in Kyoto objected to GHG emission reductions on the grounds that any efforts by the industrial countries to reduce CO_2 would be cancelled out by the pollution from the developing countries. They asked for GHG cuts by countries like India and China also, which are nowhere near the 'critical income level.'
5. Larry Summers, US Treasury Secretary, is quoted to have suggested that more high-polluting industries should move to developing countries, as people there do not live so long anyway, so multinationals would have to pay out less in compensation for any deaths caused by their pollution. http:// www.greenparty.org.uk

6. Actually there have been a number of instances where polluting industries and goods have been shifted to developing countries including India, which are quoted in this chapter.
7. For instance the growth of chemical industrial units in Jeedimetla, Patencheru and Bolarum in Hyderabad followed the ban of production of process chemicals in USA.
8. USA, the largest producer of GMOs, dominated by the giant TNC—Monsanto (responsible for 91 per cent of the GE crops grown in the world), is forcing many countries to accept them under the shield of WTO. Although the European Union and many developing countries refused to import GMOs, in 2003 the US government filed a formal complaint in the WTO against EU, for its de facto moratorium on GMO imports (Greenpeace 2003).
9. But the WTO feels that environmental implications affecting free trade and its own rules are negligible since there only about 20 out of 500 environmental rules restricting trade on environmental grounds—such as trade in ODS, hazardous wastes, endangered species. The larger implications of trade on the environment are ignored [WTO website].
10. The Indian Constitution enjoins the 'States to take measures to protect and improve the environment and to safeguard the forests and wildlife of the country.' It also states that 'it is the fundamental duty of every citizen to protect and improve the natural environment, including forests, lakes and rivers, and wildlife, and to have ecological compassion for the living creatures.'
11. These include Monocrotophos, methyl parathion and five other highly toxic and banned pesticides, even though they have to be phased out under the POPs UNEP treaty of 1995.
12. India and China rank 5th and 2nd in world's emissions of carbon dioxide. *World Resources*, 1997–98.
13. The irony of this situation is that many farmers in India are now using the colas to kill pests effectively and economically in their fields. Farmers in the Durg, Rajnandgaon and Dhamtari districts of Chhattisgarh say they have successfully used Pepsi and Coke to protect their rice plantations against pests and it reduces the costs of pesticides by Rs 55–60 per acre. [BBC news.2004.http://news.bbc.co.uk/2/hi/south_asia/3977351.stm]

REFERENCES

Annual Report 2001. UNEP.
———. 2002. Ministry of Environment and Forests, New Delhi.
———. 2003. CPCB, New Delhi.
Bhattacharya, A. 2002. 'Environmental Technology Diffusion in a FDI Regime: A Developing Country Perspective', CSI Conference, International Institutions and Multinational Enterprises: Global Markets, Global Players. Austria.
Boulding, K. 1966. 'The Coming of the Space Ship Earth,' in H. Jarrett (ed.), *Environmental Quality in a Growing Economy*. Baltimore: John Hopkins University Press.
Down to Earth, 1999. 'When Health is not Wealth', 3 January.
———. 2000. 'Global Environmental Governance', 15 March.
———. 2004. 'No Consensus', 15 June.
Energy Information Administration. 'India: Environmental Issues. http://www.eia.doe.gov.
Greenpeace. 2003. *The US War on Bio safety; Renewed Aggression by a Rogue State*. http://www.greenpeace.org.
Hettige, Hemamala, Muthukumara, Mani and David Wheeler. 1997. 'Industrial Pollution In Economic Development: Kuznets Revisited'. Development Research Group. *World Bank Papers*.
Jha, Veena. 1999. 'Investment Liberalisation and Environmental Protection: Conflicts and Compatibilities in the Case of India'. UNCTAD/DICM Project Report on Cross Border Environmental Management in Transnational Corporations.
Kothari, A. 1998. *Alternate Economic Survey, 1991–96. Seven Years of Structural Adjustment*. Delhi: Rainbow Publishers Ltd.
Low, P. and A.Yeats. 1992. 'Do Dirty Industries Migrate?', *International Trade and Environment*. Discussion Paper No. 159, Washington D.C: World Bank.
Mani, M. and D. Wheeler 1997. 'In Search of Pollution Havens? Dirty Industries in the World Economy 1960–95'. *World Bank Papers*.
Meadows, D. et al. 1972. *The Limits to Growth*. New York: Universe Books.
Panth, Prabha. 1997. 'The Structural Adjustment Programme and Sustainable Development in India', in K. Azam's *Economic Liberalisation in India, Implications for Indo-US relations*, Hyderabad: Delta Publishing House.

Perez, Oren. 2002. 'The Many Faces of the Trade-Environment Conflict: Some Lessons for the Constitutionalization Project'. *European Integration Online Papers.*

Sampson, Gary P. 2000. 'Trade, the Environment, and the WTO: A Policy Agenda'. *Overseas Development Council Policy,* Essay No. 5.

Shaman, David. 1996. 'India's Pollution Regulatory Structure and Background'. *World Bank Papers.*

SIA (FDI Data Cell). 2003. Department of Industrial Policy & Promotion. Ministry of Commerce & Industry.

Sudarshan, R. 1996. 'Better Environment & Natural Resource Management'. Discussion, India International Centre, New Delhi.

Wheeler, David. 2000. 'Racing to the Bottom? Foreign Investment and Air Quality in Developing Countries'. Development Research Group. *World Bank Papers.*

World Development Report 1992, Delhi: Oxford.

18

Environmental Protection: The Role of Regulatory System in India

P.M. Prasad

Abstract: India incorporated environmental protection measures in the Constitution to implement decisions taken at international conventions and conferences. In addition, the Bhopal Gas Tragedy necessitated for the Government of India to enact a comprehensive environmental legislation to mitigate externalities. On the basis of these, the Indian Parliament enacted the Environmental Protection Act, 1986. This is an umbrella legislation that consolidated the provisions of the Water Act of 1974 and the Air Act of 1981. Within the framework of the legislation(s), India established Pollution Control Boards (PCBs) in order to prevent, control and abate environmental pollution. The focus of this chapter is to evaluate the functioning of PCBs (in particular the State Pollution Control Board of Andhra Pradesh and the Central Pollution Control Board, New Delhi) in relation to the prevention of externalities. The analysis of the chapter is based on primary as well as secondary data. The data has been collected from published and unpublished records. Based on the review of these documents, a questionnaire was prepared to obtain the opinion of the officials of PCBs on the functioning of the Boards. The primary emphasis was on the degree to which the objective of improving environmental quality in India has been fulfilled. The inferences drawn from the empirical analysis were then critically evaluated in the light of the theory of regulatory system. This provides insights into the effectiveness of providing incentives to polluters to take precautionary measures. The study reveals that the regulatory system is unable to improve the environmental quality effectively and efficiently because of an increase in its responsibilities, and the absence of deterrence mechanisms within the PCBs for imposing fines against rogue industries.

INTRODUCTION

Environmental measures to regulate emissions of air and water pollution are important because of the limitations of market induced correctives. In addition, the liability system in India is unable to improve the environmental quality in the country because of informational asymmetries with respect to scientific knowledge, legal delays and poor monitoring of compliance (Prasad 2004). For these reasons, the Government of India established PCBs (regulatory system) at both the Central and State levels. Unfortunately, environmental degradation persists even after three decades of regulatory oversight. This necessitates an evaluation of the *ex-ante* approach in terms of internalisation as well as prevention of pollution externalities.

The state plays a major role both in the formulation as well as the enforcement of laws. According to Ogus (1994), there are four different degrees of state intervention: (1) regulation of information, (2) standards, (3) licensing, and (4) price controls. *Licensing* is the highest degree of state intervention because the firm has to take prior approval from the regulatory agency to market the product. However, state intervention also involves:

- administrative costs;[1]
- compliance costs;[2] and
- indirect costs.[3]

Regulatory agencies, under certain circumstances, also erect barriers[4] which in turn impose costs on citizens.

This is an *ex-ante* approach where parties pay a fine after violating regulatory standards and sometimes even before harm has occurred. Standards are defined by the state, which also plays a major role in the enforcement of laws. On the other hand, liability is an *ex-post* approach, where parties pay damages after the harm has occurred. Under this approach, courts set the due level of care based on the nature and the facts of the case, if harm occurs.

The theory of liability versus regulation reveals that both the systems have their advantages[5] and disadvantages[6] in providing incentives to the tortfeasor to take precautionary measures to reduce the risk of harm. Neither, however, protects citizens perfectly. Therefore, an optimal mix of regulatory and liability systems is required to internalise the externalities. They are substitutable[7] as well as complementary.[8]

It should be noted that in the case of joint use of liability and regulation, the regulatory system sets minimum standards. In addition, it adopts the probabilistic method to test the established standards. It is uncertain whether a tortfeasor can be found violating standards under this approach. Similarly, the court system may be unable to provide an incentive to the tortfeasor to reduce the risk of harm because of no case having been filed against him. Under these circumstances, a minimum regulatory standard may perhaps reduce the severity of the risk of harm. Once, the regulatory agency formulates the optimal standards, then the courts shall resolve the conflicts between the *ex-ante* and *ex-post* approaches. Thus, the optimal mix of liability and regulation should provide incentives to the parties to take precautionary measures to reduce the risk of harm.

This chapter tries to focus on the critical evaluation of the functioning of the Pollution Control Boards (regulatory system) in terms of the prevention of environmental degradation in India.

ENVIRONMENTAL PROTECTION SYSTEM IN INDIA

Environmental Laws

The Indian Constitution provides for power sharing between the federal and state governments. Parliament has the power to legislate for the whole country, while the State Legislatures are empowered to make laws only for their respective territorial jurisdictions. Under Article 246 of the Constitution, the subject areas of legislation are divided between the Union and the States into three lists: Union, State and Concurrent list. Central law prevails over a State law in the concurrent list,[9] however, State law prevails if it has received Presidential Assent. The Constitution also provides that the Centre may enact laws on State list, after receiving consent from the respective states.[10]

After the 1972 UN Conference on Environment and Human Development at Stockholm, the Indian government amended the Indian Constitution and adopted Articles 48A,[11] 51A(g)[12] and 253.[13] On the basis of these Articles, the Indian Parliament enacted the Prevention and Control of Pollution Act, 1981 (Air Act) and the Environmental Protection Act of 1986.

An outline of the environmental legislation(s) in India is given below:

The Water Act of 1974 (Amendment, 1988)

This is the first law passed in India whose objective was to ensure that the domestic and industrial pollutants are not discharged into rivers and lakes without adequate treatment. The reason is that such a discharge renders the water unsuitable as a source of drinking water and for irrigation and support of marine life.

In order to achieve its objective, PCBs at Central and State levels were created to establish and enforce standards for factories discharging pollutants into bodies of water. The State Boards are empowered to issue Consent for Establishment (CFE) whenever a firm wanted to establish a new factory and also issue Consent for Operation (CFO) for existing factories. They were also given the authority to close factories or, in the case of disconnecting power and water supply, issue directions to the concerned departments for enforcement of the Boards' standards.

The Water Act[14] provided for civil and criminal penalties for the violation of its provisions. It imposes the penalty with fine up to Rs 10,000 or with imprisonment (minimum of three months) up to six years or with both. In the case of continuous violation (after first conviction), it imposes an additional fine up to Rs 5,000 every day. In addition, Section 41 of the Act imposes, with fine, imprisonment up to seven years if the violation continues for one year.

The Air Act of 1981 (Amendment, 1987)

The objective of the Air Act of 1981 was to control and reduce air pollution. The working of this Act and the enforcement mechanisms are similar to that of Water Act. What was novel is that the Act also called for the abatement of noise pollution. The Air Act[15] provided for civil and criminal penalties for the violation of its provisions. It imposes the penalty with fine up to Rs 10,000 or with imprisonment (minimum of three months) up to six years or with both. In the case of continuous violation (after first conviction), it imposes with additional fine up to Rs 5,000 every day. In addition, Section 37 of the Act imposes, with fine, imprisonment up to seven years if the violation continues for one year.

Environmental Protection Act, 1986 (The EP Act)

The objective of the EP Act is to protect and improve the environment in the country. It is an umbrella legislation that consolidated the provisions of the Air and the Water Act. It were the environmental disasters[16] that prodded the Indian Government into passing comprehensive environmental legislation, including rules relating to storing, handling and use of hazardous waste.

The EP Act empowered the Indian Government to make necessary rules and regulations to fulfil its objectives. It is under this Act and its rules that government takes all necessary steps such as the formulation of national environmental standards, to prescribe procedures for managing hazardous substances, to regulate industrial locations, to establish safeguards for preventing accidents and to collect and disseminate information regarding environmental pollution. It also empowered the Government to set up parallel regulatory agencies to protect parts of the environment and to delegate its powers to such an agency. For example, the Government could set up an agency to protect coastal resources.

The Act[17] provided for civil and criminal penalties for the violation of its provisions. It imposes the penalty with fine up to Rs 1,00,000 or with imprisonment up to five years or with both. In the case of continuous violation (after first conviction), it imposes an additional fine up to Rs 5,000 every day. In addition, Section 37 of the Act imposes with fine, imprisonment up to seven years if the violation continues for one year.

The Water (Section 60), the Air (Section 52) and the EP (Section 24) Acts reveal overriding effects.

The Public Liability Insurance Act, 1991

The focus of this Act was to provide for the payment of immediate compensation to the victims of industrial accidents.

Environmental Protection Rules, 1986

The Rules of 1986 empowers the formulation of standards for emission of environmental pollutants. In general, the Rules were formulated by the Government of India for working and conduct of business under the Environment (Protection) Act, 1986. The formulated rules are: the Hazardous Waste (Management and Handling) Rules of 1989,[18] the Public Insurance Act of 1991 (Amendment, 1992),[19] and Biomedical Waste (Management and Handling) Rules of 1998[20] etc.

Enforcement of Environmental Laws

The established environmental rules and regulations are enforced by the concerned administrative authorities. In addition, they act upon the direction of the courts and PCBs. Thus, both the *ex-post* and *ex-ante* approaches are playing an active role in improvement of environmental quality in the country. The PCBs, in particular, try to prevent environmental degradation through formulation of standards, issuance of consents for establishment and operation and closure orders to rogue industries.

REGULATORY SYSTEM IN INDIA

The PCBs are a two-tier system, that is, the Central Pollution Control Board (CPCB) at the central level and the State Pollution Control Boards (SPCBs) at the state level.

Water Boards were established under the provisions of the Water Act of 1974[21] in order to prevent water pollution. The Boards later received the additional responsibility to control air pollution under the provisions of the Air Act of 1981. The Water Boards were then renamed as PCBs under the provisions of the Environmental Protection Act of 1986. The responsibilities of PCBs increased with the adoption of environmental protection rules in the context of prevention of water pollution, supervision of hazardous wastes, implementation of court directions, etc.

Central Pollution Control Board (CPCB)

The Central Pollution Control Board was first established in September 1974 under the provisions of the Water Act to promote cleanliness of streams and wells in India. It got additional responsibilities in terms of prevention and controlling of air pollution under the provisions of the Air Act. In its structure, the CPCB consists of twelve members[22] and six zonal offices.[23]

Functional and Structural Aspects of CPCB

The CPCB[24] as a nodal agency, tries to promote cleanliness of surface and ground water; to prevent, control and abate air pollution; to advise central government in the matters of prevention of water and air pollution; to coordinate activities of states and settle disputes; to direct and provide assistance to State Boards in prevention of water and air pollution; to formulate minimum national standards; to recognise laboratories for the analysis of samples; to submit expert reports based on the directions of the Court; and to promote research, training and dissemination of information about the prevention of water and air pollution.

Activities of CPCB

Standard Formulations: The CPCB formulates pollution standards for industries under the provisions of the Water and the Air Acts. These standards are called 'the Minimum National Standards (MINAS)' for liquid effluents and air emissions. They are approved and notified by the Ministry of Environment and Forests.[25] As a matter of fact, the State Pollution Control Boards are empowered to impose stringent standards which could be over and above the MINAS. The development of MINAS by CPCB from 1990–91 to 1998–99, can be presented with the help Table 18.1.

Table 18.1 Development of Standards: CPCB

Year(s)	*Developed*	*Developing*	*Initiation*	*Total*
1990–91	11	11	8	30
1991–92	9	8	8	25
1992–93	12	10	17	39
1993–94	3	11	11	25
1994–95	7	11	10	28
1995–96	17	10	8	35
1996–97	8	5	5	18
1997–98	6	6	6	18
1998–99	11	2	2	15

Source: Compiled from the Annual Reports, CPCB.

Table 18.1 reveals that the Board has developed eighty-four MINAS from 1990–91 to 1998–99. It has also developed thirty-seven and thirty-one categories of industrial effluent and emission standards respectively. This is in addition to ambient air, ambient noise, automobile and fuels quality specifications for petrol and diesel.

Effluent Treatment Plants (ETPs): The CPCB promotes Common Effluent Treatment Plants (CETPs) in clusters of Small-Scale Industries (SSIs) because of the fact that SSIs may have financial constraints, lack of space and installation of small effluent treatment plants at their respective units may not be viable. In addition to this, the CPCB monitors the working of the established CETPs.

Eco-labelling: Rapid industrialisation and urbanisation and changes in the production and consumption patterns may generate negative externalities. In such a situation, the activities of regulatory agencies alone are inadequate to internalise the externalities. Thus, there is a need for proactive and promotional role by the manufacturers and the consumers to prevent environmental pollution. The Eco-mark scheme signals to the consumers that the product is eco-friendly. It also provides incentives to the manufacturers to adopt green technology to produce eco-friendly products.

The Scheme on Eco-labelling, which is voluntary in nature, was launched by the Government of India in 1991 to encourage consumers to use environment friendly goods and to achieve sustainable development. The technical committee,[26] with the help of product specific subcommittees,[27] finalises the guidelines of Eco-mark for various product categories. The Ministry of Environment and Forests then notifies them in the Gazette. In fact, the guidelines encompass from the extraction of raw material for manufacturing of a product to disposal of the used product by the consumer. The Eco-mark label is awarded to consumer goods that satisfy the specified environmental criteria and the quality requirements of Indian standards. Since 1991, the Eco-mark criteria have been finalised and notified for sixteen product categories.[28]

Hazardous Waste Management: The waste generated by households, hospitals, industries and their improper disposal creates health hazards. The Hazardous Waste Rules, 1989 under the EP Act and the EP Rules direct that hazardous waste disposal sites have to be designed and managed in such a way that no harmful substances reach the biosphere and hydrosphere in an unacceptable quantity.[29] The Board has issued directions to SPCBs and District Pollution Control Committees (DPCCs) to monitor the steps taken by the municipalities for the prevention, control and abatement of pollution due to land filling.

Recognition of Labs: The Government of India gave powers to the CPCB to recognise environmental laboratories and analysts as Government analysts.[30] Since its inception up to February 2001, the CPCB has recognised forty-four laboratories in the country. In addition, the World Bank project on Industrial Pollution Control strengthened the laboratories of the PCBs.[31]

CPCB and Its Human Capital: The CPCB manpower stood at 373 by the end of March 1999. The administration got the maximum number of personnel (218) as compared to the technical and scientific personnel (155). Figure 18.1 explains the status of the Human Capital from 1990–91 to 1998–99.

Figure 18.1 CPCB's Human Capital

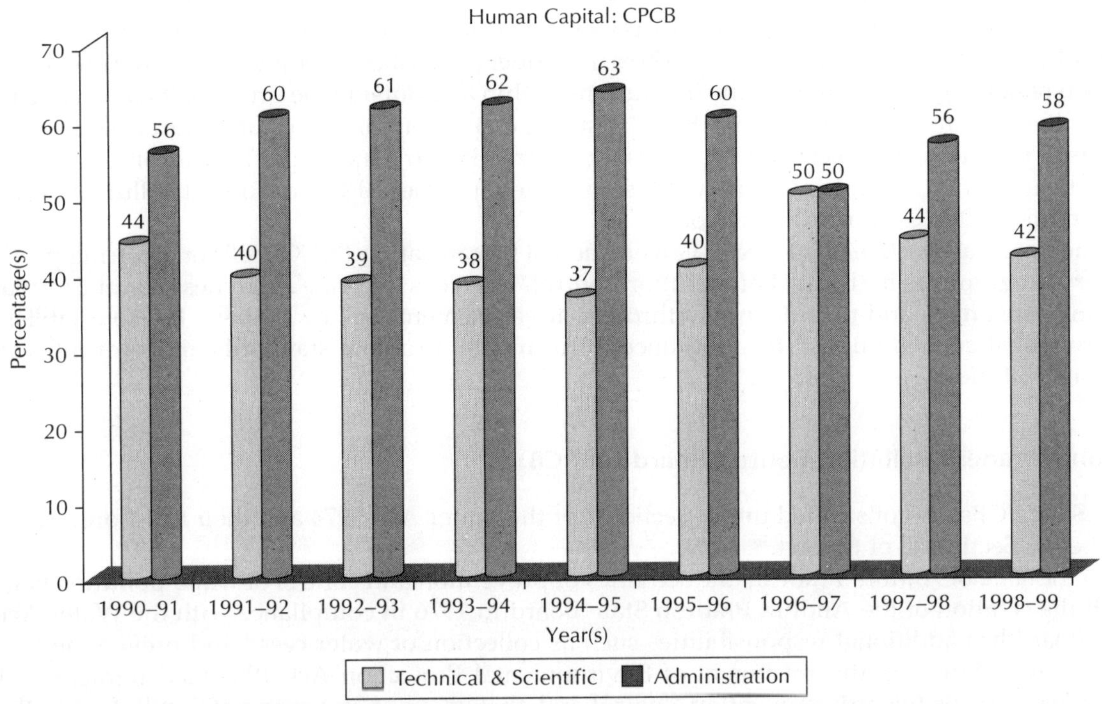

Source: Compiled from the Annual Reports, CPCB.

Usually, the CPCB requires more scientific and technical personnel than administrative personnel to perform its functions. Figure 18.1 reveals that the administrative personnel were higher in number than the scientific and technical personnel except in the year 1996–97. This is one indication that the Board is unable to fulfil its objectives. Moreover, the head of the CPCB is a civil service employee.

CPCB and Its Money Capital: The Ministry of Environment and Forest provides grants to CPCB to meet day-to-day expenses. In addition to this, the CPCB raises financial resources by carrying out various projects. On the other hand, the Board is spending its money in the forms of revenue and project expenditure while keeping some revenue in bank deposits too. Its financial structure can be explained with the help of Figure 18.2.

Figure 18.2 shows that the grant from the MoEF, over the years,[32] has been reduced from eighty per cent to thirty per cent of the total receipts. In fact, the resources generated by carrying out projects went up to fifty per cent of the total receipts. In case of payments, the CPCB has reduced its revenue expenditure as well as revenue product expenditure. However, it is keeping its resources in the form of bank deposits which varies from twenty to forty per cent of the total payments.[33]

The CPCB being a regulatory agency is more prone to regulatory capture. In addition, uncertainty prevails over its resource generation. This may be one of the reasons why the CPCB wishes to keep some of its money in the form of bank deposits.

Promotional activities: The CPCB chooses its members from different fields[34] in order to achieve widespread acceptability of its standards. The Board conducts workshops, seminars and even encourages its personnel to go for training at the institutions within and outside the country. It has established the Pollution Information Centre in 1994 to conduct exhibitions and to create awareness about the status of pollution, its effects and measures for control. It also handles public grievances and even investigates serious complaints, otherwise refers to the concerned SPCBs or District Pollution Control Committees (DPCCs) for further action.

The Board also established a Non-Governmental Organisation (NGO) cell for ensuring participatory programmes in the field of pollution control. Its role is to create awareness about Eco-mark among consumers and manufacturers through advertisements in newspapers. It also establishes international collaborations[35] for assistance to formulate, promote standards and prevent water and air pollution.

Andhra Pradesh Pollution Control Board (APPCB)

The State PCBs are constituted under Section 4 of the Water Act, 1974 and their functions are prescribed in Section 17 of the Act.[36]

In the state of Andhra Pradesh, the activities of prevention and control of water pollution began with the creation of the Andhra Pradesh State Board in 1976 in compliance with the Water Act.[37] The Board has additional responsibilities such as collection of water cess[38] and prevention of air pollution.[39] Moreover, the enactment of Environmental Protection Act, 1986 has augmented the activities of PCBs towards prevention, control and abatement of environmental pollution in their respective states. The APPCB consists of nine members[40] with five zonal[41] and seventeen regional[42] offices. The members have to meet at least once in every three months in order to deal with the requirements of the APPCB.[43] The APPCB is a two-tier system. The first tier consists of its Chairman, Member Secretary and other members (not exceeding 15). All are nominated by the State Government of Andhra Pradesh. The second one consists of appointed regular staff who run the day-to-day activities of the Board. The organisational aspects of the Board can be explained with the help of Figure 18.3.

Functional and Structural Aspects of APPCB

The main functions of the APPCB are: maintaining and restoring the wholesomeness of water, prevention and control of air pollution, formulation of standards in consultation with CPCB, advising

Figure 18.2 CPCB's Money Capital

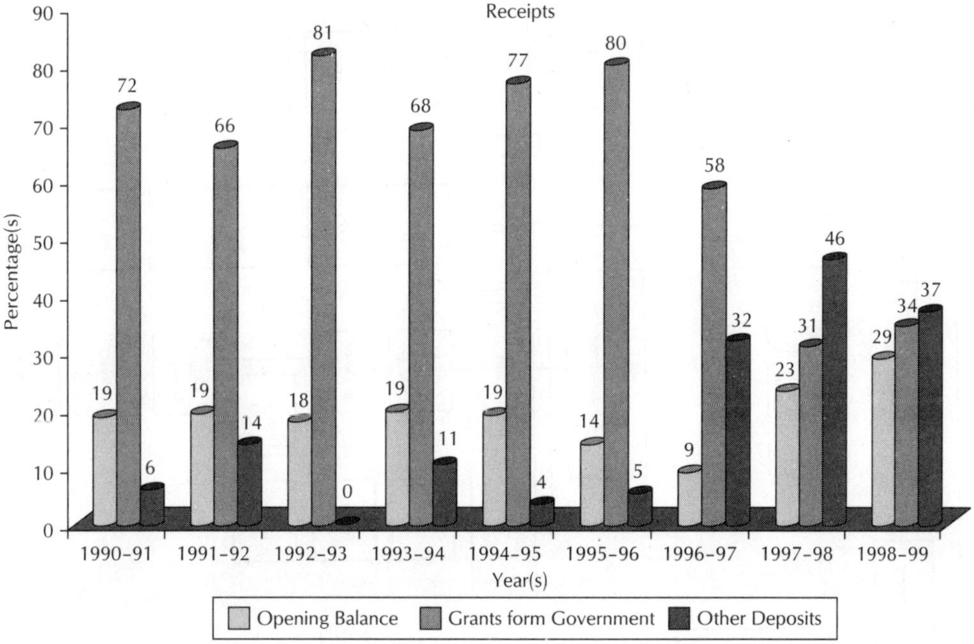

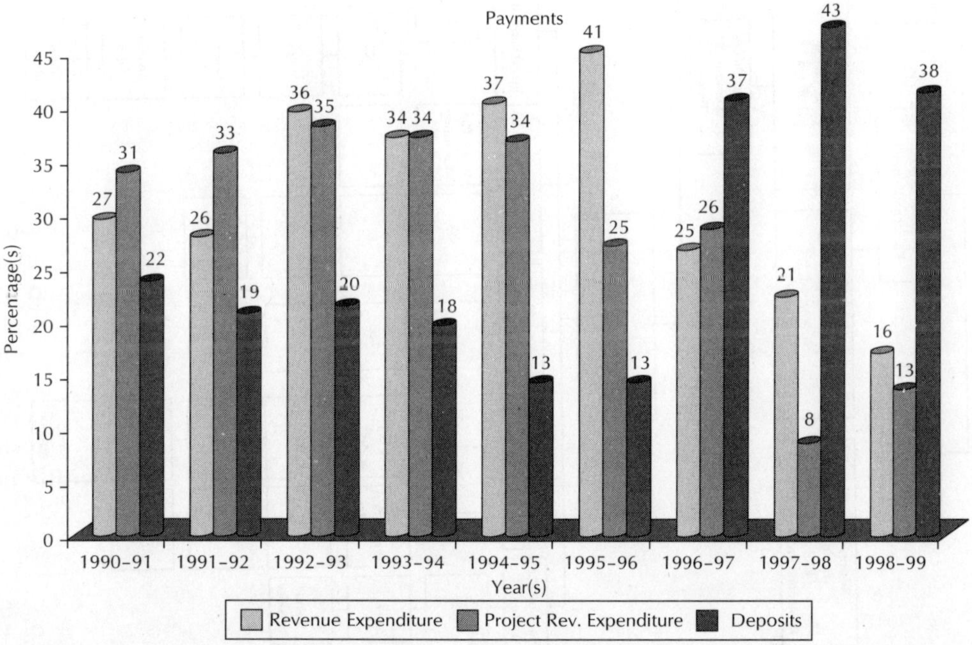

Source: Compiled from the Annual Reports, CPCB.

Figure 18.3 Organisation of the APPCB

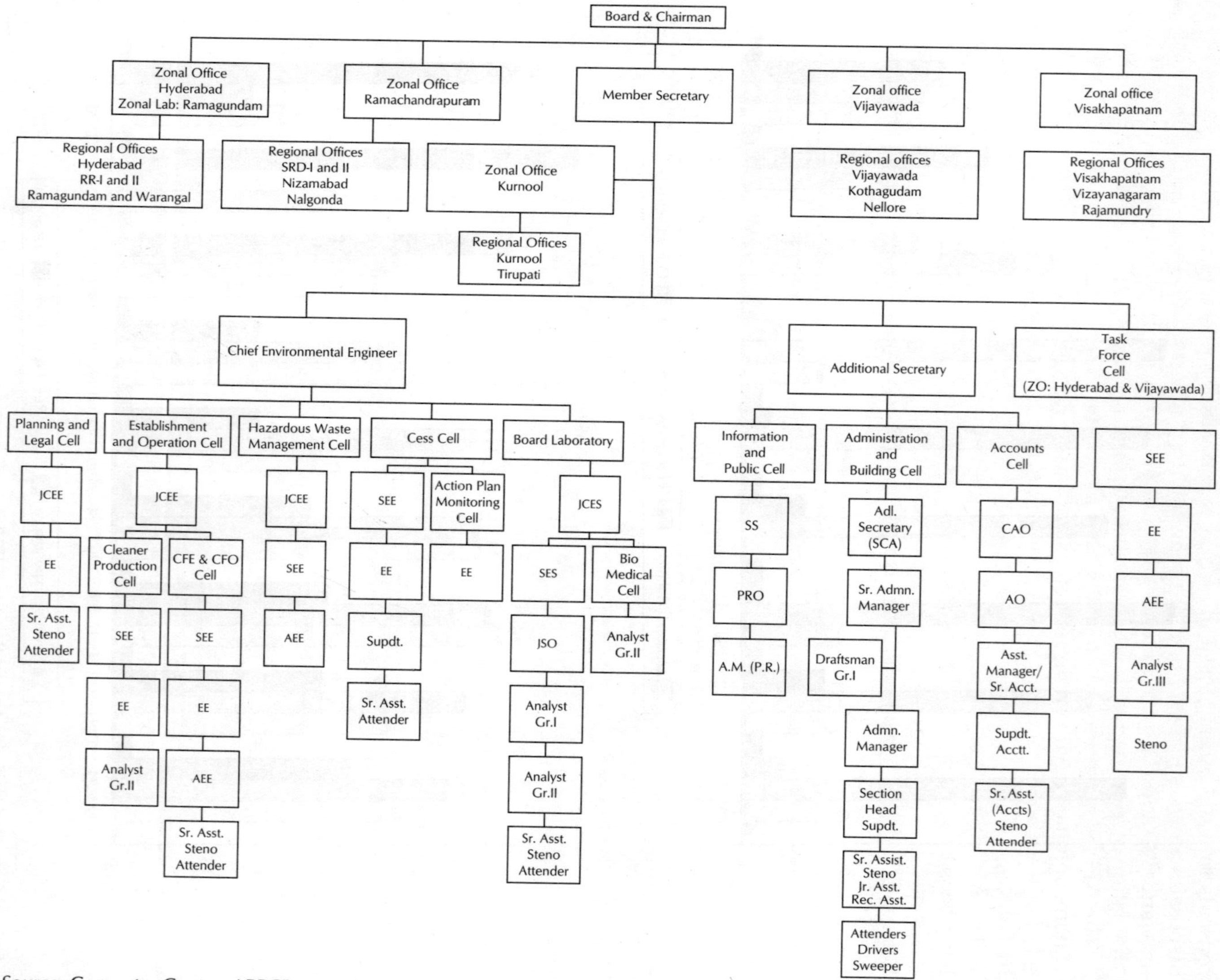

Source: Computer Centre, APPCB.

the state government in the location of industry and prevention of environmental pollution, issuing consents for establishment and operation to industries, collecting water cess, establishing waste disposal management system, controlling the improper use of consents, recognising laboratories for standardisation and environmental quality control and promoting research, training and dissemination of information in the interests of citizens.

The APPCB is empowered to collect samples[44] from industry and issue closure orders[45] in case of non-compliance with environmental standards.[46] The Water and Air Acts spell out an elaborate set of powers on inspection, regulation and punishment relating to the violation of standards fixed by Boards.

The APPCB have earmarked the environmentally sensitive areas (hot-spots) in the state. They have been marked in the map (Figure 18.4). Their focus is to prevent further degradation of the environment at these hot-spots (Figure 18.4). One can expect the PCBs to try and reduce the number of hot-spots. If that is not possible, they can at least try to limit the expansion of hot-spots in future.

Figure 18.4 Environmental Hot-Spots of Andhra Pradesh

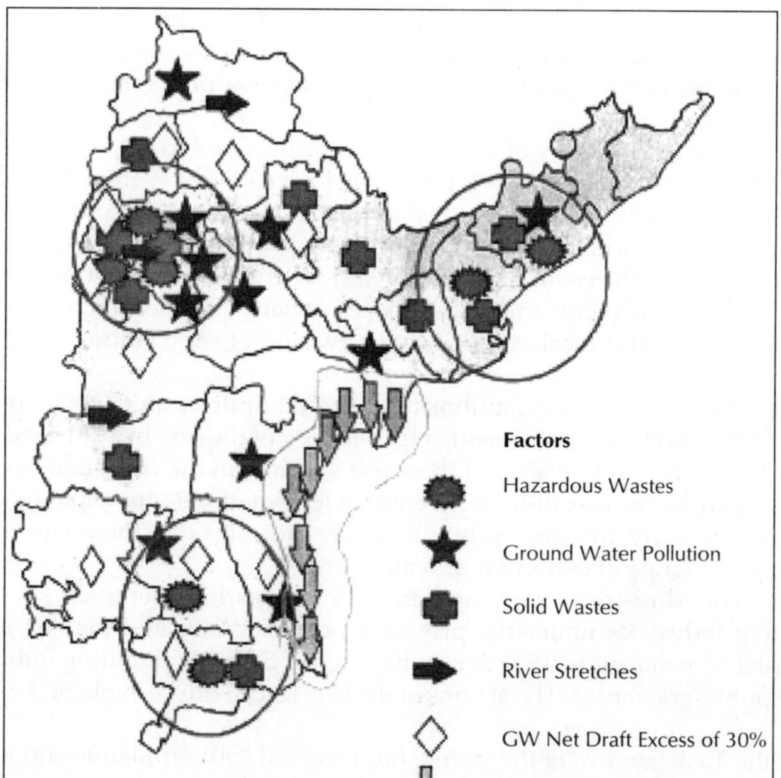

Source: Annual Report, APPCB, 1998–99, p. 2.

The Activities of APPCB

Consent for Establishment (CFE) and Consent for Operation (CFO): The industries must get CFE at the time of establishment of a new plant. Similarly, the existing industries have to get CFO[47] in order to continue with their activities. Industries have to pay a fee to obtain APPCB's consents.[48]

The APPCB grants consent to establish a new plant and may grant it subject to conditions[49] that are open for public scrutiny. These consents are reviewed once in every two years with one condition for renewal being the fulfilment of the previous years' consent conditions. In the case of issuing consent for operations, there are three categories of industries—*red*,[50] *orange*[51] and *green*.[52] The *red* category industries will be the highly polluting, the *green* industries will be the least polluting and the *orange* category of industries fall in between the red and the green. The APPCB has to record the reasons for refusal of consent in writing and issue notices to industrialists that are denied the consent. The procedural aspects of the issuance of consents by the APPCB can be explained with the help of the chart as in Figure 18.5. The issuance of consent for operation under the air and the water acts seems to be much more time consuming than the issuance of consent for establishment (Figure 18.5).

The consents issued by the Board from 1991–92 to 1999–2000 can be explained with the help of the graph as in Figure 18.6. It indicates that in the case of the APPCB, consents issued for establishment and for operation accounted for twenty-five and seventy-five per cent, respectively, of the total 12,375 consents issued.

Water Cess: The Member Secretary of the APPCB is empowered to assess and collect water cess[53] on behalf of the Government of India (GoI) under the provisions of Water Cess Act.[54] The cess is to be levied for the quantity of the water consumed by specified industries[55] and local authorities.[56] The industries as well local authorities can get a rebate of up to twenty-five percent of the cess payable if they follow certain procedures and standards laid by the Government of India.[57] The APPCB charges interest[58] for late payment, and also imposes penalty[59] for non-payment. It will take legal action[60] against industries and local authorities for evasion of cess, failure to furnish returns and false returns.

The number of industries and local authorities covered under water cess increased from 398 in 1990–91 to 919 in 1999–2000. The assessment and collection of the cess by the board can be explained with the help of Figure 18.7. It reveals that there is a shortfall in the collection of cess except in the year 1997–98. This may be because only about eight to ten out of 112 Municipalities/Municipal Corporations pay cess regularly and promptly.[61] It clearly indicates that the Board is lenient towards the local authorities in terms of collection of water cess.

Closure Orders: The Ministry of Environment and Forests has identified seventeen categories of highly polluting industries under the provisions of the Water and the Air Acts. The APPCB established a Task Force in August 1995 to monitor the problematic polluting industries and to provide a forum for public grievances. The actions of the task force[62] can be explained with the help of the Figure 18.8.

It shows that the Task Force over the years[63] has received 950 complaints and issued 186 closure orders. It initiates legal hearings, with complainants and technical experts, before taking any decision on closure of industry. The APPCB, however, is not empowered to impose fines on non-compliant industries. It has to either issue directions (such as closure, prohibition and stoppage of water and electricity services) or file a case in the court against these industries and wait for the court verdict.

Figure 18.5 Consents Procedure of APPCB

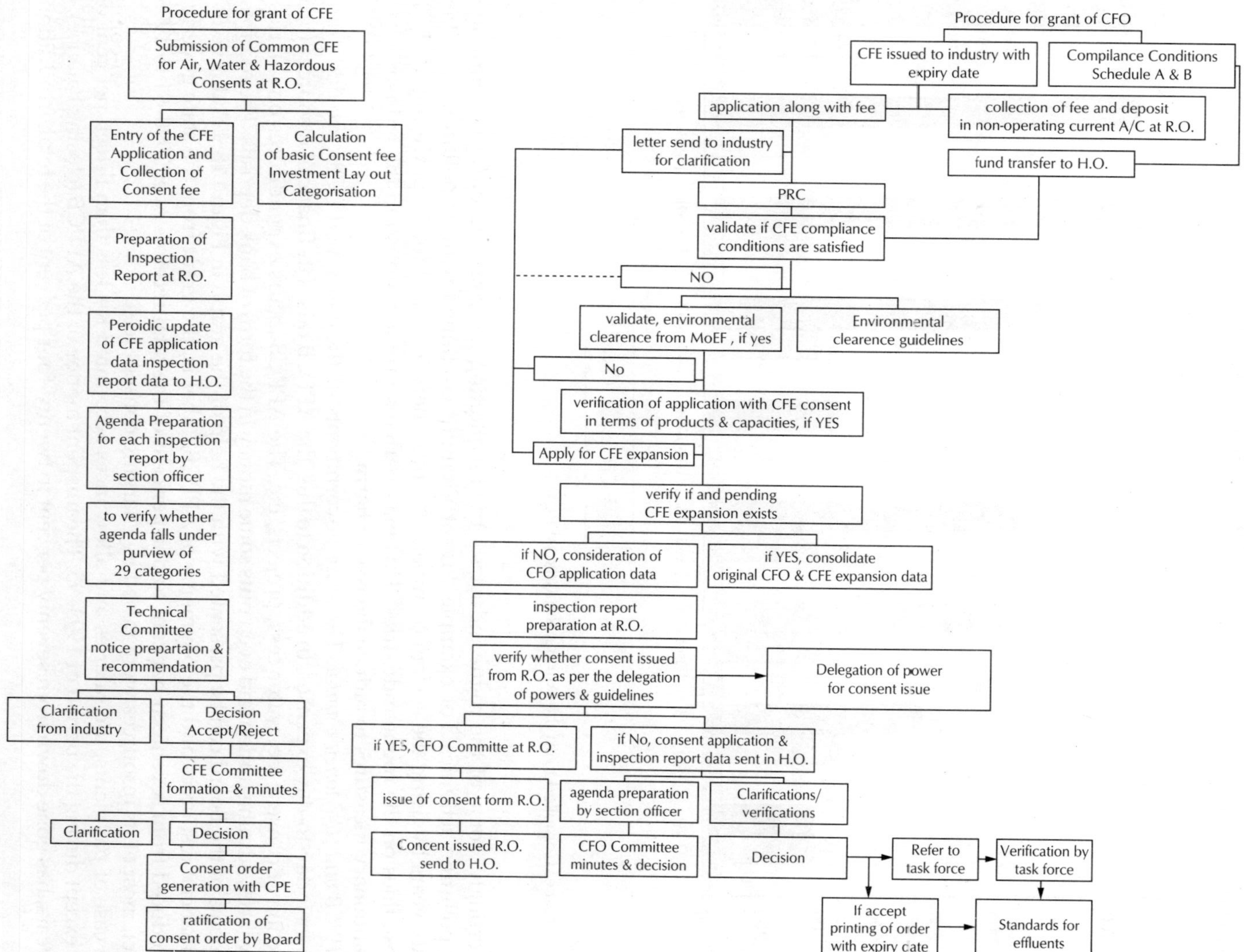

Source: Compiled from the Annual Reports, APPCB

Figure 18.6 Consents Issued by APPCB

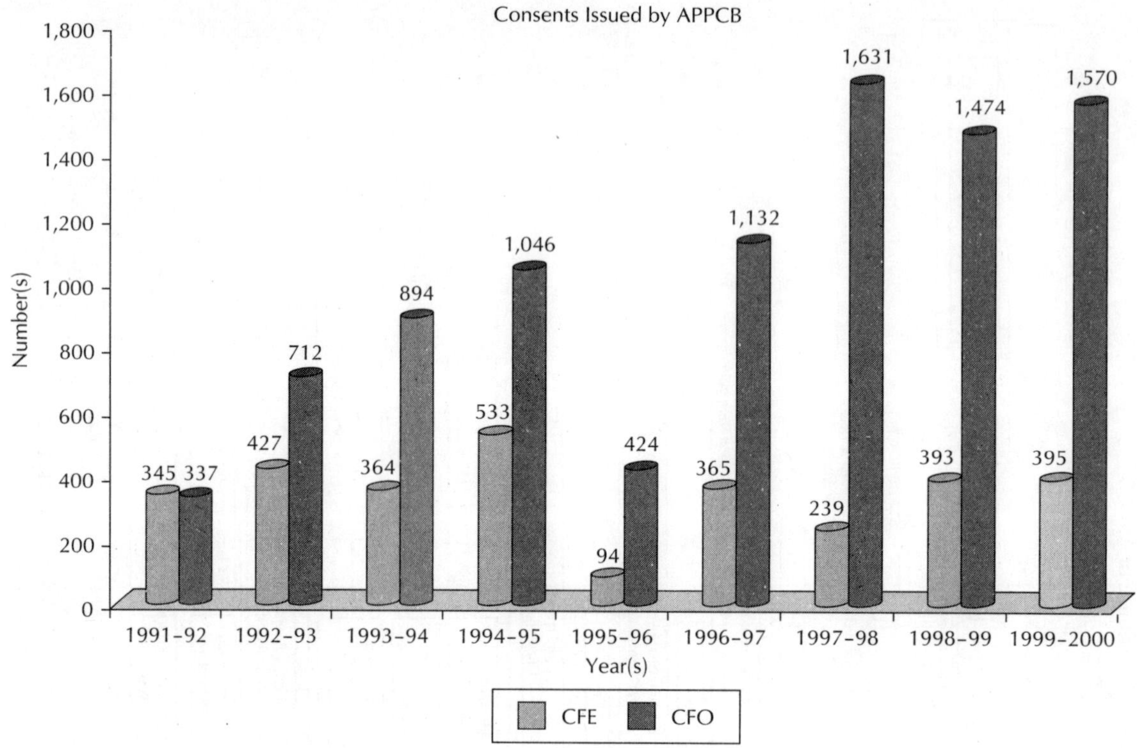

Source: Compiled from the Annual Reports, APPCB.

Interestingly, some of the rogue industries tried exploiting the courts in order to continue with their polluting activities. For example, Jayant Vitamins continued with its pollution activities for twenty years by using the system of appeals.[64] Thus, there is a need for empowering the PCBs to impose fines on the rogue industries. This may perhaps provide a push to the industries to take precautionary measures to reduce the risk of harm.

APPCB and its Money Capital: The state governments of the concerned PCBs provide grants to meet their day-to-day expenses. In addition to this, the APPCB raises its financial resources through collection of consent fee, water cess, projects, etc. The APPCB incurs expenditures on salaries and allowances, administration and even puts some money in the form of bank deposits. The receipts and payments of the Board can be presented with the help of the graph in Figure 18.9 and reveals that the financial resources and the payments of APPCB,[65] that is, the grant from the state government has reduced from seventeen per cent to one per cent of the total receipts over 1990–91 to 1996–97. In fact, most of its resources were raised by charging consent fees.

In case of payments, the salaries and allowances accounted for less than fifteen per cent of the total except during the period of 1993–95. The unspent money of the APPCB (closing balance), over the years, has gone down from seventy per cent to twenty-four per cent of the total. In addition, the

Figure 18.7 Assessment and Collection of Water Cess

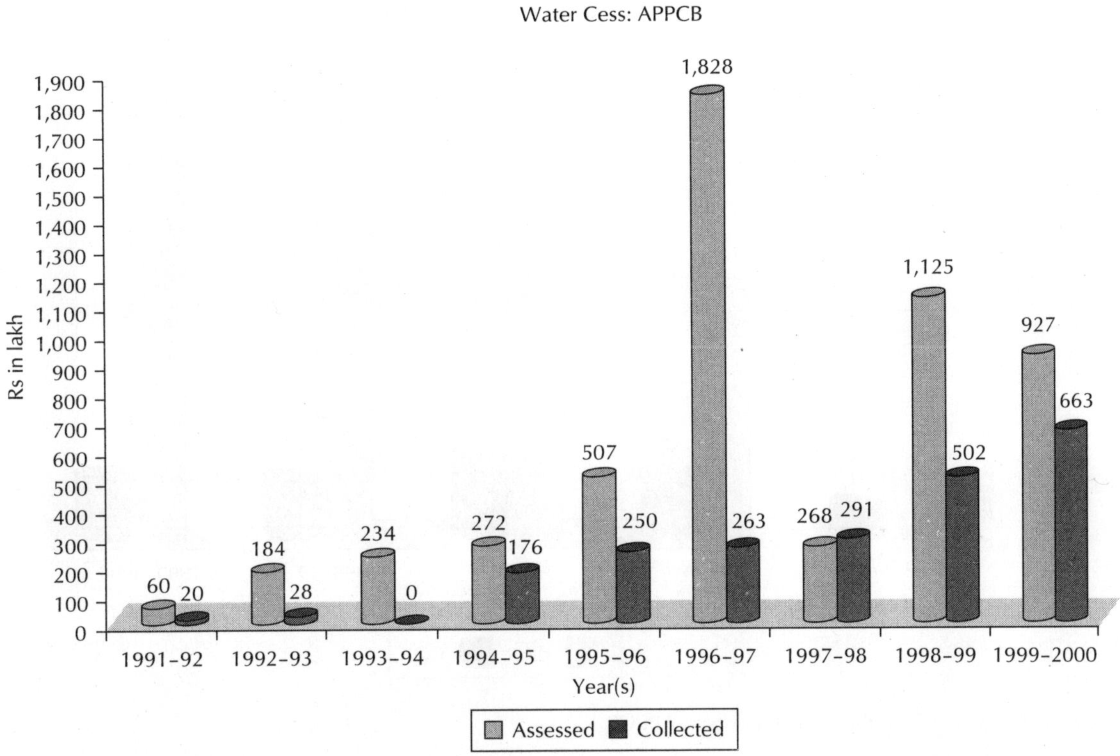

Source: Compiled from the Annual Reports, APPCB.

board is keeping some of its resources in the form of bank deposits, which accounts for forty-eight per cent of the total payments. Once we treat the closing balance and the bank deposits of the board as unspent money, then the available money for the wider activities of the board accounts for less than the fifty per cent of its total payments.

Usually, the received money is not sufficient to carry out the board's obligations. The graph (Figure 18.9) reveals that the board is not utilising its available resources effectively and efficiently in order to prevent the environmental degradation. One reason could be the prevalence of uncertainty over resource generation. There is a need for policy measures towards the strengthening of the PCBs' resource base.

Promotional and Informational Activities: The APPCB encourages its officials to attend training programmes, workshops and seminars to broaden their skills, increase interaction with experts and establish partnership with institutions. It conducts environmental awareness programmes, orientation programmes[66] and seminars on environmental issues to create awareness among stakeholders. The Documentation Centre[67] of the Board keeps track of the data on the projects and programmes conducted by the different cells of the Board. The APPCB networks with NGOs and educational institutions for collaboration in public consultation programmes.

Figure 18.8 Closure Orders

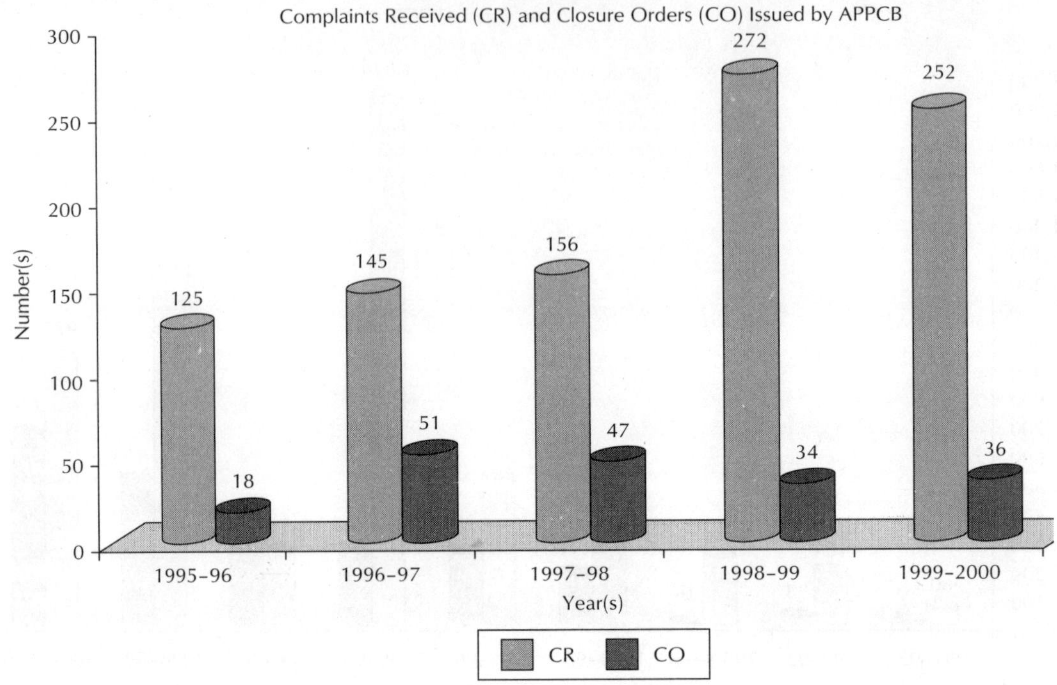

Source: Compiled from the Annual Reports, APPCB.

Hazardous Waste Management: According to MoEF, 'around five million tonnes of hazardous wastes are generated in India every year.' It is largely concentrated in four states: Andhra Pradesh, Gujarat, Maharashtra and Tamil Nadu.

The Hazardous Waste Management Cell has been established[68] along with the Cleaner Production Cell by the APPCB with the financial assistance of Australian Government.[69] The purpose is to identify, quantify and characterise hazardous waste producing industries. The Cleaner Production Cell also advises and provides technical and financial assistance to industries to minimise waste by adopting cleaner production options. The cell tries to create awareness among industrialists by releasing information bulletins[70] about the advantages of waste minimisation and adoption of cleaner technology. The cell provides incentives, such as the issuing of a three-year Consent for Operations concessions in water cess payment, etc. to industries that are practising cleaner production. As per the rules, the APPCB identified 596 industries as hazardous waste generating industries and issued authorisation for 535 industries for onsite collection and safe storage.

The cell perhaps should foresee the consequences of disposal of Municipal Solid Waste, which includes household trash and the hazardous hospital waste. A majority of the cities and the towns in our country dispose the waste without segregation which leads to long-term ecological effects.

Laboratory Testing: In 1977, the APPCB established central and regional[71] laboratories to undertake analysis of all polluting parameters.[72] Laboratory equipment is added through government grants

Figure 18.9 APPCB's Money Capital

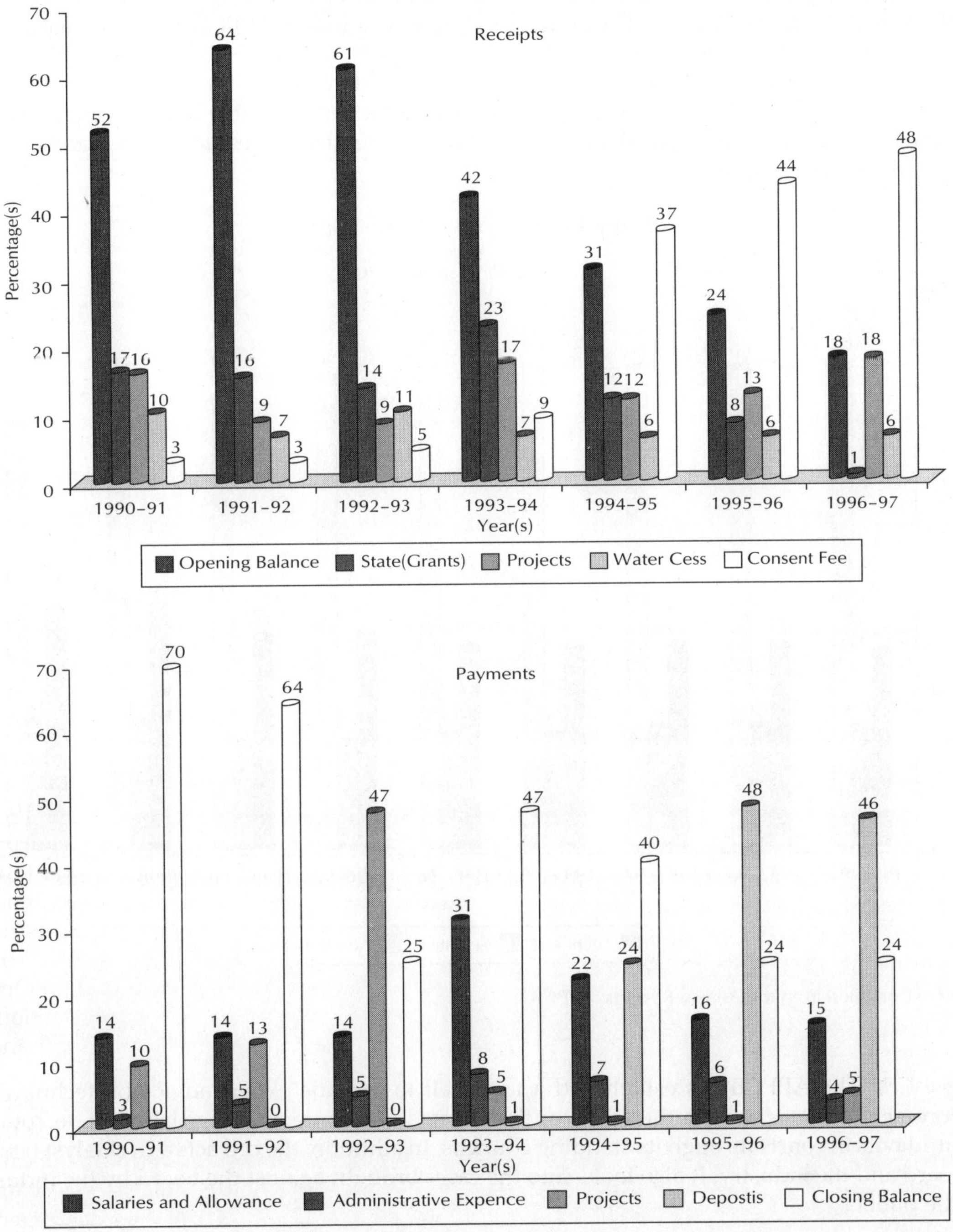

Source: Compiled from the Annual Reports, APPCB.

and projects. The Central laboratory of the APPCB, over the years,[73] has carried out forty-four per cent and fifty-six per cent of the total 41,448 sample analysis of water and air, respectively.

APPCB and its Human Capital: The status of Human Capital of APPCB can be explained with the help Figure 18.10. It indicates that the Board's manpower stood at 258 by the end of March 2000. The administration, technical, and scientific personnel accounted for fifty-three, thirty-one and sixteen per cent respectively, of the total staff. The ratio of technical persons to the number of polluted (red and orange categories) industries stood at 1:100. We can infer from these statistics that the APPCB has very restricted range of activities.

Figure 18.10 APPCB's Human Capital

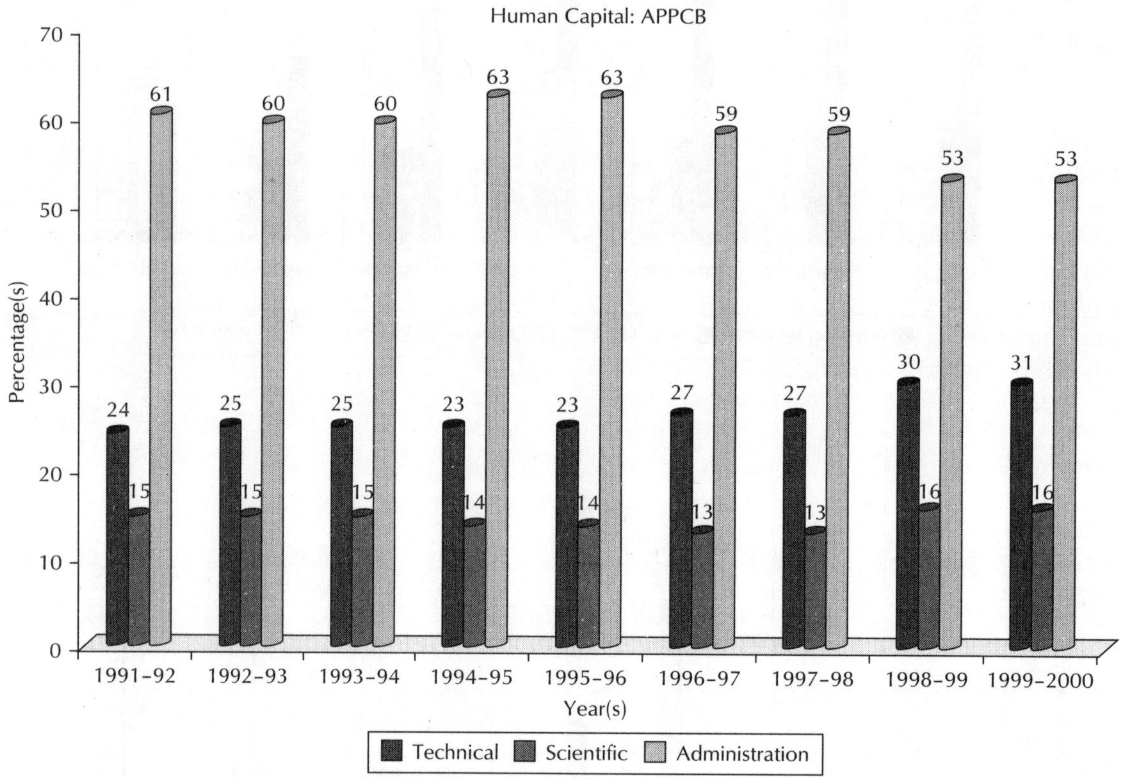

Source: Compiled from the Annual Reports, APPCB

Legal Cell: The APPCB has established a legal cell to provide expert advice on technical and environ-mental issues, within the scope of the enacted environmental legislation(s), to courts. It files affidavits in courts through its standing councils. In addition, the cell acts as a catalyst between the board and the industry. It also looks into the cases brought against the board by the industries and the public.

VIEWS OF THE OFFICIALS OF THE PCBs

We prepared questionnaires for the Boards' officials in order to obtain their opinion on the functioning of PCBs. These questionnaires were prepared after a review of the environmental PIL cases from the Supreme Court of India and the Andhra Pradesh High Court; environmental legislation that they enacted; and material collected from the CPCB and the APPCB.

Six officials were interviewed to get their opinion on the functioning of the regulatory system in the country. We asked for their opinion on the following issues:

Dissemination of Information

Although the board will provide expert opinion to the Courts about the state of affairs of environmental pollution, some of the officials conceded that their expert reports may hide factual information about the polluting industry.

The PCBs also do not provide the citizens information about the activities of polluters. For instance, disclosure of information about the industries may create panic amongst the public. In addition, they argue that the disclosure of information about the negative externalities of the polluters may be exploited by the rival/competitive industries.

Dissemination of information about polluters is the bedrock function of the PCBs. Citizens can exercise their rights and campaign against the polluting industries.

Issuance of Consents

Officials were of the opinion that in the case of issuance of the Consents, such as CFE and CFO, standards were applied uniformly irrespective of the nature of the industries (that is, public or private) involved.

Implementation of Standards

The schedule of implementation of standards varies amongst the industries. For example, Thermal Power, Integrated Iron & Steel, Oil Refineries and Mines may require larger investments and longer time to install pollution abatement machinery.

According to the officials, monitoring industry is a complex issue for the PCBs. A majority of the industries do not comply with the conditions because they are not economically viable.[74] Further, even industries that pretend to be complying, actually do not. For example, industries establish the Effluent Treatment Plant but do not run it and they operate it only at the time of the PCBs team's visit. Courts are also impotent to stop the pollution and, at times, provide incentives to the industries to pollute more.[75] As a result, a majority of the officials favour the introduction of the concept of fine against rogue industries.

Officials also felt that the decentralised regulatory system should curb pollution but the involvement of politicians and lack of honesty among the PCB's personnel led to a virtual breakdown. In addition, they face pressure from interest groups; threats; inadequate job security; lack of trained personnel, expertise, financial resources and infrastructure facilities incentives; and, last but not the least, lack of sincerity.

Moreover, they felt that the standards formulated by the PCBs are not scientifically and economically viable. That is the reason why most of industries violate the standards. As a result, a majority of the officials of the PCBs stressed the need for the establishment of separate environmental protection courts equipped with technical and scientific prowess. According to them, it would also result in timely disposal of cases, monitoring and the implementation of orders, etc.

Suggestions by the Officials

Officials were against the establishment of parallel regulatory agencies, such as Shore Area Regulatory Authorities,[76] but favoured competition among PCBs in order to achieve sustainable development in the country. They made the following suggestions:

- Establishment of the National Environmental Protection Authority in place of CPCB and similarly establishing State Environmental Protection Authority in place of SPCBs.
- All cities should have pollution monitoring stations.
- No governmental interference.
- Committee systems should be encouraged.
- Awareness programmes through the media.

CRITICAL EVALUATION OF THE PCBs IN THE LIGHT OF THEORY OF REGULATION

Enforcement of Environmental Standards

PCB personnel monitor compliance with environmental standards by industries by issuing show cause notices, legal counselling and closure orders. They are empowered to collect samples in and around the industry premises and test them to determine whether the effluents/emissions are in compliance with the standards or not. In majority of the cases, they take action against polluting industries based on complaints received from citizens. PCB's do not have any 'Consent for Establishment/Operation Manual' to carry out minimum sample tests and to ensure that the industries strictly follow the relevant standards. It is handicapped in terms of enforcing its standards and providing deterrent incentives to violating firms because the PCBs are not empowered to use punitive measures. However, the PCBs may blacklist the polluting firms. Moreover, PCBs have a greater number of administrative personnel than scientific and technical personnel. A conclusion that can be drawn is that the PCBs are handicapped in terms of carrying out its functions towards the prevention of pollution in the country.

PCBs and their Resource Mobilisation

PCBs over the years have been woefully under-funded. They have to raise their resources to meet even their daily expenditure. The SPCBs, in particular, raise income through the consent fee, No Objection Certificate (NOC) and water cess that are paid by the industrialists and the local authorities. For example, in the year 1999–2000, the APPCB raised Rs 10.6 cr. (out of the total receipts of Rs 10.9 cr.) through consent fees, NOC and water cess.

These circumstances increase the possibility that the PCBs may issue consents subject to conditions that favour the industries rather than protect the environment in the country. Thus, self-reliance of PCBs comes at a great cost and reduces their effectiveness. This is an issue that needs to be addressed because the Board has been unable to reduce the degradation of the environment in the State of Andhra Pradesh for the last three decades.

Informational Difficulty

It is difficult to get information from a regulatory agency. In the case of PCBs, even the information that they are required to disseminate is not made available to the public, often in the name of confidentiality, secrecy or not to create panic amongst the public.

Information programmes such as labelling and reporting requirements may help foster (market-oriented) solutions to environmental problems. The PCBs need to publish the information on firms' use, storage and release of hazardous chemicals. In fact, dissemination of this type of information to the public may bring awareness and also ease the task of the PCBs in monitoring the activities of polluting industries. Public scrutiny can provide incentives to firms to alter their behaviour.

Information is power and the consequence of asymmetric information further aggravates environmental pollution in the country. Moreover, information disclosure provides an opportunity to the public to carry out their statutory duty under Article 51A (g) of the constitution.[77] Hence, the PCBs should disclose information rather than restrict their activities to just conducting awareness programmes among the general public.

Influence of Interest Groups on PCBs' Activity

The theory of regulation predicts that interest groups always cast a shadow on regulatory activities because it is easy to capture. Given the increasing levels of environmental pollution, the PCBs are not free from the influence of interest groups. Further, there is indirect evidence of the validity of the theoretical arguments about capture theory. For instance, since PCBs mobilise their own resources, consents are issued with conditions that are favourable to interest groups. Otherwise it would not have been possible to mobilise ninety-five per cent of its resources through Consent Fee, NOC and Water Cess in 1999–2000.

Moreover, the lack of job security amongst the PCB employees provides an opportunity to the interest groups. Monetary bribes are theoretically possible, although they are not common because of their illegality. It is, however, personal relationships that provide incentives to the government officials to treat their industry partners kindly. The industry also exercises power by obtaining transfers of key elected officials who have influence over the agency.[78]

Jurisdictions of the PCBs

There is often overlapping jurisdiction that creates problems with the enforcement of the environmental law. For example, a regulatory agency's jurisdiction is on the territory (State boundary); but there are other parallel agencies such as Shore Area Regulatory Authority, Traffic Authority, licensing authority for small scale industries, etc., the intersection of which leads to delay and confusion as the regulatory agencies debate their respective jurisdiction.[79]

Expert Role of PCBs

Although the PCBs are playing an important role in providing expert opinion to the liability system, their work is often shoddy and has drawn strictures from the Supreme Court. For example, in the context of Jayant Vitamins Ltd, the Supreme Court was not satisfied with the pollution report submitted by the MPPCB.[80] Similarly, once, the CPCB had submitted its expert report to the Supreme Court of India by visiting the polluting industries on a day on which they were closed[81] and during the peak monsoon period when the effluent discharge is diluted by rainwater.[82]

PCBs and Issuance of their Consents

The study reveals that industries play a tactical role while obtaining consents from the PCBs. The industries initially apply for consent to produce less polluted goods but actually they produce highly polluted goods under the same consent.[83]

PCBs and Issuance of Closure Orders

Industries continue with their pollution activities even after receiving closure orders from the concerned PCB, by simply changing the name of the polluting unit. For example in *B. Sadanandam vs Government of AP & others*, W. P. No. 17148/1999 in the HC of AP:

The residents of Allwyn Colony (126 out of 2000 members) approached the HC of AP against the activities of the Hyderabad Ossein Ltd (animal bones crushing and storage unit) in a residential zone. The court issued closure orders in the year 1997 against the industry. The Trans Gel Industry, however, has taken over the Hyderabad Ossein Ltd and requested the APPCB to revoke the closure orders. The APPCB, after hearing complainant's view (fourteen out of 126 members), issued temporary revocation of closure order subject to certain conditions. The court in its order stated that the APPCB's revocation of closure of the industry ends on 30 January, 2000. It directed the APPCB to make periodical inspection and take appropriate action against the industry in case of default.

Does the Compliance with PCBs' Standards Exempt the Polluter from the Liability?

The Theory of Liability versus Regulation reveals that compliance with standards does not protect the polluting industry from liability in the event of harm. At the same time, non-compliance does not automatically lead to liability. It should be noted that even after obtaining the requisite consents, victims can go to court. Under these circumstances, the industrial activities, that are hazardous and irreversible in nature, will have to be monitored to achieve compliance with specified standards.

However, the PCBs may prevent the polluter from establishing his units and operating existing units under the provisions of the Consents for Establishment and Consents for Operation, respectively.

Since PCBs formulate only Minimum National Standards (MINAS), the courts justify that, irrespective of compliance with the standards, the polluter is liable for the harm occurred. In other words, the polluter who complies with the MINAS may not be exempted from the liability in the event of harm. For instance, under the provisions of Public Interest Litigation, the citizens of India are approaching the liability system to seek remedial measures against the rogue industries.

SUGGESTIONS FOR IMPROVING THE FUNCTIONING OF THE PCBs

Financial Assistance

PCBs require resources for the formulation and monitoring of standards, conducting exhibitions and awareness programmes on the availability of state of the art green technology and measures for abatement of pollution. We find that the Government has drastically reduced its contributions to PCBs. For instance, the Andhra Pradesh State Government provides only one percent of the total receipts of the APPCB. In addition, mobilisation of resources through water cess is inadequate because the local authorities do not pay their dues and the Board is unable to do much to recover them. Moreover, the PCBs cannot rely on NGOs and foreign funding because it is specific and temporary in nature. They, therefore, rely on fees gathered from the issuance of consents. There is, however, variability in revenue obtained from consents because it depends upon the establishment of new industries and also the existence of old industries. We also find that the PCBs are depositing their funds with financial institutions to obtain interest income rather than investing in research and development because of the uncertainty about the revenue.

As a result we have a unique situation. On the one hand, as an *ex-ante* system, the PCBs should adopt stringent measures against polluting industries to prevent environmental degradation. On the other hand, since they raise their own revenues, PCBs may be required to compromise in use of stringent measures. Thus, there is a need for providing financial assistance directly by the Ministry of Finance.

Dissemination of Information

Information asymmetry is the root cause for the breakdown of markets. We find that the PCBs are reluctant to disclose information to the public about the activities of polluting units. Information disclosure by the PCBs largely reveals their achievements in abating pollution. These claims can, however, be challenged since there is clear evidence of environmental degradation.

Disclosure of the activities of polluting industries provides a push to manufacture environment friendly products and to comply with environmental standards. Disclosure enables the public to participate in the preservation of ecology in the country. The Right to Information Act will enable the public to obtain the necessary information if the PCBs are unwilling to part with it.

Punitive Measures

Fines imposed on those violating environmental standards generate efficient deterrence. The Water, the Air and the EP Acts provided for civil and criminal penalties for the violation of their provisions. In addition, the Acts also entail overriding effects. The PCBs, may be because of these effects, are not empowered to impose fines and their activities are restricted to the issuance of consents and monitoring compliance by the industries. In the event that industries violate the consent conditions, the PCBs need to follow the principles of natural justice in order to carry out action against the industries, including approaching the judiciary. Once it is under the purview of judiciary, the industries can happily be in business till the decision of the Court and the implementation of the Court's order. It is, therefore, wise to grant PCBs, the power to impose fines, including punitive damages to prevent environmental degradation *ex-ante*.

Adopting Sample Testing Manual

The effectiveness and efficiency of the functioning of the PCBs depend upon the monitoring of established environmental standards. This function is directly dependant on the periodic collection of samples and their testing by the PCBs in laboratories. We find that the collection of samples by the PCBs is dependant on whether there is a complaint and they are not performing their duty of voluntarily collecting samples. Moreover, the PCBs do not have a minimum sample testing manual under the provisions of the Consent for Establishment and Operation given by them. Failure to test provides incentives to polluting industries not to comply with environmental standards. The PCBs, therefore, need to adopt the minimum sample testing manual to enforce their standards.

Organisational and Structural Changes

The regulatory responsibility of the PCBs is increasing because of the severity and increase in environmental pollution. In the decentralised model, with the CPCB acting a nodal agency under the Ministry of Environment and Forest, the CPCB collects information and provides it to the SPCBs. The role of the SPCBs is restricted to their respective state jurisdictions. One of the justifications for establishment of SPCBs is that these institutions facilitate greater participation by the people in local affairs; promote better planning and implementation of development and environmental programmes, and enhance the responsiveness to the needs of the people. However, the PCBs have been unable to internalise the externalities in an effective and efficient manner. There is, therefore, a need for the establishment of a separate independent statutory agency to prevent and reduce pollution in the country.

SUMMARY AND CONCLUSION

The study focused on the evaluation of the functioning of the Pollution Control Boards (PCBs) with the goal of determining whether the regulatory system is effective in preventing environmental pollution in India. Since the market and the liability systems are unable to provide incentives to the polluter to reduce pollution, there is a need for the regulatory system to prevent, control and abate environmental pollution in the country. The PCBs were established under the provisions of the Water, Air and EP Acts in order to fulfil the objectives of formulating environmental standards, monitoring them, issuing consents for the establishment and operation of industries, and advising the Courts and the Government on the science and technicalities of environmental issues.

Our study is based on primary and secondary data. The insights obtained from the data were used to prepare questionnaires that were then submitted to the officials of the PCBs to get their opinion on the functioning of the PCBs. We have critically analysed the data and the opinions of the PCB officials in the light of the theory of the regulatory system to determine the role of an *ex-ante* system in abatement of pollution in the country.

The study reveals that the role of the Board is of great importance in preventing, controlling and abating environmental pollution in the country. The decentralised system of PCBs, however, is ineffective in ensuring internalisation of environmental concerns in the process of economic development. This is mainly because the responsibilities and the roadblocks are manifold—inadequate human resources, particularly, technical and scientific staff, prevalence of uncertainty over resource base, presence of the influence of the interest groups, existence of jurisdictional problems, absence of punitive measures, non-existence of minimum sampling tests manual, lack of effective and efficient working culture, and non-disclosure of information about the activities of the hazardous industries.

Thus, there is a need to introduce policies on restructuring of the existing PCBs, establishment of competitive environment, empowerment of PCBs to impose fine against rogue industries, putting in place an incentive mechanism for the personnel, reducing the revenue generation responsibility and the provision of financial assistance directly by the Ministry of Finance.

Overall, the study emphasises the necessity for improving the functioning of the regulatory system by making necessary changes not only in substance of the law, but also in the working conditions of the PCBs so as to improve the overall environmental quality in the country.

ACKNOWLEDGEMENTS

I thank Prof. Jyoti Parikh, Chairperson of the Environmental Economic Research Committee (under EMCaB Programme) for project assistance, the Member Secretaries of both the Central Pollution Control Board and the State Pollution Control Board of Andhra Pradesh for their permission to collect data from the documents, and Manoj Dalvi and Ramamohan Rao for their efforts in substantially improving the quality of the chapter.

NOTES

1. 'They are largely borne by the regulatory agency, which has the task of formulating, monitoring and enforcing standards'. See for instance, Anthony I. Ogus, *Regulation: Legal Forms and the Economic Theory*, 1994, p. 155.
2. 'They are the capital expenditure on equipment and adoption of plant to meet the standard'. Ibid.
3. 'They fall under the category of productive inefficiency, the inhibition of technology, and allocative inefficiency. The assessment of indirect costs is problematic because relevant effects which are widespread and data is difficult to obtain'. Ibid.
4. For example, if two firms apply for a license to establish their units and only one of them gets the license. So, the other firm which hasn't obtained licence has also incurred expenditure on lobbying, leading to a waste of resources. The result may be an inadvertent limiting of competition between firms in the adoption of green technology.
5. The liability system is effective in providing incentives to the tortfeasor to take precautionary measures to reduce the risk of harm by setting the due level of care based on case by case adjudication, generating information from private parties, lower administrative costs, error correction by way of appeals, etc. Similarly, the regulatory system is effective in providing incentives to the tortfeasor to take precautionary measures to reduce risk of harm by the formulation of regulatory standards through scientific knowledge, collection of fines (which is helpful in case of inadequate wealth of tortfeasor), etc.
6. The liability system has limitations with respect to the award of non-pecuniary costs, economic consequences of full, over, and under-compensation, rational apathy, establishment of causational links, law's delay, etc. Similarly, regulatory system also has limitations such as regulatory capture, adverse effects in the case of formulations of standards for private goods, etc.
7. For example, in the case of chopping down a tree in one's yard, it is less costly to use liability to force appropriate caution than to construct a myriad of permits and regulations covering tree felling. At the same time, in another example concerning air pollution, it is less costly to promulgate well thought out standard regulations than to let each victim to take the tortfeasor to court (C. D. Kolstand et. al., 'Ex-post Liability for Harm vs Ex-ante Safety Regulation Substitutes or Complements', *American Economic Review*, 1990).
8. For instance, in the case of potential deficiencies of incompatible uses of neighbouring property, where a hospital is situated next to a noisy, dusty cement manufacturing industry, there may be possibilities of minimising externalities by zoning ordinances (which is a from of *ex-ante* approach) and at the same time exposing the externality generator to nuisance liability (which is a from of *ex-post* approach). The classic comparison of the efficiency aspects of these alternate methods of minimising this type of externality is given by R. Ellikson, 'Alternatives to Zoning: Covenants, Nuisance, and Fines as Land Use Controls', *University of Chicago Law Review*, Summer 1973, pp. 681–781.
9. Under the provisions of the Articles 251 and 254.
10. For example, The Water (Prevention and Control of Pollution) Act, 1974 was enacted by the Parliament after consent resolutions were passed by twelve State Legislatures.

11. It states that 'The State shall endeavour to protect and improve the environment and to safeguard the forests and wildlife of the country'.
12. This article imposes a responsibility on every citizen 'to protect and improve the natural environment including forests, lakes, rivers and wildlife, and to have compassion for living creatures'.
13. It (read with entry 13 of the Union list) provides power to the Centre to make laws implementing India's international obligations and also with regard to any decision made at international conference.
14. Sections 41 to 45.
15. Sections 37 to 39.
16. The Bhopal Disaster of 3 December, 1984.
17. Section 15.
18. The objective of the Rule is to control generation, collection, treatment, import, storage and handling of hazardous waste.
19. It was enacted to provide immediate relief to the victims of an accident involving a hazardous substance. The act imposes strict liability upon the owner of the hazardous substance and has to pay relief as the Act obligates every owner to take out an insurance policy covering potential liability from an accident and also must make a contribution to an Environmental Relief Fund established by the GoI. The fund is used as relief to the victims of an accident.
20. Makes it binding on the health care institutions to streamline the process of proper handling of hospital waste such as segregation, disposal, collection and treatment, enacted by the Government of India to improve the environmental quality.
21. Water is a subject in the State List under the Constitution (Entry 17, List II, Seventh Schedule). So the Act was enacted by the Parliament after consent resolutions were passed by 12 State Legislatures under Article 252 (1) of the Constitution.
22. One chairman; five members as the representatives of government; three members from industry, agriculture and trade; two members from the PSUs; and a Member Secretary—all are nominated by Central Government.
23. They are at Lucknow, Bhopal, Shillong, Kolkata, Vadodara and Bangalore.
24. Under the provisions of the Water and the Air Acts, 1974 and 1981, respectively.
25. Under Section 25 of the Environmental Protection Act, 1986. The State Pollution Control Board (SPCBs) must take into consideration these standards while issuing consents to the industries. The SPCBs can make their standards more stringent than the Central standards but not less stringent. A World Bank study, however, stated that the MINAS fixed by the CPCB have not left any flexibility for the SPCBs to make them more stringent as MINAS at their current levels require near the maximum effluent reduction that is technically achievable.
26. It consists of experts from Council for Scientific and Industrial Research (CSIR), National Test House, Bureau of Indian Standards (BIS), National Institute of Occupational Health (NIOH) and consumer organisations.
27. The subcommittees prepare the draft Eco-mark criteria in their respective product categories in consultation with the organisations and the government. The draft criteria prepared by the technical subcommittee are considered by the technical committee for its recommendation to MoEF, and then draft notification is issued for public comments. The public comments received by MOEF are again examined by the technical committee and on the basis of its recommendation the final notification is issued by the MoEF so that the relevant Indian Standards are amended accordingly and the ECOMARK scheme becomes operational.
28. Since 1991, the Eco-mark criteria has been finalised and notified for sixteen product categories like soaps and detergents, paper, food items, lubricating oils, packaging material/package, architectural paints and powder coatings, batteries, electrical/electronic goods, food additives, wood substitutes, cosmetics, aerosol propellants, textiles, plastic products, fire extinguisher and leather.
29. In a study (Inventorisation of Hazardous Waste Generation), the CPCB found that the hazardous wastes generation in eight states (Gujarat, Jammu & Kashmir, Punjab, Kerala, Andhra Pradesh, Madhya Pradesh, National Capital Region and Orissa) accounted for 19 lakh tpa.
30. Under Sections 12 and 13 of the Environmental Protection Act, 1986.
31. In order to improve the abilities of PCBs of the heavy industrial states in the country such as Gujarat, Maharashtra, Tamil Nadu and Uttar Pradesh.
32. That is from 1990–91 to 1998–99.
33. Except for two years (that is, during 1994–95 and 1995–96)
34. Such as industry, agriculture, trade, etc.
35. With the countries: Germany, the Netherlands, Norway, and EEC.

36. Section 25 and 26 of the Act provide for refusal or withdrawal of consent by the Boards. Section 33 of the Act empowers the Boards to make an application for directions to the Court of a Judicial Magistrate, where the Boards apprehend pollution. Section 33-A (53 of 1988) further empowers the Boards to direct the closure of polluted industries by regulation of supply of electricity, water or other sources.
37. The Water (Prevention and Control of Pollution) Act, 1974 (Amendment, 1988).
38. Under the Water (Prevention and Control of Pollution) Cess Act, 1977.
39. Under the Air (Prevention and Control of Pollution) Act, 1981 (Amendment, 1987)—noise pollution was added as an air pollution in 2000.
40. Special Chief Secretary from the Department of Environment, Forests, Science & Technology, the Government of Andhra Pradesh, acts as an ex-Officio Chairman; Member Secretary of APPCB acts as member convenor; one member each from the Departments of Municipal Administration & Urban development, Health, Medical & Family Welfare, Commissioner of Industries, Forests, Andhra Pradesh State Financial corporation, Andhra Pradesh Industrial Investment Corporation.

    The office of the Member Secretary supported by thirteen cells: (1) CFE & CFO Cell (2) Cleaner Production Cell (3) Legal Cell (4) Bio-Medical Cell (5) Action Plan Monitoring Cell (6) Hazardous Management Cell (7) Task Fore Cell (HO) (8) Information & Public Cell (9) Accounts (10) Administration Cell & Building Cell (11) Cess Cell (12) Board Laboratory (13) Documentation Centre.
41. They are Visakapatnam, R.C. Puram, Hyderabad, Vijayawada and Kurnool.
42. They are Rajamundry, Visakhapatnam, Vizianagaram, Medek I & II, Nalgonda, Nizamabad, Hyderabad, Ramagundam, Ranga Reddy I & II, Warangal, Nellore, Kothagudem, Vijayawada, Tirupati and Kurnool.
43. Chairman of the Board may convene a meeting whenever there is urgency—Section 8 and 10 of the Water and the Air Acts, respectively.
44. Section 21 and 26 of the Water and the Air Acts, respectively.
45. Section 33 (A) and 31 (A) of the Water and the Air Acts, respectively.
46. However, the industries can approach Appellate Tribunal and even go up to Apex Court of India in order to get remedial measures against disputed closure orders passed by the Board.
47. Section 25 and 21 of the Water and the Air Acts, respectively.
48. The fee is based on the total investment of the industry. However, it excludes working capital and expansion costs of the industry but not depreciation.
49. Conditional clearances may be given to the industries. This increases the possibility of fraud. For example, the State of Andhra Pradesh imposed certain environmental conditions for Hatcheries and Aquaculture projects. One of the conditions is that the production of shrimp shall not exceed six metric tonnes per hectare per crop and allowed only two crops per year, to maintain pollution levels at a minimum. In reality, many industries violated these conditions. For example, in Nellore District, the Mega Prawn farms have no effluent treatment plants. They discharge effluents and dump the dead and diseased shrimp into the Buckingham Canal that supplies drinking water to Chennai City (*Indian Council for Enviro-legal Action vs Union of India and others*, W. P. No. 664/1993, Supreme Court of India). According to the *Public Trust Doctrine* (International Law), natural resources are public property entrusted to the Government of India for their safe and judicious use. Any action that leads to their improper use and damage amounts to violation of the doctrine. Thus, environmental clearances given by authorities without comprehensive Environmental Impact Assessment (EIA) are illegal.
50. Every year by reviewing the fulfilment of previous year conditions.
51. Every two years by reviewing the fulfilment of previous year(s) conditions.
52. Every five years by reviewing the fulfilment of previous year(s) conditions.
53. Section 5 of the Water Cess Act of 1977.
54. Water (Prevention and Control of Pollution) Cess Act, 1977 (Amendment, 1991). It was enacted to augment the resources of the PCBs and to conserve water.
55. The major cess paying industries are Thermal Power Plants, Pulp and Paper, Heavy Water Plant, Vizag Steel Plant, etc.
56. The APPCB has been regularly assessing about 112 Municipalities (Gr. I, II, III)/Municipal operations.
57. Under the Environmental Protection Act, 1986.
58. Section 10 of the Water Cess Act.
59. Section 11 of the Water Cess Act.

60. Section 14 of the Water Cess Act.
61. The Municipalities/Municipal Corporations and Thermal Power plants were not paid dues of about Rs 4.5 and 22.4 crores, respectively.
62. The primary aim is attending to complaints received from the public as well as industries, issuing of show cause notices/ directions/closure orders to those industries which did not comply with the standards, night patrolling to prevent illegal outlets, and investigating the root cause of pollution and providing preventive measures.
63. That is from 1995–96 to 1999–2000.
64. *Hari Ram Patidar vs Union of India & others,* (in the SC, W. P. No. 330/ 1995). See Supra note 1 at p. 263.
65. We have reviewed the receipts and the payments of APPCB instead of income and expenditure for purposes of the study.
66. Like AP Children's Environmental Science and Action Congress.
67. Established in 1998.
68. Based on the Hazardous Waste (Management and Handling) Rules, 1989 (Amendment, 2000).
69. In 1990, the Government of Australia offered assistance (Australian Agency for International Development—AusAID) to the Government of India on environmental maters. The APPCB has designed a project on Hyderabad Waste Management with the assistance and the project started in the year 1996 in order to promote waste minimisation and cleaner production.
70. So far the cell has issued ten information bulletins such as waste minimisation in textiles, electroplating industries, chemical industries, bulk drugs, dye intermediate industries, etc.
71. Six regional labs at Ramagundam, Rajamundry, Vijayawada, Vishakapatnam, Tirupati, and Warangal.
72. Such as heavy metals, pesticide residues, air pollutants, organising water pollution and industrial waste surveys, establishing water quality standards, maintaining the data, etc.
73. That is, from 1991–92 to 1999–2000.
74. In the case of the effluents of Drug Industries, Patancheru, even after treating the effluents under common treatment plant, the TDS (BoD and CoD) levels are high (from 14,000 to 1,200) and need to find the place for their discharge. The Supreme Court of India asked the CPCB to look into the matter and do the needful. The Board estimated a cost of Rs 16 cores (approximately) to resolve the problem of treated effluents of the Drug Industries of Patancheru.
75. For example, in the *K. Sai Vijayendra Singh vs Andhra Pradesh State Pollution Control Board* (W. P. No. 28363/ 1997, in the High Court of Andhra Pradesh), the APPCB issued closure orders based on the court direction. However, the industry challenged the closure orders of the Board by filling a writ petition in the Court. The Court, in its interim orders, dismissed the closure orders, in spite of the Board's affidavit stating that the closure order was in the line with compliance with the Court direction. After nine months, the Court again passed the orders against the industry by stating that the industry had not complied with the APPCB standards.
76. In a Public Interest Case (*Indian Council for Enviro-legal Action vs Union of India and others*), the Supreme Court of India stated that 'considering the fact that the PCBs are not only overworked and have a limited role in effective implementation of the Notification 1991, the GoI should consider setting up State as well as National Coastal Management Authorities under Section 3 of the Environmental Protection Act, 1986'.
77. It is a fundamental duty which imposes a responsibility on every citizen 'to protect and improve the natural environment including forests, lakes, rivers and wildlife, and to have compassion for living creatures'.
78. See *Vellore Citizens Welfare Forum vs Union of India & others,* W. P. No. 914/ 1991 in the SC.
79. *The Citizens' Forum vs Government of AP and others,* W. P. No. 27917/ 1996, in the HC of AP.
80. The MPPCB in its affidavit (1992) stated that the Jayant Vitamins Ltd, Ratlam, is the chief source of the pollution and discharges about 600 cm per day of effluents into Kurel River which is a drinking water source for nearby villages and towns. The Board initially granted consent for one year in 1975 and did not renew it because of violation of its standards. However, the industry, in its affidavit, stated that the effluents are from the neighbouring industries such as Ratlam Alcohol Plant (produces alcohol form molasses), Sajan Industry (produces H. Acid), Stattar Drugs (produces Trimathoprine, Ibuprofen, Atenol and Isonieazid), Sri Ram Chemical Industry (produces Sodium and Sulphide solid), and Diesel Shed Western Railway, Ratlam (repairs its diesel engines). The MPPCB filed its affidavit in the Supreme Court of India which in its order (20 January, 1995) stated that it is not satisfied with the report of the MPPCB and directed the CPCB to inspect the industries and file a report.

81. *Tarala V. Patel & Others vs Government of Pondicherry,* W. P. No. 184/1996 in the SC.
82. *Rourkela Shramik Sangh vs UoI and others,* W. P. No. 285/1991 in the SC.
83. *Paryavaran Suraksha Sangarsh Samiti vs Union of India & others,* W. P. No. 94/1990 in the SC. Chemical industries, such as Messers. Hindustan Agro Chemicals, Silver Chemicals, Rajasthan Multi Fertilisers, Phosphate India and Jyoti Chemicals, in and around Bichhri village (GIRWA Taluk, Udaipur District) in Rajasthan were polluting the environment. The RPCB, in its affidavit indicated that the Hindustan Agro Chemicals obtained No Objection Certificate (NOC) subject to certain conditions to produce sulphuric acid and alumina sulphate. However, the industry started producing Olsum and Single Super Phosphate (SSP).

REFERENCES

APHC 1990–91 to 1998–99. Public Interest Litigation, Registrar Judiciary, High Court of Andhra Pradesh (APHC), Hyderabad.

APSPCB 1991–92 to 1999–2000. *Annual Reports.* Hyderabad: Andhra Pradesh State Pollution Control Board.

Bhaskaran, G., 1998a. *Pollution Control Acts.* Chennai: C. Sitaraman & Co.

———. 1998b. *The Environmental (Protection Act), 1998,* Chennai: C. Sitarman & Co.

CPCB 1990–91 to 1998–99. *Annual Reports.* New Delhi: Central Pollution Control Board.

Divan, S. & Rosencranz, A. 2001. *Environmental Law and Policy in India: Cases, Materials and Statutes.* New Delhi: Oxford University Press.

Ellikson, C.R. 1973. 'Alternatives to Zoning: Covenants, Nuisance and Fines as Land Use Controls', *University of Chicago Law Review,* 40(4): 681–781.

Ogus, I.A. 1994. *Regulation: Legal Form and the Economic Theory.* Oxford: Clanrendon Press.

Kolstad, C.D. et al. 1990. 'Ex Post Liability for Harm vs Ex Ante Safety Regulation: Substitutes or Complements', *American Economic Review,* 80(4): 888–911.

Prasad, P.M. 1999. 'The Relative Efficiency of Liability vs Regulation in Providing Incentives to the Tortfeasor', *Journal of the Indian Law Institute,* 41(3&4): 405–28.

———. 2004. 'Environmental Protection: The Role of Liability System in India', *Economic and Political Weekly,* 39(3): 257–69. 17–23 January.

Rose-Ackerman, S. 1992. *Rethinking the Progressive Agenda.* New York: The Free Press.

Sankar, U. *Law and Institutions Relating to Environmental Protection in India,* (Unpublished).

Shavell, S. 1984. 'Liability for Harm versus Regulation of Safety', *Journal of Legal Studies,* 13(2): 357–74.

———. 1980. 'Strict Liability versus Negligence', *Journal of Legal Studies,* 9(1): 1–25.

———. 1984. 'A Model of the Optimal Use of Liability and Safety Regulation', *Rand Journal of Economics,* 15(2): 271–80.

SCI 1990–1999. Public Interest Litigation, Public Interest Litigation Cell, Supreme Court of India (SCI), New Delhi.

19

Successful Voluntary Participation in Common Property Resource

Lekha Mukhopadhyay[1]

Abstract: This chapter examines the potential of successful voluntary participation in effort regulation programme to check over utilisation and thus degradation of Common Property Resources (CPR) in a society where CPR is nested under private property regime with inequality. With a hypothetical example of agro-pastoral village community under the threat of forest degradation and thus decline of fodder in the coming future, it shows that potential decline in marginal effectiveness of effort will be more for a 'rich' cattle owner. This cannot however lead to the Olson conclusion—the rich will have more incentive to restrain use of fodder resources. It is assumed that fodder collector's potential urge to increase per cattle effort in future decreases as marginal effectiveness of effort due to overuse of forest, decreases. In a game-theoretic framework, in the context of voluntary participation each community member individualistically tries to solve the problem of optimal allocation of present and future effort in fodder collection. The solution is path-dependent: the individual's choice of deploying effort at present depends on the remaining stock of fodder (CPR)—determined by the community's total effort in the past for fodder collection. While estimating optimal allocation of effort between two periods under threat of forest degradation in the coming future, herder's Nash optimal effort is shown to be dependent on four parameters: (1) herd size; (2) fodder stock condition in the forest; (3) effort endowment restriction effect; and (4) complementary (between effort and milk production) restriction effect. These theoretical results practically send the messages to the policy makers in the participatory forest management for wildlife conservation in forest reserves in developing countries. Open grazing is a perennial problem, having dangerous impact on wildlife conservation. These theoretical results can throw some light to indicate why various types of effort regulation measures like stall feeding, periodical closure, restricting days to enter forest etc. are not being so successful.

INTRODUCTION

Collective action for conservation of Common Property Natural resources (CPR) may take various forms: (1) development of institutions for rules and regulations for management of CPR; (2) mobilisation of private resources like effort and money for protection and maintenance of CPR; (3) coordination of activities to minimise the congestion externality in CPR field; and

(4) information sharing, like, sharing the CPR harvesting technology and so on. The practical experiences with various institutions for CPR management however have developed a set of theoretical puzzles; two of which to mention in the present context are concerned with: (*i*) relation between heterogeneity and prospective collective action in CPR management; and (*ii*) relation between physical condition of CPR and prospective collective management. Sometimes the community's non-acceptance of collective management institutions of CPR endogenously produces heterogeneity like income-wealth variability and so on. Sometimes heterogeneities themselves like locational differences (in case of head-end and tail-end farmers in irrigation system), income-wealth variability (like cattle ownership, ownership of fishing net, agricultural land holding etc.), differential time preferences (that is, preference between present and future consumption; Ostrom 1990), exogenously determine the prospective collective action (Baland and Patteau 2003). The physical condition of CPR also plays an important role in framing the community's decision to cooperate or not to cooperate in collective conservation programme. The villages with acute water scarcity exhibit less cooperative management, as studied in some villages in Mexico and South India (Bardhan and Dayton 2002). If due to degradation, yield of CPR is unpredictable and risky and collective management can generate some risk pooling and risk sharing benefit, then the possibility of cooperation increases (Runge 1981; McKloskey 1976). In villages in the Swiss Alps, in fertile lands in the lower valley, private appropriations easily occur in contrast to arid highlands used as community pasture under the management of village councils (Netting 1972, 1976, 1981). In the backdrop of ecological and economic complexity constituted by resource condition and heterogeneous benefit and cost of different resource users from common resources, collective action in CPR management can evolve successfully if overall for the community, there exists a possibility of forming at least one minimal coalition of CPR users for whom benefit from enforcement of collective management rules is greater than over all costs—(*i*) upfront cost of time and effort for devising rules, (*ii*) short-term cost for self restrained strategies, and (*iii*) long-term cost for maintenance of rules (Ostrom 1999).

In a heterogeneous society with unequal distribution of private wealth or resources, CPR users may have different impact of resource degradation over a time period on the effort deployed for harvesting common resources as well as on the output-benefit using CPR. Due to heterogeneity in Private Property Resources (PPR), used along with CPR in production of some private benefit, for different 'rich' and 'poor' CPR users, different types of effort endowment restriction (that is, effort to collect CPR per unit of PPR) and complementarity restriction (that is, maximum CPR that can be used as a complement to PPR) may work as binding constraints on their optimal choice of deploying effort in CPR field (Mukhopadhyay 2002). All these have serious implications on initiating collective regulation regime in any of the forms referred to at the beginning. In order to launch a regulated management regime on common property field, peoples' action in a non-regulated manner (that is, in a non-cooperative fashion) to bring it about is required. This is the essence of voluntary collective action. Harnessing the linkage between resource quality and wealth inequality, the main focus of this chapter is addressed to a set of questions: (*i*) Who among the 'poor and 'rich' PPR owners will be more interested in launching CPR regulation?; and (*ii*) How is one to determine the potential of success of CPR regulation in an economically heterogeneous society under the threat of resource degradation? To lend concreteness to the problem, the chapter considers the example of a community forestland (CPR), which is the only source of fodder for milk production (a private good) in a community, heterogeneous with respect to cattle holding. The chapter examines the potential of success of launching effort regulation (for collecting fodder) through peoples' voluntary response to the regulation. In the proposed analytical framework, the results from the theoretical exercises show that given resource condition, difference in responses to regulation (in terms of restraining per cattle effort in collecting fodder) occurs due to the difference in net benefit. Resource condition determines the effectiveness of effort per unit of cattle deployed for collection of fodder. This chapter shows, given the resource condition, how three important factors, that is, unequal distribution of private property resources (PPR; cattle holding), the degree of substitutability of PPR by the effort used for appropriation of CPR (fodder), and the degree of complementarity of PPR (cattle holding) to CPR (fodder), make a difference in responses to regulation vis-à-vis difference in private benefit (in terms of milk production).

The chapter has been organised as follows. Section 2 considers a general outline of the proposed model for thematic presentation. Section 3 gives a formal presentation of the model with derived propositions. Section 4, after compiling the results from Section 3, reaches the conclusion.

A GENERAL OUTLINE OF THE MODEL

More and more CPR is congested, effectiveness of deploying effort in CPR field declines. In the context of inequality in distribution of private resources (PPR) used with CPR for production of private good, effectiveness declines with the size of holding PPR. Now suppose some agent (say conservationist) external to the community, to check congestion, wants to launch effort regulation through voluntary collective action. Whether each member would like to participate in effort regulation programme or not, is to be determined by solving his problem of optimal allocation of effort between 'present' and 'future' in the CPR field, which is a path-dependent solution.

Consider an agro-pastoral community in which the forest is the only source of fodder for feeding cattle to produce milk output. The forest is restricted to the community members only. If regulation is introduced, there will be effort regulation in terms of say, restricting weekly the number of days for each member household to go to forest for collecting fodder. The more the forest-pasture is degraded for congestion the farther the herder has to go from the village to collect fodder. This implies that he has to spend more time for fodder collection for each of the cattle owned so that the effectiveness of effort for milk output through the effort for fodder collection will decline.[2] Since the regulation programme is proposed to occur through voluntary collective action, the choice to follow (contend) or not to follow (defect) regulation is open to each community member.

The theme of this chapter has been textured in terms of a two-player-two-stage (considered here as two periods, 'present' and 'future') game in community forest field with backward induction strategies. 'Players' are heterogeneous with respect to cattle holding, that is, cattle property (which is a private property resource for milk production) is unequally distributed. All members of the community are engaged in production of milk and for that, they depend on forest, the CPR, which is the only source of fodder. The pasture has a finite stock of fodder, which grows in these two periods at a constant rate of the initial stock.

Now the production of milk of each of the member of the society depends on (*i*) the size of her cattle property; (*ii*) the effort she puts for each cattle for collection of fodder; and (*iii*) some externality regarding how much fodder from the existing stock has been collected by other members in the society, given her own collection. Her production of milk (which is also considered as payoff in the model) is solely dependent on her own effort/action (that is, independent of other's effort/action in the community), as long as total effort deployed by the society for collection of fodder does not exceed the total stock. In other words, how far per cattle effort deployed in collection of fodder would be effective for production of milk depends upon the congestion externality. If the forest for fodder collection is not congested, a herder may get production of milk, as much as effort she puts upto some optimal level of fodder given the number of cattle she owns (cattle are assumed to be equally productive) and given the production technology.

The problem of congestion externality in the model has been transcribed with a dynamic perspective in two periods game. The set of players and distribution of cattle property are assumed to remain the same between these two periods.[3] Given the backward induction strategy, the individual player being at the 'present' (stage 1) anticipates the effect of her present period collection of fodder and that of her opponent, on her future period collection and the future period stock of fodder. On the basis of this anticipation, she chooses the present period's sub-game perfect Nash equilibrium strategy to maximise milk production (payoff) in these two periods. Since our model deliberately assumes non-existence of past before period 1 and non-existence of any future beyond period 2, given the finite stock of fodder, assuming equal endowment of effort for each player in each period, this finite two-stage game of perfect information shows how individual players with different size of cattle holding in a non-cooperative way solve the problem of optimal allocation of per cattle effort (vis-à-vis, per cattle collection of fodder) between these two periods.

Among the infinite number of choices of action paths of the players, this model however restricts to a few of them. It considers the phenomenon if someone wants to introduce restriction on effort for fodder collection at 'present', what would be the Nash choice of action/per cattle effort, given that there is an overall threat of breaking the rules for fodder collection in future in the community.

Regarding the allocation of effort in the model, two possible restrictions on per cattle effort have been taken into consideration. First, it is assumed that total effort endowment is fixed for each player (say twenty-four hours a day). Maximum possible effort per cattle for collecting fodder decreases with increase in the number of cattle holding. If we assume that effort cannot be hired, then for a declining stock of fodder, in the collection of it, it may not be possible to increase per cattle effort further for the 'rich' (in terms of cattle holding) player. This binding constraint for the 'rich' player on effort allocation is termed as effort endowment restriction. The second kind of restriction on per cattle effort comes out from technological complementarity of fodder (the CPR units) to milk output. If there is sufficient stock of fodder, more effort for collecting fodder per cattle increases more milk production but up to a certain point, not beyond that. This complementarity restriction in turn imposes restriction on substitutability of cattle by effort. Because of this complementarity restriction, one cannot compensate for the loss of milk production (that is, the loss of payoff) due to small size of cattle property merely by increasing per cattle effort for collection of fodder. This 'complementarity restriction' may be the binding constraint for the 'poor' player on his effort allocation problem. The consequences of both these binding constraints on players' decisions to contend or defect effort regulation have been examined in the chapter.[4]

The general conclusion we derived from our theoretical exercises is that, in a private property regime with unequal distribution of private property (the cattle property in our model) equal effort per unit of PPR (vis-à-vis., equal level of per cattle collection of fodder) doesn't necessarily lead to equal benefit (that is, equal level of production of milk) and vice versa.

A SIMPLE COMMON PROPERTY RESOURCES GAME IN COMMUNITY FOREST FOR FODDER COLLECTION WITH BACKWARD INDUCTION STRATEGIES

Model Specification

Consider a society of two players, N = {1, 2}, with a finite time horizon of two periods, T = {1, 2}, the 'present' and 'future', with no 'past'. The life span of each player is assumed to cover these two periods. There is a common property resource, say, forest, with a finite stock of fodder, which at the beginning of the game is S and grows by Δ between these two periods. $\Delta < S$, is a simple follow-up of the existing trend in the literature handling the dynamics of the natural growth of renewable resources: growth of resources is a decreasing function of the size of resource stock. (Jorgensen and Yeung (1999)).

The players collect fodder from the forest to feed the cattle since open grazing in this society is not allowed. The cattle are homogeneous in terms of productivity. There is inequality in the distribution of cattle properties (K), so that the number of cattle owned by the *ith* player, is assumed to be less than the number of cattle owned by the *jth* player (assuming, $i = 1$ and $j = 2$). For each of the *ith* player K_i is assumed to be the same across the periods and so is K_j.

The interaction of the players determines the amount of fodder that each player will collect from the forest in each period. The final outcome however is independent of players' decisions. Since this is a two period game, the move in the second period ('future') is conditioned by the outcome of the first, ('present'), that is, by the history of the game till the second stage is reached. This implies that each of the player's strategy is a complete plan of action for the whole game.

Let a_i^t effort per unit of cattle (expressed in terms of labour hours), deployed for collection of fodder, be the action variable of the *ith* player in period t. Given the number of cattle K_i fixed, and given the fixed endowment of total effort E_i for each player in each period: $a_i^t \in \{0, \frac{E_i}{K_i}\} \rightarrow \Re_+$. It is assumed that, $E_i = E_j$, so that total effort endowment of each of the player in each period is the same although the per cattle effort endowment of *ith* player is less than that of *jth* player, that is, $\frac{E_i}{K_i} < \frac{E_j}{K_j}$. At the beginning of the period 1 say, 'present' since the stock of fodder is $S^1 = S$, and between these two periods, the stock grows by Δ, in period 2, say 'future', the maximum available fodder (if nothing is used in the 'present') is, $(S + \Delta)$.

Specification of Production (or Payoff) Function

The production function (also the payoff function) of milk in this model has two parts; the first part considers the total effort used for collection of fodder ($= K_i a_i^t$) and the second part constitutes the 'effectiveness' of effort (ψ_i^t) which depends on congestion externality effect. If there is no congestion externality, that is, there is enough stock, let $\psi_i^t = 1$ and less than 1 if congestion matters:

D1. The production function (also the payoff function) of milk of the *ith* player at period *t*, Q_i^t is defined as a function of total effort, (= $K_i a_i^t$), and the 'effectiveness' function ψ_i^t, such that,

$$Q_i^t = K_i a_i^t \psi_i^t$$

where, $\psi_i^t = \dfrac{S^t}{\sum K_i a_i^t}$

$\dfrac{S^t}{\sum K_i a_i^t} \geq 1$, that is, stock is not overexploited $\psi_i^t = 1$. If otherwise, $0 \leq \dfrac{S^{\langle t \rangle}}{\sum K_i a_i^{\langle t \rangle}}) < 1; 0 \leq \psi_i^{\langle t \rangle} < 1.$[5] As $\sum K_i a_i^{\langle t \rangle}$ increases beyond $S^{\langle t \rangle}$ the effectiveness of effort $\psi_i^{\langle t \rangle}$ declines, taking the values less than one. The effectiveness curve is graphically shown in Figure 19.1.

Figure 19.1 Effectiveness Curve

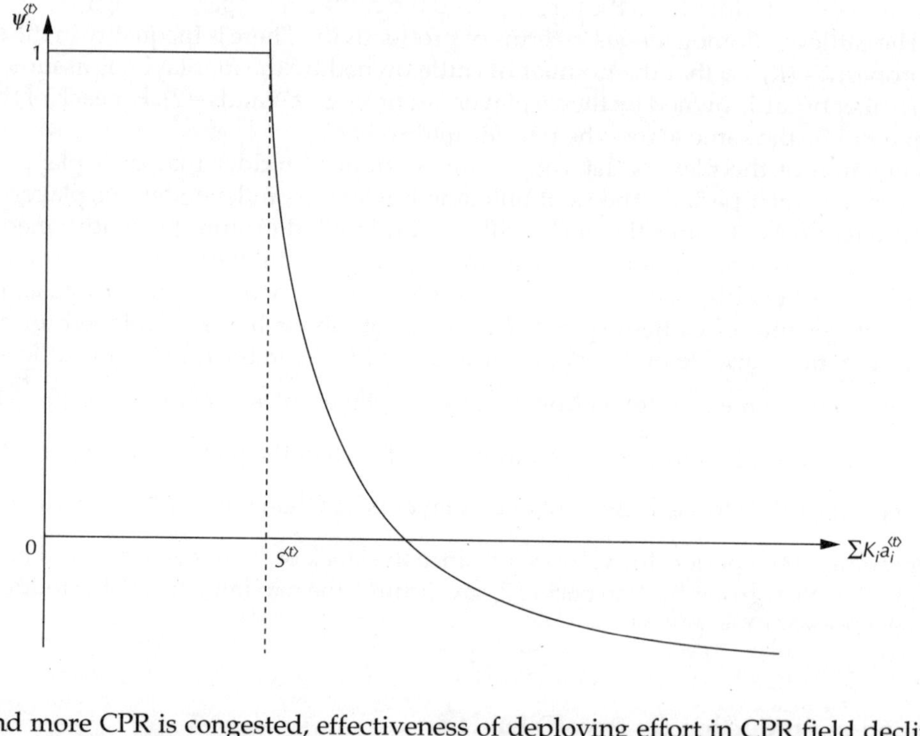

More and more CPR is congested, effectiveness of deploying effort in CPR field declines. Here the objective of each individual member in the community is optimal allocation of effort at present and at future in the CPR field, which is a path dependent solution. The effectiveness of deploying effort in CPR field has an inflictive role on players' optimal allocation problem.

Compementarity Restriction on Milk Production or Payoff Function

In the production or payoff function, per cattle effort for fodder collection, a_i^t is complementary to the milk production. Other things remaining the same, as per cattle fodder collection increases, as a complementary effect milk production also increases. But there is a limit to this complementarity. There is so far no restriction imposed on complementarity. Given the number of cattle K_i, if the effectiveness factor $\psi_i^t = 1$, that is, forest is not congested, milk production increases with a_i^t up to say, $\underline{a}$. Beyond $\underline{a}$ with ψ_i^t being equal to 1, milk production per cattle remains unchanged. If effectiveness $\psi_i^t < 1$, that is, forest is congested, more effort will be required for each cattle to collect fodder. In that case, complementarity restriction will be different; say, $\bar{a}$. Obviously, $\underline{a} < \bar{a}$.

Incorporating complementarity restriction into the production (also the payoff) function, in this way, we are now able to handle two distinct possibilities; one, where, complementarity restriction acts as a binding constraint and the other, where the complementarity restriction does not bind the individual's choice of per cattle action in the common property field. If for the ith player, com plementarity restriction acts as the binding constraint, there are two possibilities:

(C1) $Q_i^t = K_i\underline{a}$, if forest is not congested;
(C2) $Q_i^t < K_i\bar{a}$, if forest is congested; $\bar{a} > \underline{a}$.

Possibility (2) indicates that if forest is congested, and ith player has already reached the limit of complementary effect $\bar{a}$, because of congestion externality he would get milk output less than $K_i\bar{a}$. If for example, $\psi_i^t = \frac{1}{2}$, then $Q_i^t = K_i\frac{\bar{a}}{2}$ and this less than he could otherwise get through maximum complementary effect if forest was not congested, that is, $K_i\frac{\bar{a}}{2} < K_i\underline{a}$. Complementarity restriction in other words, makes the production/payoff function discontinuous for $a_i^t > \underline{a}$ (or, $a_i^t > \bar{a}$) and other things remaining the same, the marginal productivity of $a_i^t > \underline{a}$ (or $a_i^t > \bar{a}$) becomes zero.[6]

Two-period Milk Production or Payoff Function

Since community member-players in the model are solving their problem of optimal allocation of effort between two periods: 'present' and 'future', they are concerned with two period milk production or payoff function

$$
\begin{aligned}
Q_i &= Q_i^{\langle 1\rangle}(a_i^{\langle 1\rangle}, a_j^{\langle 1\rangle}) + Q_i^{\langle 2\rangle}(a_i^{\langle 2\rangle}(a_i^{\langle 1\rangle}, \hat{a}_j^{\langle 1\rangle}), a_j^{\langle 2\rangle}(a_i^{\langle 1\rangle}, a_j^{\langle 1\rangle}))\\
&= K_i a_i^{\langle 1\rangle}\psi_i^{\langle 1\rangle} + K_i a_i^{\langle 2\rangle}\psi_i^{\langle 2\rangle}\\
&= K_i a_i^{\langle 1\rangle}\frac{S^{\langle 1\rangle}}{\sum K_i a_i^{\langle 1\rangle}} + K_i a_i^{\langle 2\rangle}\frac{S^{\langle 2\rangle}}{\sum K_i a_i^{\langle 2\rangle}}
\end{aligned}
$$

where, $S^{\langle 2\rangle} = (S^{\langle 1\rangle} - \sum K_i a_i^{\langle 1\rangle}) + \Delta$

The first part of the two period milk production or payoff is the function of the set of actions of players in period 1 $(a_i^{\langle 1\rangle}, a_j^{\langle 1\rangle})$, while the second part is the function of actions of period 2, $(a_i^{\langle 2\rangle}(a_i^{\langle 1\rangle}, \hat{a}_j^{\langle 1\rangle}), a_j^{\langle 2\rangle}(a_i^{\langle 1\rangle}, a_j^{\langle 1\rangle})$. The choice of action in period 2 is contingent upon actions chosen in period 1.

DECENTRALISED NASH SOLUTION IN COMMUNITY FOREST GAME

The Nash solution in the two-stage community forest game with backward induction strategies gets solved by:

$$\underset{a_i^1}{Max}[Q_i^{\langle 1\rangle}(a_i^{\langle 1\rangle}, a_j^{\langle 1\rangle}) + Q_i^{\langle 2\rangle}(a_i^{\langle 2\rangle}(a_i^{\langle 1\rangle}, \hat{a}_j^{\langle 1\rangle}), a_j^{\langle 2\rangle}(a_i^{\langle 1\rangle}, a_j^{\langle 1\rangle}))]$$

that is, by

$$\underset{a_i^1}{Max}[K_i a_i^{\langle 1\rangle}\frac{S^{\langle 1\rangle}}{\sum K_i a_i^{\langle 1\rangle}} + K_i a_i^{\langle 2\rangle}\frac{S^{\langle 2\rangle}}{\sum K_i a_i^{\langle 2\rangle}}] \quad (1)$$

The equation (1) is used as generic equation in the model. In period 2, sub-game perfect equilibrium $[a_i^{\langle 2\rangle}(\hat{a}_i^{\langle 1\rangle}, \hat{a}_j^{\langle 1\rangle}), a_j^{\langle 2\rangle}(\hat{a}_i^{\langle 1\rangle}, \hat{a}_j^{\langle 1\rangle})]$ is obtained by solving each of the following equations:

$$\frac{\partial}{\partial a_i^{\langle 2\rangle}}[k_i a_i^{\langle 2\rangle}\psi_i^{\langle 2\rangle}(.)] = 0 \quad (2)$$

$$\frac{\partial}{\partial a_j^{\langle 2\rangle}}[k_j a_j^{\langle 2\rangle}\psi_j^{\langle 2\rangle}(.)] = 0 \quad (3)$$

Solving (5), we get:

$$\psi_i^{\langle 2\rangle}(.) + a_i^{\langle 2\rangle}\frac{\partial\psi_i^{\langle 2\rangle}(.)}{\partial a_i^{\langle 2\rangle}} = 0 \quad (3.1)$$

$\frac{\partial\psi_i^{\langle 2\rangle}}{\partial a_i^{\langle 2\rangle}}$ in the second part of L.H.S of (6.!) shows the marginal effectiveness of effort in period 2, if forest is congested, given the action of period 1 (that is, 'present'). Since, $\frac{\partial\psi_i^{\langle 2\rangle}}{\partial a_i^{\langle 2\rangle}} = -\frac{\psi_i^{\langle 2\rangle}}{\sum K_i a_i^{\langle 2\rangle}}K_i$, and $\frac{\partial}{\partial\psi_i^{\langle 2\rangle}}(\frac{\partial\psi_i^{\langle 2\rangle}}{\partial a_i^{\langle 2\rangle}}) = -\frac{K_i}{\sum K_i a_i^{\langle 2\rangle}}$ we can develop Proposition 1.

Proposition 1 Given the action of period 1 (that is, 'present'), more and more forest is congested; effectiveness of effort in period 2 (that is, 'future') for collection of fodder (that is, marginal effectiveness) declines in general and it declines more for 'rich' (in terms of cattle property) compared to 'poor'.

Corollary 1.1 Lesser the marginal effectiveness of effort in period 2 for the rich $(= -\frac{K_i}{\sum K_i a_i^{\langle 2\rangle}})$, lesser will be the per cattle effort for fodder collection

This is because from (3.1) we get $a_i^{\langle 2\rangle} = \frac{\psi_i^{\langle 2\rangle}(\Sigma K_i a_i^{\langle 2\rangle})^2}{S^{\langle 2\rangle} K_i}$, which decreases with the size of K_i (Proof in Appendix A)

With an hypothetical example of the values of effectiveness of effort ranging between 0 and 1 in a community with two players *P* and *R* heterogeneous in terms of cattle property (shown in Table 19.1) the marginal effectiveness curves for 'rich' and 'poor' have been plotted graphically in Figure 19.2.

Table 19.1 Marginal Effectiveness of 'Rich' (*R*) and 'Poor' (*P*) in a Hypothetical Example

$\psi^{(2)}_{i=P.R}$	$\Sigma K_i a^{(2)}_{i=P.R}$	K_P	K_R	$\frac{\psi_i^{<2>}}{\sum K_i a_i^{<2>}} K_P$	$\frac{\psi_j^{<2>}}{\sum K_i a_i^{<2>}} K_R$
0.1	100	20	30	0.020	0.030
0.2	90	20	30	0.040	0.067
0.3	80	20	30	0.075	0.113
0.4	70	20	30	0.011	0.171
0.5	60	20	30	0.167	0.250
0.6	50	20	30	0.240	0.360
0.7	40	20	30	0.350	0.525
0.8	30	20	30	0.533	0.800
0.9	20	20	30	0.900	1.350
1.0	10	20	30	2	3

Potentiality of Voluntary Acceptance of Effort Regulation in Community Forest at 'Present' When There is Threat of Breaking Regulation in 'Future'

The practical experiences with voluntary collective management of community forest and other types of common property resources show a large number of instances where the collective regulation has been broken in phase 2 after successful launching of the regulation programme in phase 1. It is plausible to assume that while introducing voluntary collective regulation of effort for fodder collection from community forest, the community itself and each constituent member faces threat of breaking rules and regulations in future. In our proposed theoretical framework this implies that we are considering only the action path in the two-stage community forest game with backward induction strategies, which is characterised by congestion in the period 2 (that is, 'future') that is, $\psi_i^{\langle 2\rangle} < 1$, $\frac{\partial \psi_i^{\langle 2\rangle}}{\partial a_i^{\langle 2\rangle}}$, < 0 and also $\frac{\partial}{\partial \psi_i^{\langle 2\rangle}} (\frac{\partial \psi_i^{\langle 2\rangle}}{\partial a_i^{\langle 2\rangle}}) < 0$; but $\psi_i^{\langle 1\rangle} = 1$, which means forest is not congested in period 1 (that is, 'present') due to regulation.

Since the strategy of the game is backward induction, sub-game perfect equilibrium solution in stage 2 will be rolled back into the stage 1 game. In stage 2 there are two possible cases: Case (1): Neither by complementarity restriction nor by effort endowment restriction, per cattle effort for fodder collection is bounded; and, Case (2): Per cattle effort for fodder collection is bounded by either of these restrictions for some player.

Figure 19.2 Marginal Effectiveness Curve of 'Rich' and 'Poor' Cattle Owner

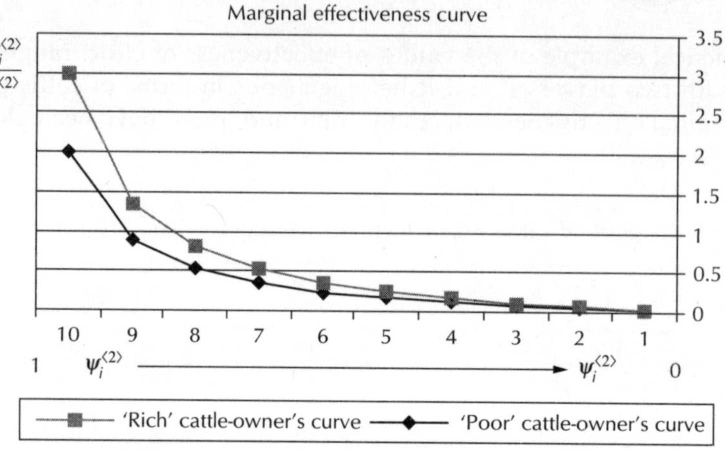

Case 1: Per cattle effort for fodder collection is not bounded by any restriction

Solving (6.1) for ith and similarly for jth players we get following two reaction equations:

$$\psi_i^{\langle 2\rangle} - \frac{K_i a_i^{\langle 2\rangle}}{(\sum K_i a_i^{\langle 2\rangle})^2} S^{\langle 2\rangle} = 0$$

$$\psi_j^{\langle 2\rangle} - \frac{K_j a_j^{\langle 2\rangle}}{(\sum K_i a_i^{\langle 2\rangle})^2} S^{\langle 2\rangle} = 0$$

$\psi_i^{\langle 2\rangle} = \psi_j^{\langle 2\rangle}$; $\frac{K_i}{(\sum K_i a_i^{\langle 2\rangle})^2} S^{\langle 2\rangle}$ in the second part of the reaction equation shows the slope of reaction curves. Since $\frac{S^{\langle 2\rangle}}{(\sum K_i a_i^{\langle 2\rangle})^2}$ is the common part in both the equations, slopes of reaction curves vary according to the size of the cattle property $K_i(K_j)$.

As a sub-game perfect solution in stage 2, we get:

$$K_i a_i^{\langle 2\rangle} = K_j a_j^{\langle 2\rangle} \tag{4}$$

which means; $a_i^{\langle 2\rangle} = \frac{K_j}{K_i} a_j^{\langle 2\rangle}$; By the model specification $\frac{K_j}{K_i} < 1$. To distinguish between two players in terms of cattle property, let us denote 'rich' ith player now and later by the suffix r and 'poor' by p and thus $a_r^{\langle 2\rangle} < a_p^{\langle 2\rangle}$.

Plugging the value from (7) i.e $\sum K_i a_i^{\langle 2\rangle} = 2K_i a_i^{\langle 2\rangle}$, into the generic equation (1), we get the decentralised Nash solution in per cattle effort:

$$\tilde{a}_r^{\langle 1\rangle} = \frac{1-\frac{1}{2}\psi_r^{\langle 2\rangle}(1+\psi_r^{\langle 2\rangle}\frac{\partial a_r^{\langle 2\rangle}}{\partial a_r^{\langle 1\rangle}})}{K_r} S^{\langle 1\rangle} \quad (5.1)$$

$$\tilde{a}_p^{\langle 1\rangle} = \frac{1-\frac{1}{2}\psi_p^{\langle 2\rangle}(1+\psi_p^{\langle 2\rangle}\frac{\partial a_p^{\langle 2\rangle}}{\partial a_p^{\langle 1\rangle}})}{K_p} S^{\langle 1\rangle} \quad (5.2)$$

$\tilde{a}_r^{\langle 1\rangle}$ and $\tilde{a}_p^{\langle 1\rangle}$ must satisfy the effort endowment restriction; that is, $\tilde{a}_r^{\langle 1\rangle} \le \frac{E}{K_r}$ and $\tilde{a}_p^{\langle 1\rangle} \le \frac{E}{K_p}$. From (8.1) and (8.2), the results in Nash Solution, we can set a number of propositions:

Proposition 2: The Nash optimal level of per cattle effort for fodder collection from community forest in 'present' under the threat of forest degradation in 'future' will

(*i*) increase with present stock of fodder; and
(*ii*) decrease with greater anticipated rate of forest degradation (that is, decrease in effectiveness of effort) in future

$\frac{\partial a_r^{\langle 2\rangle}}{\partial a_r^{\langle 1\rangle}}$ (and $\frac{\partial a_p^{\langle 2\rangle}}{\partial a_p^{\langle 1\rangle}}$) in the Nash solution 5.1 (and 5.2) indicates the respective player's choice to change per cattle effort in period 2 ('future') in response to change in effort in period 1 ('present'). Degradation of forest due to congestion and therefore decline in effectiveness of effort is already presumed, so that, sign of $\frac{\partial a_r^{\langle 2\rangle}}{\partial a_r^{\langle 1\rangle}}$ (and $\frac{\partial a_p^{\langle 2\rangle}}{\partial a_p^{\langle 1\rangle}}$) must be positive if no complementarity restriction works as binding restriction on per cattle effort for fodder collection in period 1. It is shown in corollary 1 above that due to congestion in period 2, marginal effectiveness of effort is lesser for 'rich' player compared to 'poor' player and $a_r^{\langle 2\rangle} < a_p^{\langle 2\rangle}$. If for 'rich' per cattle effort in period 2 is lesser than that of the 'poor', we can make the following assumption:

Assumption: $0 \le \frac{\partial a_r^{\langle 2\rangle}}{\partial a_r^{\langle 1\rangle}} < \frac{\partial a_p^{\langle 2\rangle}}{\partial a_p^{\langle 1\rangle}}$, that is, while increasing per cattle effort due to congestion externality in period 2, 'rich' player will increase less compared that of 'poor'.

Comparing between (5.1) and (5.2), we see that the ratio of two numerators $\frac{1-\frac{1}{2}\psi_r^{\langle 2\rangle}(1+\psi_r^{\langle 2\rangle}\frac{\partial a_r^{\langle 2\rangle}}{\partial a_r^{\langle 1\rangle}})}{1-\frac{1}{2}\psi_p^{\langle 2\rangle}(1+\psi_p^{\langle 2\rangle}\frac{\partial a_p^{\langle 2\rangle}}{\partial a_p^{\langle 1\rangle}})} > 1$

which implies $K_r\tilde{a}_r^{\langle 1\rangle} > K_p\tilde{a}_p^{\langle 1\rangle}$; that is, total effort for fodder collection and thus total benefit from community forest of the 'rich' player will be greater than that of the 'poor'.

Now from the whole game plan in the optimal choice per cattle effort of 'rich' player will be greater than that of 'poor' that is, $\tilde{a}_r^{\langle 1\rangle} > \tilde{a}_p^{\langle 1\rangle}$ if:

$$\frac{1-\frac{1}{2}\psi_r^{\langle 2\rangle}(1+\psi_r^{\langle 2\rangle}\frac{\partial a_r^{\langle 2\rangle}}{\partial a_r^{\langle 1\rangle}}}{1-\frac{1}{2}\psi_p^{\langle 2\rangle}(1+\psi_p^{\langle 2\rangle}\frac{\partial a_p^{\langle 2\rangle}}{\partial a_p^{\langle 1\rangle}}} > \frac{K_r}{K_p} \tag{6.1.1}$$

Choice of per cattle effort of 'rich' player on the other hand, will be lesser than that of 'poor' that is, $\tilde{a}_r^{\langle 1\rangle} < \tilde{a}_p^{\langle 1\rangle}$ if:

$$1 < \frac{1-\frac{1}{2}\psi_r^{\langle 2\rangle}(1+\psi_r^{\langle 2\rangle}\frac{\partial a_r^{\langle 2\rangle}}{\partial a_r^{\langle 1\rangle}}}{1-\frac{1}{2}\psi_p^{\langle 2\rangle}(1+\psi_p^{\langle 2\rangle}\frac{\partial a_p^{\langle 2\rangle}}{\partial a_p^{\langle 1\rangle}}} < \frac{K_r}{K_p} \tag{6.1.2}$$

Now effort regulation through voluntary participation will be potentially successful if regulator fixes effort at a^*, such that:

$a^* \geq \tilde{a}_r^{\langle 1\rangle} > \tilde{a}_p^{\langle 1\rangle}$, in case (6.1.1) occurs, or
$a^* \geq \tilde{a}_p^{\langle 1\rangle} > \tilde{a}_r^{\langle 1\rangle}$, if (6.1.2) occurs

Irrespective of greater ness of per cattle effort optimally chosen by 'rich' or 'poor', proportion of effort to the total community effort and therefore the milk output (benefit) of the 'rich' player will be greater than that of 'poor' player; that is, $\frac{K_r\,\tilde{a}_r^{\langle 1\rangle}}{\sum K_i a_i^{\langle 1\rangle}} > \frac{K_p\,\tilde{a}_p^{\langle 1\rangle}}{\sum K_i a_i^{\langle 1\rangle}}$.

From the derived results now we can develop the following proposition:

Proposition 3: If CPR (here, community forest) is nested in private property regime with inequality in distribution of PPR (here cattle property):

(*i*) Per cattle effort for fodder collection of the 'rich' cattle owner will be greater than that of the 'poor' if the output benefit using community forest of the 'rich' in proportion to that of the 'poor' is greater than the cattle holding of the 'rich' in proportion to that of the 'poor';
(*ii*) Per cattle effort for fodder collection of the 'poor' cattle owner will be greater than that of the 'rich' if the output benefit using community forest of the 'rich' in proportion to that of the 'poor' is lesser than the cattle holding of the 'rich' in proportion to that of the 'poor'; and
(*iii*) CPR regulation through voluntary participation will be successful if regulation sustains inequality in benefit from CPR. (Proof is given in the Appendix).

The proposition 3(i) and 3(ii) however cannot rule out effort endowment restriction, that is, $\tilde{a}_r^{\langle 1\rangle} \leq \frac{E}{K_r}$ and $\tilde{a}_p^{\langle 1\rangle} \leq \frac{E}{K_p}$ and, also $\frac{E}{K_r} < \frac{E}{K_p}$. If in any case, effort endowment restriction becomes the binding constraint for 'rich' player, that is, $\tilde{a}_r^{\langle 1\rangle} = \dfrac{E}{K_r}$ then case (3.1): will not arise; that is, in spite of greater potential output benefit from the forest, the 'rich' player will not be able to deploy greater per cattle effort. In this case, it would be his potential loss. This is made under the assumption that labour hiring is not possible. If in that case he can hire labour and the cost of hiring labour is less than or equal to $(K_r\tilde{a}_r^{\langle 1\rangle} - E)$, then he can still invest greater effort for fodder collection to reap greater benefit

Case 2: Per cattle effort for fodder collection is bounded by complementarity restriction.
Complementarity restriction in period 1: Let in period 1 complementarity restriction ($\bar{a}$) work as binding constraint. $\bar{a} \in \Re_+$, that is, $\bar{a}$ is any real number which is exogeneously given. Now we may consider the following four logical possibilities in each of the cases (6.1.1) and (6.1.2) with different ranges of values that $\bar{a}$ can take in relation to the first stage sub-game values:

Case (6.1.1): $\tilde{a}_r^{\langle 1\rangle} > \tilde{a}_p^{\langle 1\rangle}$
Possibility 1.1: $\tilde{a}_r^{\langle 1\rangle} > \tilde{a}_p^{\langle 1\rangle} \geq \bar{a}$
Possibility 1.2: $\tilde{a}_r^{\langle 1\rangle} \geq \bar{a} > \tilde{a}_p^{\langle 1\rangle}$
Possibility 1.3: $\bar{a} \geq \tilde{a}_r^{\langle 1\rangle} > \tilde{a}_p^{\langle 1\rangle}$

Case (6.1.2): $\tilde{a}_r^{\langle 1\rangle} < \tilde{a}_p^{\langle 1\rangle}$
Possibility 2.1: $\tilde{a}_p^{\langle 1\rangle} > \tilde{a}_r^{\langle 1\rangle} \geq \bar{a}$
Possibility 2.2: $\tilde{a}_p^{\langle 1\rangle} \geq \bar{a} > \tilde{a}_r^{\langle 1\rangle}$
Possibility 2.3: $\bar{a} \geq \tilde{a}_p^{\langle 1\rangle} > \tilde{a}_r^{\langle 1\rangle}$

Now to make effort regulation through voluntary participation successful if regulator fixes effort at a^*, such that: $a^* \geq \tilde{a}_r^{\langle 1\rangle} > \tilde{a}_p^{\langle 1\rangle}$, in case (6.1.1), in possibility 1.1 this implies $a^* \geq \tilde{a}_r^{\langle 1\rangle} > \tilde{a}_p^{\langle 1\rangle} \geq \bar{a}$. In possibility 1.2, it leads to $a^* \geq \tilde{a}_r^{\langle 1\rangle} \geq \bar{a} > \tilde{a}_p^{\langle 1\rangle}$ and in possibility 1.3, it is little bit tricky because it is possible that either, $a^* \geq, or \leq \bar{a}$.

In possibility (1.1), since complementarity restriction becomes the binding constraint for both the players, the potentiality for successful voluntary participation is maximum.

In possibility (1.2), since complementarity restriction becomes the binding constraint for 'rich' player, the potentiality for successful voluntary participation still exists.

In possibility (1.3), if in particular, $a^* < \bar{a}$ evolving successful voluntary participation in effort regulation becomes extremely difficult.

Similar types of inferences can be made in case (6.1.2) occurs.

SUMMARY OF RESULTS

The results from the exercises under the proposed analytical framework have been expressed in terms of a set of propositions that can be summarised as follows:

Given the action of period 'present' when more and more forest is congested, effectiveness of effort in 'future' for collection of fodder (that is, marginal effectiveness) declines in general and it declines more for 'rich' (in terms of cattle property) compared to 'poor'. Lesser the marginal effectiveness of effort in 'future' for the rich, lesser will be the per cattle effort for fodder collection.

The Nash optimal level of per cattle effort for fodder collection from community forest in 'present' under the threat of forest degradation in 'future' will: (*i*) increase with present stock of fodder; and (*ii*) decrease with greater anticipated rate of forest degradation (that is, decrease in effectiveness of effort) in future.

If CPR (here, community forest) is nested in private property regime with inequality in distribution of PPR (here cattle property): (*i*) Per cattle effort for fodder collection of the 'rich' cattle owner will be greater than that of the 'poor' if the output benefit using community forest of the 'rich' in proportion to that of the 'poor' is greater than the cattle holding of the 'rich' in proportion to that of the 'poor'; (*ii*) Per cattle effort for fodder collection of the 'poor' cattle owner will be greater than that of the 'rich' if the output benefit using community forest of the 'rich' in proportion to that of the 'poor' is lesser than the cattle holding of the 'rich' in proportion to that of the 'poor'; and (*iii*) CPR regulation through voluntary participation will be successful if regulation sustains inequality in benefit from CPR.

Finally, the results in the proposed analytical framework have far-reaching implications on policy matters. Under different conditions, it shows different upper bands of effort (conformable to Nash equilbria) restriction that regulator can fix up in a community with CPR, for conservation of CPR. Since these bands are obtained from Nash equilibria and take into account complementarity and effort endowment restrictions (whenever they are binding constraints), they are also implementable through voluntary participation.

These theoretical results practically send the messages to the policy makers in the participatory forest management for wildlife conservation in forest reserves in developing countries. Open grazing is a perennial problem in all wild reserves in developing countries. It has a dangerous impact on wildlife conservation including reducing amount of fodder, spread of diseases, disappearing species and mammals. To combat these problems, forest departments of national governments are taking various policy measures. In an ongoing field study under SANDEE (South Asian Network for Development of Environmental Economics) by the author in Buxa Tiger Reserve, in north Bengal, India, it was observed that to combat open grazing problem the Government has introduced a programme to provide loans at cheap rate of interest for purchasing highly productive 'Jersey' variety of milking cows, for which only stall feeding is recommended. This, however, is not getting much favourable response from the households. The theoretical results of this chapter can throw some light to indicate why various types of effort regulation measures like stall feeding, periodical closure (allowing cattle to graze only during cultivation time and not after harvesting is over), restricting number of days to enter forest, restricting number of head loads for fodder collection, etc. are not being so successful.

CONCLUSION

With the help of a simple model of agro-pastoral village society producing milk with cattle (private property, which is unequally distributed) and fodder (which is collected from forest, a common property), this chapter examines the potential of successful voluntary participation in effort regulation programme. For the purpose of conservation of common property resources, most of the community based organisations in various forms like periodical closure of CPR field (like that in pasture land), restricting the number of head loads, or the number of days for fodder collection, restricting the size of fishing net or number of boats, etc., make attempt to introduce effort regulation programme as part of management of common property resources. If in a community there is inequality in private property resources which is used along with CPR to produce some private benefit, it affects individual's optimal choice for allocation of effort in the common property field and individual's as well as society's benefit from the common property resources. In that case, reactions on the effort regulation programme and the choices to contend or defect the rules vary from individual to individual due to inequality. It is due to inequality that reaction to resource degradation also varies among individuals. In a generalised version of a two-player, two-stage game for effort allocation for collecting fodder from the forest with backward induction strategies, the inequality issue has been handled here in two different contexts. First, it has considered a situation where inequality itself is manifested through unequal per cattle effort endowment, where per cattle effort endowment for the 'rich' cattle owner is less compared to the 'poor' owner. Second, it has considered a situation where in conjunction with per cattle effort endowment, there exists a limit on per cattle action (or effort) beyond which milk production cannot be increased, which is called complementarity restriction in the model. In this context, sub-game perfect Nash solution to the problem of individual's allocation of effort has been derived and characterised in terms of the four parameters in the model, namely, cattle property, effort endowment, stock condition of fodder in the forest in comparison with the total community effort for harvesting from the stock, and complementarity restriction.

ACKNOWLEDGEMENTS

I express my deepest gratitude to Prasanta Pattanaik for his sincere guidance, discussion and comments on this work.

NOTES

1. This chapter has been principally developed from the Working Chapter (no. 02–07), Department of Economics, University of California, Riverside, which is part of the research supported by the Ministry of Environment and Forest, Government of India, through Overseas Post-doctoral Fellowship under World Bank India Capacity Building Programme in Environmental Economics.
2. This proposition is similar to that made by Jodha (W B Discussion Chapter 1992).
3. Dasgupta 1987, 1996; Jodha, 1986, 1990 have described how poor rural folk with the help of some degree of substitutability of capital by common pool/property resources manage to survive. CPR in that context play some remissive role on inequality.
4. In order to lend concreteness to the problem, fodder collection game in community forest has been metaphorically used. The results will not significantly change if it could be otherwise the problem of effort allocation with different sizes of fishing nets (or boats) or agricultural land holding in case of coastline fishing or ground water collection for irrigation, and so on.
5. Mukhopadhyay (2002) has expressed ψ_i^t explicitly ψ_i^t as an exponential function which is as follows: $\psi_i^t = e^{x^t - 1}$ where, $x^t - 1 = \frac{S^t}{K_i a_i^t + K_j a_j^t} - 1 = \frac{S^t - (K_i a_i^t + K_j a_j^t)}{K_i a_i^t + K_j a_j^t}$, where x^t is the proportion of forest stock over the total effort of the community for collection of fodder in period t.
6. This type of complementarity restriction may provide with an alternative explanation as to why in a small society over extraction of natural resources doesn't take place. One of the possible reason is that the size of the capital (which comes from private resources) in that society is so small that over extraction is not economically profitable.

REFERENCES

Baland, J.M and J.P. Platteau. 2003. 'Economics of Common Property Management Regimes in K.G Maler and J. Vincent, (eds), *Handbook of Environmental Economics, Vol. 1*, pp. 127–90, Elsevier.

Bardhan, P.K. and J. Dayton-Johnson, 2002. 'Inequality and the Governance of Water Resources in Mexico and South India', in J.M Baland, P. Bardhan and B. Samuel (eds), *Inequality, Cooperation and Sustainability*, Santa Fe Institute.

Dasgupta, M. 1987. 'Informal Security Mechanism and Population Retention in Rural India', *Economic Development and Cultural Change*, 36(1): 101–20.

Dasgupta, P. 1993. *An Enquiry into Well Being and Destitution*. Oxford: Clarendon Press.

Jodha, N.S. 1986. 'Common Property Resources and Rural Poor in Dry Regions of India, *Economic and Political Weekly*, 21(27): 1169–82.

——. 1990. Rural Common Property Resources: Contributions and Crisis. Foundation Day Lecture, New Delhi: Society for Promotion of Wasteland Development.

——. 1992. 'Common Property Resources: A Missing Dimension of Development Strategies'. *World Bank Discussion Paper*, No. 168.

Jorgensen, S. and D. Yeung. 2001. 'Intergenerational Cooperative Solution of a Renewable Resource Extraction Game', in L.A. Petrosjan and V.V Mazalov (eds), *Game Theory and Applications VI*.

McKloskey, D.N. 1976. English Open Fields as Behaviour Toward Risk in P. Uselding (ed.), *Research in Economic History*, Volume 1, Greenwich, Conn.

Mukhopadhyay, L. 2002. 'Common Property Resources in Private Property Regime with Inequality'. *Working paper, 02–07*, Riverside: Department of Economics, University of California.

Netting, R. McC. 1972. 'Of Men and Meadows: Strategies of Alpine land use', *Anthropological Quarterly*, 45(3): 132–44.

——. 1976. 'What Alpine Peasants in Common: Observations on Communal Tenure in a Swiss Village', *Human Ecology*, 4: 135–46.

——. 1981. *Balancing on an Alp: Ecological Change and Continuity in a Swiss Mountain Community*, Cambridge: Cambridge University Press.

Olson, M. 1965. *The Logic of Collective Action*. Cambridge, Massachusetts: Harvard University Press.

Ostrom, E. 1990. *Governing the Commons: The Evolutions of Institutions for Collective Action*. Cambridge, Massachusetts: Harvard University Press.

——. 1999. 'Self-Governance and Forest Resources'. Occasional paper 20. Jakarta. Indonesia: Centre for International Forestry Research.

Runge, C.F. 1981. 'Common Property Externalities: Isolation, Assurance and Resource Depletion in a Traditional Grazing Context', *American Journal of Agricultural Economics*, 63(4).

APPENDIX 1

The Nash solution in the two-stage CPR game with backward induction strategies gets solved by:

$$\underset{a_i^1}{Max}[Q_i^{\langle 1\rangle}(a_i^{\langle 1\rangle}, a_j^{\langle 1\rangle}) + Q_i^{\langle 2\rangle}(a_i^{\langle 2\rangle}(a_i^{\langle 1\rangle}, \hat{a}_j^{\langle 1\rangle}), a_j^{\langle 2\rangle}(a_i^{\langle 1\rangle}, a_j^{\langle 1\rangle}))]$$

that is, by
$$\underset{a_i^1}{Max}[K_i a_i^{\langle 1\rangle}\psi_i^{\langle 1\rangle}(\frac{S^{\langle 1\rangle}}{\sum K_i a_i^{\langle 1\rangle}}) + K_i a_i^{\langle 2\rangle}\psi_i^{\langle 2\rangle}(\frac{S^{\langle 2\rangle}}{\sum K_i a_i^{\langle 2\rangle}}) \tag{1}$$

The choice of action in period 2 is contingent upon actions chosen in period 1. In period 2, sub-game perfect equilibrium $[a_i^{\langle 2\rangle}(\hat{a}_i^{\langle 1\rangle}, \hat{a}_j^{\langle 1\rangle}), \hat{a}_j^{\langle 2\rangle}(\hat{a}_i^{\langle 1\rangle}, \hat{a}_j^{\langle 1\rangle})]$ is obtained by solving each of the following equations:

$$\frac{\partial}{\partial a_i^{\langle 2\rangle}}[k_i a_i^{\langle 2\rangle}\psi_i^{\langle 2\rangle}(.)] = 0 \tag{5}$$

$$\frac{\partial}{\partial a_j^{\langle 2\rangle}}[k_j a_j^{\langle 2\rangle}\psi_j^{\langle 2\rangle}(.)] = 0 \tag{6}$$

Solving (5), we get:

$$\psi_i^{\langle 2\rangle}(.) + a_i^{\langle 2\rangle}\frac{\partial \psi_i^{\langle 2\rangle}(.)}{\partial a_i^{\langle 2\rangle}} = 0$$

Here in particular,

$$\psi_i^{\langle 2\rangle} = \frac{S^{\langle 2\rangle}}{\sum K_i a_i^{\langle 2\rangle}} = \frac{S^{\langle 1\rangle} - \sum K_i a_i^{\langle 1\rangle} + \Delta}{\sum K_i a_i^{\langle 2\rangle}}$$

$$\frac{\partial \psi_i^{\langle 2\rangle}}{\partial a_i^{\langle 2\rangle}} = -\frac{S^{\langle 1\rangle} - \sum K_i a_i^{\langle 1\rangle} + \Delta}{(\sum K_i a_i^{\langle 2\rangle})^2}K_i$$

$$= -\frac{\psi_i^{\langle 2\rangle}}{\sum K_i a_i^{\langle 2\rangle}}K_i$$

(5) and (6) generate following two reaction equations:

$$\psi_i^{\langle 2\rangle} - \frac{K_i a_i^{\langle 2\rangle}}{(\sum K_i a_i^{\langle 2\rangle})^2}S^{\langle 2\rangle} = 0$$

$$\psi_j^{\langle 2\rangle} - \frac{K_j a_j^{\langle 2\rangle}}{(\sum K_i a_i^{\langle 2\rangle})^2}S^{\langle 2\rangle} = 0$$

$\psi_i^{\langle 2\rangle} = \psi_j^{\langle}$; $\frac{K_i}{(\sum K_i a_i^{\langle 2\rangle})^2}S^{\langle 2\rangle}$ in the second part of the reaction equation shows the slope of reaction curves. Since $\frac{S^{\langle 2\rangle}}{(\sum K_i a_i^{\langle 2\rangle})^2}$ is the common part in both the equations, slope of reaction curves vary according to the size of the cattle property $K_i(K_j)$.

Solving them we get:

$$K_i a_i^{\langle 2\rangle} = K_j a_j^{\langle 2\rangle} \tag{7}$$

Since the strategy of the game is backward induction the above sub-game perfect equilibrium solution will be rolled back into the stage 1 game. The game is solved by:

$$\frac{\partial}{\partial a_i^{\langle 1\rangle}}[K_i a_i^{\langle 1\rangle} \psi_i^{\langle 1\rangle}(\frac{S^{\langle 1\rangle}}{\sum K_i a_i^{\langle 1\rangle}})] + \frac{\partial}{\partial a_i^{\langle 1\rangle}}[K_i a_i^{\langle 2\rangle} \psi_i^{\langle 2\rangle}(\frac{S^{\langle 2\rangle}}{\sum K_i a_i^{\langle 2\rangle}})] = 0$$

Plugging the value from (7) into the above equation, that is, $\sum K_i a_i^{\langle 2\rangle} = 2K_i a_i^{\langle 2\rangle}$, considering $\psi_i^{\langle t\rangle} = \frac{S^{\langle t\rangle}}{\sum K_i a_i^{\langle t\rangle}}$, we get: $\frac{\partial}{\partial a_i^{\langle 1\rangle}}[K_i a_i^{\langle 1\rangle}(\frac{S^{\langle 1\rangle}}{\sum K_i a_i^{\langle 1\rangle}})] + \frac{\partial}{\partial a_i^{\langle 1\rangle}}[\psi_i^{\langle 2\rangle}.\frac{S^{\langle 2\rangle}}{2}] = 0$ The first part of the equation is:

$$\frac{\partial}{\partial a_i^{\langle 1\rangle}}[K_i a_i^{\langle 1\rangle}(\frac{S^{\langle 1\rangle}}{\sum K_i a_i^{\langle 1\rangle}})] = K_i \frac{S^{\langle 1\rangle}}{\sum K_i a_i^{\langle 1\rangle}} - K_i^2 a_i^{\langle 1\rangle} \frac{S^{\langle 1\rangle}}{(\sum K_i a_i^{\langle 1\rangle})^2}$$

$$= K_i \frac{S^{\langle 1\rangle}}{\sum K_i a_i^{\langle 1\rangle}} - K_i^2 a_i^{\langle 1\rangle} \frac{\psi_i^{\langle 1\rangle}}{\sum K_i a_i^{\langle 1\rangle}} \tag{8}$$

The second part of the equation is:

$$\frac{\partial}{\partial a_i^{\langle 1\rangle}}[\psi_i^{\langle 2\rangle}.\frac{S^{\langle 2\rangle}}{2}] = \frac{1}{2}[S^{\langle 2\rangle}\frac{\partial \psi_i^{\langle 2\rangle}}{\partial a_i^{\langle 2\rangle}}\frac{\partial a_i^{\langle 2\rangle}}{\partial a_i^{\langle 1\rangle}}. + \psi_i^{\langle 2\rangle}\frac{\partial S^{\langle 2\rangle}}{\partial a_i^{\langle 1\rangle}}]$$

$$S^{\langle 2\rangle} = S^{\langle 1\rangle} + \Delta - \sum K_i a_i^{\langle 1\rangle}$$

$$\frac{\partial S^{\langle 2\rangle}}{\partial a_i^{\langle 1\rangle}} = -K_i$$

$$= -\frac{1}{2}K_i \psi_i^{\langle 2\rangle}[1 + \frac{S^{\langle 2\rangle}}{\sum K_i a_i^{\langle 2\rangle}}\frac{\partial a_i^{\langle 2\rangle}}{\partial a_i^{\langle 1\rangle}}]$$

$$= -\frac{1}{2}K_i \psi_i^{\langle 2\rangle}[1 + \psi_i^{\langle 2\rangle}\frac{\partial a_i^{\langle 2\rangle}}{\partial a_i^{\langle 1\rangle}}] \tag{9}$$

Combining (8) and (9) we get:

$$\frac{S^{\langle 1\rangle}}{\sum K_i a_i^{\langle 1\rangle}} - K_i\ a_i^{\langle 1\rangle}\frac{\psi_i^{\langle 1\rangle}}{\sum K_i a_i^{\langle 1\rangle}} - \frac{1}{2}\psi_i^{\langle 2\rangle}[1 + \psi_i^{\langle 2\rangle}\frac{\partial a_i^{\langle 2\rangle}}{\partial a_i^{\langle 1\rangle}}] = 0$$

Or,

$$\psi_i^{\langle 1\rangle} - K_i\ a_i^{\langle 1\rangle}\frac{\psi_i^{\langle 1\rangle}}{\sum K_i a_i^{\langle 1\rangle}} - \frac{1}{2}\psi_i^{\langle 2\rangle}[1 + \psi_i^{\langle 2\rangle}\frac{\partial a_i^{\langle 2\rangle}}{\partial a_i^{\langle 1\rangle}}] = 0$$

Or,

$$\frac{K_i\ a_i^{\langle 1\rangle}}{\sum K_i a_i^{\langle 1\rangle}} = \frac{\psi_i^{\langle 1\rangle} - \frac{1}{2}\ \psi_i^{\langle 2\rangle}(1 + \psi_i^{\langle 2\rangle}\frac{\partial a_i^{\langle 2\rangle}}{\partial a_i^{\langle 1\rangle}})}{\psi_i^{\langle 1\rangle}}$$

Similarly for the *j*th player:

$$\frac{K_j\ a_j^{\langle 1\rangle}}{\sum K_j a_j^{\langle 1\rangle}} = \frac{\psi_j^{\langle 1\rangle} - \frac{1}{2}\ \psi_j^{\langle 2\rangle}(1 + \psi_j^{\langle 2\rangle}\frac{\partial a_j^{\langle 2\rangle}}{\partial a_j^{\langle 1\rangle}})}{\psi_i^{\langle 1\rangle}}$$

Proof of corollary 1
From (6.1) we get,

$$\psi_i^{\langle 2\rangle} + a_i^{\langle 2\rangle}\frac{\partial\,\psi_i^{\langle 2\rangle}}{\partial a_i^{\langle 2\rangle}} = 0;$$

Or, $$a_i^{\langle 2\rangle} = \frac{-\psi_i^{\langle 2\rangle}}{\frac{\partial\,\psi_i^{\langle 2\rangle}}{\partial a_i^{\langle 2\rangle}}} = \frac{-\psi_i^{\langle 2\rangle}}{\frac{-\psi_i^{\langle 2\rangle}}{\sum K_i a_i^{\langle 2\rangle}}K_i} = \frac{\sum K_i a_i^{\langle 2\rangle}}{K_i}$$, which decreases with K_i

20

Integrating Watershed Management: Stakeholders, Dynamics and Institutions

Saravanan V. Subramanian

Abstract: The important challenge for institutional research in future lies in going beyond emphasising the importance of institutions towards explaining how institutions matter in managing natural resources such as water. Institutions have been called to realign along the regional boundaries of watersheds or river basins for integrated management for ecologically consistent human development. Though such regional units are physical in nature, institutions, understood as patterned social behaviour, evolved over time which is essential for their management and do not strictly follow their physical boundaries. Institutions interact in diverse action arenas to facilitate or constrain actors involved in managing watersheds. These arenas may be location-specific or generic, formal or informal, and are naturally evolved or deliberately created. Diverse institutions operate and interact in these action arenas where all decisions are related to the use and management of the resources in the watershed. This chapter aims to examine how watershed institutions can be integrated by exploiting the interactive nature of institutions across action arenas and the interlinked nature of actors' actions.

INTRODUCTION

Watershed or a river basin is often considered as an appropriate unit for integrated management of natural resources for ecologically consistent human development. It is largely assumed to integrate relevant sectoral institutions at state, district or at village level.[1] Such attempts are assumed to identify and bring together various stakeholders at one level or to negotiate their diverse concerns in managing watersheds. Though such attempts have made an impact on economic and social fronts (Chopra 1999; GOI 1999; Ninnan 1997), sustainability of these attempts have come under scrutiny in recent years (Saravanan 2002). First, external institutions attempt to bring about changes by directly involving the stakeholders and their behaviour pattern. Such an approach not only creates new stakeholder groups, but also authorises existing resource use when they are highly dynamic depending on the context (Mosse 1997; Saravanan 1998). Second, such an approach is expensive, requires specialised skills and knowledge and cannot in any way assume that the sensitivity of the people and their livelihoods will not be affected. Third, the approach assumes that stakeholders are easy to identify and their unequal capabilities can be negotiated when brought onto a common forum. What makes this approach significant is the emphasis on interactive function of stakeholders, but if this is supplemented with indirect approaches, it can play a significant role in bringing about institutional changes.

In the real world scenario, stakeholders do not always stake their claim over resources but rather depend on endowments they possess, the characteristics of resources at that particular period of time and institutional rules in a particular 'strategic context' as 'actors'. These actors participate in managing watershed, using diverse actions by integrating diverse water management institutions. Here, integration of institutions takes place not at various administrative (state/district/village) or physical (watershed/river basin) jurisdictions, rather in various action arenas. These arenas are location-specific or generic, formal or informal, and are naturally evolved or deliberately created in a strategic context, where actors share a common understanding of an issue and have a shared vision to overcome. The actors use diverse forms of participation—ranging from passive submission to debate and negotiation. Though various factors shape the decisions in this arena, institutions remain the cross-cutting influence over actors. Diverse institutional rules and actors operate and interact in these action arenas where all decisions are related to the use and management of the resources in the watershed. This chapter aims to examine the interactive nature of institutions across action arenas and the interlinked nature of actors' strategies. The chapter has three objectives to address the research gap: (*i*) to understand how interactive institutions across various arenas influence management of water resources at hamlet level, and their linkages with poverty; (*ii*) to analyse the actions of actors who are directly affected by a particular problem in accessing different action arenas; and (*iii*) to examine the interactive nature of diverse institutional rules in facilitating and constraining decisions at action arenas in managing water resources. These objectives are examined with an empirical application of the 'agent-actor-crowd' model to a core water-related issue applicable in each of the four socio-economically and hydrologically distinct hamlets selected from two watersheds in Himachal Pradesh, India. The information regarding the interactive nature was collected using diverse research methods (participatory methods, structured interviews, semi-structured interviews and intuitive observation) to capture the complexity. The data was collected from samples of actors in different action arenas using a combination of investigations.

The study reveals the complexity of institutions in sharing and management of resources. It illustrates the dynamic nature of stakeholders who, depending on their endowments, prevailing institutional rules and resources in context, take part in water-related decisions as 'actors' and 'agents'. Unlike the contemporary approach emphasising on the collaborative model, the study traces the prevalence of agents who play an important role in integrating institutions and negotiating diverse concerns within and between action arenas. Facilitating these agents, the study offers options for participatory dialogue process among actors, provides channel for information dissemination and evolves cost effective options for institutional change. The study, though a piecemeal attempt to examine the role of institutions, emphasises the need to strengthen sectoral approach of managing natural resources and identifies issues for devolving power to various institutions for polycentric governance.

The chapter is organised into six sections. The first section provides the conceptual background and methodology adopted to capture the complexity of water resource management. The third section provides an overview of the area of study. The fourth section depicts the complexity and clutter of institutions interacting and influencing resource management by examining their role in creating 'virtual' scarcity, in authorising resources' distribution and in building the capability of households to access. The fifth section examines the options available for actors in addressing water resource management at local level. Different decision-making arenas up to district levels are examined to understand the interactive nature of institutions and their role in facilitating and constraining the agents. The final section concludes by identifying key insights for institutional change and opportunities for improving the institutions in the arena to predict models of institutional change.

CONCEPTUAL BACKGROUND AND METHODOLOGY

Institutions and actors interact among each other in taking decisions related to water management. Such interactions have been recognised in action arena[2] (hereafter as arena) (Ostrom et al. 1994). Arena represents a complex system (refer to, special issue of Ecological Modelling 2002; Railsback 2001) that characterises openness, diversity of actors, non-linear fashion of interaction and heterogeneity. In spite of this, these arenas are characterised as involving emergent properties, multi-scale interactions, unexpected behaviours and self-organisation capacity, which make them a 'complex adaptive system'. Though a number of factors (physical, social and cultural) influence the arena, institutions, understood as patterned behaviour of social group over a period of time, constitute a cross-cutting factor and a particular driving force in the decision-making process (Young 1999) (Figure 20.1). These arenas are hierarchically placed and influenced by situational variables, which interact in arenas to have their impact. Contextual variables influence the panarchy at various points in time.

Action arena is a social practice ordered across space and time (Giddens 1984). It may be location-specific or generic, formal or informal, and naturally evolved or deliberately created in a strategic context, where actors are involved in performing diverse actions—may even be strategic or communicative (Alexander 2001). It is the capability of few actors, that act as agents in accessing other action arenas and drawing upon the modalities.

Though these two actions combine in complex forms in a 'strategic context'[3] of the action arena, it is the capability of few actors, who act as 'agents' in accessing other action arenas by drawing upon the modalities of existing institutions in the reproduction of systems of interactions, by the same token reconstituting their properties (Giddens 1984: 2). Using social network approach of following 'agents,' the interactive nature of institutions are explored. Though institutions are complex and diverse, they often overlap with a number of forces to constrain and facilitate the management of water resources in diverse action arena (Dorcey 1986).

The institutional arrangements in action arenas consist of structures, components and rules (Figure. 20.2). The institutional structures consist of public, private and user groups (Meinzen-Dick and Rosegrant 1997; Bruns and Meinzen-Dick 2003). Each of these structures have three components (Saleth and Dinar 1999)—policy, administrative and legal. These components have various institutional rules, broadly relate to seven types (Ostrom et al. 1994). Very little is known about the complexity of interaction and, consequently, the mixture of rules and principles involved in action arena (Cars et al. 2003; Lubell 2004; Mehta 2002; Ostrom 2001; Pahl-Wostl 2002; Pradhan et al. 1997). To analyse the complexity of interaction among institutions controlling individuals' access to water, a case study approach is necessary (Neuman 2003) as it enables us to capture the complexities of and the relationship between humans and environment (Young 1999). This type of approach provides insights for understanding contextual factors influencing institutional phenomenon in a selected watershed, where micro level or the actions of individual people connects the macro level or large scale social structure and processes (Neuman 2003: 33).

Figure 20.1 Framework for Analysing Institutional Integration for Watershed Management

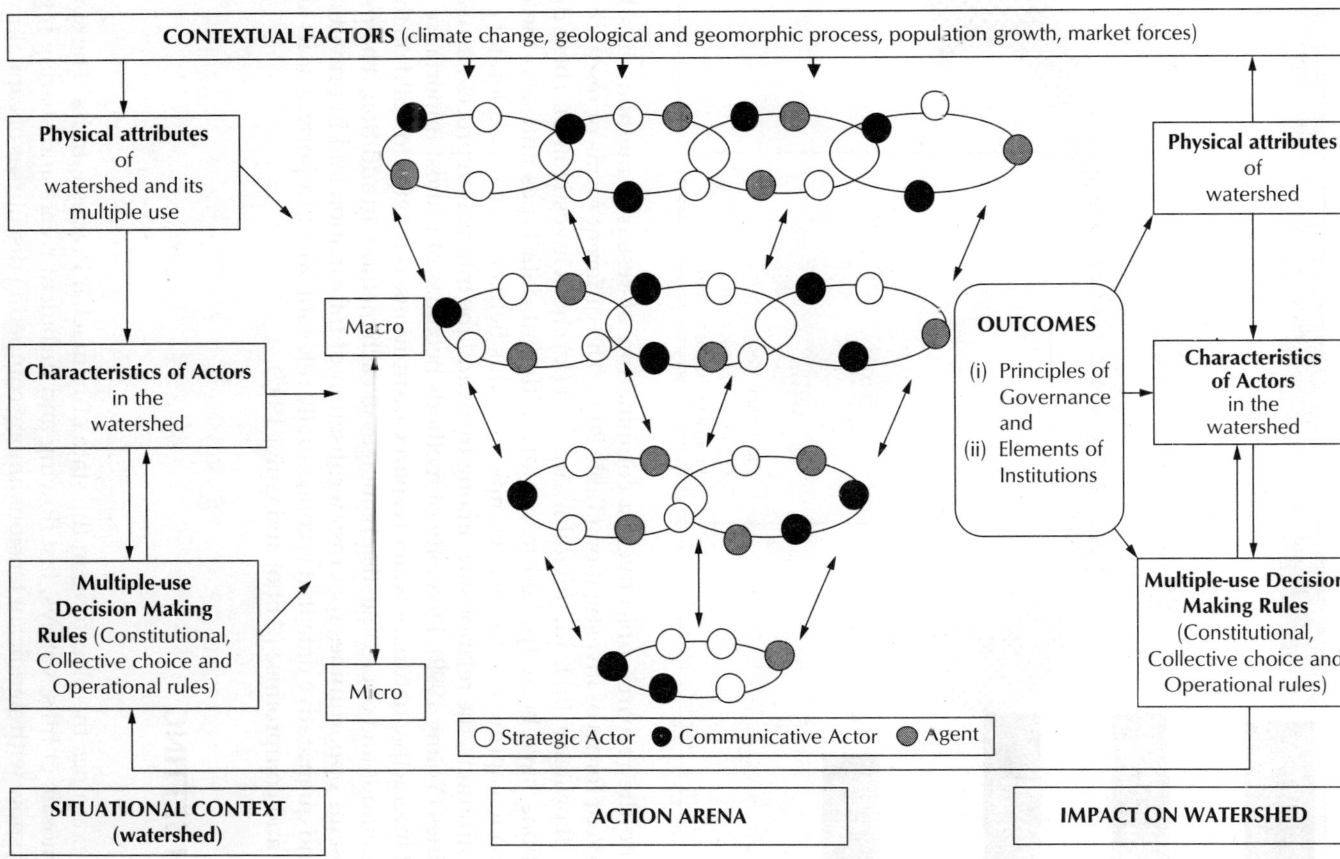

Figure 20.2 Institutional Arrangements in Action Arena

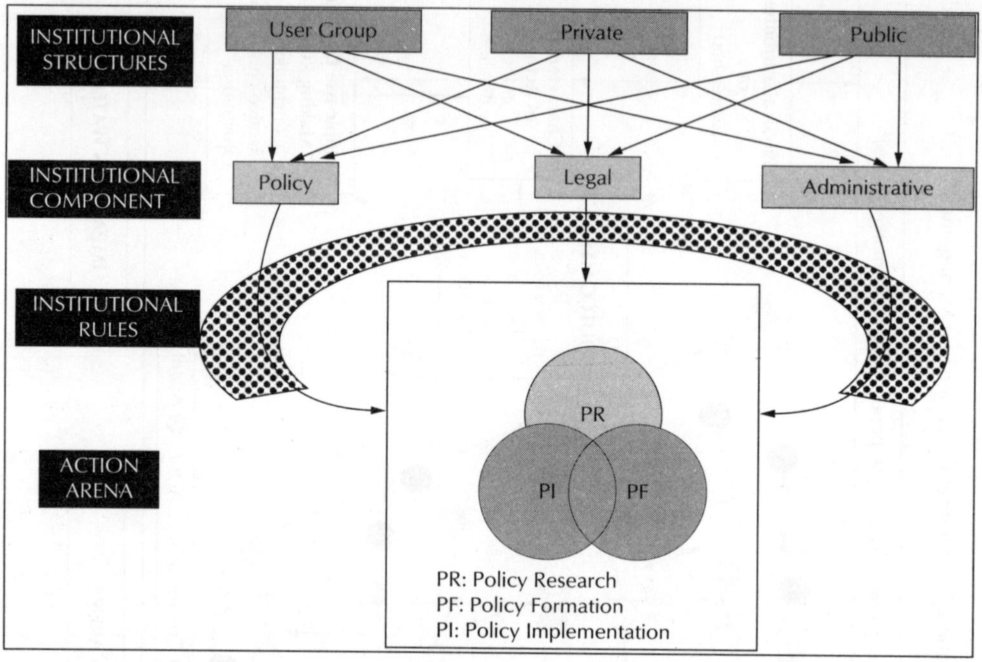

To examine the research objectives, a combination of research methods has been applied through different forms of investigation (Table 20.1). The different forms of research investigation adopted are: (*i*) primary; (*ii*) lead; (*iii*) follow-up; and (*iv*) check investigation. These different forms of investigations have been applied to capture different data types and also the contextuality of information to capture institutional complexity. This helps to remain exploratory in describing the role of institutions, the relationship among them and the interaction process in order to capture the complexities (Young 1999). Diversity of methods becomes of utmost importance in examining complex and interactive nature of water resource institutions as it helps to build on the advantages of different methods and overcome disadvantages of each method. In addition, it helps in validation and also presents opportunities for cross-fertilisation of information and in providing a balanced qualitative and quantitative data that is contextually relevant. More important, is its ability to build on creativity and compromise (Abbot and Guijit 1997).

RESEARCH SETTING

The chapter examines the objectives in the state of Himachal Pradesh, India. The state represents an intricate mosaic of hills, valleys, fast flowing and turbulent rivers and soaring high mountains covered with snow with significant tensions among competing discourses of capital-intensive forms of economic development, environmental conservation and participatory forms of eco-development

Table 20.1 Diverse Research Methods and Forms of Investigation

Research Objectives	*Methods*	*Sampling*	*Strengths*	*Limitations*	*Type of Investigation*
Examine the interactive nature of institutions influencing water management in the watershed.	Secondary documents	-	Provides information on past baseline data of the region (Such as Census of India, Watershed Socio-Economic plans and agriculture research information).	It is less contextual and can be used as baseline information.	Primary
Analyse the actions adopted by actors to access water	Participatory research methods	Transect mapping, wealth mapping, resource mapping and social calendar.	Gives an overall picture of the case study area and general perception in the region on issues (quality of land, ownership of land, cropping pattern, social class and annual calendar). Further helps in building rapport with people	Less Contextual and information biased against the poor.	Primary and lead
	Semi-structured interviews	(i) 5 randomly selected community leaders, village elders, educated adults and irrigation functionaries. (ii) Government functionaries related to water resources, like Dept. of Forest, Irrigation and Rural Development.	One-on-one interview will elicit in-depth and truthful information from informants not wiling to share in-group interview setting, particularly when rapport is well established (History of village and irrigation management, change in cropping, problems in managing water and reasons for these problems).	Time consuming and highly individual opinion.	In-depth
	Focus Group Discussions	2 FGDs with 3-member group consisting of poor members in the village.	Generates a common view of the poor, their problems and reasons. Explores new issues and various factors that influence their perceptions.	Time consuming and do not provide practical insights on individual actions.	In-depth and cross-checking.

(*Table 20.1 continued*)

(Table 20.1 continued)

Research Objectives	*Methods*	*Sampling*	*Strengths*	*Limitations*	*Type of Investigation*
	Structured Interviews	About 160 households were stratified based on landholdings and other occupations from 4 hamlets.	Helpful in generating necessary, quantifiable, non-controversial and non-sensitive information. Easy to code and tabulate.	Less time consuming and not suitable for sensitive and controversial information.	Primary
Analyse the capability of agents to integrate institutions to manage water resources at local level.	Structured Interviews	About 160 households were stratified based on landholdings and other occupations from 4 hamlets.	Helped in (*i*) identifying different strategies adopted by various users in accessing or managing water and (*ii*) identifying agents at hamlet who negotiate with other action arenas outside the hamlet.	Did not help in identifying other action arenas accessed for managing water.	Primary
	Semi-structured interviews	With agents identified by the sample households at hamlet level (Total agents–6 from 4 hamlets).	One-on-one interview with the agent helped in eliciting information on the institutions and their rules that facilitate and constrain the agents' action, the arenas he accesses and the reasons.	Time consuming, and gives only formal means of negotiation process.	In-depth
		With agents in arenas outside hamlet level.	One-on-one interview with the agent helped in eliciting information on the institutions and their rules that facilitate and constrain the agents action, the arenas he accesses and the reasons.	Time consuming.	In-depth
	Semi-structured interviews.	With actors in each arena.	As the interview is held with actors on the functioning of agents, it helps to illustrate the actual negotiation process and the institutions' roles in the arena.	Time consuming.	Cross-checking

(Coward 2003; Baker and Saberwal 2003). Taking a case of the most backward district Sirmaur in the southern part of the state (Annexure 1), four socio-economically and hydrologically distinct hamlets were selected from two watersheds representing different agro-climatic conditions. This district is an ideal candidate in order to examine the actions of actors in accessing water due to the presence of diverse agro-climatic conditions within the district (suitable for comparability), scarce availability of water, socio-economic backwardness and the existence of diverse irrigation system.

Two watersheds were selected, one each from the low hills sub-tropical (Shiwalik) zone and the mid-hills sub-humid zone. A comparative picture of the watersheds illustrates the former as located in low altitude zone (600–1,000 msl), easily accessible to plains and well-off, while the latter in the mid hills (1,000–2,000 msl) as relatively remote and backward (Table 20.2). The two watersheds were selected based on: (*a*) competing claims over water resources in the region—this is identified by an irrigation source benefitting more number of hamlets/villages, number of overlapping irrigation sources (such as *Khuls*, canals and wells), different cropping patterns and any conflicts over water use; (*b*) ecological characteristics of vulnerability; (*c*) willingness of the people to support the proposed research study; and (*d*) access to transportation facilities (as the researcher had to coordinate the research work in two watersheds in different agro-ecological regions). The hamlets in each watershed were selected through discussion with gatekeepers and village leaders on their location in watershed (upstream and downstream), economic backwardness of that particular hamlet and scarce availability of water (for irrigation).

The two hamlets (Khairwala and Pipalwala) selected out of ten in Khairi-Ka-Kala watershed in the low hills sub-tropical zone (hereafter referred to as low hills), are relatively (compared to other hamlets) backward in the watershed. The first hamlet, Khairwala, is located upstream of the watershed and has a high population of Muslim Gujjars (Scheduled Tribes) and less number of Rajputs (forward caste). These people, though agriculturists, supplement their livelihood by selling milk and engaging in other part time non-agricultural labour outside the hamlet. They are remotely located from the main group of hamlets, due to their occupation and backwardness. The hamlet has irrigation facility through lift method from the nearby river Markhanda, through which they grow maize, wheat and fodder grasses. The people of Pipalwala hamlet, though agriculturists, depend on employment (formal and informal) from nearby towns in Himachal and Haryana for their living. The *Khul* (diversion-based) irrigation systems that entail drawing of water by gravity from the river Markhanda are the only source of irrigation. This enables them to cultivate maize, wheat, fodder grass and vegetables for domestic consumption.

Compared to their counterparts, the people of hamlets in Rajana Watershed located in the mid-hills sub-humid zone (hereafter referred to as mid-hills) are agriculturists (with limited employment opportunities) and are economically backward due to remoteness. Here, there are two major castes—the Rajputs (forward caste) and Kohlis (Scheduled Caste), with Brahmans and *Chamars* (another class of Scheduled Caste) being in the minority. Further these hamlets being close by have the same socio-cultural characteristics. However, being apart from each other by about 100 metres in altitude makes a great difference in their agriculture pattern. The hamlet Uppala Rajana (located upstream) grows tomato and ginger (also has the potential to grow other vegetables) in rainfed conditions (with limited irrigation in May) and organically. It also has a very good soil condition. In contrast, hamlet Nichala Rajana (downstream) is unable to grow tomato and ginger successfully due to problems of pests and unsuitable soil conditions, in spite of having *Khul* based irrigation

Table 20.2 Physiography and Socio-Economic Background of the Case Study Hamlets

	Khairi-Ka-Kala Watershed		*Rajana Watershed*	
Revenue Village	Bikram Bagh	Pipalwala	Rajana	Rajana
Case Study Hamlet	Khairwal	Pipalwala	Uppala Rajana	Nichala Rajana
Agro-climatic zone	Low hills sub-tropical (Shiwalik) zone		Mid-Hills sub-humid zone	
Physiography	Moderate steep to steep low hills of Shiwaliks.		Steep to very steep high hills of Lesser Himalayas.	
Altitudinal Location (in metres)	400–600		1,000–1,200	
Rainfall	About 1,000 mm		About 1,200–1,500 mm	
Intensity of soil Erosion	Severe		Moderate	
Slope	Moderate		Steep	
Soil type	Loamy	Sandy to loamy	Loamy to clayey	Sandy to loamy
Location in Watershed	Upstream	Downstream	Upstream	Downstream
Population (as in 2002) (Households)	307 (Muslim Gujjars—Scheduled Tribe) 96 (Rajputs—Forward Caste) (57)	270 (Forward caste) 101 (Scheduled Caste) 11 (Scheduled Tribe) (64)	357 (Kohli—Scheduled Caste)247 (Rajputs & Brahmins—Forward Caste) (75)	393 (Kohli—Scheduled Caste) 250 (Rajputs—Forward Caste) (80)
Average Household Size	7	6	8	8
Caste	Muslim Gujjars (Scheduled Tribe) and Rajputs	Multi-caste	Rajputs, Brahmins and Kohlis (Scheduled Caste)	Rajputs and Kohlis (Scheduled Caste)
Main source of Drinking Water Facilities	Handpump	Handpump	*Bavdi* (stored spring water)	Spring
No. of households having access to toilet facilities at home (% of total pop.)	None	10 (23%)	None	3 (9%)
Dominant Economy	Labourers, regular employment (formal) and marketing of milk.	Regular employment in formal institutions, labourers and marketing of milk.	Agriculturists and regular employment (formal).	Regular employment (formal) and agriculture.

(*Table 20.2 continued*)

(*Table 20.2 continued*)

	Khairi-Ka-Kala Watershed		*Rajana Watershed*	
Irrigation Type	Lift Irrigation System	*Khul* Irrigation System	Rainfed	*Khul* Irrigation system
Major Food crops	Maize and Wheat	Maize and Wheat	Maize and Wheat	Maize and Wheat
Other crops (including cash crops)	Fodder grass	fodder grass, mangoes (only large landowners) and minor vegetables.	Ginger, tomato and vegetables	Ginger and vegetables.
Average annual income (both cash and non-cash) (in Indian[4] Rs) of Household	48.199	70,393	50,906	46,078
Social Class (main indicators from wealth ranking)				
Rich	Land holdings more than 2 acres, concrete house and good number of cattles.	Households supplementing their agriculture with pension from defence or government departments.	Landholdings more than 2 acres.	Landholdings more than 5 acres.
Upper middle	-	-	Landholding size between 1 to 2 acres and regularly employed in Mining industries.	Landholdings between 3–5 acres.
Middle	Good agriculture land and regular employment in formal ard informal institutions.	Households supplementing their agriculture with regular income from formal and informal institutions.	Landholding size between 0.4 to 1 acre.	-
Lower middle	-	-	Landholding size between 0.2 to 0.4. acre.	Landholdings between 1–3 acres.
Poor	Landholding of less than an acre and working as labourer.	Households supplementing their agriculture with income from labour employment.	Landholding size less than 0.2 acre and employed as labour (mainly from SC community). They have landholding in tail-end location.	Landholdings less than 1 acre.
Very poor	6 households (having un-irrigated land uphill.	2 Households having un-irrigated land.	-	-

Source: Field survey, 2004.

facilities. Examining water resource management in these diverse settings offers a range of insights for understanding the management and the options stemming from them.

Water Management in a 'Strategic Context'

To understand the strategic context in which institutions and actors interact, various problems faced in the hamlet were identified through semi-structured interviews with key persons in the hamlets. Of various problems, the core water related problem was taken into account and various actors and their roles were identified. Examining problems across the case study hamlet helps in identifying the perceived problems by the respondents, and also in identifying the core water related problem in the hamlet. In three of the (out of four) hamlets, 'distribution of water' was a problem, which is normally cited to outsiders as a problem of 'less water more land to irrigate'. In the fourth hamlet, the problem was of non-availability of any irrigation facilities.

WATER RESOURCE MANAGEMENT AND INSTITUTIONS

Managing water resources in the case study hamlets require understanding the history of the hamlets, their management pattern and the role of current institutional arrangements affecting water resources. Semi-structured interviews were conducted with people directly affected by this particular problem and with key officials in relevant departments on the role of institutions in influencing water management. Broadly, it could be illustrated in terms of their role in perceiving water availability, the way it is distributed and in building the capability of actors in accessing water. In each of these categories, institutions (both micro and macro) coalesce in diverse arenas at various time periods in shaping the management of water resources. It also reveals how the poor are being marginalised.

CONSTRUCTING VIRTUAL SCARCITY

Availability of water is often considered to be infinite and naturally available (through rain). In recent decades, the finite nature is only understood in relevance to surface and groundwater and, therefore, the emphasis on harvesting rainwater (which is assumed to be infinite). The study demonstrates how external institutions perceive water availability and in the process construct virtual scarcity (Figure 20.3).

Early settlers (as in the case of Khairi-Ka-Kala Watershed)—the princely rulers of Sirmaur district—King Shamsher Singh constructed a *Khul* (a diversion based irrigation system) from the river Markhanda to cultivate their orchards about 3 Kilometres downstream. The princely ruler solely managed it as it was a private property. Later, the ruler, due to close acquaintance with the people of Daduwala (upstream hamlet), extended irrigation rights. After independence, the Public Works Department of the then Union Territory of Himachal Pradesh took over the maintenance

Figure 20.3 Institutions Constructing Water Scarcity

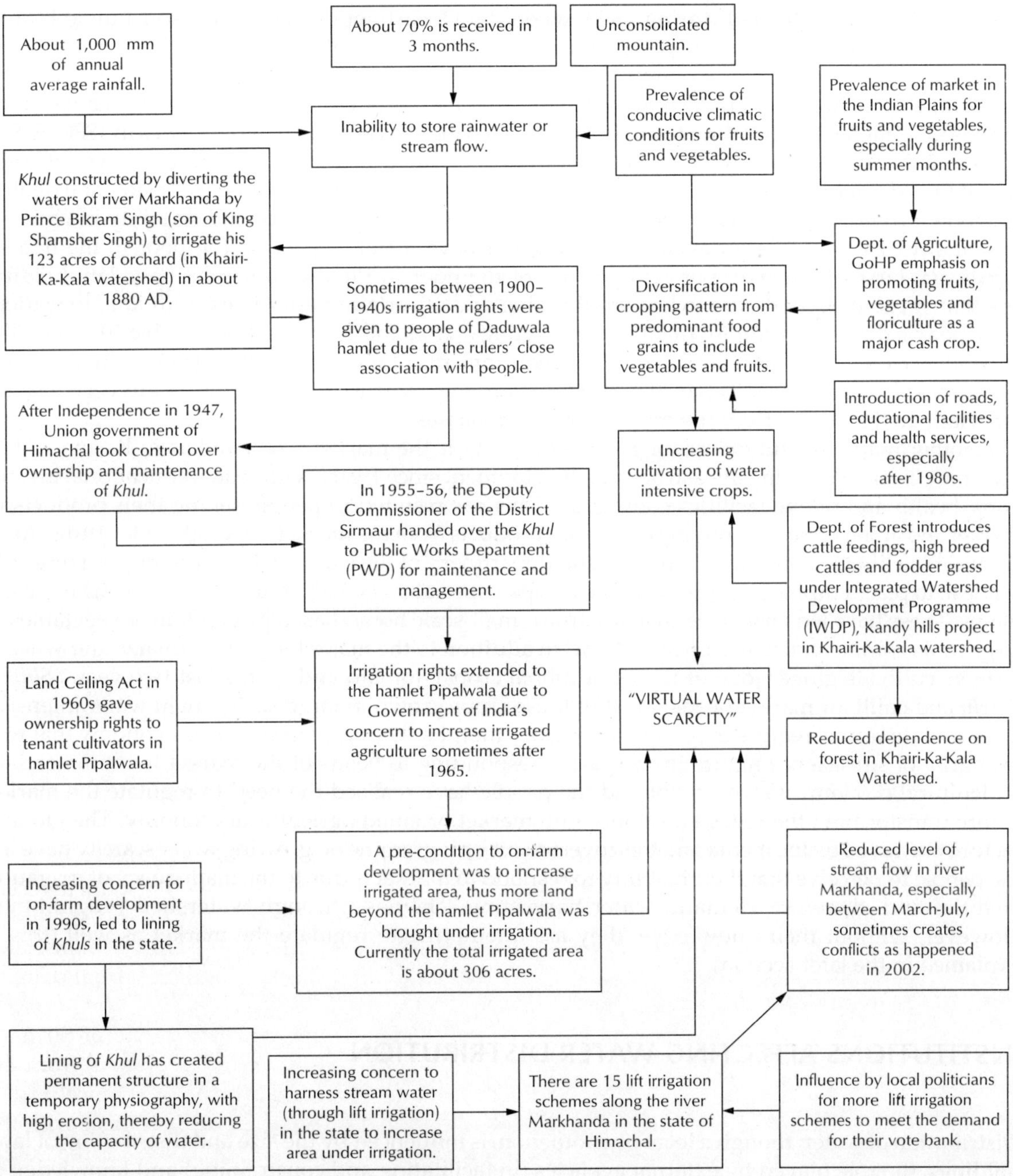

and management of *Khul* from the princely rulers and extended irrigation rights to the downstream hamlets (one of them being Pipalwala) during 1960s with the intention of increasing irrigated area. Once again it was extended during 1990, when the Department of Irrigation and Public Health (DoIPH) lined the *Khul* in the name of on-farm development. Though the lining might have improved efficiency of water; it did create scarcity in two ways. First, it had to comply with the directive of the Government of Himachal Pradesh, which states that if *Khuls* are lined, the irrigated area has to be increased[5]. This led to extending the irrigated area beyond its capacity from 123 acres in 1880s to 306 acres in 2003. Second, the lining created a permanent structure in a very temporary physical landscape[6] thus demanding regular de-silting and channelising. The unregulated extension of irrigated area only provides superficial hopes to the people rather than assurance and certainty in terms of availability of water. These developments did not have major impact on the poor in the hamlet Pipalwala. First, about sixty per cent (25/44) of the poor in Pipalwala hamlet have landholding less than an acre. Second, due to less landholding and uncertainty associated with *Khul* irrigation, these people depend on employment and marketing of milk for their livelihood (Table 20.3). Finally, dependence on agricultural land is only for requirements of food grains that grow even in rainfed conditions. It is clear from Table 20.3 that the economic returns from agriculture are very meagre especially to the people from the middle and poor classes.

Another major institution creating virtual scarcity is the market. The hamlets in Rajana watershed have been witnessing infrastructural development since 1980s, with roads, educational institutions, health and phone facilities coming up. This has resulted in people selling their products in market and at the same time buying consumer products from the market, especially after 1990s. Also, favourable climatic conditions have led the department of agriculture, Government of Himachal Pradesh to place emphasis on growing cash crops, especially vegetables and fruits for markets in the plains. These have led farmers to increase from small-scale home-based production of vegetables to large-scale commercial market needs. Now, in addition to the major food crops (maize and wheat), farmers cultivate ginger (one of their traditional crops), tomato and, in the last two years, *Shimla mirchi* and chilli for market needs. Most of these crops grown in large-scale, are in water intensive areas and therefore, require irrigation during dry months. The agriculture based economy that was primarily of subsistence nature in the past is responding to needs of the market for commercial/agricultural economy. Government and the people have realised the need to regulate the market before transforming the village economy into market oriented agricultural economy. The gloomy picture portrayed by the media and the government's programme on growing water scarcity have led the people to perceive that their inability to respond to market is due to the inadequacy of irrigation facilities and, therefore, demand water harvesting schemes (through watershed programme). However, without their knowledge, they are attempting to regulate the market as well (this is explained in the later section).

INSTITUTIONS AFFECTING WATER DISTRIBUTION

Distribution of water, though a local phenomenon, is influenced by the size and distribution of land holdings, the role played by external agencies (in facilitating and constraining) and knowledge of users (Figure 20.4). Distribution of landholdings and their size is primarily influenced by institutional

Table 20.3 Different Sources of Income of Sampled Households

Income Class/ Social Class	*Agriculture* % of Cash Income	*Agriculture* % of Non-Cash Income	*Diary* % of Cash Income	*Diary* % of Cash Income	Income from Employment (Formal and Informal) (in %)	Annual Average Household Income (Indian Rs)
Pipalwala Hamlet						
Rich	5	24	9	10	52	67,553
Middle	1	10	16	14	59	1,00,615
Poor	1	15	15	14	56	45,556
Khairwala Hamlet						
Rich	5	28	17	8	42	56,839
Middle	0	22	20	13	45	42,892
Poor	0	9	11	1	80	50,173
Uppala Rajana						
Rich	43	19	0	17	20	95,920
Upper Middle	19	23	0	33	25	55,735
Middle	36	17	0	39	8	51,767
Lower Middle	18	6	0	54	23	29,532
Poor	9	15	0	32	45	25,738
Nichala Rajana						
Rich	43	5	0	46	6	52,840
Upper Middle	21	6	2	54	17	35,464
Lower Middle	4	2	0	21	72	1,12,110
Poor	7	2	4	37	50	36,829

Source: Field Survey, 2004.

evolution of the hamlets through history. For instance, the Rajputs (early settlers) who occupied lands in Rajana watershed took control and ownership of all lands. In order to meet their labour requirements, the Rajput community brought in the Kohli (Scheduled Caste) community to work as tenant cultivators on their land. It was the Land Reforms Act of Government of India in 1960s that gave ownership rights to these tenant cultivators. Unfortunately, the decision to part away with the land rested with the Rajputs, who often gave away poor quality and tail-end located lands.

Being early settlers and landlords, they had the right to decide (even now) on matters pertaining to the village administration. Water distribution in *Khul*-irrigated areas were not an exception. Being large landholders and head farmers, the distribution was tailored—as land-based distribution—to benefit the Rajputs than the Kohlis. Though the inadequacies of such distribution was shared in private by the Kohli community, none of them was able to openly question the Rajputs due to the cultural bond of subordination that existed. However, few Kohli community members do break these norms independently using strategic actions—take water directly from *Khul* channels through tubes or pipes. Being categorised as a 'private *Khul*'[7] by the government of Himachal Pradesh, the department of irrigation and public health (DoIPH) rarely supervises the inefficiency of the irrigation practice, in a way facilitating the inefficiency of water distribution.

In contrast to water distribution in Rajana watershed where external institutions influenced local distribution practice, in Pipalwala hamlet the water distribution is totally influenced by external institutions on the assumption that people are knowledgeable and efficient in distributing water. The

Figure 20.4 Institutions Affecting Water Distribution

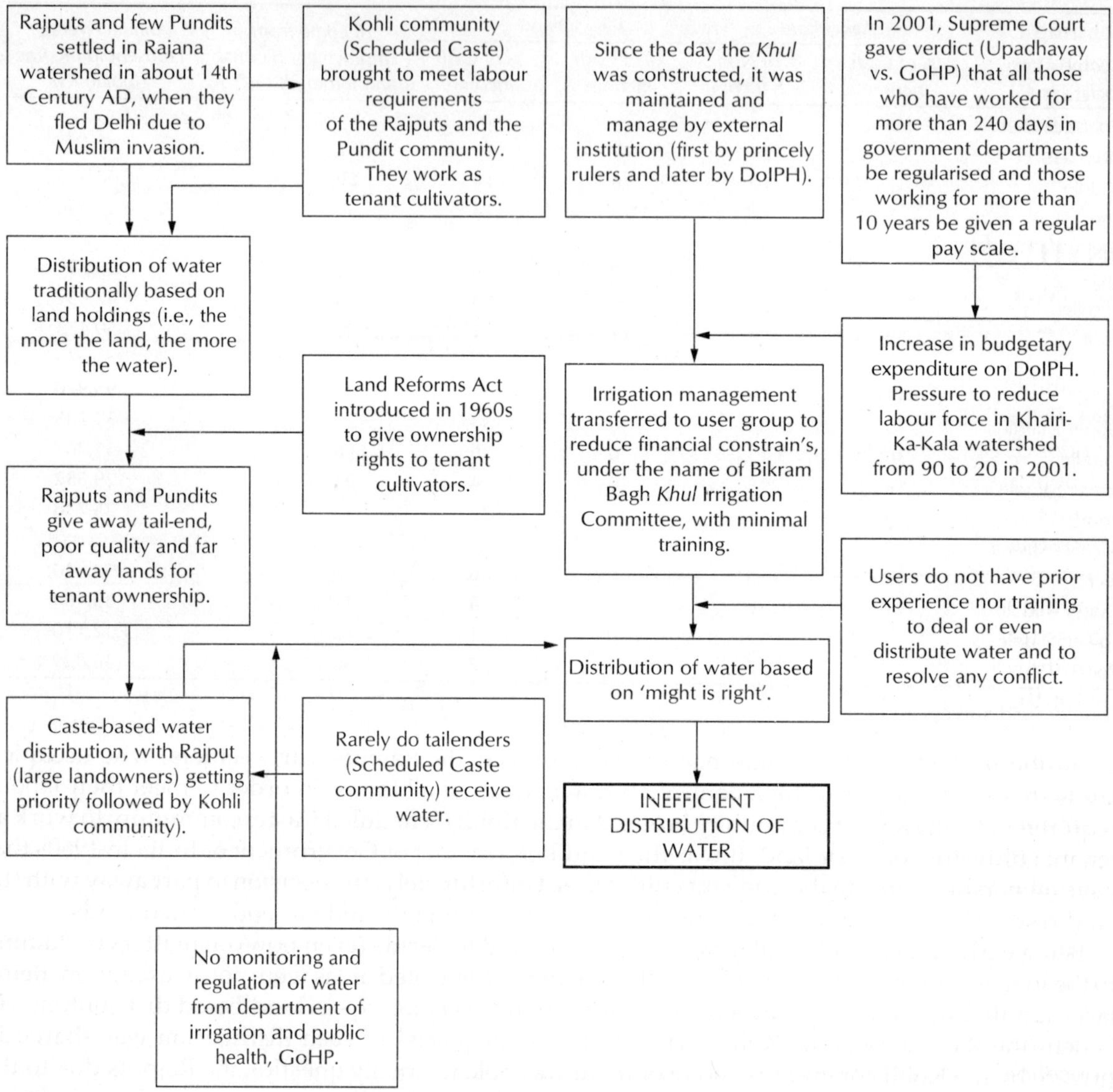

distribution of *Khul* irrigation in Pipalwala hamlet was in the past carried out by the people appointed by the princely ruler, who distributed water first to the rulers' orchards and then to the people. The distribution was primarily based on the first-come-first-serve basis irrespective of the location of the field in the command area. After independence, the water bodies were taken over by the Public Works Department (PWD) and later by the Department of Irrigation and Public Health (DoIPH). The DoIPH employed water distributors, though there was no major change in the distribution pattern. It was in 2001 that a Supreme Court directive made DoIPH to regularise all daily waged employees with various other benefits. This led to increasing the financial burden on the department, leading DoIPH to transfer the maintenance and management of *Khul* to the user groups, who did not have any previous experience nor were given any training. Initially these users followed the pattern of distributing water as done by the department staff, but unfortunately due to social bonds of preferential treatment for some and impartiality for others, the distribution has gone awry. Now the distribution is primarily through the 'might is right' principle, leading to wastage of water.

INSTITUTIONS AFFECTING THE CAPABILITY OF ACTORS TO ACCESS WATER

Capability of actors to access or utilise water depends on the various endowments each of the households has. Prominent among them include: the type of land available for cultivation, household size and gender differentials within the households. Type of land available for cultivation is one of the factors influencing a household's capability to access water. The landholding size matters the most in all the hamlets. With 30–50 per cent (varying across caste studies) of the sampled households having less than an acre of cultivable land (either in *Khul* command area or in unirrigated land), the returns from this is not significant for the poor to invest time and energy in accessing water. In Rajana watershed, in addition to landholding size, the location and quality matters for enhancing or constraining actors' access to water. More than ninety per cent of the land owned by the Scheduled Caste Kohli community is located in the tail-end. Due to the inefficient distribution of water and also the distance factor to monitor wild animals encroaching on the land (for unirrigated lands in Uppala Rajana), dependence on this particular land becomes expensive and meaningless. In contrast, the rich people in Rajana watershed (the Rajput community) have better access to irrigation facilities and also have the ability to monitor the land from wild animals, which contributes to more than forty per cent of their annual income. The uncertainty in the availability of water, inefficient distribution and less returns from cultivable land leads the poor and middle class households to depend on employment (formal and informal) that contributes to 50–70 per cent of their income (Table 20.3).

Household size matters for getting adequate returns from cultivating the land, especially in the Rajana watershed. Being remotely located, the households have to depend on their family labour for cultivating their land. It is notable (Table 20.4) that the poorer the family, the smaller the family size.

Sex ratio of these households also matters in utilising the productivity of the land. The richer the household, the higher the sex ratio (Table 20.4). This is normally found among the Rajput families. It is also found that among these families, the females work more (in the fields, cattle yard and at home) than males who spend time travelling to towns and engage in village works. A sample study of male and females in four families (two from Nichala Rajana and two from Uppala Rajana) of the Rajput community indicates that females work for about 17–18 hours a day. Such differentials in the work pattern also reflect the need for female children for family labour.

The inadequacy of the existing institutional structure does not have any major impact on the poor. The cultivable land is the lowest among the poor. Even if the land is available, the quality is poor and located in tail-end thereby getting less access to water. In addition, these households are less literate,

Table 20.4 Distribution of Household Size among Sampled Population (in per cent)—Rajana Watershed

	Household Size				
Social Class	*Less than 4*	*5–8*	*9–10*	*11 and above*	*Sex Ratio*
Uppala Rajana					
Rich	20	–	20	60	1,760
Upper Middle	8	30	31	31	1,133
Middle	30	70	–	–	889
Lower Middle	14	58	–	28	1,000
Poor	38	62	–	–	1,050
Nichala Rajana					
Rich	–	–	30	70	1,000
Upper Middle	17	55	5	23	1,278
Lower Middle	–	62	–	38	1,375
Poor	28	44	–	28	952

Source: Field Survey, 2004.

disabled and have a single member or a large family. The inadequacy of getting sufficient returns from agriculture has led many of these poor families to depend on employment (formal or informal) in other sectors in and around the hamlets. It is these sources that contribute between 50–70 per cent of their monthly income. Any effort at addressing the problems of the poor requires a more comprehensive approach rather than sector specific ones.

Institutions influencing water resource management are diverse, ranging from micro-macro institutions that coalesce at diverse action arenas. Though both external and internal institutions influence management of water resources, it is the formal external institutions that play a major role in initiating change at micro level by portraying the resource as scarce or surplus. It is interesting to note that new actors are entering (like Supreme Court of India, market, District Rural Development Agencies, GOHP) the grid in making claims over resources.

ACTORS, ACTIONS AND ARENAS

The inadequacy of existing institutions in managing water resources is recognised by each actor at the hamlet level who attempts diverse actions to modify the existing institutions, creating new ones or even accessing diverse other institutions. Broadly these actions are classified as strategic and communicative (Table 20.5) (Alexander 2001). The former represents actions taken for the realisation of particular self-interested goals (coercive power), while the latter aims at achieving collective decisions (enabling power). Though both these actions indicate the inadequacy of existing institutional structures in diverse forms of collective actions, it is the communicative actions that aim to strengthen or empower the existing institutional structure or attempt to overcome the inadequacy

Table 20.5 Diverse Actions of Households to Access Water

Hamlets (Interview Question)	*Strategic*	*Communicative*
Pipalwala		
If you do not get water from Khul *irrigation systems as per your turn, what do you do?*	1. Wait, wait and wait... 2. Take directly from *Khul* channel. 3. Buy water from others. 4. I do get if there is sufficient water. 5. I use my might (fight) to get water. 6. I do not depend on this *Khul* water for my income.	1. Inform President of the irrigation committee and get water. 2. Get water by negotiating with the person irrigating at the moment. 3. If I do not get water as per turn, I investigate and take water.
Khairwala		
If you do not get water from lift irrigation systems as per your turn, what do you do?	1. Wait, wait and wait... 2. If water is available I get them. 3. I use my might (fight) to irrigate. 4. I take water directly by opening the gate wall, as it is close to my field.	1. If I do not as per turn, I inform the water operators of the DoIPH. 2. I try to solve the problem through negotiation, if I do not get water as per turn. 3. I investigate and take water. 4. Inform the President of the irrigation committee and irrigate the field.
Nichala Rajana		
Whom do you contact to access water from Khul *irrigation system?*	1. I do not contact anyone to get water. 2. I take water from *Khul* systems directly. 3. I do not depend on this water for my income.	1. We contact the Rajputs to irrigate our field.
Uppala Rajana		
Who told you to cultivate tomato crop and why?	1. I do not have time to spend on cultivating tomato (as the persons are employed elsewhere). 2. There is not enough labour force in family so that we can cultivate tomato. 3. As our fields are located near the forest, it is difficult to cultivate tomato as wild animals destroy them.	1. The Village leader influenced me. 2. I was influenced to cultivate by a schoolmaster. 3. I decided to cultivate myself (by looking at others). 4. The villagers started growing them, so did I.

through democratic principles of consensus seeking. This does not mean that strategic actions are less important as examining them will offer insights into ways to overcome the inadequacy. For the purpose of research (with limited time and cost factor), communicative actions are examined for their role in promoting water resource for local development.

A simplistic analysis (Table 20.6) of the actions adopted by diverse households indicates that it is mostly the middle class which adopts communicative actions (except from the hamlet Pipalwala) while the poor and the rich adopt strategic actions. It is interesting to note that the rich households steal water and use their might to access water while the poor use the action of 'wait and watch' and depend on employments. In communicative actions, actors communicate with others for collective decisions. In this arena at the hamlet level, not all actors take a lead role as 'agents' but only those who have the capability to draw upon the modalities of existing institutions in modifying or reconstituting their properties by accessing supra-arenas.

Table 20.6 Percentage of Households using Diverse Actions to Access Water

	Actions	
Social Class	*Strategic*	*Communicative*
Pipalwala Hamlet		
Rich	5	7
Middle	23	16
Poor	14	35
Khairwala Hamlet		
Rich	14	11
Middle	16	34
Poor	16	9
Uppala Rajana		
Rich	7	11
Upper Middle	15	19
Middle	3	5
Lower Middle	3	17
Poor	14	5
Nichala Rajana		
Rich	0	9
Upper Middle	2	52
Lower Middle	12	3
Poor	12	10

Source: Field Survey, 2004.

OPTIONS FOR WATER RESOURCE MANAGEMENT

There are diverse options available for hamlet-level agents to address the inadequacy of existing institutional arrangements (Table 20.7) that include: approaching the relevant government departments (here it is the Department of Irrigation and Public Health—DoIPH), the political representatives and the markets. Each of these arenas is accessed for some specific reasons. It is clear for these agents that for technical problems, it is the departments (like in case of Khairwala) that matter. But in case they require new irrigation schemes, they approach both DoIPH for technical clearance and the member of legislative assembly (MLA) for seeking additional funds (like in case of Pipalwala) (Figure 20.5).

Though the DoIPH can also mobilise additional funds through sectoral allocations every year, there is greater dependence on the political representatives as they are committed and are also easily accessible to people. While in both hamlets of Rajana watershed, the agent proposes to address through two different options (Figure 20.6): (*i*) increase availability of water through various water harvesting measures (as he has been told during watershed training programme); and (*ii*) improve infrastructure facilities (transporting and seeking better markets in plains) for marketing their cash crops—ginger and tomato.

Table 20.7 Different Action Arenas Accessed by Agents

	Khairi-Ka-Kala Watershed			*Rajana Watershed*	
Watershed Hamlet	*Khairwala*		*Pipalwala*	*Nichala Rajana*	*Uppala Rajana*
Agents	Mr *Nazim Ali*, President of LIS;	Mr *Sundar Das*, Member in LIS	Mr *Parem Singh* President of KIC; Mr *Sher*, Vice-President, KIC	Mr *Charan Singh*, Chairman of WDC	Village leader,
Problems Perceived by Agents	Inadequacy of Water	Problem of Water distribution	Inadequacy of water	Inadequacy of water	Problem in getting better price for products
Arenas Accessed	DoIPH	DoIPH	MLA, DoIPH	DRDA, DoF	Market
Purpose of Accessing these Arenas	Seeking additional lift irrigation scheme	For DoIPH to take over distribution	For a new lift irrigation scheme	Enhance water availability (harvesting measures) through watershed development	Improve infrastructure for marketing of ginger and tomato

INSTITUTIONAL INTEGRATION IN ARENAS

Agents interact among each other to take decisions within and among diverse arenas. In each of them, institutions integrate in diverse and complex ways to facilitate and constrain agents' decisions. Understanding the institutions involved will enable to deconstruct the complexity and grasp the interactions among institutions in diverse arenas. Though different types of institutions interact in the arena, they consist of three basic components that enable agents to take decisions: (*i*) policy institutions that provide guidelines on who should enter the arena, what position they should hold and how the outcomes have to be; (*ii*) legal institutions which authorise agents to take decisions; and (*iii*) administrative institutions that enable the agents to transform their decisions into actions and their actions into outcomes in a cost effective manner.

Examining the institutional integration in arenas indicates how agents emerge and the institutions that facilitate their decisions. In both the watersheds under study, informal institutions set the policies for hamlet level agents (Table 20.8). Of these, the role of social network is important. Though this makes them eligible, the legal authority for taking decisions is provided by the external institutions (DoIPH, DRDA, and DoF). This enables them to access administrative institutions to implement their decisions.

Of the three agents at the hamlet level, the agent at Rajana, Charan Singh, offers an example. He had been a village leader for the past decade and also the *Nambardar* (village revenue collector), but it was only for three years that he has been active as an agent. The credit goes to the watershed development

Figure 20.5 Agents in Different Action Arenas—Pipalwala, Khairi-Ka-Kala Watershed

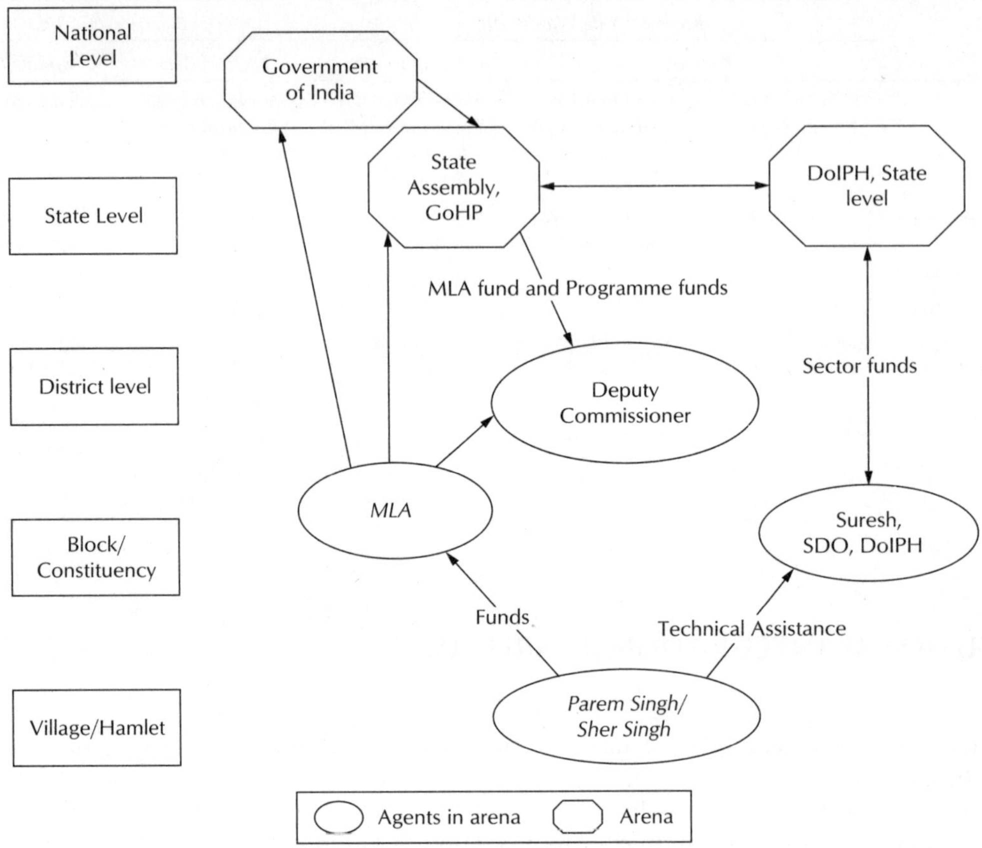

programme implemented under the Integrated Wasteland Development Programme (IWDP) of District Rural Development Agency (DRDA). Under this programme, he had been appointed as the Chairman of the Watershed Development Committee. Being the Chairman, he got the opportunity to meet the bureaucrats of various departments and also to know about their programmes. This also serves his self-interest—the need of earning a livelihood by taking these programmes to his villages. These agents play an important role in bringing development programmes to the village, but the challenge lies in monitoring and regulating these agents and their actions to address the concern of water resource upgradation for local development. In contrast, as the agents move higher-up, the role of informal institutions in setting policies gets reduced. However, for all the agents it is only the formal institutions that provide legal authority and administrative support in implementing their decisions.

Options to integrate institutions from other arenas are limited to government officials rather than political representatives. The demands made by the people pertain to technical, managerial and financial matters. The line departments are able to address the technical and managerial matters, but not the financial ones. Though they could forward such requests to the District Development Committee or

Figure 20.6 Agents and Arenas—Rajana Watershed

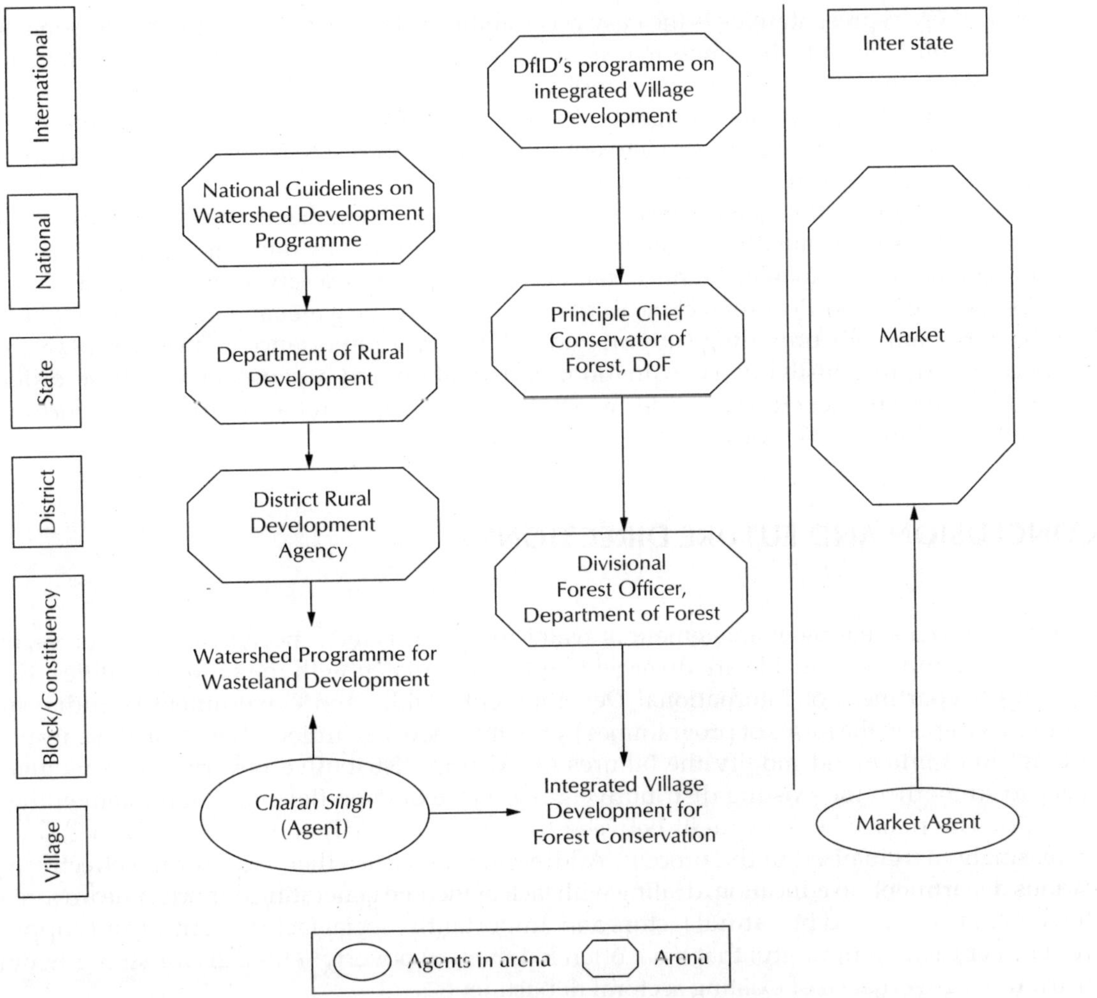

Table 20.8 Types of Institutions Facilitating Agents

Agents	*Policy*	*Legal*	*Administrative*
Hamlet Level	Village Institutions	DoIPH, IC 73rd Amendment	DoIPH, Political parties
Block Level (MLA/SDO)	Village Institutions and DoIPH	GoHP, DoIPH, Vote bank	DoIPH
District Level (PO)	DRD and PR	GoHP, DRD and PR	DRDA
Market Agent	Market and Village institutions	State Government	GoHP

to the Deputy Commissioners, the limitation imposed by the respective organisations hinders them from doing so (sometimes, the officials also reject the demand). This makes the hamlet level agents seek other arenas such as the political representatives. These representatives have access to diverse sources of funds—the state legislative assembly for including the demand in sectoral allocation, the district development committee for programme funds and the MLA Funds.[8] Another advantage of seeking these representatives is the easy accessibility—the language he speaks, personal access, simplicity in outlook and willingness to hear and solve their problems. The most important one is the trust that the politician builds with the people. This makes a lot of difference to the people, though he only forwards the plea made by agents to various departments. In fact, if one goes to meet him, his office functions like a helpline service centre. This is in contrast to Deputy Commissioner's office or even the simple government department.

Agents' decisions in the arena are influenced by the perception that they have on the attributes of governance. These attributes help agents in pursuing their goals by integrating diverse institutions (Table 20.9). Equity for hamlet agents is said to be *Khudrat ke diya* (given by God) and can only be managed. While the DoIPH uses technical criteria to approve the water and irrigation schemes, MLA uses his vote banks for providing support and DRDA gives importance to 'peoples' plan'. Similar is the case with responsibility, coordination, participation and accountability. These differences illustrate the different conditions under which the agent exists and also provides opportunity for designing institutions in accordance.

CONCLUSION AND FUTURE DIRECTIONS

The study reveals that the management of water resources is influenced by diverse forces, but the institutional options available are divergent and do not match with the ground realities. External agencies (Department of International Development—DFID and Government of India) impose various concepts in the forms of programmes by creating new institutions. Rarely do these institutions attempt to examine and modify the failures of existing distributive governance. This gives less space to strengthen the existing distributive governance or even flexibility in implementing these concepts. The poor who are caught between the macro (formal) and micro (informal) are being increasingly marginalised in the process. Addressing the issues then requires the effective role of various departments in education, dealing with lack of income generating opportunities, overcoming the constraints imposed by natural factors and, importantly, social factors (control and suppression from upper caste community) that have often led them to poverty. This calls for strengthening the distributive governance of existing sectoral departments.

This chapter provides an opportunity for utilising agents in facilitating development programmes. Agents at the hamlet and block levels emerge due to the village level informal institutions. However, the legal authority to take decisions is provided by the external formal institutions. This offers an opportunity to build on these agents by providing opportunities for existing agents to come forward and create opportunities for new agents at the hamlet level. This does not require new institutions to emerge but rather calls for existing government departments or NGOs to be interactive with the micro level reality, to share information about various ongoing and future programmes and provide regular advisory support. This calls for existing line departments (specially field level officials) to be proactive, to visit villages and discuss issues. This does not mean that government officials have to create user group in villages, rather they have to try to interact with the people to understand the

Table 20.9 Attributes of Governance and Arena

		Village	Block		District	
Agents		*Hamlet Agents*	*SDO (DoIPH)*	*MLA*	*PO, DRDA*	*Market*
Attributes						
Equity	Type	Inequity is given by God, it can only be managed.	Based on technical feasibility	More the vote bank, more the support	Better the user group, more the support	Better pricing
	Institutions Facilitating	Village institutions	DoIPH/Village Institutions	Political party	DRDA—watershed Guidelines	Village institution/ Market
Responsibility	Type	Assumed/ Assigned	Assigned	Assumed	Assigned	Assumed
	Institutions Facilitating	Village institutions/ Government Department	Institution based and Village institutions	Vote Bank	DLWDC/DoRD	Market
Coordination	Type	Authority-based	Need-based	Authority-based	System-based	Authority-based
	Institutions Facilitating	Village institutions and Irrigation committee	DoIPH	Vote Bank	DRDA	Market/Social network
Participation	Type	Authority-based	Rules and Regulation	Authority-based	Structure-based	Pricing based
	Institutions Facilitating	Village institution	DoIPH	Power	DRDA	Market/Village institution
Accountability	Type	Authority-based	Rules and Regulation	No accountability	DRDA/ DLWDC/User group	No-Accountability
	Institutions Facilitating	Village Institution	DoIPH	-		-

impact of contemporary programmes on people and their livelihood. This will offer opportunities to identify or create agents at micro level. Facilitating these agents can offer opportunities to bring in desired institutional change for water resources management.

Agents above the hamlet level are largely from the formal institutions. However, unclear roles of these (political representatives and limited role of bureaucrats) seem to be of concern in the case study region. It is not clear as to whom the political representative is responsible and accountable. Very often people are made to take up the burden (during election by voting a right candidate), but are unaware about mechanisms that are in place to oversee their decisions that can be respected and monitored. In contrast, bureaucrats have too much of accountability problem but limited autonomy to take decisions. Very often they seek opinion of their higher-ups for a decision or have to bow down to political interferences. These agents are constrained from taking independent decisions due to interferences from political bosses or higher-ups. Too much or too little constraints, make these agents have different perceptions in addressing the attributes of governance (equity, responsibility, coordination, participation and accountability). Examination of more of these attributes could serve as major guidelines for policy and programme interventions for necessary institutional change for managing water resources.

The study is only a piecemeal attempt as part of the research programme. It offers opportunities on two fronts. The first one lies in further improving the institutions in each action arena. Some of the areas for examination lie in identifying different types of integration that are in place, examining the interaction between formal and informal rules and applying the design principles of Institutional and Analysis Development framework as a heuristic tool. On the theoretical front, this will contribute towards blending the institutionalist approach, emerging from common property theories, with planning theories to predict models of institutional arrangements. The second one lies in moving forward with this small piece of preliminary research to examine the feasibility of providing guidelines for policies and programmes at district level.

ACKNOWLEDGEMENTS

The author is grateful to the University of Queensland—International Postgraduate Research Scholarship in Australia, and to International Water Management Institute (IWMI) for doctoral fellowship for the field work in India. I am indebted to Keith Richards, D.B. Gupta, George Verghese, Kanchan Chopra, Ramesh Chand, Chetan Singh and Joel Ruet for their initial encouragement. Comments on earlier drafts from Bruce Mitchell, Keith Richards, Ruth Meinzen-Dick, Dipak Gyawali, Kanchan Chopra, Ramesh Chand, Chetan Singh and Ajaya Dixit helped in refining research questions. Special thanks to Walter Coward, N.S Bisht and Chetan Singh for their support in clarifying and sorting out problems during the field work in Himachal Pradesh. My supervisors—Geoff McDonald, Basil van Horen, David Ip and Maria Saleth, played a major role in shaping and encouraging the research. I am indebted to district level officials, especially M.L. Sharma, Rajesh Maria and Arvin Pande for ensuring that my research reflects reality. The people of Rajana and Bikram Bagh for information, love and affection during nine months of stay in the field.

NOTES

1. Attempts to integrate institutions at state level have been witnessed in the past by the formation of Water Resource Organisations (WRO) under the influence of World Bank; at district level through the formation of Watershed Development Programme such as in Doon Valley and Indo-Changer projects; and at village level through watershed development committee.
2. Few term this as 'forums' (Bruns and Meinzen-Dick 2000; Moench et al. 2003) or 'platforms' (Chamala 1995; Steins and Edwards 1999). However, 'action arenas' is appropriate as it describes action.
3. The strategic context considers a wide spectrum of issues, involves a wide range of actors having a shared vision and understanding in making well-informed strategic choices that shapes their future and, more importantly, considers the ability of these actors to administer and enforce these decisions.
4. The conversion rate of Indian currency to one $ US is Rs 45.00.
5. Personal communication from Suresh Kumar, Sub-Divisional Officer, Nahan Division, Department of Irrigation and Public Health (DoIPH), Government of Himachal Pradesh, 15 October 2004.
6. The *Khul* in the study is channeled along the mountain ranges, which due to unconsolidated landscape has high erosion. This often leads to silt accumulation in the *Khul*, thus reducing its capacity. This calls for desilting the *Khul* very often, sometimes thrice a year, which people are unable to do along the three kilometre belt.

7. *Khul* irrigation systems are classified in revenue records as private (when it is managed and maintained by people) and government (if it is maintained and managed by DoIPH).
8. Allocated Rs 24 lakh every year to MLA for development works in his/her constituency.

REFERENCES

Abbot, J. and I. Guijit. 1997. 'Methodological Complementarity: Creativity and Compromise', *PLA Notes,* (28): 27–32.

Ahmed, Q.K., A.K. Biswas, R. Rangachari, and M.M. Sainju. 2001. 'A Framework for Sustainable Development of Ganga-Brahmaputra-Meghna Basin', in Q.K. Ahmed, A.K. Biswas, R. Rangachari and M.M. Sainju (eds), *Ganga-Brahmaputra-Meghna Region: A Framework for Sustainable Development.* Dhaka: The University Press.

Alexander, E.R. 2001. 'The Planner-Prince: Interdependence, Rationalities and Post-Communicative Practice', *Planning Theory and Practice,* 2(3): 311–24.

Baker, M. and V. Saberwal. 2003. Introduction: Unraveling the Tangle State-Society Negotiations over Natural Resources in Himachal Pradesh, *Himalayan Research Bulletin,* 21(2): 1–3.

Bruns, B.R. and R.S. Meinzen-Dick. 2000. 'Negotiative Water Rights: Implications for Research and Action', in B.R. Bruns and R.S. Meinzen-Dick (eds), *Negotiating Water Rights,* Washington: International Food Policy Research Institute, and New Delhi: Vistaar Publications.

Cars, G., P. Healey, A. Madanipour and C.de Magalhaes, (eds) 2003. *Urban Governance, Institutional Capacity and Social Milieux.* Hampshire: Ashgate Publishing Limited.

Castells, M. 1997. *The Power of Identity. The Information Age: Economy, Society and Culture. Volume 2.* Malden MA: Blackwell.

Chamala, S. and K. Keith. (eds) 1995. *Participative Approaches for Landcare: Perspectives, Policies and Programs.* Brisbane: Australian Academic Press.

Chopra, Kanchan. 1999. 'Evaluation of Watershed Programmes in India—A Review', in John Farrington, Cathryn Turton and A.J. James (eds), *Participatory Watershed Development: Challenges for the Twenty-First Century,* pp. 30–48. New Delhi: Oxford University Press.

Coward, W.J. 2003. 'The Kullu Valleys in Motion', *Himalayan Research Bulletin,* 21(2): 5–14.

Dorcey, A.H.J. 1986. *Bargaining in the Governance of Pacific Coastal Resources: Research and Reform.* British Columbia: Northburn Printers and Stationers Ltd. University of British Columbia Press.

Ecological Modelling 2002. *Modelling Complex Ecological Systems: An Introduction.* 153, pp. 1–6.

Edwards, V.M. and N.A. Steins. 1998. 'Developing An Analytical Framework for Multiple-Use Commons', *Journal of Theoretical Politics,* 10(3): 347–83.

Giddens, A. 1984. *The Constitution of Society.* Cambridge: Polity Press.

GOI 1993. National Policy for Integrated Development in Himalaya—Report of the Expert Group. New Delhi: Planning Commission, Ministry of Human Resources Development, Government of India.

———. 1997 Legal Instruments on Rivers in India—Volume 1. *Constitutional Provisions, Central Laws and Notifications.* New Delhi: Central Water Commission, Ministry of Water Resources, Government of India.

———. 1999. *National Commission for Integrated Water Resources Development Plan.* New Delhi: Ministry of Water Resources. September.

———. 2002. *National Water Policy – 2002.* New Delhi: Ministry of Water Resources, Government of India.

———. 2002. *National Human Development Report 2001.* New Delhi: Planning Commission, Government of India.

Hofwegen, P.J.M.V. and G.W. Jaspers. 1999. 'Analytical Framework for Integrated Water Resource Management: Guidelines for Assessment of Institutional Framework'. International Institute of Infrastructural, Hydraulic and Environmental Engineering (IHE), Monograph 2. Rotterdam: AA Balkema Publishers.

Ingram, H.M., D.E. Mann, G.D. Weatherford and H.J. Cortner. 1984. 'Guidelines for Improved Institutional Analysis in Water Resources Planning', *Water Resources Research,* 20(3): 323–34.

Jodha, N.S. 1998a. 'Poverty-Environmental Resource Degradation Links: Questioning the Basic Premises', *Issues in Mountain Development,* No. 1. Kathmandu: International Centre for Integrated Mountain Development (ICIMOD).

Lane, M.B. and G. McDonald. 2002. 'Crisis, Change, and Institutions in Forest Management: The Wet Tropics of Northern Australia', *Journal of Rural Studies,* 18: 145–256.

Lubell, M. 2004. 'Collaborative Environmental Institutions: All Talk and No Action?', *Journal of Policy Analysis and Management.*, 23(3): 549–73.

McDonald, G.T., T. Morrison and M. Land. 2003. Integrating Natural Resource Management Systems for Better Environmental Outcomes. *Australian Water Summit*, Sydney.

Mehta, L. 2002. 'Water for the 21st Century: Challenges and Misconceptions'. *IDS Working Paper 111*, Sussex: IDS.

Mitchell, B. (ed.) 1990. *Integrated Water Management: International Experiences and Perspectives*. London: Belhaven Press.

Moench, M., A. Dixit, S. Janakarajan, M. Rathore and S. Mudrakartha. 2003. 'The Fluid Mosaic: Water Governance in the Context of Variability, Uncertainty and Change'. *A Synthesis Paper*. Kathmandu: NWCF and Colorado: ISET.

Mollinga, P. 2000. 'Introduction' in P. Mollinga (ed.), *Water For Food and Rural Development: Approaches and Initiatives in South Asia*. New Delhi: Sage Publishers.

Mosse, D.M. 1997. 'The Symbolic Making of Common Property Resources: History, Ecology and Locality in a Tank Irrigated landscape in South India', *Development and Change*, 28: 467–504.

Neuman, L.W. 2003. *Social Research Methods: Qualitative and Quantitative Approaches*. Boston: Allyn and Bacon.

Ninnan, K.N. 1998. 'An Assessment of European Aided Watershed Development Projects in India from the Perspective of Poverty Reduction and the Poor'. *Centre for Development Research (CDR) Working Paper 98.3*, p. 31. Denmark: CDR.

North, D.C. 1990. *Institutions, Institutional Change and Economic Performance* Cambridge: Cambridge University Press.

Ostrom, E. 1985. 'A Method of Institutional Analysis', in F.X. Kaufman, G. Majone and V. Ostrom (eds), *Guidance, Control, and Evaluation in the Public Sector – The Bielefeld Interdisciplinary Project* (Berlin, New York, Walter, De Gruyter).

———. 1986. 'An Agenda for the Study of Institutions', *Public Choice*, 48: 3–25.

———. 1990. *Governing the Commons – The Evolution of Institutions for Collective Action*. Cambridge: Cambridge University Press.

———. 1998. 'The Institutional Analysis and Development Approach', in E.T. Loehman and D.M. Kilgour (eds), *Designing Institutions for Environmental and Resource Management*. Massachusetts: Edward Edgar Publsihing Limited.

Ostrom, E. 2001. 'Vulnerability and Polycentric Governance Systems', *International Human Dimensions Programme on Global Environmental Change Update*, 3(1): 1–4.

Ostrom, E., R. Gardner and J. Walker. 1994. *Institutional Analysis and Common-Pool Resources*. Ann Arbor: The University of Michigan Press.

Pahl-Wostl, C. 2002. 'Participative and Stakeholder-based Policy Design, Evaluation and Modeling Processes', *Integrated Assessment*, 3(1): 3–14.

Pradhan, U. 1997. 'Address by Dr Ujjwal Pradhan', Programme Officer, Ford Foundation, in R. Pradhan, V.F. Benda-Beckman, V.K. Benda-Beckman, J.H.L. Spiertz, S.S. Khadka and A.K. Haq (eds), *Water Rights, Conflict and Policy: Proceedings of a Workshop Held in Kathmandu, Nepal, January 22–24, 1996*. Colombo, Sri Lanka: International Irrigation Management Institute.

Railsback, S.F. 2001. 'Concepts from Complex Adaptive Systems as a Framework for Individual-based Modelling', *Ecological Modelling.*, 139: 47–62.

Richards, K. 2004. Comments Received Through Email on the Draft Proposal. 1 February.

———. (forthcoming) 'Ethical Grounds for an Integrated Geography', in S. Trudgill (ed.), *Contemporary Meaning in Physical Geography*.

Sadeque, S.Z. 2000. 'Poverty Assessment, Poverty Reduction, and Sustainable Livelihoods: How Poverty Mapping, Institutional Analysis, and Participatory Governance can Make a Difference'. *Issues in Mountain Development, No. 5*. Kathmandu: International Centre for Integrated Mountain Development (ICIMOD).

Saleth, M.R. and A. Dinar. 1999. 'Evaluating Water Institutions and Water Sector Performance'. *World Bank Technical Paper No. 447*. Washington DC: The World Bank.

Saleth, M.R., M. Samad, D. Molden, and I. Hussain. 2003. 'Water, Poverty and Gender: An Overview of Issues and Policies', *Water Policy*, 5: 385–98.

Saravanan, V.S. 1998. 'Sustaining Community-Based Resource Management: Conditions for Collective Action – Case Study of Two Tank Irrigation Societies in Tamil Nadu, India'. M. Phil Thesis (unpublished), Department of Geography, Cambridge: University of Cambridge.

———. 2002. 'Institutionalising Community-based Watershed Management in India: Elements of Institutional Sustainability', *Water, Science and Technology*, 45(11): 113–24.

———. 2004. 'Integrating Watershed Management Institutions: Examining What, Where and How?' Mimeo, New Delhi: International Water Management Institute (IWMI).

Steins, N.A. and V.M. Edwards. 1999. 'Collective Action in Common-Pool Resource Management: The Contribution of a Social Constructivist Perspective to Existing Theory', *Society and Natural Resources*, 12(6): 539–57.
Steins, N., N.G. Roling and V.M. Edwards. 2000. 'Re-'designing' the Principles: An Interactive Perspective to CPR Theory'. *8th Conference of the International Association for the Study of Common* Property, *1–4 June 2000*. Indiana: Bloomington.
UN-WWAP 2003. Water for People—Water for Life: Executive Summary. The United Nations-World Water Development Report. Barcelona: Berghahn Books.
Verghese, B.G. 1999. *Waters of Hope: From Vision to Reality in Himalaya – Ganga Development Cooperation*. Dhaka: The University Press Limited.
Young, O.R. 1999. Institutional Dimensions of Global Environmental Change. *International Human Dimensions Programme Report 9*. Bonn.
———. 2002. *The Institutional Dimensions of Environmental Change*. Cambridge, Massachusetts: Massachusetts Institute of Technology.

ANNEXURE 1

Annexure 1 District-wise Selected Indicators of Development—Himachal Pradesh

District	*Area (sq km)*	*Population (2001)*	*Decadal Growth Rate (1991–2001)*	*Sex Ratio*	*CBR 1991*	*IMR 1991*	*% of Habitation with Safe Drinking Water*	*Metalled Roads per sq km*	*Per Capita Income (1999–2000) at 1990–91 Prices*	*% of Rural Population below Poverty Level*	*Literacy 2001*
Chamba	6,528	4,60,499	17	961	35	104	98	7	6,058	62	64
Kinnuar	6,401	83,950	18	851	31	123	100	4	7,930	27	NA
Kullu	5,503	3,79,865	26	928	33	102	98	7	6,098	19	73
Lahaul & Spiti	13,835	33,224	6	804	28	59	100	2	12,559	38	73
Shimla	5,131	7,21,745	17	898	29	104	92	20	8,304	34	80
Sirmaur	2,825	4,58,351	21	901	34	94	89	2	5,650	23	71
Mandi	3,950	9,00,987	16	1,014	30	69	98	25	5,313	–	76
Bilaspur	1,167	3,40,375	15	992	28	71	100	47	7,547	27	79
Hamirpur	1,118	4,12,009	12	1,102	25	65	100	47	4,243	24	83
Kangra	5,739	13,38,536	14	1,027	28	77	97	32	5,736	24	81
Solan	1,936	4,99,380	31	853	30	84	96	36	11,231	27	77
Una	1,540	4,47,967	18	997	29	82	100	47	4,480	19	81
Himachal Pradesh	55,673	60,77,248	18	970							

Source: GoHP. (2002) Himachal Pradesh Human Development Report–2002. Planning Commission, GoHP, Shimla.

21

Interactive Goal Linear Programming in Land Use Planning in Haryana[1]

A.K. Vasisht[2] and *S.S. Kumar*

Abstract: The ever increasing nature of our country's population growth presents prodigious demand on food supplies. This coupled with frequent natural hazards and calamities continue to disturb the balance between the food requirements and production reserves. Food security is the availability and accessibility of sufficient food of the desired quality at all times and is the outcome of food supply and demand, modified by the socio-economic characteristics that determine prices and purchasing power. The food security of the people is ensured only when the factors like availability of adequate supplies, access to sufficient food, stability of supplies and cultural acceptability, act in harmony. Besides having sufficient capacity for food production, issues such as employment, capital, infrastructure and diet of rural population also contribute directly or indirectly to the food security of the rural areas. Although there is pressure to increase production to meet higher demand, lately the growth in cultivated area, production and yield has slowed significantly. The annual rate of growth in food production and yield peaked during the early years of the Green Revolution, but has declined since the 1980s. The perceived gradual increase in environmental degradation, early signs of which are becoming visible in rural areas that benefited largely from the Green Revolution technologies, is further compounding the problem. There is a great concern now about declining soil fertility, changes in water table depth, rising salinity, resistance of harmful organisms to many pesticides and degradation of irrigation water quality.

INTRODUCTION

Food is a basic need of any society and safeguarding the food supply has been a major consideration in policy development. Food demand of humans has increased dramatically over the last fifty years. The explosive increase in food production in the western world has led to a situation of supply exceeding demand. These surpluses of food were absorbed by the developing countries during 1950s but, in the 1960s, imports by the latter shrank because of the Green Revolution. After enjoying self-sufficiency in food during the last three decades, many Asian countries are once again at the crossroads, facing tremendous new challenges because of continued population growth, globalisation, environmental degradation and stagnation in farm productivity in intensive farming areas. Rapidly increasing population necessitates that the productivity of the land be further increased. This has to be achieved without increasing environmental degradation while maintaining or increasing the farmer's income. From an ecological point of view, land use/land cover change is a major factor affecting the health and stability of an ecosystem. Therefore, economically viable optimal solutions for land use can be determined by the use of a systems approach whereby the biophysical potential of the resources available and the socio-economic constraints, which are often inherently conflicting in nature, are considered to determine the consequences and trade-offs of different sets of policy aims on agriculture.

Planning for sustainable food security requires an integrated assessment of biophysical, socio-economic, political and environmental conditions. Several studies have been carried out in the past to determine the potential of different kinds of land, but most of these have focused on one or two aspects only. Such land evaluation methodologies do not relate biophysical criteria to crop productivity, intensity of input use, socio-economic conditions and environmental impact. The market forces also control land use patterns at any given time, besides it being determined by the biophysical potential of the land. At the same time, market-driven land use patterns often lead to unsustainable use of land and thus land degradation and decreasing profits of the rural masses in the long run.

Haryana, in northwestern India, has considerable potential to withdraw agricultural land from cultivation, without affecting basic food production and income at the aggregate level. This would require that the small farmers, who cannot use alternative, efficient and capital intensive technologies, do not cultivate their land and that their water and other resources be made available to other farmers who presumably could use these more efficiently. Alternatively, technologies that are affordable and can be applied on small farms should be developed.

In this chapter, a methodology has been developed for exploratory land use analysis and planning using Interactive Multiple Goal Linear Programming (IMGLP) approach. The use of this Decision Support System (DSS) has been illustrated for the State of Haryana. The results revealed that Haryana has ample opportunities to increase food production and agricultural income compared with current levels, provided additional water resources could be made available. The current natural constraints of land and water, limit maximum food production to seventeen million tonnes. Here, the model assumes that all water and capital within a land unit can be shared. It implies that groundwater resources available within a farm can be transported to other farms without cost, irrespective of distance involved. This does not look feasible, even with the current policy of an almost free water supply in the region.

METHODOLOGY

Regional land use analysis and planning for food security should be oriented towards maximisation of the welfare function of society from the non-renewable resource land. It should recognise land as a resource that provides space, is indestructible and can be viewed as a source of flow of production/consumption services whose composition depends on the use to which the space is allotted. This spatial pattern is variable over time depending on human activity and, therefore, intertemporal allocations of these services have their consequences. Land use planning is thus an interdisciplinary task that needs both biophysical and land economics evaluation.

The approach of Multiple Goal Linear Programming (MGLP) used for the current study provides such a framework for considering biophysical and socio-economic resources and constraints. An optimisation framework, consisting of linear programming or other techniques, represents a normative approach that is often used to search for the best solution with limited resources. In this approach, an objective function is maximised or minimised by selecting from different possible activities and subject to several regional constraints. Prior knowledge of the decision makers' choices has prime importance in formulating objective functions. Their preferences are expressed as objective functions and targets in the model. Decision making for many real-world problems is often the responsibility of a group of individuals, each with its own goals and aspirations, rather than of a single individual. Besides, in any society, preferences of the people are likely to be multidirectional. Therefore, it is necessary to develop a land use planning model for food security in a multi-objective framework. Moreover, the existing methodologies for evaluating optimal land use plans lack flexibility to rapidly respond to continuously changing policy environments with multiple and often conflicting goals. The research results based strictly on biophysical evaluation bereft of socio-economic considerations have not always been successful. In agro-ecosystems that are equal to or bigger than a farm, socio-economic processes become equally important.

The MGLP approach has been used in several studies for land use analysis and planning at the farm level (Schans 1991), village level (Huizing and Bronsveld 1994), sub-regional and regional level (Schipper et al. 1995; Veeneklaas et al. 1991) and even at the continental level (WRR 1992). It requires decision makers to specify maximum allowable levels for the (n-l) objectives to solve the n-dimensional multi-objective problem. This method can be used to generate the non-inferior set for all types of objectives. The result of each iteration is presented to decision makers to seek their preferences and then articulated back to the model through modified values of objective functions and targets. The process continues till the decision makers are satisfied with their choices and an optimal solution is obtained. This implies that this approach needs a series of iterations to arrive at the desired output. In the first iteration, all targets are set to a minimum value, resulting in an optimal solution that satisfies the entire minimum requirement simultaneously. This process is repeated sequentially for all objective functions, which will result in the definition of technically feasible objectives, targets and constraints. Moreover, the maximum attainable value for each objective function is also achieved. In the next step, the target values are further tightened, reflecting the aspirations of the decision makers. This will reduce the technically feasible solution space. The process continues till the decision makers reach a Pareto optimal solution, that is no further feasible solution can be achieved with the same or better performance for all criteria under consideration.

THE MULTIPLE GOAL LINEAR PROGRAMMING MODEL (MGLP) FOR HARYANA

The aim of IMGLP is to quantify the upper limits of production of food and other commodities in the State of Haryana and to identify production systems that are both economically viable and agronomically efficient and have a minimal impact on the environment. Rice and wheat, commonly grown in double cropping rotation, are the major cereal crops of this region and their average productivity ranges between three to five tonnes/ha.

The MGLP model for Haryana covers sixteen districts (as per the 1991 census database), which can be viewed as a combination of various land units. A land unit is delineated overlaying agro-ecological units and district boundaries. The model contains five resources: land, water, labour, capital and fertiliser. Land and water resources have been defined in two dimensions—administrative and agro-ecological—because of the distinct heterogeneity in different properties of land units in the same district. Since the district is the basic planning and production unit, labour, capital and fertiliser resources have been defined at the district level.

Various production functions have been specified through input-output relations for fifteen land use types at five technology levels. Land use types represent different farming regimes (irrigated versus non-irrigated). These are summarised in Table 21.1. A specific technology level through its uniqueness of input-output combinations characterises each land use type. Input-output combinations are determined by several factors related to land use and technology level.

Table 21.1 Districts, Land use Types, Technologies, Products and Farm Types in Haryana used for Land use Analysis

Districts	*Land use Types*	*Technologies*	*Products*	*Farm Types*
Ambala	Rice-rice-wheat	Current	Rice	Small
Bhiwani	Rice-wheat	Potential	Basmati Rice	Medium
Faridabad	Basmati rice-wheat	Current +	Wheat	Medium-Large
Gurgaon	Rice-mustard	25% yield gap	Sugar	Large
Hissar	Cotton-wheat	Current +	Mustard	Very Large
Jind	Maize-chickpea	50% yield gap		Pearl Millet
Kaithal	Maize-mustard	Current +		Cotton
Kamal	Maize-potato-wheat	75% yield gap		Maize
Kurukshetra	Sugarcane-wheat			Gram
Mohindergarh		Irrigated pearl millet-wheat		Potato
Panipat	Rain-fed pearl millet-wheat			Milk
Rewari				Fallow-wheat
Rohtak				Fallow-chickpea
Sirsa				Fallow-mustard
Sonipat				Pearl millet-fallow
Yamunanagar				

Milk is also an important product related to land use in Haryana. Therefore, besides cropping activities, livestock activities with three animal types, cow, buffalo and hybrid cow, are also considered in the model.

These land use types result in eleven products, including milk from each animal type.

The behaviour of the producers is described by assuming that they aim at maximum returns from the land unit under existing resource constraints. Five farm types varying in the size of landholding are considered. This is used as a proxy variable to represent the technology adoption capability of producers.

Since the livelihood of most of the population of Haryana basically depends on agriculture, it was assumed in all analyses that at least 98 per cent of the land has to be used for agriculture.

The market for agricultural products is assumed to be unaffected by producers' decisions at the district level. Irrespective of the quantities, all products can be sold or purchased at a fixed price for a district. This may not always be true but this assumption allows us to keep the model simple and explore all possible opportunities for the future irrespective of trade scenarios so that finally a limited policy environment can be explored in different scenarios.

Table 21.3 shows the indices and abbreviations used in the equations of the MGLP model.

In irrigated land units, all fifteen land use types (luts) were considered, whereas in rain-fed land units, only five luts were considered.

Land Use Activities

Two types of activities are included in the MGLP model for Haryana: cropping activities and livestock activities. For each activity, only those items of input-output that are needed for objective functions and constraints considered in the model are quantified.

Cropping activities are expressed as land use types (lut) applied at a certain technology level (t). We defined fifteen land use types for Haryana (Table 21.1). Inputs and outputs of these cropping activities are differentiated by land unit (u) and technology (t) and they also may vary by month or season. Inputs required for cropping activities are fertilisers, labour force, water and capital. Outputs from cropping activities are main products and by products of the crop and residues are used as feed for animals.

Animal types specify livestock activities. Inputs required for livestock are feed and capital. Livestock activities are linked to cropping activities through the availability of crop residues for feed in each land use type.

Because both cropping activities and livestock activities generate outputs for objective functions, a land use activity is defined as a combination of a cropping activity (lut, t) and a livestock activity (a, at). The variable $\text{LU–Area}_{du,lut,t,a,at}$ used in the MGLP model is the area allocated to each land use activity in each land unit (du).

$\text{LU-Promising}_{du,lut,t}$ is applied in the MGLP model as a promising land use indicator, which enables the model to handle different policy scenario analyses in a simple way and improves efficiency by reducing the size of the matrix. The value of this indicator is switched between 1 and 0 to identify whether a land use type (lut) can be applied in a land unit (du) or not.

Objective Functions

Objective functions for the model were formulated considering social, economic and environmental aspects of development for Haryana. These objective functions are as follows:

1. Social objective functions: Food grain production and employment.
2. Economic objective function: Income.

3. Environmental objective functions: Agricultural area, water use and biocide residue index and N leaching.

Each objective function comprises six cases each of which is characterised by a combination of constraints:

1. Land resource, which is always a constraint.
2. Land + water resources.
3. Land + technology adoption levels applicable by farm size groups.
4. Land + technology adoption + water resources.
5. Land + technology adoption + water resources + capital availability.
6. Land + technology adoption + water + capital + labour availability.

Table 21.2 shows the number of combinations of land units, land use types and technology levels. Table 21.3 shows the indices and abbreviations used in the equations of the MGLP model.

Table 21.2 Number of Combinations Related to Land use in Haryana

Item	*Abbreviation*	*Size*
Number of agro-ecological units	NAE	58
Number of land units (District agro-ecological combinations)	NDU	257
Number of land unit-land use type combinations	NDULut	2,855

Table 21.3 Indices and Abbreviations used for Defining Land use Types and Input/Output Relationships in the MGLP Model

Index	*Description*	*Classes*
U	Agro-ecological units	58 agro-ecological units
D	District	16 districts
Du	Land unit	257 land units (combinations of district, agro-ecological units and irrigated/unirrigated areas)
Lut	Land Use Type	15 land use types
P	Product	11 products, including milk from each animal species
T	Technology Level	Five technology levels
M	Months	12 months
A	Animal	Three types of animals: cows, buffaloes, hybrid cows
At	Combinations of an animal and livestock technology level	Two technology levels (current and improved) for each animal
F	Type of Fertiliser	Three types of fertiliser: N, P, K
S	Season Code	Three seasons: summer, *kharif* (monsoon), *rabi* (winter)

Note: * Land Unit (du) is used as a basic unit in the model, but a variable can vary either by district (d) or by agro-ecological unit (u) of this combination (du).

Social Objective Functions: Food Grain Production and Employment

Haryana is one of the major food-producing states in India and it contributes significantly to the public food distribution system of the federal government. Therefore, food grain production (Food) is one of the social objective functions to be maximised:

$$\text{Food} = \Sigma_{du}\Sigma_{lut}\,\Sigma_{t}\Sigma_{a}\Sigma_{at}\,(\text{Productivity}_{u\ in\ du,lut,t} \times \text{LU–Promising}_{du,lut,t} \times \text{LU–Area}_{du,lut,t,a,at}) \tag{1}$$

where $\text{Productivity}_{u\ in\ du,lut,t}$ is the yield of grains (rice, *basmati* rice, summer rice and wheat) in each land unit by various land use types at different technology levels.

Creating more gainful employment in the agricultural sector is essential for sustaining the development of the state. To realise this objective, we selected employment as another social objective function to be maximised:

$$\text{Employment} = \Sigma_{du}\Sigma_{lut}\,\Sigma_{t}\Sigma_{a}\Sigma_{at}\,(\text{Labour}_{u\ in\ du,lut,t,a,at} \times \text{LU–Promising}_{–du,lut,t} \times \text{LU–Area}_{du,lut,t,at}) \tag{2}$$

where $\text{Labour}_{u,lut,t,a,at}$ is the total labour required in a year for land use activities calculated from the labour requirement in each month.

$$\text{Labour}_{u.lut,t,a,at} = \Sigma_{m}\Sigma_{p}\ \text{MonthlyLabour}_{u,lut,t,p,m} \times \text{Lut-Product}_{lut,p} \tag{3}$$

However, the labour input for livestock activity was not considered because in Haryana, this homestead activity is generally taken care of by the family members in their spare time.

Economic Objective Function: Income

Income from agriculture is a major factor that determines crop and technology selection. This was selected as an objective function to be maximised to express the goal of economic development of the farmers and the region:

$$\text{Income} = \Sigma_{du}\Sigma_{lut}\Sigma_{t}\Sigma_{a}\Sigma_{at}\,(\text{Income–Ha}_{du,lut,t,a,at} \times \text{LU–Promising}_{lut,p} \times \text{LU–Area}_{du,lut,t,a,at}) \tag{4}$$

where Income–Ha is the net revenue from both cropping and livestock activities and is equal to the total revenue from the sale of all products, including milk, after subtracting the production cost of all inputs.

Income–Ha was calculated from operational costs and gross returns per hectare. Operational cost per ha does not include the fixed cost of the land and was derived by the following expression:

$$\begin{aligned}\text{Operational Cost}_{du,lut,t,a,at} &= \Sigma p[(\text{VariableCost}_{u\ in\ du,lut,t,a,at,p} + \text{PumpCost}_{du,lut,t,a,at,p}) \times \text{Lut–Product}_{lut,p}]\\ &+ (\text{NoAnimal}_{u\ in\ du,lut,t,a,at} \times \text{MilkCost}_{a,at})\end{aligned} \tag{5}$$

In the model, the cost of pumping water (PumpCost) is separated from other input costs because it varies over seasons and across crops depending on the amount of water pumped:

$$\text{PumpCost}_{du,lut,t,a,at,p} = \Sigma m\ \text{Month–Pump}_{u\ in\ du,lut,t,a,at,p,m} \times \text{Month–Pump–Price}_{du,m} \tag{6}$$

Month–Pump $_{u,lut,t,a,at,p,m}$ is the amount of water pumped for irrigation for a specific crop and month and Month–Pump–Price$_{du,m}$ is the unit cost of pumping water in a month. VariableCost$_{u,lut,t,a,at,p}$ is the cost for crops excluding the costs of water pumping and rearing livestock, NoAnimal$_{u,lut,t,a,at}$ is the number of animals per hectare and MilkCost$_{a,at}$ is the annual cost of producing milk from one animal. This leads to

$$\text{GrossReturn}_{du,lut,t,a,at} = (\text{NoAnimal}_{u\ in\ du,lut,t,a,at} \times \text{MilkIncome}_{a,at}) + \Sigma p\,((\text{Productivity}_{u\ in\ du,lut,t,p} \times \text{FGPrice}_{u\ in\ du,lut,t,p} \times \text{PriceAdjust}_{dindu,p}) \times \text{Lut–Product}_{lut,p}) + (\text{RevResidue}_{u\ in\ du,lut,t,a,at}) \times \text{Lut–Product}_{lut,p}) \quad (7)$$

where Productivity $_{u,lut,t,p}$ is the yield level of a product, FGPrice $_{u,lut,t,p}$ is the farm-gate price of a product and PriceAdjust$_{d,p}$ is a factor used to adjust the price across districts for different products. This price difference occurs mainly because of changes in market accessibility. RevResidue $_{u,lut,t,a,at}$ is the income from crop residues except for wheat and pearl millet (which have been used for livestock).

Net income is calculated as the difference between gross returns and costs:

$$\text{Income–Ha}_{du,lut,t,a,at} = \text{GrossReturn}_{du,lut,t,a,at} - \text{Operational Cost}_{du,lut,t,a,at} \quad (8)$$

Environmental Objective Functions: Agricultural Area, Water use and N Leaching

The pressure on land is increasing because of the increase in population, industrialisation and the requirements for various other non-agricultural activities. Moreover, there is concern that, ideally, about one-third of the land should be left for forest for environmental sustainability. Therefore, agricultural area in Haryana is considered as an objective function to be minimised:

$$\text{AgriArea} = \Sigma_{du}\Sigma_{lut}\Sigma_{t}\Sigma_{a}\Sigma_{at}\,(\text{LU–Promising}_{du,lut,t} \times \text{LU–Area}_{–du,lut,t,a,at}) \quad (9)$$

There are also concerns in Haryana about sustainability as the state moves into the post-Green Revolution era. The environmental goals for agricultural development in Haryana are to minimise two other environmental objective functions – water use and Nitrogen Leaching:

$$\text{WaterUse} = \Sigma_{du}\Sigma_{lut}\Sigma_{t}\Sigma_{a}\Sigma_{at}\,(\text{ET}_{u\ in\ du,lut,t} \times \text{LU–Promising}_{lut,p} \times \text{LU–Area}_{–du,lut,t,a,at}) \quad (10)$$

where ET $_{u,lut,t}$ is the total water needed in a year for each land use activity calculated from its monthly water requirement. Drinking water required for animals is a relatively low amount compared with the water required for crops and has therefore been ignored.

$$\text{ET}_{du,lut,t} = \Sigma_{m}\,\text{MonthlyET}_{du,lut,t,m}$$

The model provides total nitrogen leached out (NLoss) at different levels of nitrogen application:

$$\text{NLoss} = \Sigma_{du}\Sigma_{lut}\Sigma_{t}\Sigma_{a}\Sigma_{at}\Sigma_{p}\,(\text{Nleaching}_{u\ in\ du,lut,t,p} \times \text{LU–Promising}_{–du,lut,p,t} \times \text{LU–Area}_{du,lut,t,a,at}) \quad (11)$$

where Nleaching $_{u,lut,t,p}$ is leaching of nitrate–N below 150 cm of the soil profile.

Constraints

Many biophysical characteristics and socio-economic factors constrain regional land use. These can be broadly grouped into natural resource constraints and external input constraints. In the model, a target of development, such as total production of certain products to satisfy the demand of the local population, has the same formulation as a constraint.

Natural Resource Constraints: Land and Water Resources

As mentioned earlier, the land resource has been defined with two dimensions—agro-ecological unit (u) and district (d)—to enable the model to capture biophysical homogeneity at the land unit level and homogeneity in socio-economic variables at the district level. The first constraint in land resource is that the total area of all land use types in each land unit ($DUArea_{du}$) should not be greater than the available land resource ($AvLand_{du}$):

$$DUArea_{du} = \Sigma_{lut}\Sigma_{t}\Sigma_{a}\Sigma_{at}\ (LU\text{-}Promising_{du,lut,t} \times LU\text{-}Area_{du,lut,t,a,at}) <= AvLand_{du} \quad (12)$$

where $AvLand_{du}$ is the available land in all land units (du).

In Haryana, 20.4 per cent of the land is made up of small holdings (< 2 ha) and 35.5 per cent of the holdings are from 2 to 5 ha (Table 21.4).

Table 21.4 Categories of Farmers in Haryana by Area and Size of Landholding

Category	*Size of Land-holding (ha)*	*Percentage of Land-holdings (%)*	*Area (ha)*	*Area (%)*
Small	< 2	60.5	7,57,731	20.4
Medium	2–5	27.5	13,18,110	35.5
Medium-Large	5–10	9.0	9,25,968	25.0
Large	10–20	2.5	4,76,677	12.8
Very Large	> 20	0.5	2,32,729	6.3

Only 6.3 per cent of the holdings are larger than 20 hectares. Resource availability can greatly vary depending upon the size of the landholding and other production resources of farmers. Since household modelling is not directly considered in our model, we have restricted, as a surrogate, the land area that can be used for different technologies depending upon the size of the landholdings. Thus the entire area of Haryana, irrespective of the size of landholding, can use 1st (current) and 2nd levels of technologies. The adoption of higher technologies requires more capital and a larger knowledge base. It was assumed that small farmers cannot adopt the 3rd, 4th and 5th level of technologies, whereas large and very large farmers can adopt the 4th level of technology. Only very large farmers can adopt the 5th level of technology (Table 21.5). The share in total area in Table 21.5 is used to estimate the maximum land resource available to each technology level ($AvTechLand_{du,t}$):

$$AvTechLand_{du,t} = AvLand_{du} \times CF_{d,t} \quad (13)$$

where $CF_{d,t}$ is the share of a technology level in the total area.

Table 21.5 Capability of the Farmers of Haryana to Adopt Different Technologies

Technology Level	*Farmers*	*Total Area (%)*
1	Small, Medium, Medium-Large, Large and very Large	100
2	Small, Medium, Medium-Large, Large and very Large	100
3	Medium, Medium-Large, Large and very Large	79.6
4	Large and very Large	19.1
5	Very Large	6.3

Thus, another land constraint is that the total area of all land use types by each technology level (DUTArea $_{du,t}$) should not be greater than the land resources available for that level (AvTechLand $_{du,t}$):

$$\text{DUTArea}_{du,t} = \Sigma_{lut}\Sigma_{a}\Sigma_{at}(\text{LU–Promising}_{du,lut,t} \times \text{LU–Area}_{du,lut,t,a,at}) <= \text{AvTechLand}_{du,t} \quad (14)$$

Water Resources

Both groundwater and surface water are considered when estimating total water available for irrigation. The model assumes that different land use types within it can share the water available within a land unit.

Total water use in a year in each land unit (Water$_{du}$) should not be greater than the available water resources in that land unit (AvWater$_{du}$):

$$\text{Water}_{du} = \Sigma_{lut}\Sigma_{t}\Sigma_{a}\Sigma_{at}(\text{ET}_{du,lut,t} \times \text{LU–Promising}_{du,lut,t} \times \text{LU–Area}_{du,lut,t,a,at}) <= \text{A vWater}_{du} \quad (15)$$

where ET$_{du,lut,t}$ is the total water requirement of a land use type in a year aggregated from water requirements in each month.

Socioeconomic Constraints: Labour, Capital and Input Supply

Similar to water, the constraint in labour availability by month is considered. The following constraint is applied for labour:

Labour use (Labour$_{dist,m}$) in each district in each month should not be greater than the available labour force (AvLabour$_{dist,m}$)

$$\text{Labour}_{dist,m} = \Sigma_{du}\Sigma_{lut}\Sigma_{t}\Sigma_{a}\Sigma_{at}((\text{MonthlyLabour}_{u,lut,t,a,at,m} \times \text{LU–Promising}_{du,lut,t} \times \text{LU–Area}_{du,lut,t,a,at} \text{ I with d in du = dist}) <= \text{A vLabour}_{dist,m} \quad (16)$$

where MonthlyLabour $_{u,lut,t,a,at,m}$ is the labour requirement in each month.

It was assumed that capital could be shared or borrowed within the district. The constraint in capital was therefore formulated as the total capital requirement (Capital $_{dist}$) should not be greater than the available capital (AvCapital $_{dist}$):

$$\text{Capital}_{dist} = \Sigma_{du}\Sigma_{lut}\Sigma_{t}\Sigma_{a}\Sigma_{at}(\text{Capital-Ha}_{du,lut,t,a,at} \times \text{LU–Promising}_{du,lut,t} \times \text{LU–Area}_{du,lut,t,a,at} \text{ I with d in du = dist}) <= \text{A vCapital}_{dist} \quad (17)$$

where Capital-Ha $_{du,lut,t,a,at}$ is the total cost for land use activity.

Fertiliser availability is also considered as a major constraint to agricultural production. Therefore, the total fertiliser requirement ($Fertiliser_{dist,f}$) should not be greater than the available fertiliser ($AvFertiliser_{dist,f}$):

$$Fertiliser_{dist,f} = \Sigma_{du}\Sigma_{lut}\Sigma_{t}\Sigma_{a}\Sigma_{at}(Fertiliser\text{-}Ha_{uin\ du,lut,t,f} \times LU\text{-}Promising_{du,lut,t} \times LU\text{-}Area_{du,lut,t,a,at} <= A\ vFertiliser_{dist,f} \quad (18)$$

where $Fertiliser\text{–}Ha_{u,lut,t,f}$ is the total fertiliser required for a land use activity.

RESULTS AND DISCUSSIONS

Since the majority of the population of Haryana depends on agriculture for its basic livelihood, the model was forced to assume that all agricultural land of the state is cultivated in all cases except in the scenario in which agricultural land use was minimised. The upper limits of different objective functions were determined by optimising each one separately and deriving the 'extreme points' to identify the feasible solution space under the specified restrictions. Thus, the model first calculates the value of each objective function by imposing land as a constraint plus the lower bounds for production of the different commodities, defined as the production figures for 1996–97 (Table 21.6, last column). Subsequently, all other constraints were introduced successively in the subsequent rounds of optimisations to evaluate the effect of each constraint on the feasible solution space. In the final run, all constraints and current targets for other crops were imposed concurrently. In this chapter, the analysis has been presented for two objective functions, namely, maximisation of food and minimisation of water use.

Maximising Food Grain Production

The results of maximisation of food showed that the maximum attainable food production (rice + wheat) in Haryana was 39.1 million tonnes when land was the only constraint and the current targets for other products were met (Table 21.6). Corresponding milk production was 6.8 billion litres. To produce this, however, Haryana would need, besides arable land, 56.4 billion cubic metres of water, 1.5 million tonnes of N fertiliser, 666 million labour days and 114.2 billion rupees of capital for operational costs. These requirements are several times higher than what is currently (1996–97 level) available in the state. This indicates that, if such resources were made available, farmers could generate an income of 109.9 billion rupees per annum. The associated land use would result in a loss of 61.4 thousand tonnes of N through leaching.

This case provides information on the maximum food production possibilities in Haryana. However, it is not considered a feasible solution because of the extremely high amount of resources needed to produce these levels. These resources are neither currently available nor do they appear to become available in the next ten to twenty years.

The availability of irrigation water was imposed as the next constraint, in addition to land, to determine the maximum possible food production in Haryana with only the natural resources

Table 21.6 Production of Different Commodities, Income, Resource Requirements and Environmental Impact at an Aggregated Level when Maximising Food Production in Haryana

		Constraints						
Item	*Unit*	*Land*	*Land+ Water*	*Land + Tech*	*Land+ Water+ Tech*	*Land+ Water+ Tech+ Capital*	*Land + Water + Tech + Capital + Labour*	*Current Level (1996–97)*
Food**	Million tonnes	39.1	17.4	28.0	11.4	11.4	11.1	10.5
Rice	Million tonnes	27.3	5.1	19.0	2.8	2.7	2.5	2.5
Wheat	Million tonnes	11.8	12.2	9.0	8.6	8.7	8.6	8.0
Oilseed	Million tonnes	0.99	0.99	0.99	0.99	0.99	0.99	0.99
Chickpea	Million tonnes	0.28	0.28	0.28	0.28	0.28*	0.28	0.28
Cotton	Million bales	1.53	1.53	1.53	1.53	1.53	1.53	1.53
Sugar (jaggery)	Million tonnes	0.90	0.9	0.90	0.90	0.90	0.90	0.90
Milk	Billion litres	6.8	6.3	5.5	5.4	4.5	4.6	4.2
Income	Billion rupees	109.9	73.8	77.8	54.3	56.3	54.9	46.1
Land used	%	100	100	100	100	100	100	100
Irrigation	Billion metre cube	56.4	17.8	51.2	16.3	16.2	15.5	18.2
N fertiliser	Million tonnes	1.51	0.79	1.25	0.64	0.64	0.61	0.65
Employment	Mill. labour days	666	384	674	364	361	347	387
Capital	Billion rupees	114.2	56.9	92.1	54.1	53.7	52.0	56.4
N loss	Thousand tonnes	61.4	37.6	62.5	39.6	39.1	37.4	31.6
Biocide index		95	94	97	132	129	125	81

Notes: * Each bale of cotton = 170 kg; ** Objective function maximised.

as constraints. Food grain production in the second case decreased to 17.4 million tonnes. Rice production, being the largest consumer of water, dropped to 5.1 million tonnes from 19.0 million tonnes. Production of other commodities was maintained at their minimum demand level (Table 21.6). These results indicate that the spatial and temporal availability of water is now the major limiting factor for increasing food grain production in Haryana. In spite of this drastic reduction in food production, milk production decreased only marginally to 6.3 billion litres. To realise these levels of production, all land available for agriculture was used and 17.8 billion cubic metres of water was needed. It is interesting to note that 2 per cent of the water available now was still not used. The available water in the *kharif* season was completely used, whereas that of the *rabi* and summer seasons was not fully used. With food production as the main goal, the model allocated all area to rice in the *kharif* season, the only food grain crop in that season whenever water availability allowed. Since the minimum targeted demand of less water-consuming crops, such as chickpea and mustard in the *rabi* season had to be fulfilled as well, a considerable area was allocated to these land use systems and hence some water remained unused.

Fertiliser, labour and capital requirements as well as farm income also decreased drastically. A reduction in nitrogen loss could be observed compared to the first case. This is the result of a drastic shift in cropping pattern from rice-rice-wheat, the cropping system that consumes the highest amount of nitrogen fertiliser, to rice-wheat and fallow-wheat.

In the third case, in addition to land, the constraint of technology adoption was introduced to take into consideration the limited capacity of small and medium farmers to adopt capital intensive technologies. Water availability was not included as a constraint in this scenario. Optimal food grain production decreased to twenty-eight million tonnes and corresponding milk production to 5.5 billion litres. Production of all other commodities was at their 1996–97 levels. Relative to the land constraint, the requirements of water, fertiliser and capital decreased and total farm income decreased by 30 per cent.

When land, water and technology adoption were simultaneously introduced as constraints in the fourth case, food grain production decreased further to 11.4 million tonnes. Rice production declined to 2.8 million tonnes, which was very close to the minimum targeted demand. For wheat, the situation was almost the same. Production of other commodities was maintained at their minimum demand level. To achieve this level of production, all land available for agriculture was used and 16.3 billion litres of water was used. Almost 10 per cent of the available water remained unused, largely in the *rabi* and summer seasons, possibly because the technology adoption constraint limits the use of higher level technologies that efficiently use water.

Fertiliser, labour and capital requirements also decreased drastically and were lower than their current (1996–97) level of use in the state. This is perhaps because now the primary goal of farmers is to maximise income and not necessarily food production, as aimed at in this scenario.

The introduction of capital and labour availability as additional constraints in the fifth case resulted in similar total food grain production (11.4 million tonnes), but milk production dropped to 4.5–4.6 billion litres. The use of all inputs for production as well as outputs remained similar to the third case.

The results indicate that at the aggregate state level, even with all constraints (land, technology, water, capital and labour) imposed in the sixth case, production and income could be somewhat higher than what are currently (1996–97) achieved.

Minimising Water Use

The earlier scenario analyses revealed that restricted availability of water was the major constraint to increasing food production in Haryana. Therefore, in this scenario, a minimum water requirement was determined to produce current levels of food grains, oilseed, pulses, cotton and sugar. Results showed that, if the land resource was the only constraint, the current levels of production in Haryana could be attained with only 9.9 billion cubic metres of water, which is almost half of the current water use. This scenario still generates higher milk production and income than the 1996–97 baseline, but drastically reduces employment opportunities in the agricultural sector. At the same time, resource requirements in terms of capital and N fertiliser also decreased. N loss was maintained at the same level, but the biocide residue index declined drastically because only 6.8 per cent of the area was allocated to cotton-wheat and maize-potato-wheat, the two most biocide-consuming land use systems (Tables 21.7 and 21.8). Fallow-wheat occupied 60.6 per cent of the arable area of the state. Other important cropping systems were rice-wheat, maize-mustard and pearl millet-wheat.

Table 21.7 Production of Milk, Income, Resource Requirements and Environmental Impact at the Aggregate Level when Minimising Water use in Haryana

		Constraints						
Item	*Unit*	*Land*	*Land+ Water*	*Land+ Tech*	*Land+ Tech+ Water*	*Land+ Tech+ Water+ Capital*	*Land + Tech + Water + Capital + Labour*	*Current Level (1996–97)*
Milk	Billion litres	4.9	4.3	4.9	5.0	4.3	4.4	4.2
Income	Billion rupees	58.5	52.3	55.0	50.1	51.8	51.6	46.1
Land used	%	100.0	100.0	100.0	100.0	100.0	100.0	100.0
Irrigation*	Billion metre cube	9.9	12.3	11.4	13.7	13.7	13.8	18.2
N fertiliser	Million tonnes	0.56	0.59	0.59	0.57	0.57	0.57	0.65
Employment	Million labour days	236	310	301	341	341	341	387
Capital	Billion rupees	46.0	47.7	47.7	50.3	50.4	50.4	56.4
N loss	Thousand tonnes	31.9	27.6	33.3	34.6	34.6	35.1	31.6
Biocide index	–	31	93	77	122	122	121	81

Notes: Production of all crops was at their 1996–97 level in all scenarios and hence is not shown.
* Objective function minimised.

Table 21.8 Area (per cent of Agricultural Land) under Different Land use Types when Water use was Minimised

	Constraints					
Land use Type	*Land*	*Land + Water*	*Land + Tech*	*Land + Tech + Water*	*Land + Tech + Water + Capital*	*Land + Tech + Water + Capital + Labour*
Rice-rice-wheat	0.0	0.0	0.0	1.2	1.2	1.3
Rice-wheat	6.8	5.01	8.8	9.5	9.5	9.5
Basmati rice-wheat	0.0	0.0	0.0	0.0	0.0	0.0
Rice-mustard	2.4	6.55	3.1	6.5	6.5	6.4
Cotton-wheat	6.8	7.02	11.7	20.0	20.0	19.2
Maize-mustard	6.0	0.17	7.5	7.2	7.2	7.2
Maize-chickpea	2.3	0.0	3.0	3.6	3.6	3.5
Maize-potato-wheat	0.0	0.0	0.0	0.0	0.0	0.0
Sugarcane-wheat	2.7	2.45	3.4	3.8	3.8	3.9
Irrigated pearl millet-wheat	1.4	0.0	2.5	5.4	5.4	5.5
Rain-fed pearl millet-wheat	11.1	18	10.3	7.1	7.1	7.1
Fallow-wheat	60.6	51.17	49.8	35.7	35.7	36.3
Fallow-chickpea	0.0	0.0	0.0	0.0	0.0	0.0
Fallow-mustard	0.0	9.68	0.0	0.0	0.0	0.0
Pearl millet-fallow	0.0	17.98	0.0	0.0	0.0	0.0

When other constraints were gradually added in this scenario, water use still remained below 75 per cent of the current use, while maintaining the current level of production of different commodities and income (Table 21.7). This was attained by the predominance of fallow-wheat, cotton-wheat, rice-wheat, maize-mustard and pearl millet-wheat cropping systems.

CONCLUSIONS

The results presented in this chapter are exploratory in nature. The main purpose of this chapter was to develop a methodology for exploratory land use analysis and planning using Interactive Multiple Goal Linear Programming (IMGLP) approach. It reveals that Haryana has ample opportunities to increase food production and agricultural income compared with current levels, provided additional water resources could be made available. The current natural constraints of land and water limit maximum food production to seventeen million tonnes. Here, the model assumes that all water and capital within a land unit can be shared. It implies that groundwater resources available within a farm can be transported to other farms without cost, irrespective of distance involved. This does not look feasible, even with the current policy of an almost free water supply in the region.

Haryana has considerable potential to withdraw agricultural land from cultivation, without affecting basic food production and income at the aggregate level. This would require that the small farmers who cannot use alternative, efficient and capital intensive technologies not cultivate their land and that their water and other resources be made available to other farmers who presumably could use these more efficiently. Alternatively, technologies that are affordable and can be applied on small farms should be developed.

NOTES

1. This chapter draws heavily from the project report entitled, 'Land Use Analysis and Planning for Sustainable Food Security: with an Illustration for the State of Haryana, India', by P.K. Aggarwal et al. (eds), 2001. India: IARI, Philippines: IRRI and the Netherlands: WUR.
2. Corresponding author.

REFERENCES

Conway, G.R., 1987. 'The Properties of Agroecosystems'. *Agricultural Systems* 24, 95–117.

Hazel, P.B.R and R.D. Norton, 1986. Mathematical Programming for Economic Programming in Agriculture. New York, USA: Macmillan.

Huizing, H. and K. Bronsveld, 1994. 'Interactive Multiple-Goal Analysis for Land Use Planning'. *ITC Journal* 4, 366–73.

Schans, J., 1991. 'Optimal Potato Production Systems with Respect to Economic and Ecological Goals'. *Agricultural Systems*, 37, 387–97.

Schipper, RA., D.M. Janssen and J.J. Stoorvogel, 1995. 'Sub-Regional Linear Programming Models in Land Use Analysis: A Case Study of the Neguev Settlement, Costa Rica. Netherlands' *Journal of Agricultural Sciences*, 43, 83–109.

Stomph, T.J., La. Fresco and H. Van Keulen, 1994. Land Use System Evaluation: Concepts and Methodology. *Agricultural Systems*, 44, 243–55.

Stoorvogel, J.J., RA. Schipper and D.M. Jansen, 1995. USTED: A Methodology for a Quantitative Analysis of Land Use Scenarios. Netherlands, *Journal of Agricultural Sciences*, 43, 19–29.

Veeneklaas, F.R, S. Cisse, P.A. Gosseye, N. Van Duivenbooden and H. Van Keulen, 1991. 'Competing for Limited Resources: The Case of the Fifth Region of Mali', Report 4. *Development Scenarios*. CABO-DLO, Wageningen, The Netherlands.

WRR-Netherlands Scientific Council for Government Policy, 1992. 'Ground for Choices: Four Perspectives for the Rural Areas in the European Community'. Reports to the Government no. 42, Sdu, the Hague, The Netherlands, p. 149.

About the Editors and Contributors

THE EDITORS

Pushpam Kumar, Institute of Economic Growth, Delhi, India.
B. Sudhakara Reddy, Indira Gandhi Institute of Development Research, Mumbai, India.

THE CONTRIBUTORS

P. Balachandra, Indian Institute of Science, Bangalore, India.

Sumana Bandyopadhyay, University of Kalyani, Nadia, India.

P.R. Bhat, Indian Institute of Science, Bangalore, India.

Rabindra N. Bhattacharya, Jadavpur University, Kolkata, India.

Brent Bleys, Vrije Universiteit Brussel Pleinlaan, Brussels, Belgium.

H.N. Chanakya, Indian Institute of Science, Bangalore, India.

J. Chowdhury, Jadavpur University, Kolkata, India.

Saudamini Das, Swami Shardhanand College, Delhi, India.

J.V. Dhameliya, Foundation for Ecological Security, Anand, India.

Anita Chattopadhyay Gupta, Muralidhar Girls' College, Kolkata, India.

Kausik Gupta, University of Burdwan, Burdwan, India.

Zakir Husain, Rabindra Bharti University, Kolkata, India.

M. Zulfiquar Ali Islam, University of Rajshahi, Rajshahi, Bangladesh.

S. Iyengar, Centre for Social Studies, Surat, India.

P.K. Jain, Indian Agricultural Research Institute, New Delhi, India.

Raman Khaddaria, Jadavpur University, Kolkata, India.

Ranjit Kumar, Indian Agricultural Research Institute, New Delhi, India.

S.S. Kumar, Indian Agricultural Research Institute, New Delhi, India.

A.K. Mangal, Indian Agricultural Research Institute, New Delhi, India.

Dibyendu Mondal, Foundation for Ecological Security, Anand, India.

Lekha Mukhopadhyay, Jogamaya Devi College, Kolkata, India.

Ramakrishna Nallathiga, Centre for Good Governance, Hyderabad, India.

K. Narayanan, Indian Institute of Technology, Mumbai, India.

Prabha Panth, P.G. College, Secunderabad, India.

M.S.V. Prasad, College of Management Studies GITAM, Vishakapatnam, India.

P.M. Prasad, Indian Institute of Technology, Kanpur, India.

A. Ramanathan, Indian Institute of Technology, Mumbai, India.

V. Ramprasad, Indian Institute of Science, Bangalore, India.

R. Jagannatha Rao, Indian Institute of Science, Bangalore, India.

N.H. Ravindranath, Indian Institute of Science, Bangalore, India.

Jyothis Sathyapalan, Centre for Social Studies, Surat, India.

Amita Shah, Gujarat Institute of Development Research, Ahmedabad, India.

Alka Singh, Indian Agricultural Research Institute, New Delhi, India.

S. Singh, Foundation for Ecological Security, Anand, India.

S.T. Somashekara Reddy, Indian Institute of Management, Bangalore, India.

B.S. Sri, Mrs. A.V.N. College, Vishakapatnam, India.

Subhash Chandra Srivastava, National Institute of Rural Development, Hyderabad, India.

Saravanan V. Subramanian, University of Brisbane, Brisbane, Australia.

P. Sudha, Indian Institute of Science, Bangalore, India.

A.K. Vasisht, Indian Agricultural Research Institute, New Delhi, India.

Index